The Mainstream of
Algebra and Trigonometry

The Mainstream of Algebra and Trigonometry

A.W. Goodman

Houghton Mifflin Company   Boston

Atlanta   Dallas   Geneva, Illinois   Hopewell, New Jersey   Palo Alto

DEDICATED TO THE MEMORY OF MY PARENTS

Printed in the United States of America

Library of Congress Catalog Card Number: 72–5246

ISBN: 0–395–16004–9

Preface

The Mainstream of Algebra and Trigonometry provides the necessary background material for the student who wishes to study calculus. This book is also suitable as a text in algebra and trigonometry for the student who will not go on to calculus. The two subjects covered are intrinsically interesting and are often useful in fields far removed from calculus.

In writing this book every effort was made to provide clear and readable explanations, a proper balance between intuition and rigor, examples which illustrate the concepts and point out the pitfalls, and problems that are carefully graded to allow the student to progress gradually from the routine mechanical ones to those that offer a real challenge. The arrangement of the material in this text differs somewhat from that used by most of the textbooks on algebra and trigonometry. The arrangement is the logical consequence of two facts: (1) The amount of time devoted to algebra and trigonometry varies from a short course of one quarter to a long course of three quarters or two semesters. (2) Students entering a course in algebra and trigonometry may have widely diverse backgrounds.

Although it is assumed that the student has had some previous contact with algebra and geometry the book begins with the most elementary concepts. Thus, a well-prepared class with only one quarter to devote to this material could begin with Chapter 5 or 6, bypass all of the starred sections, and end with Chapter 13, Section 6. Such a program would include all of the material that is strictly essential for calculus but would omit many interesting topics.

If the students are not well prepared, they should be granted more time. Such a class should cover the first five chapters slowly and go as far beyond Chapter 13 as time permits.

The teacher will want to create his own syllabus to suit the local climate and conditions. Some possibilities are set forth in the following table, but this table should be regarded merely as a guide.

Starred sections contain material that is either difficult or not essential for calculus. When time is limited, such material can be omitted to make way for the essential items. The choice of sections to be starred is purely personal, and one cannot hope for any universal agreement. For example, a star is placed on "Bounded Sets and the Completeness Axiom" and not on "Determinants," but many teachers would reverse this.

Trignometry is introduced as early as possible (Chapters 7, 8, 9, 10, and 12) so that even in a short course this important topic will not be slighted. The modern trend is to emphasize the theory of the trigonometric functions and to push the applications to the background. Following this concept, "solving triangles" is covered last, in Chapter 12. The teacher who prefers the older order will have no difficulty presenting Sections 1 and 2 of Chapter 12 before Chapter 7. The entire chapter on solving triangles (Chapter 12) can be covered immediately after the introductory material (Chapter 7) and logarithms (Chapter 11).

Time Available	Number of class hours	Suggested chapters to be covered	Number of exercises
One quarter, 5 credits	$5 \times 10 = 50$	1 and 6 through 13	44 (a)
One semester, 4 credits	$4 \times 16 = 64$	1 and 5 through 16 or 1 through 13	58 (a) 58 (a)
One semester, 5 credits	$5 \times 16 = 80$	1 and 5 through 19 or 1 through 16	68 (a) 67 (a)
Two quarters, 4 credits	$4 \times 20 = 80$		
Two semesters, 3 credits	$3 \times 32 = 96$	1 and 6 through 19 or 1 through 16	77 (a) 80
Three quarters, 4 credits	$4 \times 30 = 120$	1 through 19	91
Two semesters, 4 credits	$4 \times 32 = 128$		

(a) Omit those sections that are starred, together with the associated exercises.

Many students have trouble understanding inverse functions. Consequently, this topic is postponed until it is absolutely essential (Chapter 10). Since the logarithm function is easily grasped as the inverse of an exponential function, logarithms are covered in Chapter 11. The teacher who feels that this relation is unnecessary can take up the study of logarithms immediately after Chapter 6 or at any later point. Interpolation is first explained in Chapter 11 on logarithms and is used again in Chapter 12 (Triangles). Consequently, in any rearrangement Chapter 11 should precede Chapter 12.

If a teacher wishes to cover more algebra before taking up trigonometry, Chapters 13, 14, 15, and 16 (Systems of Equations, Mathematical Induction, The Binomial Theorem, and Permutations, Combinations, and Probability) can be moved forward to follow Chapter 6.

Chapter 17 on vectors presents several applications of trigonometry. The sum of two vectors is carefully explained here so that it can be used in Chapter 18 for the sum of two complex numbers. Those who skip Chapter 17, in order to spend more time with complex numbers, can supply the definition of vector addition when it is needed in Chapter 18.

Chapter 19 on polynomials is placed after the chapter on complex numbers because the polynomials under consideration are permitted to have complex zeros and complex coefficients. Interpolation is used in the section on the approximation of irrational zeros of a polynomial. With only a few changes these two items can be omitted. Then the chapter on polynomials can be taken up at any time after Chapter 6.

The following schematic diagram displays the relations between the chapters. It can be used as a guide by the teacher who wishes to reorder the topics.

With no essential loss we have omitted polar coordinates, conic sections, limits, and continuity, because these topics are always covered thoroughly in the standard analytic geometry and calculus course.

The text contains 91 exercises, not including the exercises in the appendices. Answers to all odd-numbered problems are given at the end of the book. The appendices are not considered a part of the normal course work. Chapter 2 is essentially a review of arithmetic. Consequently, for all of the problems in Chapter 2 and the appendices, the answers are supplied at the end of the book. Answers for the remainder of the even-numbered problems and full solutions for selected problems are available to the instructor in a separate solutions manual prepared by Cynthia (Mrs. Norman) Mansour.

Most of the exercises contain more problems than are needed. Here the author has intentionally erred on the generous side because it is much easier

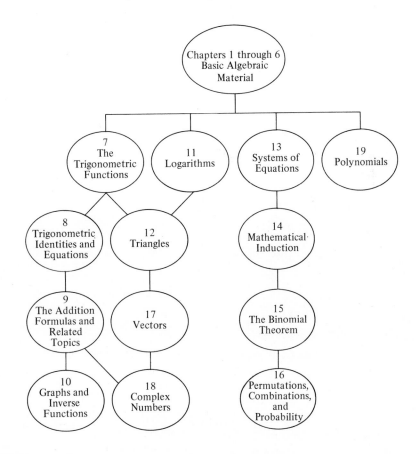

for the teacher to select a subset from a long set for an assignment, than to construct a superset by adding his own problems.

There are a few pedagogical devices that are worth noting. Each chapter closes with a set of review questions that emphasizes the imporant items in that chapter. Important formulas are set in color. In the opinion of the author these are the formulas that the student should memorize.

The symbol ▌ is used to mark the end of a proof. The symbol ▲ marks the end of the solution of an example. Whenever a proof is omitted, the theorem is labeled either PLE ("the proof is left as an exercise") or PWO ("the proof will be omitted"). These symbols shold help orient the student. Further discussion of these special symbols will be found in Chapter 1, Section 8.

The author is indebted to many friends for help with this book. Particular thanks go to Professor Edwin Comfort, who called my attention to the modern treatment of synthetic division that is presented in Chapter 19. I am very grateful to Professor Elmer Tolsted, Professor Herbert Gross, and Miss Deborah Hughes Hallett for carefully reading the original manuscript. Their many comments, criticisms, and suggestions were most welcome and quite helpful. Special mention is due to my student assistant, Mr. Larry Lewis, who read the entire manuscript, checked every example, and worked every problem. Because of his diligence, patience, and care, this book is as free of error as is humanly possible. Finally, I express my thanks to Mrs. Janelle Fortson and Miss Wanda Balliet who shared equally the burden of typing the manuscript.

A. W. GOODMAN

Table of Contents

Some Preliminary Items

1 We assume that the student has had some contact with algebra and geometry. Consequently, an exhaustive study of the basic concepts is not necessary. In this chapter we review only a few items, choosing the most important, together with those that may have caused the reader some trouble in his previous work. The strong student may skip this chapter (and perhaps the next) returning to them only if the need should arise, or if his curiosity passes the critical point.

1.1 THE USE OF SYMBOLS

The liberal use of symbols in mathematics is often a stumbling block for the student. But, much of the power of mathematics lies in the fact that very complex ideas can be expressed with great ease when suitable symbols are available. For example, consider the problem: Find a real number x such that

$$x^3 - 3x^2 + 7 = 0. \tag{1}$$

Only a few symbols are used in equation (1). If the reader will try to restate this problem in words using only the symbols for the numbers three, seven, and zero, he will certainly have difficulty.

Once we realize that the use of symbols shortens the labor of writing, adds clarity to our ideas, and speeds the thinking process, we may conclude that the

1

more symbols we have available, the better off we are. To some extent, this is true. However, new symbols must be introduced gradually. The student must be given ample opportunity to practice using each new set of symbols before he is confronted with another set. For the convenience of the student, we give a list of the important symbols just before the index.

1.2 A TWO-MINUTE HISTORY OF SYMBOLS

In antiquity mathematical ideas were expressed by long complicated sentences in the language then in use (Greek, Arabic, Latin, etc.). Gradually mathematicians realized that the sentences could be shortened by using symbols to replace words and often they merely chose the first letter of the key word. If such a system were in use today,

$$\text{"Solve the equation } x^3 - 3x^2 + 7 = 0\text{"}$$

might read,

$$\text{"Find u so that ututumthtutupseqz."}$$

Here u represents the unknown, t stands for times, p stands for plus, and m stands for minus. The letter pair *th* would represent the numeral three to avoid confusion with t for times. Clearly s and z represent seven and zero, respectively, and eq means equals.

Although such a system is bad, it is not completely unmanageable. Indeed, despite the handicaps imposed by a poor notation, these early geniuses made remarkable discoveries. What made matters worse was the fact that communication was slow, and libraries were rare so that each society developed its own system of symbols. Consequently, when an archeologist finds a new manuscript, a knowledge of mathematics and the language of the manuscript may not be sufficient for reading the treasure. He must also know the symbols used by the writer before he can distinguish a valuable mathematical document from an account of a military expedition or the records of a collector of taxes.

The symbols we use today were introduced gradually and all of them within the past 500 years. Table 1 gives more detailed information. Once introduced, the symbols were not immediately adopted. On the contrary, often 50 years or more elapsed before a new symbol was accepted and used by the majority of mathematicians.

Today mathematicians are in a more fortunate position. Most of the symbols in Table 1 and a great many other symbols used in higher mathematics are used throughout the civilized world, always with the same meaning. As a result, new mathematical discoveries made and published in any part of the world can be immediately read and understood by mathematicians everywhere.

TABLE 1			
Symbol	Meaning or Name	Date Introduced	Inventor
$+$	Plus	1486	Unknown
$-$	Minus	1486	Unknown
$\sqrt{}$	Square root	1526	Christoff Rudolff
$(\,)$	Parentheses	1556	Nicolo Fontana Tartaglia
$=$	Equals	1557	Robert Recorde
.	Decimal point	1617	John Napier
$>$	Greater than	1631	Thomas Harriot
$<$	Less than	1631	Thomas Harriot
$\times$	Multiplication	1631	William Oughtred
.	Multiplication	1631	Thomas Harriot
AB	Multiplication by juxta-position	1637	René Descartes
x, y, z	Letters near the end of the alphabet for unknown quantities	1637	René Descartes
a, b, c	Letters near the begining of the alphabet for known quantities	1637	René Descartes
$\div$	Division	1659	Johann Rahn
$\leqq$	Less than or equal to	1734	Pierre Bouguer
$\geqq$	Greater than or equal to	1734	Pierre Bouguer
$\neq$	Not equal	1740(?)	Leonhard Euler(?)
a^n	The exponent notation		See below
$a^1, a^2, a^3, \ldots$	n a positive integer	1637	René Descartes
$a^{-1}, a^{1/2}, \ldots$	n a negative integer or a fraction	1659	John Wallis
a^n	n any real number	1676	Isaac Newton
π	Ratio of circumference to diameter in a circle	1706	William Jones

Details of the history of mathematics are available in the following books:

W. W. R. Ball. *History of Mathematics*, 4th ed. New York: Dover Publications, Inc., 1960.

E. T. Bell, *The Development of Mathematics*, 2d ed. New York: McGraw-Hill Publishing Co., 1945.

Florian Cajori. *A History of Mathematical Notation*. Vols. 1 and 2. LaSalle, Illinois: Open Court Publishing Co., 1928.

Howard Eves. *An Introduction to the History of Mathematics.* 2d ed. New York: Holt, Rinehart, and Winston, 1963.

Vera Sanford. *A Short History of Mathematics.* Boston: Houghton Mifflin Co., 1930.

D. J. Struik, *A Source Book in Mathematics.* Cambridge, Mass.: Harvard University Press, 1969.

1.3 EQUALITY AND IDENTITY

The equal sign ($=$) has only one meaning, but it arises in a variety of ways, and this may cause confusion. The equation $A = B$ means that both A and B represent the same object. For example, in the equation $17 - 3 - 4 = 16 + 11 - 17$ the expressions on each side of the equal sign both represent the same number, ten. In contrast, the equation

$$(2) \qquad\qquad 17x^2 - 3x - 4 = 16x^2 + 11x - 17$$

also states that the expressions on each side represent the same number, but in this example the statement is not always true. In this case the statement (2) is true if $x = 1$ or $x = 13$ and is false for all other values for x.

The equal sign may arise in a third way, as illustrated by the equation

$$(3) \qquad\qquad (1+x)^2 = 1 + 2x + x^2.$$

In this example, the statement is true when x is replaced by any number. Although the equal sign has the same meaning in equations (2) and (3), there is an essential difference in the two equations. We call (3) an *identity* because it is true for all values of x. We call (2) a *conditional equation* because it is true only on the condition that $x = 1$ or $x = 13$.

The equal sign may arise in a fourth way. This is illustrated by the statement: Let

$$(4) \qquad\qquad \mathbf{N} = \text{the set of positive integers.}$$

In equation (4) the symbol on the left is defined by the material on the right. Thus both sides of (4) represent the same object. But in this example there is nothing to prove because (4) is a definition. Equation (4) resembles equation (3) in that both are always true, but it differs from equation (3) in that equation (3) is a statement of a (very simple) theorem that must be proved (or has already been proved).

We have seen that the $=$ sign can arise in at least four different situations. We are justified in using the same sign in all four situations, because in each case it is asserted that the expressions on each side represent the same object. However, as illustrated in equation (2), the assertion of equality may be false.

If we want to emphasize that an equation is an identity, we will use the symbol $\equiv$. Thus in (3) we may (if we wish) write

(5) $$(1+x)^2 \equiv 1 + 2x + x^2,$$

and in (4) we may write

(6) $$\mathbf{N} \equiv \text{the set of positive integers.}$$

The symbol $\equiv$ is called the *identity symbol*. In (5), one may read $\equiv$ as "is identically equal to." In (6), one may read $\equiv$ as "is defined to be."

1.4 SUBSCRIPTS AND SUPERSCRIPTS

Suppose that we have a problem involving the areas of four triangles (the exact nature of the problem is unimportant). It is natural to use the letter A to represent the area of the first triangle, and to use a for the area of the second triangle. For the third area we might use α (Greek letter alpha) because it corresponds to the English a. But now we are without a suitable choice of a symbol for the area of the fourth triangle. The solution is quite simple. We return to the letter A and put little numbers called *subscripts* just below the letter, thus: A_1, A_2, A_3, A_4, and use these to represent the areas of the four triangles. These symbols are read: A sub-one, A sub-two, etc. If we are in a hurry we may say A-one, A-two, etc. Clearly the device of adding subscripts greatly enlarges the numbers of symbols available for our use.

We can also use *superscripts*. Thus we might write $A^{(1)}$, $A^{(2)}$, $A^{(3)}$, and $A^{(4)}$ to denote the altitudes of the triangles. Here we want to avoid A_1, because presumably in the problem at hand it has already been assigned the meaning of area. The symbols $A^{(1)}$, $A^{(2)}$, ... are read: A upper-one, A upper-two, etc. The superscripts are enclosed in parentheses to distinguish them from powers. Thus $A^{(2)} \neq A^2$, because the latter is AA or A squared, whereas $A^{(2)}$ is merely a symbol; in this example it is the symbol for the altitude of the second triangle.

A subscript or superscript is also called an *index*. The indices may be represented by a letter such as k. As a matter of shorthand we can indicate the four altitudes $A^{(1)}$, $A^{(2)}$, $A^{(3)}$, $A^{(4)}$ by writing $A^{(k)}$, $(k = 1, 2, 3, 4)$.

In some situations we may wish to consider a finite number of triangles, where the number of triangles involved is unknown. In this case we let k (or some other suitable letter) denote the number of triangles. Then the areas of the triangles would naturally be denoted by $A_1, A_2, A_3, ..., A_k$. Here the three dots between A_3 and A_k represent the missing items. If $k = 4$, there are no missing items. If $k = 9$, then the three dots represent A_4, A_5, A_6, A_7, and A_8. If $k = 100$ it would be silly to write individual symbols for the area of each triangle so some notation such as the three dots is necessary. We will say more about this notation in Section 1.9.

1.5 AXIOM SYSTEMS

The student has already met the axioms for Euclidean geometry. For example, "Through two distinct points one and only one line can be drawn," and "Two distinct straight lines meet in at most one point" are two of the axioms.

The concept of a set of axioms is not restricted to Euclidean geometry. Indeed, any mathematical subject may have a suitable set of axioms associated with it.

DEFINITION 1

A Set of Axioms *Let* $\mathbf{S} = \{S_1, S_2, ..., S_k\}$ *be a set of statements about the elements of a certain set* $\mathbf{M}$, *where each* S_i *is an individual statement for* $i = 1, 2, ..., k$. *If it is agreed to accept these statements about the elements of* $\mathbf{M}$ *as true without proof, then the set* $\{S_1, S_2, ..., S_k\}$ *is called a set of axioms for* $\mathbf{M}$.

At first glance this definition may seem unpleasant. The student may well ask "Why not prove the statements in **S**?" But a moment of thought will convince him that in order to prove these statements he must use another set of statements

$$\mathbf{T} = \{T_1, T_2, ..., T_l\},$$

and if he wishes to prove the statements in **T**, then he must use another set of statements

$$\mathbf{V} = \{V_1, V_2, ..., V_m\}.$$

Clearly this process of going "backward" cannot continue indefinitely. We must agree to start "somewhere" and an axiom system is just a formalization of our agreement on our starting place.

If we can agree to start wherever we wish, why not just start by stating all of the difficult theorems of algebra as axioms, and save the labor of proving them. If the selection of an axiom system is completely wild, this can be done. But we ourselves, as conscientious mathematicians, impose certain natural and aesthetic limitations on the choice of the statements in **S**, such as (**I**) simplicity, (**II**) economy, (**III**) utility, and (**IV**) consistency.

(**I**) *Simplicity* Each axiom in the set **S** should be as simple as possible.

(**II**) *Economy* The set **S** should have as few statements as possible, while preserving the simplicity and utility. If we can prove S_j using the other statements in **S**, then S_j should be dropped as an axiom.

(**III**) *Utility* The set **S** should be selected so that it is most convenient for proving the theorems that we have in view, while preserving the simplicity of the set.

(**IV**) *Consistency* The statements in **S** must be consistent. If we can prove that one of the statements in **S** is false, using the set **S**, then the axiom system is inconsistent and must be revised.

The real number system is basic for any book on algebra. Because it is rather complicated, a detailed account of the development and structure of this system would only impede our progress in learning algebra and trigonometry. Briefly, the system of real numbers is constructed from the natural numbers (the positive integers). A set of five axioms (the Peano axioms) for the natural numbers is given in Appendix 2. From these axioms, all the theorems about the real numbers can be derived. Although the derivation is straightforward and natural, it is neither short nor easy. Consequently the student should postpone this study until he has attained more mathematical maturity.

A system of fifteen axioms for the real numbers is presented in Chapters 2 and 3. Although this latter system is more numerous than the Peano axioms, and presumably more complicated, we use it because we can obtain the results we need more rapidly.

1.6 WHAT IS A THEOREM?

A theorem is a statement that can be proved by a valid logical argument using definitions, previously proved theorems, and axioms. In some cases it may not be necessary to use all three of these items in the proof.

The assertion "$2+2=4$" is a true statement, and it can be proved by a logical argument, beginning with the three definitions $1+1=2$, $2+1=3$, and $3+1=4$ and using the associative law of addition. Still we do not wish to call $2+2=4$ a theorem, because it is neither very interesting nor very important for the development of mathematics.

Mathematics is populated with millions of facts, formulas, laws, and propositions, each one provable by a valid logical argument. To call every such item a theorem would be ridiculous. The fascinating and important facts would be swallowed up and lost in a sea of dull and trivial items. The honorable title of "Theorem" is reserved for those facts, formulas, laws, or propositions that are either very interesting or very important, or both. It must be realized that personal taste and judgement enter into the selection of those statements that receive the title of theorem. Consequently, one man's theorem may be another man's proposition, while a third person may call the same statement a formula or a law. The essential item is that (no matter what the title) the assertion can be proved by a valid argument.

The term "lemma" is also used in place of theorem, proposition, etc. The title "lemma" is used to indicate that the statement is an important step in the proof of a theorem; but the "lemma" by itself is not interesting enough or important enough to be called a theorem.

In some cases one man's axiom may be another man's theorem. This can occur if the two men have selected a different set of axioms as the starting point on which to base their chain of arguments. The student who is confronted by such an apparent inconsistency may suffer some distress. However, he should recognize that either organization of the material may be correct, and he should select that set of axioms that appeals to him (always reserving the right to switch as his taste changes).

1.7 THE FORM OF A THEOREM

As an example consider

Theorem 1   *If a_0, a_1, a_2, and a_3 are real numbers and $a_0 \neq 0$, then the equation*

$$(7) \qquad\qquad a_0 x^3 + a_1 x^2 + a_2 x + a_3 = 0$$

has at least one real root.

This theorem has the form of

Theorem A. *If H, then C.*

In this form H represents a statement (or a series of statements) called the *hypothesis*, (or *hypotheses*) and C represents a statement (or series of statements) called the *conclusion*. To prove the theorem, one must give a correct argument showing that if H is true, then C is true. The direction of the argument is often indicated by the symbol $\Rightarrow$ (read, "implies"). With this symbol Theorem A can be put in the form

$$(8) \qquad\qquad H \Rightarrow C,$$

(read, "H implies C").

In Theorem 1, the hypothesis is: a_0, a_1, a_2, and a_3 are real numbers, and $a_0 \neq 0$. The conclusion is: Equation (7) has at least one real root (is true for at least one real number).

The English language offers a rich variety of ways of expressing $H \Rightarrow C$ and to avoid the boredom of repetition most authors use the many permissible variations in stating theorems. Whenever such a variation is confusing, the reader should put the theorem in the form $H \Rightarrow C$ before proceeding with the proof. Another source of annoyance may be the form of

Theorem B. *H if and only if C*

This assertion is really two theorems combined into one for efficiency. These are Theorems B_1 and B_2.

Theorem B_1. *H, if C.*

This means if C then H, or $C \Rightarrow H$.

Theorem B₂. *H only if C.*

This means if H, then C, or $H \Rightarrow C$. Hence, to prove Theorem B, we must prove two simpler theorems, i.e. we must prove B_1 and B_2.

1.8 SOME UNUSUAL SYMBOLS

The reader is already familiar with many of the standard symbols, such as $+$, $-$, $\times$, $\sqrt{\ }$, etc. Other symbols will be explained as they appear for the first time. A few special symbols are treated here.

■ **The proof is completed.** It is convenient to have a mark to signal the end of a proof. Thus if the reader has trouble understanding the proof, he can at least locate the end of the proof, and then reread the proof until it does become clear. In the past this place was often indicated by the letters Q.E.D., which abbreviate "Quod Erat Demonstrandum," the Latin phrase for "which was to be demonstrated." In recent times it has become the custom to use the symbol ■ with exactly the same meaning. In this book we will use ■.

▲ **The solution of the example is completed.** It is also convenient to have a symbol marking the end of the solution of an example. An example may also be a theorem, and a theorem may also serve as an example. Nevertheless there may be some advantage in distinguishing between the two; hence, we will use the different symbols ■ and ▲.

PWO **The proof will be omitted.** Frequently there are good reasons for omitting a proof. Perhaps the proof is long, uninteresting, or uses advanced methods and concepts beyond the scope of this text. In any one of these cases we label the theorem *PWO* to advise the reader that the proof will be omitted.

PLE **The proof is left as an exercise.** We also label a theorem *PLE* to indicate that the proof will be omitted. But in this case the proof of the theorem is within the capacity of the student, and he is expected to supply the proof for himself.

In order to avoid overloading the student with too many symbols, we shall avoid the use of certain popular ones that are often used in other texts. But your instructor may wish to use them in his lectures, or you may meet these symbols when you read other books, so we list them here for your convenience

$\wedge \equiv$ and	$\ni \equiv$ such that
$\vee \equiv$ or	$\Rightarrow \equiv$ implies
$\exists \equiv$ there exists	$\Leftrightarrow \equiv$ implies and is implied by
$\forall \equiv$ for all (for every)	

**1.9   THE THREE DOTS
NOTATION**

In Chapter 14 we will prove the very pretty formula

$$(9) \qquad\qquad 1 + 3 + 5 + \cdots + (2n-1) = n^2.$$

To follow the proof we must first understand what the formula says, and hence we must understand the meaning of the three dots in equation (9). These dots mean that the terms in the sum continue in the manner indicated by the first few written until the last term $2n-1$ is reached. For example, if $n = 8$, then $2n-1 = 2 \times 8 - 1 = 15$ and equation (9) states that

$$1 + 3 + 5 + 7 + 9 + 11 + 13 + 15 = 8^2 = 64.$$

In this case the three dots represent the terms $7+9+11+13$. If $n = 3$, then $2n-1 = 5$ and in this case equation (9) means $1+3+5 = 3^2 = 9$. If $n = 1$, the left-hand side of equation (9) has only the single term 1.

Why do we bother with the three dots notation? In the first place it is shorter. If $n = 10,000,000$, then equation (9) would have 10,000,000 terms. It would take about 116 days to write all of the terms of the sum, writing at the rate of one term per second and not pausing to eat or sleep.

Further, equation (9) is true for any positive integer n. It would be impossible to convey this idea in a formula if we did not have the three dots notation or some other equivalent device.

1.10   THE GREEK ALPHABET

As the use of symbols grew, certain symbols were reserved for specific ideas. Thus, r meant radius, h meant height, t meant time, d meant distance, r meant rate—but already the alphabet is being crowded, for here we have r with two different meanings, radius and rate. As mathematics expanded and more new concepts were added, it became obvious that no single symbol could represent one concept only. Either some symbols must do double duty, or new symbols must be introduced to relieve the congestion. The early scholars all knew Greek, so it was natural for them to introduce Greek letters, and these letters are still with us today in mathematics, though the study of the Greek language has dwindled almost to zero.

For ready reference we give the complete Greek alphabet, but the student is advised *not* to learn it. Instead he can learn a few letters at a time, referring to this alphabet as the need arises. He already knows π, which has been reserved the world over for the ratio "circumference/diameter" in a circle and is rarely used for anything else. In this book we shall use only those letters that are marked with boldface. As the reader goes further in mathematics more Greek letters will be used, but they will be introduced gradually and should not cause the student any trouble.

Alpha	A, α	Iota	I, ι	**Rho**	P, ρ
Beta	B, β	Kappa	K, κ	**Sigma**	Σ, σ
Gamma	Γ, γ	Lambda	Λ, λ	Tau	T, τ
Delta	Δ, δ	Mu	M, μ	Upsilon	Υ, υ
Epsilon	E, ε	Nu	N, ν	**Phi**	Φ, ϕ
Zeta	Z, ζ	Xi	Ξ, ξ	Chi	X, χ
Eta	H, η	Omicron	O, o	Psi	Ψ, ψ
Theta	Θ, θ	**Pi**	Π, π	Omega	Ω, ω

1.11 THE ABUSE OF SYMBOLS

In a modern textbook we may find an expression such as

3.1.10 $$\forall a \in \mathbf{R} \wedge \forall b \in \mathbf{R} \ \exists x \in \mathbf{R} \ni a + x = b.$$

This looks quite formidable, but with a little effort it will become relatively harmless. In the first place the "decimal" 3.1.10 merely means that this is item 10 in Section 1 of Chapter 3. Further translating some of the symbols we obtain:

3.1.10 *For every real number a and every real number b there is a real number x such that $a + x = b$.*

Presumably the alert reader knows which chapter and section he is reading so the "decimal" is superfluous. The statement can be simplified still further to yield

Theorem 10 *If a and b are any two real numbers, then the equation $a + x = b$ always has a unique solution.*

It is true that the form 3.1.10 is shorter than Theorem 10, but it saves only a small amount of space, and at a cost of much time and painful effort on the part of the poor reader.

In this text we prefer to use the expanded version because this form is easier for the student. Further, in numbering equations the positive integers are quite satisfactory. But in referring to results obtained in other chapters the "decimal" system may have some merit. Thus, Theorem 3.7 means Theorem 7 in Chapter 3, and Section 1.11 means Section 11 in Chapter 1.

1.12 THE SYMBOL FOR INFINITY

The standard symbol for infinity is ∞. The reputation of this symbol is somewhat tarnished because on some occasions it has been used as a number. For example, the statements $\infty - \infty = 0$, and $\infty/\infty = 1$ are pure nonsense and can lead to erroneous results. But the fault does not lie with the symbol, but rather with those who misuse it. The symbol ∞ is very useful and must not be rejected entirely. Each time that a new use for ∞ is introduced, the meaning should be explained. It is perfectly proper to use ∞ with any meaning that has previously been agreed upon, provided that ∞ is not used as a number. For example, we may use the symbols $1/0 = \infty$ to indicate that $1/0$ is not a number or that $1/0$ is undefined. Such a notation has some value because it serves to remind us that when x is close to zero and positive, $1/x$ is very large. However, if we use $1/0 = \infty$ to deduce that $0 \cdot \infty = 1$, then we are treating ∞ as a number. If we permit such manipulations, then it is an easy matter to prove that $2 = 1$.

The Real Numbers

2

Since the set **R** of real numbers will be in constant use throughout this book, some knowledge of the real numbers is absolutely essential. However, if we pause to give a detailed development of the real number system, we will delay our study of algebra and trigonometry.

Our compromise is to list the axioms for the real numbers and the basic properties that can be derived from them. In Appendix 2 we discuss some of the axioms and prove as theorems some of the rules of operation that are listed here. With this organization the student is not tempted to spend too much time on the foundations. Nevertheless, this logically important material is available if the curious student wishes to see it, or the teacher wishes to use it.

We assume that the student is familiar with the basic elements of set theory, and has had some experience with the notations, $a \in S$, $\subset$, $\cap$, etc. Those who need a quick review, will find the required material in Appendix 1. We use bold letters such as $\mathbf{A}, \mathbf{B}, \mathbf{C}, ..., \mathbf{Q}, \mathbf{R}, \mathbf{S}, \mathbf{T}$, to denote sets.

2.1 ADDITION AND MULTIPLICATION

Let **R** denote the set of all real numbers. This set contains the natural numbers,

$$\mathbf{N} \equiv \{1, 2, 3, 4, ...\},$$

together with 0 (zero) and the negatives of the natural numbers,

$$\{-1, -2, -3, \ldots\}.$$

Further **R** contains the set **Q** of all rational numbers. These are the numbers that can be expressed as the ratio of two integers. Consequently, the set **Q** includes such numbers as $1/2$, $-2/3$, $111/16$, $-109/64$, and $5/1$. Finally, **R** contains roots such as $\sqrt{3}$, $\sqrt[3]{5}$, $\sqrt{2}+\sqrt{7}$, $\sqrt[3]{11}-\sqrt{59}$, together with such bizarre numbers as π, $\sqrt{2}^{\sqrt{3}}$, and $(\sqrt{13}-\sqrt[3]{7})^{\sqrt{73}}$. We will have more to say about some of these numbers later on. For the moment, these numbers are to be regarded as examples that illustrate the nature of **R**.

The familiar operation of adding two real numbers satisfies the following five axioms.

AXIOM 1

The Closure Property for Addition *For every pair a, b in* **R**, *there is a uniquely determined element in* **R** *called the sum of a and b and denoted by a+b.*

The word "closure" means that when we add two elements in **R**, the sum is again in **R**. We say that the set **R** is *closed under addition.*

Examples: $\quad 3 + 7 = 10 \in \mathbf{R}$,

$$9 + (-3) = 6 \in \mathbf{R},$$

$$\frac{3}{4} + \frac{11}{4} = \frac{14}{4} = \frac{7}{2} \in \mathbf{R},$$

$$\sqrt{2} + \sqrt[5]{3} = \text{a number in } \mathbf{R}.$$

Notice that the last example is somewhat different from the first three. Indeed there is apparently no reasonable way of simplifying the expression $\sqrt{2}+\sqrt[5]{3}$. Axiom 1 states that the sum is in **R**.

AXIOM 2

The Commutative Law for Addition *For every pair a, b in* **R**

(1) $$a + b = b + a.$$

This axiom states that we can interchange (commute) the order of the elements to be added without changing their sum.

Examples: $\quad 3 + 7 = 7 + 3$,

$$9 + (-3) = (-3) + 9,$$

$$\sqrt{2} + \sqrt[5]{3} = \sqrt[5]{3} + \sqrt{2}.$$

AXIOM 3

The Associative Law for Addition *For every* a, b, c *in* **R**

(2)
$$(a+b) + c = a + (b+c).$$

Here, and in many other places, the parentheses are used to indicate the order in which operations are to be performed. On the left side of equation (2) we are to add a and b first, and then to this sum we must add c. On the right side of (2) we are to add a to the sum of b and c. Axiom 3 states that in either case, the final result is the same. In other words, the sum of three real numbers does not depend on the way the terms are grouped (associated). Consequently, we may drop the parentheses and write $a+b+c$ for the quantity on either side of (2).

Example: $(7+11) + 19 = 7 + (11+19)$

$ 18 \quad + 19 \doteq 7 + \quad 30.$

AXIOM 4

There is a real number called zero (denoted by 0*) in* **R** *such that for every* a *in* **R**

(3)
$$a + 0 = a.$$

The element 0 is called the *identity element for addition*. Normally one should be careful to specify whether zero is added on the left or on the right of a in (3), but for real numbers such care is unnecessary because by the Commutative Law for Addition (Axiom 2)

$$0 + a = a + 0 = a.$$

AXIOM 5

For each a *in* **R** *there is an element in* **R** *denoted by* $-a$ *or by* $(-a)$ *such that*

(4)
$$a + (-a) = 0.$$

This element $(-a)$ is called the *additive inverse* of a, or the *negative* of a. Again the Commutative Law tells us that $(-a)+a = 0$. Hence, the order in (4) is unimportant.

Examples: $7 + (-7) = 0,$ and $(-13) + [-(-13)] = 0.$

DEFINITION 1

Subtraction *The difference* $b-a$ *is defined by*

(5)
$$b - a = b + (-a).$$

In other words to subtract a from b, we add the additive inverse of a to b.

The familiar operation of multiplying two real numbers satisfies five axioms which are similar to the five axioms for addition. Further, there is one important axiom that relates addition and multiplication.

AXIOM 6

The Closure Property for Multiplication *For every pair* a, b *in* **R** *there is a uniquely determined element in* **R** *called the product of a and b, and denoted by* ab *(or by* $a \cdot b$ *or* $a \times b$*).*

Examples: $7 \cdot 19 \ \sqrt{3} \sqrt[5]{17}, \quad \sqrt{11}\,\pi, \quad \pi^2 \quad$ are in **R**.

AXIOM 7

The Commutative Law for Multiplication *For every* a, b *in* **R**

(6)
$$ab = ba.$$

Example: $13 \times 57 = 741 = 57 \times 13.$

AXIOM 8

The Associative Law for Multiplication *For every* a, b, c *in* **R**

(7)
$$(ab)c = a(bc).$$

Example: $(7 \times 13) \times 19 = 7 \times (13 \times 19)$

$$91 \quad \times 19 = 7 \times \quad 247.$$

AXIOM 9

There is a real number called one (denoted by 1*) in* **R** *such that for every a in* **R**

(8)
$$a \cdot 1 = a.$$

The element 1 is called the *identity element for multiplication.* Again one should be careful to specify whether 1 multiplies a on the left or on the right in (8), but again such care is unnecessary because by the Commutative Law for Multiplication (Axiom 7) $1 \cdot a = a \cdot 1 = a.$

AXIOM 10

For each a in **R**, *such that* $a \neq 0$, *there is an element in* **R** *denoted by* $1/a$ $\left(or\ by\ \dfrac{1}{a}\right)$ *such that*

(9)
$$a\left(\frac{1}{a}\right) = 1.$$

The product in (9) may be written as $a \cdot (1/a)$ or $a(1/a)$. The element $1/a$ is called the *multiplicative inverse* (or the *reciprocal*) of a. Again the Commutative Law tells us that $(1/a)a = a(1/a) = 1$. Hence the order of the terms in (9) is unimportant.

Examples: $(27)(1/27) = 1, \quad (-16)\big(1/(-16)\big) = 1.$

DEFINITION 2   ***Division***   *If $b \neq 0$, the quotient a/b is defined by*

(10)
$$\frac{a}{b} = a\left(\frac{1}{b}\right).$$

In other words, to divide a by b, we multiply a by the reciprocal of b. If $b = 0$, then the quotient a/b is not defined.

Finally, we need an axiom that relates addition and multiplication. This is

AXIOM 11   ***The Left Distributive Law***   *For every a, b, c in* **R**

(11)
$$a(b+c) = ab + ac.$$

Example: $3(9+17) = (3)(9) + (3)(17)$

$\qquad\qquad (3)(26) \;=\; 27 \;+\; 51$

$\qquad\qquad\qquad 78 = 78.$

We prefer to indicate the product of 3 and 9 by $(3)(9)$ or by $3(9)$. Juxta-position would give 39 which looks like thirty-nine. The dot $3 \cdot 9$ might be confused with the decimal 3.9. The symbol $\times$ resembles the variable x.

The order of terms in (11) is unimportant because with the Commutative Law for Multiplication we can prove a Right Distributive Law: *For every a, b, c in* **R**

(12)
$$(a+b)c = ac + bc.$$

Any set **S** with two operations such as $+$ and $\times$ that satisfies the Axioms 1 through 11 (and has at least two elements) is called a *field*. Consequently the real numbers form a field. But this is not the only field. It is possible to have a field with only a finite number of elements. The curious reader is referred to Appendix 2 for further information about fields.

2.2 PROPERTIES OF THE REAL NUMBERS

We first look at the phrase "*there is a uniquely determined element in* **R**." This phrase occurs in Axiom 1 (for addition) and again in Axiom 6 (for multiplication). Since the sum of two numbers is unique we know that if

(13)
$$a = b,$$

then (by Axiom 1)

(14)
$$a + c = b + c.$$

We refer to the process of going from equation (13) to equation (14) as "*adding c to both sides of the equation.*"

In the same way it follows from Axiom 6 that we can "*multiply both sides of an equation by c.*" Thus if

(13) $$a = b,$$

then (because the product is unique)

(15) $$ac = bc.$$

Notice that in (14), c may be negative, so we can subtract the same quantity from both sides of an equation. Further in (15), c may be $1/d$ so we can divide both sides of an equation by d, if $d \neq 0$.

We now list some of the properties of the real numbers that follow logically from the axioms, the definitions, and the two operations that we have just described. We group these properties under several headings, and we illustrate a few of these properties with numerical examples. Here $a, b, c, \ldots$ are arbitrary real numbers, with the one exception that if a letter appears in the denominator, then we assume it is not zero.

Rule for zero:

(16) $$0 \cdot a = a \cdot 0 = 0.$$

Rules of operation for the negative sign:

(17) $$-(-a) = a,$$

(18) $$-(a+b) = (-a) + (-b),$$

(19) $$-(a-b) = b - a,$$

(20) $$(-a)b = a(-b) = -(ab),$$

(21) $$(-a)(-b) = ab,$$

(22) $$\frac{-a}{b} = \frac{a}{-b} = -\frac{a}{b}, \qquad b \neq 0,$$

(23) $$\frac{-a}{-b} = \frac{a}{b}, \qquad b \neq 0.$$

Rules of operation for fractions:

(24) $$b\left(\frac{a}{b}\right) = a, \qquad b \neq 0,$$

(25) $$1/(1/a) = a, \qquad a \neq 0,$$

(26) $$\frac{1}{a} \cdot \frac{1}{b} = \frac{1}{ab}, \qquad a \neq 0, \qquad b \neq 0,$$

(27) $$\frac{a}{d} + \frac{b}{d} = \frac{a+b}{d}, \qquad d \neq 0,$$

$$\frac{13}{37} + \frac{19}{37} = \frac{32}{37},$$

(28) $$\frac{a}{b} \cdot \frac{c}{d} = \frac{ac}{bd}, \qquad b \neq 0, \qquad d \neq 0,$$

$$\frac{3}{4} \cdot \frac{5}{7} = \frac{15}{28},$$

(29) $$\frac{ad}{bd} = \frac{a}{b}, \qquad b \neq 0, \qquad d \neq 0,$$

$$\frac{45}{95} = \frac{9(5)}{19(5)} = \frac{9}{19},$$

(30) $$\frac{a}{b} + \frac{c}{d} = \frac{ad+bc}{bd}, \qquad b \neq 0, \qquad d \neq 0,$$

$$\frac{3}{4} + \frac{5}{13} = \frac{3(13)+4(5)}{4(13)} = \frac{39+20}{52} = \frac{59}{52},$$

(31) $$\frac{a/b}{c/d} = \frac{ad}{bc}, \qquad b \neq 0, \qquad c \neq 0, \qquad d \neq 0,$$

$$\frac{3/4}{5/13} = \frac{3(13)}{4(5)} = \frac{39}{20}.$$

Rules for the expansion of products:

(32) $$a(b+c+d) = (b+c+d)a = ab + ac + ad,$$

(33) $$(a+b)(c+d) = ac + ad + bc + bd,$$

(34) $$(a+b)^2 = a^2 + 2ab + b^2,$$

(35) $$(a+b)^3 = a^3 + 3a^2b + 3ab^2 + b^3,$$

(36) $$(a-b)(a+b) = a^2 - b^2,$$

(37) $$(a-b)(a^2+ab+b^2) = a^3 - b^3,$$

(38) $$(a-b)(a+b)(a^2+b^2) = a^4 - b^4,$$

(39) $$(a+b)(a^2-ab+b^2) = a^3 + b^3.$$

Rules of implication: We state these as theorems.

Theorem 1 If $a+b=0$, *then*

(40) $$a = -b, \quad and \quad b = -a.$$

Theorem 2 If $ab = 1$, *then*

(41) $$a = \frac{1}{b}, \quad and \quad b = \frac{1}{a}.$$

Theorem 3 If $ab = 0$, *then either* $a = 0$, *or* $b = 0$, *or both.*

Theorem 4 If $a+c = b+c$, *then* $a = b$.

Theorem 5 If $ca = cb$ and $c \neq 0$, *then* $a = b$.

Theorem 6 If $b \neq 0$, *and* $d \neq 0$, *and*

(42) $$ad = bc,$$

then

(43) $$\frac{a}{b} = \frac{c}{d}.$$

Conversely, if (43) *is true, then* (42) *is also true.*

Note that Theorem 5 is a cancellation law for products. Equation (29) is a cancellation law for fractions. It is often helpful to use equation (29) in the reverse direction, that is, we may insert a nonzero factor d in the numerator and the denominator of a fraction whenever we wish.

Exercise 1

In problems 1 through 12 prove the given assertion using only the axioms, the definitions, and the properties of the real numbers covered so far. *Hint:* Most of these are just special cases or slight extensions of equations (16) through (39).

1. $-0 = 0$

2. $a(b-c) = ab - ac$

3. If $a \neq 0$, then $\dfrac{a}{a} \neq 1$.

4. $\dfrac{a}{1} = a$

5. If $a \neq 0$ and $b \neq 0$, then $\dfrac{a}{b} \cdot \dfrac{b}{a} = 1$.

6. If $d \neq 0$, then $\dfrac{a}{d} + \dfrac{b}{d} + \dfrac{c}{d} = \dfrac{a+b+c}{d}$.

7. If $d \neq 0$, then $\dfrac{a}{d} - \dfrac{b}{d} = \dfrac{a-b}{d}$.

8. If $b \neq 0$ and $c \neq 0$, then $\dfrac{a/b}{c} = \dfrac{a}{bc}$.

9. If $b \neq 0$ and $c \neq 0$, then $\dfrac{a}{b/c} = \dfrac{ac}{b}$.

10. $(a-b)^2 = a^2 - 2ab + b^2$

11. If $b \neq 0$ and $d \neq 0$, then $\dfrac{a}{b} - \dfrac{c}{d} = \dfrac{ad-bc}{bd}$.

12. If $b \neq 0$, $d \neq 0$, and $f \neq 0$, then $\dfrac{a}{b} + \dfrac{c}{d} + \dfrac{e}{f} = \dfrac{adf + cbf + ebd}{bdf}$.

In problems 13 through 28 justify the indicated computation by referring to a suitable axiom, rule, theorem, or earlier problem.

13. $3(-7) = -21$

14. $(-3)(-5) = 15$

15. $5(11-7) = 55 - 35$

16. $\dfrac{3}{4} \cdot \dfrac{7}{11} = \dfrac{21}{44}$

17. $\dfrac{44}{55} = \dfrac{4}{5}$

18. $\dfrac{7}{11} = \dfrac{119}{187}$

19. $\dfrac{1}{3} + \dfrac{5}{7} = \dfrac{22}{21}$

20. $\dfrac{218}{153} = \dfrac{11}{17} + \dfrac{7}{9}$

21. $\dfrac{3}{8} + \dfrac{5}{16} = \dfrac{11}{16}$

22. $\dfrac{1}{2} + \dfrac{2}{3} + \dfrac{3}{5} = \dfrac{53}{30}$

23. $(-1)a = a(-1) = -a$

24. $\dfrac{-1}{1} = -1 = \dfrac{1}{-1}$

25. $\dfrac{2/3}{5/7} = \dfrac{14}{15}$

26. $\dfrac{20/9}{35/81} = \dfrac{36}{7}$

27. $13(7) = (10+3)(10-3) = 100 - 9$

28. $13^2 = (10+3)^2 = 100 + 60 + 9$

In problems 29 through 49 complete the indicated computation.

29. $52/13$

30. $-49/119$

31. $11(23-32)$

32. $-96/(-32)$

33. $\dfrac{100/25}{5}$

34. $\dfrac{100}{25/5}$

35. $\dfrac{63/25}{7}$

36. $\dfrac{63}{25/7}$

37. $\dfrac{3}{4} + \dfrac{4}{5} + \dfrac{5}{7}$

38. $\dfrac{3}{4} - \dfrac{4}{5} + \dfrac{5}{7}$

39. $\dfrac{3}{4} - \dfrac{4}{5} - \dfrac{5}{7}$

40. $29^2 = (30-1)^2$

41. $18^3 = (20-2)^3$

42. $\dfrac{2/3}{4/5}$

43. $\dfrac{3/7}{9/28}$

44. $\dfrac{\dfrac{1}{2}}{\dfrac{2}{3} + \dfrac{3}{4}}$

45. $\dfrac{\dfrac{2}{3} - \dfrac{1}{4}}{\dfrac{5}{6} - \dfrac{1}{2}}$

46. $\dfrac{\dfrac{1}{2}}{\dfrac{2+3}{3+4}}$

47. $\dfrac{\dfrac{2-1}{3-4}}{\dfrac{5-1}{6-2}}$

48. $\dfrac{1-2}{\dfrac{2}{3} - \dfrac{3}{4}}$

49. $\dfrac{3\left(\dfrac{2}{3} - \dfrac{4}{5}\right)}{4\left(\dfrac{1}{6} - \dfrac{5}{8}\right)}$

2.3 RATIONAL AND IRRATIONAL NUMBERS

Any field **F** contains the number 1 (by Axiom 9). Since it is closed under addition, **F** also contains $1+1$, it contains $(1+1)+1$, and so on. Either this process of repeatedly adding 1 always produces a new element, or it may happen that at some stage, the addition of 1 gives an element of **F** that has appeared earlier. For example, if **F** has only two elements we will find that

$$1 + 1 = 0, \qquad (1+1) + 1 = 0 + 1 = 1.$$

In any field with only a finite number of elements, such a repetition must occur but in the real number system the process of adding 1 may be continued indefinitely without repetition. At each stage the addition of 1 produces a number not encountered before. We need symbols for these numbers and of course we adopt the ones we have been practicing with since early childhood. But we must emphasize that these are definitions. Thus, by definition

$$2 \equiv 1 + 1,$$

$$3 \equiv 2 + 1 = (1+1) + 1,$$

$$4 \equiv 3 + 1 = ((1+1) + 1) + 1, \text{ etc.}$$

The set $\mathbf{N} = \{1, 2, 3, 4, \ldots\}$ obtained in this way has an infinite number of elements and is called the set of *natural numbers*, or the set of *positive integers*. When we adjoin 0 and the negatives of the elements in **N** we get the set

$$\mathbf{Z} \equiv \{0, 1, -1, 2, -2, 3, -3, \ldots\},$$

which is also denoted by $\mathbf{Z} \equiv \{0, \pm 1, \pm 2, \pm 3, \ldots\}$. The set **Z** is called the set of *integers*, and each element of **Z** is called an *integer*.

DEFINITION 3 | ***Rational Number*** *A number x is said to be a rational number if it can be put in the form $x = a/b$ where a and b are integers and $b \neq 0$.*

The set of all rational numbers is denoted by **Q** (to remind us of the word quotient). If a is an integer, then $a = a/1$ and, consequently, a is also a rational number. Clearly we have the hierarchy of sets

(44) $$\mathbf{N} \subset \mathbf{Z} \subset \mathbf{Q} \subset \mathbf{R}.$$

The set of rational numbers is sufficient for all of the usual applications of mathematics to accounting, business, etc. However, **Q** is aesthetically unsatisfactory because it is incomplete. If we try to find $\sqrt{2}$ in **Q** we will certainly fail.‡ Any number in **R** that is not in **Q**, is said to be an *irrational number*. Numbers such as $\sqrt{2}$, $\sqrt[3]{17}$, $\sqrt[5]{1001}$, π, $\sqrt{7}+\pi^2$ are irrational numbers. However $\sqrt[3]{24,389}$ is rational, because $\sqrt[3]{24,389} = 29$.

The next step is a careful development of the decimal system for representing real numbers. To shorten our labor we will omit this step, and we will assume that the student is already familiar with the decimal system. We recall that any decimal expression is either periodic (repeating) or nonperiodic. For example,

$$\frac{1}{4} = 0.25000..., \qquad \frac{101}{11} = 9.181818...,$$

$$\frac{12}{37} = 0.324324..., \qquad \frac{1}{7} = 0.142857142857...,$$

are periodic decimals. The first example is said to be a *terminating* decimal. We usually drop the zeros and write $1/4 = 0.25$. The other examples are periodic and nonterminating. It is customary to indicate the periodic part with a bar; thus

$$\frac{101}{11} = 9.\overline{18}, \qquad \frac{12}{37} = 0.\overline{324}, \qquad \frac{1}{7} = 0.\overline{142857}.$$

For our purpose the main item is

Theorem 7
PWO

A number is rational if and only if its decimal representation is periodic.

Now it is easy to give a decimal expression that represents an irrational number. We merely indicate a sequence of digits that is not periodic. For example, let

$$x = 0.1010010001000010000010...,$$

where after each 1 we insert a sequence of zeros longer than any previously used. By Theorem 7, we see that x is an irrational number.

We conclude with a diagram that shows the composition of the set of real numbers.

In Chapter 18 we will see that there is a still larger field **C**, the field of complex numbers, that contains the field **R**.

‡ This is proved in Appendix 4.

FIGURE 2.1

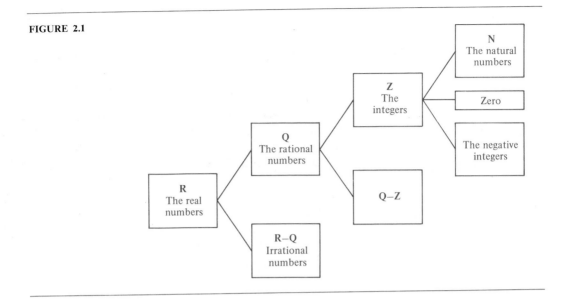

Review Questions

Try to answer the following questions as accurately as possible before consulting the text.

1. State the eleven axioms for a field. *Hint:* Five of the axioms involve addition only, five involve multiplication only, and one axiom relates the two operations.

2. Explain the meaning of the words (a) commutative, (b) associative, and (c) distributive.

3. Explain the difference between the Left Distributive Law and the Right Distributive Law.

4. Given $ab = 0$, what can you conclude? (See Theorem 3.)

5. State the rules for products and quotients that involve negative signs. (See equations (20) through (23).)

6. State the rules for the sum, product, and quotient of two fractions. (See equations (30), (29), and (31).)

7. State the test for the equality of two fractions. (See Theorem 6.)

8. Define the sets **N**, **Z**, and **Q**.

Inequalities

3

In this chapter we give four more axioms for the real numbers. These additional axioms together with the eleven axioms discussed in Chapter 2 are sufficient for a complete description of the real number system. The first three axioms are basic for the study of inequalities. We also give a very useful geometric interpretation of the real numbers as coordinates of points on a line. Throughout the remainder of the text the symbols a, b, c, etc., represent real numbers, unless otherwise stated. With this agreement we can avoid endless repetition of the statements $a \in \mathbf{R}$, $b \in \mathbf{R}$, etc.

3.1 THE ORDER AXIOMS

These axioms depend heavily on the idea of a positive number. Hence we first look at positive numbers in an intuitive and casual manner. We are accustomed to thinking of the set $\mathbf{N} = \{1, 2, 3, \ldots\}$ as a set of positive numbers and we regard each integer in $\mathbf{N}$ as a positive integer. Further the quotient a/b is regarded as a positive rational number if a and b are positive integers. Finally, an irrational number is positive if it lies "between" two positive rational numbers. For example, $\sqrt{2}$ is certainly positive since it lies "between" 1.41 and 1.42. Thus, we have an intuitive feeling about the existence of a set $\mathbf{P}$ of positive numbers. We now reformulate these ideas by giving a suitable set of three axioms for the set $\mathbf{P}$. The concept of "between" is suppressed in these axioms but will reappear in due time. Further, the axioms do not tell us explicitly which elements are positive numbers. This information is supplied in part by Theorems 1 through 7, and their corollaries.

AXIOM 12

The Trichotomy Axiom   *There is a set* **P** *called the set of positive numbers such that for each real number a exactly one of the following three assertions is true:*

$$(\text{I})\ a \in \mathbf{P}, \qquad (\text{II})\ a = 0, \qquad (\text{III})\ -a \in \mathbf{P}.$$

Each element of **P** is called a *positive number*. The set of all real numbers a for which $-a \in \mathbf{P}$ is denoted by $\mathbf{P}^{(-)}$ and is called the set of negative numbers. Each element of $\mathbf{P}^{(-)}$ is called a *negative number*. As the term "trichotomy" suggests this axiom partitions the set of real numbers into three mutually disjoint sets: the positive numbers, the negative numbers, and the set consisting of the single number 0.

For the set **P** of positive numbers there are two closure axioms.

AXIOM 13

If $a \in \mathbf{P}$ and $b \in \mathbf{P}$, then $a+b \in \mathbf{P}$.

According to this axiom, the sum of any two positive numbers is a positive number.

AXIOM 14

If $a \in \mathbf{P}$ and $b \in \mathbf{P}$, then $ab \in \mathbf{P}$.

According to this axiom, the product of any two positive numbers is a positive number.

From these axioms we can prove the following seven theorems. Since these results merely confirm our feelings about the positive numbers, the proofs are relegated to Appendix 5.

Theorem 1   *If $a \neq 0$, then $a^2 \in \mathbf{P}$.*

COROLLARY $1 \in \mathbf{P}$ *and* $-1 \in \mathbf{P}^{(-)}$.

Theorem 2   *Every natural number is in* **P**.

Theorem 3   *The product of a positive number and a negative number is a negative number.*

Theorem 4   *The product of two negative numbers is a positive number.*

Theorem 5   *The quotient of two positive numbers is a positive number.*

COROLLARY   *If a and b are natural numbers, then a/b is a positive number.*

Theorem 6   *The quotient of two negative numbers is a positive number.*

Theorem 7   *If a is a positive number and b is a negative number, then a/b and b/a are both negative numbers.*

3.2 INEQUALITIES

Now that we have an understanding of the set **P** of positive numbers, we proceed to the really important task of defining the inequality relation and the associated symbols $<$ and $>$.

DEFINITION 1   *The number a is said to be less than b, and we write*

(1) $a < b$ *(read, "a is less than b")*

if and only if the difference between $b - a \in \mathbf{P}$. Under these conditions we also write

(2) $b > a$ *(read, "b is greater than a").*

The relations (1) and (2) are called *inequalities*. From Definition 1 we see that $0 < b$ if and only if b is a positive number. Further $a < 0$ if and only if a is a negative number. As trivial examples we have

$$186 < 199, \qquad \text{because } 199 - 186 = 13 \in \mathbf{P};$$

$$-1{,}000 < 1, \qquad \text{because } 1 - (-1{,}000) = 1{,}001 \in \mathbf{P};$$

$$\frac{21}{34} < \frac{55}{89}, \qquad \text{because } \frac{55}{89} - \frac{21}{34} = \frac{1}{3{,}026} \in \mathbf{P}.$$

On the other hand, it is not at all obvious that

$$\sqrt{83} + \sqrt{117} < \sqrt{99} + \sqrt{101},$$

and it will take some effort to prove this. (See problem 7 of Exercise 6 in Chapter 5.)

Since (1) and (2) are equivalent, one of the symbols $<$ and $>$ is really unnecessary and could be dropped. However, it is convenient to have both of them available for use. It is also convenient to have a compound symbol $a \leqq b$, which means that either $a < b$ or $a = b$. Similarly, $a \geqq b$ means that either $a > b$ or $a = b$.

Theorem 8   *If a and b are any two real numbers, then exactly one of the following three relations holds*

$$\text{(I) } a < b, \qquad \text{(II) } a = b, \qquad \text{(III) } b < a.$$

Proof: By Axiom 12, either $b - a \in \mathbf{P}$, or $b - a = 0$, or $-(b-a) = a - b \in \mathbf{P}$, and these three cases are mutually exclusive. ∎

Theorem 9   *If $a < b$ and c is any positive number, then $ca < cb$.*

Thus, an inequality is preserved when multiplied on both sides by the same positive number.

EXAMPLE   If $\sqrt{7} < 2.65$, then $3\sqrt{7} < 7.95$.

Proof of Theorem 9:   Since $b - a \in \mathbf{P}$ and $c \in \mathbf{P}$, we have $c(b-a) \in \mathbf{P}$. Hence, $cb - ca \in \mathbf{P}$. Consequently, $ca < cb$. ∎

Theorem 10
PLE   *If $a < b$ and c is any negative number, then $ca > cb$.*

Thus, an inequality is reversed when multiplied on both sides by the same negative number.

EXAMPLE   If $8 < 9$, then $-8 > -9$.

Theorem 11   **The Transitive Property**   *If $a < b$ and $b < c$, then $a < c$.*

Proof: By definition $c - b \in \mathbf{P}$ and $b - a \in \mathbf{P}$. By Axiom 13 we have $(c-b) + (b-a) \in \mathbf{P}$. Hence, $c - a \in \mathbf{P}$ and $a < c$. ∎

Theorem 12
PLE   *If $a < c$ and $b < d$, then $a + b < c + d$.*

Theorem 13
PLE   *If $a < b$ then $a + c < b + c$.*

Theorem 14   *If $0 < a < b$, then $\dfrac{1}{a} > \dfrac{1}{b} > 0$.*

Thus, reciprocation reverses the inequality sign, when both members are positive.

EXAMPLE   Since $0 < 2 < 3$, we have $\dfrac{1}{2} > \dfrac{1}{3} > 0$.

Proof of Theorem 14:   Multiply both sides of $a < b$ by the positive number $1/ab$ and use Theorem 9. ∎

Theorem 15   *If $0 < a < b$ and $0 < c < d$, then $ac < bd$.*

Thus, multiplication of the corresponding terms of an inequality preserves the inequality. Of course all terms should be positive, and the inequality sign must be in the same direction in all three inequalities.

EXAMPLE   If $\sqrt{7} < 2.65$ and $\sqrt{3} < 1.74$, then

$$\sqrt{7}\sqrt{3} < (2.65)(1.74) = 4.611.$$

Proof of Theorem 15:   By Theorem 9, the inequality $a < b$ yields $ac < bc$. Similarly, $c < d$ yields $bc < bd$. Since $ac < bc$ and $bc < bd$, Theorem 11 gives $ac < bd$. ∎

In the next chapter we will discuss powers and roots in some detail. Here we assume that the reader is familiar with the elementary properties. Then we can prove

Theorem 16   *If $0 < a < b$ and n is any positive integer, then*

(3) $a^n < b^n.$

Proof: Apply Theorem 15, $n-1$ times with $c = a$ and $d = b$. ∎

Theorem 17   *If $0 < a < b$ and n is any positive integer, then*

(4) $\sqrt[n]{a} < \sqrt[n]{b},$

where, if n is even, the symbol $\sqrt[n]{}$ means the positive nth root.

EXAMPLE   Since $7 < 7.0225$, we have $\sqrt{7} < \sqrt{7.0225} = 2.65$. Hence $\sqrt{7} < 2.65$.

Proof of Theorem 17:   The proof is a little complicated because it uses the method of contradiction. By Theorem 8 there are only three possibilities:

(I) $\sqrt[n]{a} < \sqrt[n]{b}$,   (II) $\sqrt[n]{a} = \sqrt[n]{b}$,   (III) $\sqrt[n]{b} < \sqrt[n]{a}$.

We prove that the first case must hold, by showing that (II) and (III) are impossible. In each of the latter two cases we take the nth power of both sides. In case (II) we assume that $\sqrt[n]{a} = \sqrt[n]{b}$. Then we find that $a = b$. But this is impossible because, by hypothesis, $a < b$. In case (III) we apply Theorem 16 to $\sqrt[n]{b} < \sqrt[n]{a}$ and find that $b < a$. Again, this is contrary to the hypothesis that $a < b$. Since each of the cases (II) and (III) leads to a contradiction, the only case that can occur is (I). ■

In most of these theorems we can allow the equality sign to occur in the hypotheses as long as we make suitable modifications in the conclusion. In this way we obtain a large number of new theorems that vary only slightly from those already proved. There is no need to list them all, because they are really obvious.

As an illustration of the type of theorem to be expected, we give the following variations on Theorems 13 and 15.

Theorem $\overline{13}$
PLE

If $a \leq b$, then $a + c \leq b + c$.

Theorem $\overline{15}$
PLE

If $0 < a \leq b$ and $0 < c < d$, then $ac < bd$.

EXAMPLE 1

Find the set of all x such that $4x - 3 < -17$.

Solution: Let **S** be the desired set. Using Theorems 13 and 9 we see that $x \in S$ if and only if:

$$4x < -17 + 3 = -14$$

$$x < -\frac{14}{4} = -\frac{7}{2}.$$

Hence $\mathbf{S} = \{x \mid x < -7/2\}$. ▲

EXAMPLE 2

Find the set of all x such that $x^2 + 6x + 2 \leq 29$.

Solution: Again let **S** be the desired set. We observe that if we add 7 to the left side of the given inequality, we obtain the square of $x + 3$. Thus, $x \in S$ if and only if

$$x^2 + 6x + 2 + 7 \leq 29 + 7$$

$$x^2 + 6x + 9 \leq 36$$

$$(x + 3)^2 \leq 6^2.$$

Now the square of a negative number is also positive. Consequently, the last inequality is true if and only if

$$-6 \leqq x + 3 \leqq 6$$
$$-6 - 3 \leqq x \leqq 6 - 3.$$

Hence $S = \{x \mid -9 \leqq x \leqq 3\}$. ▲

Exercise 1

1. In each of the following, find the larger of the two given numbers.

    (a) $-8,$ -11 (b) $\sqrt{5}\sqrt{3},$ $\sqrt{7}\sqrt{2}$

    (c) $\sqrt[3]{11}\sqrt[3]{12},$ $\sqrt[3]{9}\sqrt[3]{15}$ (d) $18 - 29,$ $-15 + 7$

    (e) $\dfrac{3}{11},$ $\dfrac{5}{18}$ (f) $\dfrac{8}{5},$ $\dfrac{21}{13}$

    (g) $\dfrac{9}{7} - \dfrac{2}{3},$ $\dfrac{4}{5} - \dfrac{2}{11}$

In problems 2 through 10 find the set S of all x for which the given inequality is satisfied

2. $x + 5 \leqq 17$ 3. $2x - 3 \leqq 5$

4. $3x + 2 \leqq 1$ 5. $x^2 + 2x \leqq 3$

6. $x^2 - 4x + 1 \leqq 22$ 7. $x^2 + 10x \leqq 25$

★8. $2x^2 + 8x + 7 \leqq 127$ ★9. $9x^2 + 6x \leqq 63$

★10. $25x^2 + 40x + 8 \leqq -7$

11. Prove Theorems 10, 12, and 13.

In problems 12 through 18 prove the given assertion.

12. If $a < b$, then $a - c < b - c$.

13. If $a \leqq b$ and $b < c$, then $a < c$.

14. If $a \leqq c$ and $b \leqq d$, then $a + b \leqq c + d$.

15. If $a + c < b + c$, then $a < b$.

16. If $a < b$, then $-a > -b$, and conversely.

17. If $ac < bc$ and $c > 0$, then $a < b$.

18. Let $b, d \in \mathbf{P}$. Then $\dfrac{a}{b} < \dfrac{c}{d}$ if and only if $ad < bc$.

19. Convert Theorems 3 and 7 into theorems about inequalities.

20. Repeat problem 19 for Theorems 4 and 6.

21. Show that if we omit the conditions $0 < a$ and $0 < c$, then the assertion of Theorem 15 is false.

3.3 THE ABSOLUTE VALUE

It is convenient to have a special symbol $|a|$ (read, "the absolute value of a" or "the numerical value of a") that denotes a if $a \geq 0$, and denotes $-a$ (which is positive) if $a < 0$. Thus, by the definition of the symbol,

(5)
$$|a| = \begin{cases} a, & \text{if } a \geqq 0, \\ -a, & \text{if } a < 0. \end{cases}$$

Clearly either $a \geqq 0$ or $a < 0$. Consequently, this definition includes all cases.

Examples:

$$|19| = 19, \qquad |-19| = -(-19) = 19,$$

$$|x| = |-x|, \qquad \left| \pi - \frac{22}{7} \right| = \frac{22}{7} - \pi,$$

$$\left| \frac{2}{3} - \frac{3}{4} \right| = \left| \frac{2(4) - 3(3)}{3(4)} \right| = \left| \frac{8-9}{12} \right| = \left| \frac{-1}{12} \right| = \frac{1}{12}.$$

The definition gives immediately

Theorem 18
PLE

For any real number a

(6)
$$0 \leqq |a|, \qquad \text{and} \qquad -|a| \leqq a \leqq |a|.$$

Further, $|a| = 0$ if and only if $a = 0$.

Theorem 19
PLE

If a and b are real numbers, then

(7)
$$|ab| = |a||b|,$$

(8)
$$|a-b| = |b-a|,$$

(9)
$$\left| \frac{a}{b} \right| = \frac{|a|}{|b|}, \qquad b \neq 0.$$

A careful consideration of the various cases will prove (7). Then (8) and (9) follow immediately from (7).

Theorem 20
PLE

If a and b are real numbers, then

(10) $$|a+b| \leq |a|+|b|,$$

and

(11) $$|a-b| \geq |a|-|b|.$$

3.4 COORDINATES ON A LINE

The real number system can be presented without reference to geometry. However, a geometric interpretation of the real numbers is often an excellent aid to understanding relations among real numbers. To this end, we review the concept of coordinates on a line and perhaps go a little deeper.

Let L be a line. On this line we select: (a) a point O called the origin, (b) a positive direction indicated by an arrow, and (c) a unit of measurement. Such a triple is called a *directed line*. Other terms such as *number line*, *coordinate line*, and *axis* are in common use and may be used interchangeably. It is customary (but not mandatory) to place the line horizontally with the arrow to the right as indicated in Figure 3.1. When L has this position it is

FIGURE 3.1

also called the *x-axis*, because of the custom of using x (any letter will do) to indicate a real number associated with a point‡ P on L. The real number associated with P is called the *coordinate* of P and is obtained as follows. If $x = n$ a positive integer, we measure n units to the right of the origin to obtain the point whose coordinate is n. (See Figure 3.1 where the points with coordinates 1, 2, 3, and 4 are indicated.) If n is a negative integer, the procedure is the same except that we move to the left of the origin. The coordinate of the origin is 0. Suppose that $x = a/b$ where a and b are integers and $b > 0$.

‡ When we wish to distinguish between the coordinates (real numbers) and their corresponding points on L, we will use small letter $a, b, c, ..., x, ...$ for the coordinates and capitals $A, B, C, ..., P, ...$ for the corresponding points. Frequently there is no need to make this distinction, and we will often say "the point b" instead of "the point B whose coordinate is b."

We divide the unit into b equal parts to obtain smaller segments and using $|a|$ of these segments (a might be negative) we locate a point whose co-ordinate is a/b (to the right of O if $a > 0$, and to the left if $a < 0$). Hence for each rational number x, there is a unique point P on L whose coordinate is x. Further, by our construction, the distance from the origin to the point P is $|x|$ whenever x is a rational number. The point P is to the right of O if $x > 0$, and P is to the left of O if $x < 0$.

Suppose that x is an irrational number. One can argue rather persuasively that there is a point on L with the coordinate x but one cannot really give a mathematical proof (there is no foundation on which to base a proof). The proper approach is to make this assertion an axiom because indeed it is an assumption about the nature of the line L. Thus we assume that for each point P on L there is exactly one real number x that is the coordinate of P, and for each real number x there is exactly one such point. This biunique pairing is called a *one-to-one correspondence*. This assumption is not an axiom about the real numbers, but a geometric axiom about the nature of a line. Hence we give it the special title

AXIOM G

Let L be a directed line. Then there is a one-to-one correspondence between the points P on L and the set $\mathbf{R}$ of real numbers. If the number x corresponds to P, then x is called the coordinate of P. If x_1 and x_2 are coordinates of P_1 and P_2, respectively, then P_2 lies to the right of P_1 if and only if the numbers satisfy the inequality

$$(12) \qquad x_1 < x_2.$$

The last part of Axiom G serves to "locate" the "irrational points." Suppose that P_1, P_2, and P_3 have the coordinates 1.41, $\sqrt{2}$, 1.42, respectively. Since $1.41 < \sqrt{2} < 1.42$, we see that P_2 lies to the right of P_1 and to the left of P_3. Thus, the point with coordinate $\sqrt{2}$ lies "between" P_1 and P_3.

We will use the symbol $|P_1 P_2|$ to denote the length of the line segment whose endpoints are P_1 and P_2. This is also the distance between the points P_1 and P_2. If the coordinates x_1 and x_2 are rational, then it is easy to argue that:

$$(13) \qquad |P_1 P_2| = x_2 - x_1, \quad \text{if } x_1 < x_2 \quad (P_2 \text{ is to the right of } P_1)$$

and

$$(14) \qquad |P_1 P_2| = x_1 - x_2, \quad \text{if } x_2 < x_1 \quad (P_1 \text{ is to the right of } P_2).$$

In any case we have $|P_1 P_2| = |x_2 - x_1|$. If P_1 or P_2 has an irrational coordinate, then we can no longer appeal to direct measurements. The simple (and correct) way is to make

DEFINITION 2   *Distance (Length of Line Segment)*   *If P_1 and P_2 are points on L with co-ordinates x_1 and x_2, respectively, then*

(15) $$|P_1 P_2| = |x_2 - x_1|.$$

An immediate and trivial consequence of this definition is

Theorem 21   *If x is the coordinate of P, then the distance from the origin to P is $|x|$.*

Proof: From equation (15) we have $|OP| = |x - 0| = |x|$.   ∎

EXAMPLE 1   Find the distance between P_1 and P_2 in each of the following cases:

(a) $x_1 = 2,$ $x_2 = 5$ (b) $x_1 = -7,$ $x_2 = 4$

(c) $x_1 = -9,$ $x_2 = -15$ (d) $x_1 = 13/21,$ $x_2 = 8/13$

(e) $x_1 = 10,$ $x_2 = -10$ (f) $x_1 = \sqrt[3]{46},$ $x_2 = \sqrt{13}.$

Solution: By definition

(a) $|P_1 P_2| = |x_2 - x_1| = |5 - 2| = |3| = 3$

(b) $|P_1 P_2| = |x_2 - x_1| = |4 - (-7)| = |4 + 7| = |11| = 11$

(c) $|P_1 P_2| = |x_2 - x_1| = |-15 - (-9)| = |-15 + 9| = |-6| = 6$

(d) $|P_1 P_2| = |x_2 - x_1| = \left| \dfrac{8}{13} - \dfrac{13}{21} \right| = \left| \dfrac{168 - 169}{273} \right| = \left| \dfrac{-1}{273} \right| = \dfrac{1}{273}$

(e) $|P_1 P_2| = |x_2 - x_1| = |-10 - 10| = |-20| = 20$

(f) $|P_1 P_2| = |x_2 - x_1| = |\sqrt{13} - \sqrt[3]{46}|.$

In this last case we need to know which coordinate is larger before we can remove the absolute value sign. If we guess that $\sqrt{13} < \sqrt[3]{46}$ and raise both sides to the sixth power, we obtain $13^3 < 46^2$ or $2197 < 2116$. Since this is false, we have $\sqrt{13} > \sqrt[3]{46}$. Hence in case (f) we have $|P_1 P_2| = \sqrt{13} - \sqrt[3]{46}$.   ▲

The reader should draw a directed line, locate the two points, and measure the distance in each of the first five cases.

Theorem 22
PLE   *Let P_1 and P_2 be any points on L. Then*

(I) $|P_1 P_2| \geq 0,$

(II) $|P_1 P_2| = 0$ *if and only if the two points coincide,*

(III) $|P_1 P_2| = |P_2 P_1|.$

We say that P_2 lies *between* P_1 and P_3 if either

(16) $$x_1 < x_2 < x_3$$

or

(17) $$x_3 < x_2 < x_1.$$

Theorem 23
PLE

If P_2 lies between P_1 and P_3, then

(18) $$|P_1 P_3| = |P_1 P_2| + |P_2 P_3|.$$

3.5 A GEOMETRIC REPRESENTATION OF INEQUALITIES

Now that we have established a one-to-one correspondence between the set **R** of real numbers and the points on a directed line, it is easy to visualize the set of numbers that satisfies an inequality in one variable as a certain distinguished set of points on the line.

EXAMPLE 1

Draw a picture showing each of the following:

$$\mathbf{A} = \{x \mid -4 < x < -2\}, \quad \mathbf{B} = \{x \mid 1 \leqq x \leqq 2\}, \quad \mathbf{C} = \{x \mid 5 < x \leqq 8\}.$$

Solution: The sets are shown in color in Figure 3.2. Each of these sets is an example of an interval. In the figure we use parentheses on the set **A** to indicate that the points $x = -4$ and $x = -2$ are not in **A**. When the extreme points of the interval are in the set, as in **B**, this is indicated by brackets. The figure shows clearly that $x = 5$ is not in **C**, but $x = 8$ is in **C**. ▲

FIGURE 3.2

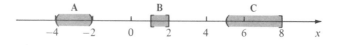

DEFINITION 3

Interval *A set of the form $\{x \mid a < x < b\}$ is called an open interval. The set $\{x \mid a \leqq x \leqq b\}$ is called a closed interval. The sets $\{x \mid a < x \leqq b\}$ and $\{x \mid a \leqq x < b\}$ are called half-open (or half-closed) intervals. Any one of these sets is an interval. For each of these intervals the points $x = a$ and $x = b$ are called the endpoints.*

For brevity, these intervals are often indicated by the symbols (a, b), $[a, b]$, $(a, b]$, and $[a, b)$, respectively. If we are not afraid of the symbol ∞, we can use it to indicate "unbounded" intervals. Thus by definition:

$$(a, \infty) \equiv \{x \mid a < x\}, \qquad [a, \infty) \equiv \{x \mid a \leqq x\},$$
$$(-\infty, b) \equiv \{x \mid x < b\}, \qquad (-\infty, b] \equiv \{x \mid x \leqq b\}.$$

Any set of this type is called a *ray*.

EXAMPLE 2   Draw a picture showing the rays $\mathbf{D} = (9, \infty)$ and $\mathbf{E} = (-\infty, 3]$.

Solution: Of course it is impossible to show the entire ray, but Figure 3.3 indicates clearly the nature of the sets $\mathbf{D}$ and $\mathbf{E}$.   ▲

FIGURE 3.3

EXAMPLE 3   Let $\mathbf{I}$ be the set of all x such that $|x - 3| \leqq 5$. Give a simpler description of this set (a) analytically and (b) geometrically.

Solution: From the definition of absolute value, $|x - 3| \leqq 5$ if and only if

(19) $-5 \leqq x - 3 \leqq 5.$

Adding 3 to each member of the inequality (19), we have

(20) $-2 \leqq x \leqq 8.$

This inequality describes the set $\mathbf{I}$ analytically. Clearly $\mathbf{I}$ is just the closed interval $[-2, 8]$.

Let P_0 and P be the points with coordinates 3 and x, respectively. Then $|x - 3| \leqq 5$ states that the distance from P_0 to P does not exceed 5. Geometrically, $\mathbf{I}$ is the set of all points P such that $|P_0 P| \leqq 5$. This set is shown in color in Figure 3.4.   ▲

FIGURE 3.4

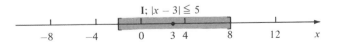

Theorem 24
PLE

Let $b > 0$. Then the set of all x such that $|x - a| \leqq b$ is a closed interval with center at the point $x = a$ and length $2b$.

EXAMPLE 4

Describe geometrically the set **J** of all x such that $|x^2 - 17| < 8$.

Solution: The given inequality is equivalent to $-8 < x^2 - 17 < 8$. This is equivalent to $9 < x^2 < 25$. Consequently, either $3 < x < 5$ or $-5 < x < -3$. Hence **J** is the union of the two open intervals $(-5, -3)$ and $(3, 5)$. The set **J** is shown in color in Figure 3.5. ▲

FIGURE 3.5

Exercise 2

In problems 1 through 6 find the indicated absolute value

1. $\left| \dfrac{1}{3} - \dfrac{1}{2} \right|$

2. $\left| \dfrac{4}{7} - \dfrac{5}{8} \right|$

3. $|\sqrt{15} - 2\sqrt{3}|$

4. $|\sqrt{17} - 3\sqrt{2}|$

5. $|5\sqrt{3} - 3\sqrt{10}|$

6. $|\sqrt{8} - \sqrt[3]{23}|$

7. Illustrate Theorem 23, by composing your own examples. In other words, select three points at random on a line and determine their coordinates. Compute the distances in accordance with Definition 2, and check that equation (18) holds.

8. Prove Theorems 19, 22, 23, and 24.

In problems 9 through 24 a set is specified by an inequality. In each case represent the set as an interval or union of intervals (if possible) and make a drawing showing the set.

9. $|x - 5| \leqq 8$

10. $|x - 7| < 2$

11. $|x + 5| < 3$

12. $|x + 100| \leqq 10$

13. $|x - 71| \leqq 0$

14. $|x - 5| \leqq -1$

15. $x^2 < 7$

16. $(x - 1)^2 < 9$

17. $1 < x^2 \leqq 16$

18. $1 \leqq x^4 < 81$

19. $-5 < x$

20. $49 \leqq x^2$

21. $|x - 3| < 1/10$

22. $|x - 3| < 1/100$

23. $|x+3| > 100$ 24. $|x+3| > 1000$

25. Let $A = \{x \mid |x-c| < a\}$ and let $B = \{x \mid |x-c| < b\}$, where $a > 0$, and $b > 0$. Prove that $A \subset B$ if and only if $a \leq b$.

26. Let $C = \{x \mid |x-a| > c\}$, and let $D = \{x \mid |x-a| > d\}$, where $c > 0$ and $d > 0$. Prove that $C \subset D$ if and only if $c \geq d$.

27. Prove that $[a,b] = (-\infty, b] \cap [a, \infty)$. Is this still true if $b < a$?

28. Let $I = [3, 10]$ and let $J = \{x \mid 3 \leq x \leq 10, \ x \in \mathbf{Q}\}$. Is it possible to make an accurate drawing of these two sets that will show that $I \neq J$?

29. Describe a geometric method of locating a point whose coordinate is (a) $\sqrt{2}$ (b) $\sqrt{5}$. *Hint:* Draw a suitable right triangle and use the Pythagorean Theorem $c^2 = a^2 + b^2$.

30. Use the results obtained in problem 29 to give a geometric method of locating a point whose coordinates are: (a) $\sqrt{3}$, (b) $\sqrt{6}$, (c) $\sqrt{7}$.

★ 31. Let A, B, and C be points whose coordinates are a, b, and $a+b$, respectively. Interpret the inequality (10) of Theorem 20 as a relation among the distance $|OA|$, $|OB|$, and $|OC|$. Show that with this interpretation the inequality is obviously true. Consider the possibility that a, or b, (or both) may be negative.

★ 32. Prove the first half of Theorem 20. *Hint:* Add the two inequalities $-|a| \leq a \leq |a|$ and $-|b| \leq b \leq |b|$.

★ 33. Complete the proof of Theorem 20 by showing that the inequality (11) follows from (10). *Hint:* $|a| = |(a-b) + b| \leq |a-b| + |b|$.

3.6 DIRECTED DISTANCES

It is often convenient to attach a sign to the distance $|P_1 P_2|$ to indicate the direction of travel from P_1 to P_2. When we do this, we obtain the directed distance which we denote by $\overline{P_1 P_2}$ (we omit the absolute value signs and add a bar). Thus by definition

$$(21) \quad \overline{P_1 P_2} = \begin{cases} |P_1 P_2|, & \text{if the direction from } P_1 \text{ to } P_2 \text{ coincides with that of L} \\ -|P_1 P_2|, & \text{if the direction from } P_1 \text{ to } P_2 \text{ is opposed to that of L} \\ 0, & \text{if } P_1 \text{ and } P_2 \text{ coincide.} \end{cases}$$

Theorem 25
PLE

Let P_1 and P_2 be two points on L with coordinates x_1 and x_2, respectively. Then

$$(22) \quad \overline{P_1 P_2} = x_2 - x_1.$$

FIGURE 3.6

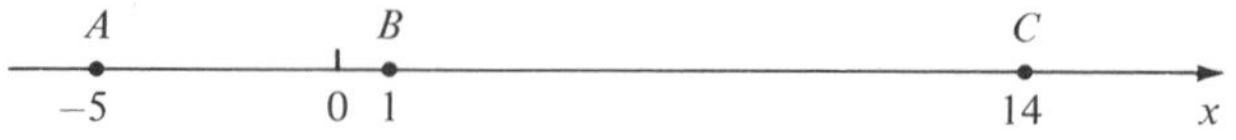

EXAMPLE 1   Let A, B, and C be the points indicated in Figure 3.6. Compute the six different directed distances that can be formed with these three points.

Solution: By Theorem 25 (or directly from the figure)

$$\overline{AB} = 1 - (-5) = 6, \qquad \overline{BA} = -5 - 1 = -6,$$
$$\overline{BC} = 14 - 1 = 13, \qquad \overline{CB} = 1 - 14 = -13,$$
$$\overline{AC} = 14 - (-5) = 19, \qquad \overline{CA} = -5 - 14 = -19. \quad \blacktriangle$$

In Theorem 23 we assume that P_2 lies between P_1 and P_3. When we use directed distances, this restriction can be dropped.

Theorem 26   Let P_1, P_2, and P_3 be any three points on a directed line. Then

(23) $$\overline{P_1 P_3} = \overline{P_1 P_2} + \overline{P_2 P_3}.$$

Proof: If x_1, x_2, and x_3 are the coordinates of these three points, then (23) is equivalent to the identity

(24) $$x_3 - x_1 = (x_2 - x_1) + (x_3 - x_2). \quad \blacksquare$$

EXAMPLE 2   Let A, B, and C be the three points indicated in Figure 3.6. There are six different ways of selecting these as P_1, P_2, and P_3. List all six ways. Check numerically that in each case equation (23) is true.

Solution: Two of the six possible selections of A, B, and C for P_1, P_2, and P_3 in (23) are

(25) $$\overline{BA} = \overline{BC} + \overline{CA} \qquad \text{and} \qquad \overline{CB} = \overline{CA} + \overline{AB}.$$

Using the directed distances just computed, we find that these are equivalent to

$$-6 = 13 + (-19) \qquad \text{and} \qquad -13 = -19 + 6.$$

The student should list the other four possible selections and in each case make a similar numerical check.

Exercise 3

1. Check Theorems 25 and 26 by composing your own numerical examples. In other words, draw a directed line L, select coordinates at random, and locate the points. Then check that equation (22) does give the directed distances, and that (23) is always satisfied.

2. *Midpoint formula.* Let P_1 and P_2 have the coordinates x_1 and x_2, and let M have the coordinate $(x_1 + x_2)/2$. Prove that M is the midpoint of the line segment joining P_1 and P_2 by proving that $\overline{P_1 M} = \overline{MP_2}$.

★ 3. *Trisection points.* With the notation of problem 2, let T_1 and T_2 have the coordinates $(2x_1 + x_2)/3$ and $(x_1 + 2x_2)/3$, respectively. Prove that T_1 and T_2 are trisection points by proving that $\overline{P_1 T_1} = \overline{T_1 T_2} = \overline{T_2 P_2}$.

★ 4. *Generalization.* Using the results of problems 2 and 3, guess at formulas for the coordinates of $Q_1, Q_2, Q_3, ..., Q_{n-1}$ if they divide the segment joining P_1 and P_2 into n equal parts. Check your guess by proving that $\overline{P_1 Q_1} = \overline{Q_1 Q_2} = \cdots = \overline{Q_{n-1} P_2}$.

5. Find the midpoint for the line segment joining P_1 and P_2 with coordinates x_1 and x_2 if:

   (a) $x_1 = 3$, $\qquad$ $x_2 = 7$ $\qquad$ (b) $x_1 = -5$, $\qquad$ $x_2 = 13$

   (c) $x_1 = -9$, $\qquad$ $x_2 = -3$ $\qquad$ (d) $x_1 = \frac{7}{13}$, $\qquad$ $x_2 = \frac{13}{7}$

   (e) $x_1 = 4 - \sqrt{13}$, $x_2 = 11 + \sqrt{13}$. (f) $x_1 = -2 - \sqrt{18}$, $x_2 = 22 + \sqrt{2}$.

★ 6. Find the trisection points for the line segment joining P_1 and P_2 with coordinates x_1 and x_2 if

   (a) $x_1 = 14$, $\quad$ $x_2 = 23$ $\qquad$ (b) $x_1 = -7$, $\qquad$ $x_2 = 32$

   (c) $x_1 = 2/5$, $\quad$ $x_2 = 5/2$ $\qquad$ (d) $x_1 = -14 - 3\sqrt{2}$, $\quad$ $x_2 = -5 + 6\sqrt{2}$.

★ 7. Find the points that divide the line segment from $x_1 = 2/3$ to $x_2 = 33/2$ into five equal parts.

★3.7 BOUNDED SETS AND THE COMPLETENESS AXIOM

DEFINITION 4 $\quad$ ***Bounded Set*** $\quad$ *A set* S *of real numbers is said to be bounded above if there is a number* u *such that for every* x *in* S *we have* $x \le u$. *The set* S *is said to be bounded below if there is a number* ℓ *such that for every* x *in* S *we have* $x \ge \ell$. *The set* S *is said to be bounded if it is bounded above and bounded below.*

Geometrically these concepts are simple. We represent the set of numbers by a corresponding set of points on a directed line. If all of the points of S are to the left of the point $x = u$, then S is bounded above. If all of the points of S are to the right of $x = \ell$, then S is bounded below. If the set S lies in an interval $[\ell, u]$, then the set S is bounded. A bounded set is pictured in Figure 3.7, where the points of S are represented by the dots.

FIGURE 3.7

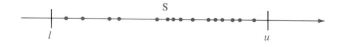

The number u in Definition 4 is called an *upper bound* for S, and the number ℓ is called a *lower bound* for S. Once we find an upper bound u for S, then any number larger than u is also an upper bound for S. The objective is to find the least upper bound for S. This number is denoted by $\ell.u.b.$ S. Similarly if S is bounded below, we want to find the greatest lower bound for S. This number is denoted by $g.\ell.b.$ S.

EXAMPLE 1

For each of the following sets determine if it is bounded above or bounded below. In each case find the $\ell.u.b.$ and $g.\ell.b.$ if they exist.

(a) N
(b) Z

(c) $\mathbf{C} = \{x \mid x = n/(n+7),\ n \in \mathbf{N}\}$
(d) $\mathbf{D} = \{x \mid x = n^2/(n+7),\ n \in \mathbf{N}\}$

(e) $\mathbf{E} = \{x \mid \sqrt{3} \leq x \leq \sqrt{17},\ x \in \mathbf{Q}\}$
(f) $\mathbf{F} = \{x \mid x = -1/(a^2 + 2b^2),\ a, b \in \mathbf{N}\}$

Solution: The reader should check carefully each of the following assertions.

(a) The set N is not bounded above. It is bounded below and $g.\ell.b.$ $\mathbf{N} = 1$.
(b) The set Z of all integers is not bounded above and is not bounded below.
(c) $\mathbf{C} = \{1/8, 2/9, 3/10, \ldots, 100/107, \ldots\}$. C is a bounded set, $\ell.u.b.$ $\mathbf{C} = 1$, and $g.\ell.b.$ $\mathbf{C} = 1/8$.
(d) $\mathbf{D} = \{1/8, 4/9, 9/10, \ldots, 10{,}000/107, \ldots\}$. D is not bounded above, but it is bounded below, and $g.\ell.b.$ $\mathbf{D} = 1/8$.
(e) $\ell.u.b.$ $\mathbf{E} = \sqrt{17}$, and $g.\ell.b.$ $\mathbf{E} = \sqrt{3}$.
(f) If $a, b \in \mathbf{N}$, then $3 \leq a^2 + 2b^2$. On the other hand, we can make $a^2 + 2b^2$ arbitrarily large. Hence,

$$0 < \frac{1}{a^2 + 2b^2} \leq \frac{1}{3},$$

where equality on the right side can occur when $a = 1$ and $b = 1$ but the equality on the left side cannot occur for any a, b in $\mathbf{N}$. Multiplying by -1 we see that $-1/3 \leq -1/(a^2 + 2b^2) < 0$. Consequently, we have that $g.\ell.b.\ \mathbf{F} = -1/3$ and $\ell.u.b.\ \mathbf{F} = 0$. ▲

In some cases the least upper bound is in the set and in other cases it is not. The reader should check the following statements about the sets of Example 1.

(a) $g.\ell.b.\ \mathbf{N} = 1 \in \mathbf{N}$.

(c) $g.\ell.b.\ \mathbf{C} = 1/8 \in \mathbf{N}$ and $\ell.u.b.\ \mathbf{C} = 1 \notin \mathbf{C}$.

(d) $g.\ell.b.\ \mathbf{D} = 1/8 \in \mathbf{D}$.

(e) $g.\ell.b.\ \mathbf{E} = \sqrt{3} \notin \mathbf{E}$ and $\ell.u.b.\ \mathbf{E} = \sqrt{17} \notin \mathbf{E}$.

(f) $g.\ell.b.\ \mathbf{F} = -1/3 \in \mathbf{F}$ and $\ell.u.b.\ \mathbf{F} = 0 \notin \mathbf{F}$.

The material covered so far is interesting and useful in its own right. However, our primary object is the necessary preparation for the last axiom for the real number system. This last axiom is

AXIOM 15

The Completeness Axiom *If* $\mathbf{S}$ *is a nonempty set of real numbers that is bounded above, then* $\mathbf{S}$ *has a least upper bound that is in* $\mathbf{R}$ *(the least upper bound is a real number).*

Any set that satisfies Axioms 1 through 15 is called a *complete ordered field*. Axioms 1 through 11 are the ones for a field. Axioms 12 through 14 order the elements (see Theorems 8 and 11), and Axiom 15 assures us that the set is complete (no holes). For example, the set $\mathbf{Q}$ satisfies the first fourteen axioms, but does not satisfy the completeness axiom. The set $\mathbf{Q}$ is not "complete" because it does not contain such numbers as $\sqrt{2}, \sqrt[5]{3}, \sqrt{5} - \sqrt{11}, \pi, \pi^2$, etc. Much of algebra and trigonometry can be studied without really using the completeness axiom. Nevertheless, it is an important property of the real numbers, and it is essential for a rigorous presentation of calculus.

Review Questions

Try to answer the following questions as accurately as possible before consulting the text.

1. State the three axioms for the set $\mathbf{P}$ of positive real numbers. These are the order axioms.

2. State the definition of $a < b$.

3. State as many theorems as you can recall that involve algebraic manipulations with inequalities. (See Theorems 9 through 17.)

4. State as many properties as you can for the absolute value symbol. (See Theorems 18, 19, and 20.)

5. Explain the difference between distance and directed distance.

6. Give the formula for the coordinate of the midpoint of a line segment. (See Exercise 3, problem 2.)

7. Define (a) a set that is bounded above, (b) a set that is bounded below, and (c) a bounded set.

8. Give an example of a set that is bounded above and not bounded below.

9. State the completeness axiom (Axiom 15).

Algebraic Expressions

|4| Manipulations with algebraic expressions may seem dull and pointless to the student. Perhaps it will help if he makes an analogy between mathematics and music. To play beautiful pieces of music the student must learn how to read the notes and he must practice scales, which in themselves may also seem dull and pointless. In mathematics practice with the manipulation of symbols corresponds to the practice of scales in music. As in music, mathematics also has its fascinating, beautiful, and deep creations, but they are harder to understand and harder to appreciate.

We have already had a taste of manipulation in Chapters 2, and 3. In this chapter the work is a little more complicated. The reader is invited to practice diligently with these mathematical exercises. Even if he fails to go far enough to see some of the exciting results in mathematics, he will probably find that what he has learned is useful.

4.1 INTEGRAL EXPONENTS

If n is a positive integer, the notation x^n is defined by

(1) $$x^n = x \cdot x \cdot x \cdots x$$

where the factor x occurs n times on the right side of (1). Of course if $n = 1$, then no multiplication is involved. In this case $x^1 = x$ by definition. In (1) the number x is called the *base*, and n is called the *exponent*. As examples we have $3^4 = 3 \cdot 3 \cdot 3 \cdot 3 = 81$ and $(-4)^3 = (-4)(-4)(-4) = -64$. The expression $(xy)^3$ means $(xy)(xy)(xy)$, but the expression xy^3 means the product $x(y^3) = xyyy$. From the definition we have

Theorem 1
PLE

Let m and n be positive integers. Then

(2)
$$x^m x^n = x^{m+n},$$

(3)
$$(x^m)^n = x^{mn},$$

(4)
$$(xy)^n = x^n y^n.$$

Further if $x \neq 0$, then

(5)
$$\frac{x^m}{x^n} = \begin{cases} x^{m-n}, & \text{if } m > n, \\ 1, & \text{if } m = n, \\ \dfrac{1}{x^{n-m}}, & \text{if } m < n, \end{cases}$$

and

(6)
$$\left(\frac{y}{x}\right)^n = \frac{y^n}{x^n}.$$

For example, to prove (2) we note that

$$x^m \cdot x^n = \underbrace{(x \cdot x \cdot \cdots \cdot x)}_{m \text{ factors}} \underbrace{(x \cdot x \cdot \cdots \cdot x)}_{n \text{ factors}} = \underbrace{x \cdot x \cdot \cdots \cdot x}_{m+n \text{ factors}} = x^{m+n}.$$

The other items in Theorem 1 are proved similarly.

EXAMPLE 1

Simplify each of the following expressions.

(a) $(3x^4 y^5)(5x^6 y)$,   (b) $(2xy^2 z^3)^4$,

(c) $\left(\dfrac{3y^2}{x^3}\right)^4 \left(\dfrac{4x}{y^2}\right)^5 \left(\dfrac{yx^3}{2}\right)^2.$

Solution: We employ the associative and commutative laws of multiplication repeatedly to justify rearrangements of the factors, but we no longer pause to give the details. From Theorem 1,

(a) $(3x^4y^5)(5x^6y) = 3 \cdot 5 \cdot x^4 \cdot x^6y^5 \cdot y = 15x^{10}y^6.$

(b) $(2xy^2z^3)^4 = (2)^4(x)^4(y^2)^4(z^3)^4 = 16x^4y^8z^{12}.$

(c) $\left(\dfrac{3y^2}{x^3}\right)^4 \left(\dfrac{4x}{y^2}\right)^5 \left(\dfrac{yx^3}{2}\right)^2 = \dfrac{3^4y^8 4^5 x^5 y^2 x^6}{x^{12} y^{10} 2^2}$

$$= \dfrac{3^4 \cdot 4^5 x^{11} y^{10}}{4x^{12} y^{10}} = \dfrac{(12)^4}{x}. \quad \blacktriangle$$

Notice that in (b) we use 16 in place of 2^4, while in (c) we prefer $(12)^4$ over 20,736.

If we use the law of exponents

$$\frac{x^m}{x^n} = x^{m-n}$$

when $n = m$, we get x^0. If $n > m$, then we obtain a negative exponent on the right side. For example,

$$\frac{x^4}{x^4} = x^{4-4} = x^0 \qquad \text{and} \qquad \frac{x^7}{x^{13}} = x^{7-13} = x^{-6}$$

But $x^4/x^4 = 1$, and $x^7/x^{13} = 1/x^6$. Hence these two examples suggest

DEFINITION 1

If $x \neq 0$ and n is any positive integer, then

(7) $$x^0 \equiv 1 \qquad \text{and} \qquad x^{-n} \equiv \frac{1}{x^n}.$$

The point of this definition lies in

Theorem 2
PWO

Let m and n be any pair of integers and suppose that $x \neq 0$, $y \neq 0$. Then equations (2) through (6) hold.

Later we will consider x^n when $n = p/q$ a rational number, and we will extend the definition in such a way that equations (2) through (6) still hold.

The proof of Theorem 2 requires the examination of a large number of cases. As an example consider equation (3) when the first exponent is positive and the second one is negative. Then equation (3) can be written in the form $(x^m)^{-n} = x^{m(-n)}$ with $m, n \in \mathbf{N}$. Using the definition we have

$$(x^m)^{-n} = \frac{1}{(x^m)^n} = \frac{1}{x^{mn}} = x^{-(mn)} = x^{m(-n)}.$$

EXAMPLE 2   Simplify each of the following expressions

(a) $(x^{-3}y^5)^{-2}$,   (b) $\dfrac{5y^2z^{-3}}{10y^{-3}z^2}$,   (c) $\dfrac{a^{-1}+b^{-2}}{a^{-3}b^{-4}}$.

Solution: Using equations (2) through (6) we have

(a) $(x^{-3}y^5)^{-2} = (x^{-3})^{-2}(y^5)^{-2} = x^{(-3)(-2)}y^{5(-2)} = \dfrac{x^6}{y^{10}}$.

(b) $\dfrac{5y^2z^{-3}}{10y^{-3}z^2} = \dfrac{5y^2/z^3}{10z^2/y^3} = \dfrac{1}{2}\cdot\dfrac{y^2}{z^3}\cdot\dfrac{y^3}{z^2} = \dfrac{y^5}{2z^5} = \dfrac{1}{2}\left(\dfrac{y}{z}\right)^5$.

(c) $\dfrac{a^{-1}+b^{-2}}{a^{-3}b^{-4}} = \dfrac{\dfrac{1}{a}+\dfrac{1}{b^2}}{\dfrac{1}{a^3}\cdot\dfrac{1}{b^4}}\cdot\dfrac{a^3b^4}{a^3b^4} = a^2b^4 + a^3b^2$. ▲

We observe that there is more than one way to work each problem. For example in (a) we could write

$$(x^{-3}y^5)^{-2} = \left(\frac{y^5}{x^3}\right)^{-2} = 1\Big/\left(\frac{y^5}{x^3}\right)^2 = 1\Big/\frac{y^{10}}{x^6} = \frac{x^6}{y^{10}}.$$

The answer to part (c) may be written as $a^2b^2(b^2+a)$. This latter form is just as simple as $a^2b^4+a^3b^2$. Consequently, there may be several different ways of simplifying a given expression. In general, to simplify a given expression means:

(a) Remove all zero and negative exponents.
(b) Combine terms whenever possible ($x^2x^3 = x^5$).
(c) Factor expressions whenever possible.
(d) Complete any indicated numerical computations.

Exercise 1

In problems 1 through 29 simplify the given expression.

1. $(-3/2)^5$

2. $(4/3)^3$

3. $\dfrac{4^{13}}{2^{25}}$

4. $\dfrac{(12)^{18}(3)^7}{(16)(4)^{16}}$

5. $2^0\cdot 4^2\cdot 8^3\cdot 16^{-4}$

6. $a^2a^4(-a)^8$

7. $(ab^2)^3(-a^2b)^{-1}$

8. $(a^{-2}b^{-3})(a^{-1}b^{-4})^{-2}$

9. $\dfrac{c^{13}d^{24}}{(c^6d^{13})^2}$

10. $4\left(\dfrac{a^3b}{a^5b^9}\right)\left(\dfrac{a^7b^5}{2a^2b^4}\right)$

11. $(-2ab^3)^5\left(\dfrac{a^0}{4b^2}\right)^3$

12. $\left(\dfrac{a^2b^3}{ab^2}\right)^4 \Big/ \left(\dfrac{a^4b^5}{a^6b^7}\right)^2$

13. $\left(\dfrac{x^3y^4}{z^5}\right)^2 \Big/ \left(\dfrac{x^2z^4}{y^5}\right)^3$

14. $\left(\dfrac{u^2v^{-2}}{w}\right)^3 \Big/ \left(\dfrac{w}{u^{-3}v^3}\right)^2$

15. $\dfrac{6x^{-2}y^0z^2}{3x^2yz^{-3}}$

16. $\left(\dfrac{u^{-3}v^4}{u^{-4}v^5}\right)^{-6}$

17. $\left(\dfrac{-c^{-3}d^{-2}e^{-1}}{c^{-4}d^{-5}e^{-6}}\right)^{-2}$

18. $\dfrac{(3^3p^4r^{-5}s^{-6})^2}{(9p^3r^{-4}s^{-2})^3}$

19. $(a^{-1}+a^{-2})a^3$

20. $(b^{-1}+b^{-2})b^{-3}$

21. $\dfrac{x^{-1}y^{-1}}{x^{-1}+y^{-1}}$

22. $\dfrac{u^{-4}+v^{-4}}{(uv)^{-4}}$

23. $\left(\dfrac{r^{-1}+s^{-1}}{r^{-2}+s^{-2}}\right)^{-1}$

24. $\left(\dfrac{r^{-2}s^{-3}}{r^{-4}+s^{-5}}\right)^{-2}$

25. $x^kx^{2k+1}(x^{3k+2})^{-1}$

26. $\dfrac{(x^{3k+2}x^{4k-3})^2}{x^{14k}}$

27. $\dfrac{(a^kb^{k+1})^2}{(a^{2-k}b^{3-k})^{-2}}$

28. $\dfrac{(x^{2k+3}y^{3k+11})^3}{(x^{k+1}y^{k+4})^6}$

29. $-x^1y^2z^3(x^2y^3z^4)^2(-x^3y^4z^5)^3(-x^7y^{10}z^{13})^{-2}$

4.2 SOME TERMINOLOGY

An expression of the form

(8) $$P_0 = 3ax^2y^3z^4 + 2bxy^4z^4 + cxy^3z^5$$

is composed of three products. Each product is called a *term* of P_0. The letters $a,b,c,\ x,y$ and z represent elements from some set. In this text they always represent real numbers (unless otherwise explicitly stated) but on some occasions we may restrict the set to be an interval. A symbol that represents an arbitrary element from a set **S** is called a *variable*. However, if the set **S** contains only one element, the symbol is called a *constant*. If x represents an arbitrary element from **S**, then **S** is called the *domain* of x or the *replacement set* for x.

It is customary to use letters near the end of the alphabet, such as x, y, and z for variables and letters near the beginning, such as a, b, and c as constants. However, there are some situations where x is a constant and a is a variable. When we select a particular number to replace a variable (letter), that number is called the *value of the variable*.

An expression such as P_0 that involves only a finite number of products and sums of the variables is called a *polynomial*. If the expression has only one term, it is called a *monomial*. If it has two terms, it is called a *binomial* etc. Obviously the polynomial P_0 given by (8) is a *trinomial*. In this case P_0 is a polynomial in the three variables x, y, and z. The factors $3a$, $2b$, and c are called *coefficients*. The numbers 2 and 3 are *numerical coefficients*, while a, b, and c are *literal coefficients*. Similar terms in a polynomial can be combined as illustrated in

EXAMPLE 1   Simplify the polynomial

$$P_1 \equiv 2x + 3y + 5z^2w + 3x - y - 7z^2w.$$

Solution: We can apply the various algebraic laws covered in Chapter 2 because the letters represent real numbers. Thus,

$$P_1 = (2x+3x) + (3y-y) + (5z^2w-7z^2w)$$
$$= (2+3)x + (3-1)y + (5-7)z^2w$$
$$= 5x + 2y - 2z^2w.$$

This is a simpler form for P_1. ▲

EXAMPLE 2   Simplify the polynomial.

$$P_2 = (x+2y) - (y+3z) + (2x-4z) - (2y-z).$$

Solution: We recall that subtraction is defined in terms of the additive inverse. Thus we replace $-(y+3z)$ by $+(-y-3z)$. It may be convenient to arrange the work in a column, just as in the addition of real numbers.

$$
\begin{array}{lll}
x & +2y & \\
 & -y & -3z \\
2x & & -4z \\
 & -2y & +z \\
\hline
\end{array}
$$

$$P_2 = (1+2)x + (2-1-2)y + (-3-4+1)z = 3x - y - 6z. \quad ▲$$

EXAMPLE 3

Simplify the expression for P_0 given in equation (8).

Solution: There are no like terms, but there are common factors. We observe that x is present as a factor in each term. By the Distributive Law

(9) $$P_0 = x(3axy^3z^4 + 2by^4z^4 + cy^3z^5).$$

Notice that y^3 is present in each term of the second factor, so again by the Distributive Law

(10) $$P_0 = xy^3(3axz^4 + 2byz^4 + cz^5).$$

Finally we can "factor out" z^4 and obtain

(11) $$P_0 = xy^3z^4(3ax + 2by + cz).$$

This is certainly simpler than (8). ▲

The sharp student will see all of the factors x, y^3, and z^4 at once and pass immediately from (8) to (11). The slower reader will gradually pick up speed with practice.

The process of going from (8) to (11) is called *factoring* the expression. Each of the expressions x, y, z, and $(3ax + 2by + cz)$ is called a *factor* of P_0. If we reverse the process and go from (11) to (8), then we are *expanding* the expression. Equation (11) is the factored form of P_0, while equation (8) is the expansion of P_0. Sometimes simplification requires both operations. This is illustrated in

EXAMPLE 4

Simplify the polynomial

$$P_4 \equiv 2y(x+3z) + 6z(2u-y) - 3z(4u-x).$$

Solution: We first expand the terms in P_4 and place like terms in the same column.

$$
\begin{array}{lllll}
2y(x+3z) = & 2xy & + 6yz & & \\
6z(2u-y) = & & - 6yz & + 12uz & \\
-3z(4u-x) = & & & - 12uz & + 3xz \\
\hline
P_4 \quad\;\; = & 2xy & + 0 & + 0 & + 3xz = x(2y+3z). \;\; \blacktriangle
\end{array}
$$

Expansion and factoring with a view to simplification are important techniques which the student must master.

Before going further with factoring we need additional terminology. The *degree* of a polynomial in one variable is the greatest exponent that appears in the polynomial. For example the polynomial

(12) $$P_5 \equiv 36 + 13x^2 + x^4 - 10x^3 - 40x$$

is of 4th degree. As a matter of neatness and organization one usually arranges the terms of the polynomial so that the exponents are decreasing. Thus, P_5 should be written as

(13) $$P_5 = x^4 - 10x^3 + 13x^2 - 40x + 36.$$

A nonzero constant c may be regarded as a polynomial of degree zero because $c = cx^0$. If the polynomial involves several variables, then it has a degree in each variable. For example the polynomial P_0 given by equation (8) is of second degree in x, fourth degree in y, and fifth degree in z. There is a systematic method for arranging the terms of a polynomial in several variables, but we do not need it here. P_2 is of first degree in each variable. Such a polynomial is called a *linear* polynomial.

The number 17 can be written as a product in many ways, thus

$$17 = 3 \cdot (17/3) = 11 \cdot (17/11) = \sqrt{29} \cdot (17/\sqrt{29}), \text{ etc.}$$

However when we speak of the factorization of an integer n into a product $q_1 q_2$, we usually mean that q_1 and q_2 are both integers. Similarly if a polynomial P is factored into the product $Q_1 Q_2$ we understand that Q_1 and Q_2 are also polynomials and that neither Q_1 nor Q_2 is a constant. A polynomial is said to be a *prime* if it cannot be factored into a product of two polynomials neither of which is a constant. A polynomial is completely factored if it is represented as a product of prime polynomials.

We consider some examples. Direct expansion of

(14) $$P_5 = (x-1)(x^2+4)(x-9)$$

will show that this is a factorization of the polynomial given by (12) or (13). If $x > 0$, then we can represent $x - 9$ as the product

$$x - 9 = (\sqrt{x} - 3)(\sqrt{x} + 3),$$

but this is not considered a factorization since the expressions $\sqrt{x} - 3$ and $\sqrt{x} + 3$ are not polynomials. Clearly, $x - 9$ is a prime factor of P_5.

The polynomial $x^2 + 4$ can not be factored using only real numbers. However if we introduce the complex number i, defined so that $i^2 = -1$, then

$$x^2 + 4 = (x + 2i)(x - 2i).$$

We describe such an event by saying that $x^2 + 4$ is *irreducible* over the field of real numbers, but it is *reducible* over the field of complex numbers. Similarly,

$$P_6 \equiv x^2 - 7 = (x + \sqrt{7})(x - \sqrt{7})$$

is irreducible over the field of rational numbers, but is reducible over the field of real numbers.

The factorization of P_5 given in equation (14) is a decomposition into a product of primes, if only real coefficients are allowed. If complex numbers may be used as coefficients, then

$$P_5 = (x - 1)(x + 2i)(x - 2i)(x - 9)$$

is the decomposition of P_5 into prime factors.

4.3 FACTORING POLYNOMIALS

The simplest type of polynomial to factor is the quadratic polynomial (polynomial of second degree) in one variable:

(15) $$P = Ax^2 + Bx + C, \qquad A \neq 0.$$

EXAMPLE 1   Factor $P_1 \equiv x^2 - 9x + 18$.

Solution. Assume that P_1 is the product

$$P_1 = (x + p)(x + q).$$

Expanding $(x + p)(x + q)$ we find that

$$P_1 = x^2 + px + qx + pq = x^2 + (p + q)x + pq.$$

This will coincide with $x^2 - 9x + 18$ if and only if

(16) $$p + q = -9 \qquad \text{and} \qquad pq = 18.$$

We now try to solve this pair of equations by inspection. (A systematic method will be explained in Chapter 5.) The simplest approach is to try integers for p and q. All pairs with product 18 are

$$1 \cdot 18, \qquad 2 \cdot 9, \qquad 3 \cdot 6, \qquad (-1)(-18), \qquad (-2)(-9), \qquad (-3)(-6).$$

Fortunately there is a pair $(-3, -6)$ such that the sum is -9. Consequently (16) is satisfied, and we have the factorization

$$P_1 \equiv x^2 - 9x + 18 = (x - 3)(x - 6). \quad \blacktriangle$$

With practice the student will learn to factor quadratics with ease. Note that $A = 1$ in Example 1. We now look at the general quadratic. Suppose that

(17) $\qquad Ax^2 + Bx + C = (ax+p)(bx+q) = abx^2 + (aq+bp)x + pq.$

Consequently we have a factorization of $Ax^2 + Bx + C$ if and only if

(18) $\qquad ab = A, \qquad pq = C, \qquad \text{and} \qquad aq + bp = B.$

Suppose now that all of the constants are integers. We list all pairs (a, b) for which $ab = A$, and all pairs (p, q) for which $pq = C$. If the numbers A and C are not large, there will not be many pairs, and it will be an easy matter to select two pairs such that $aq + bp = B$, if such a selection is possible.

EXAMPLE 2 $\qquad$ Factor $P_2 \equiv 15x^2 + 47x + 6$.

Solution: Here $A = 15$, $B = 47$, and $C = 6$. The pairs of factors are

$$ab = A = 15: \quad (\pm 15, \pm 1), \qquad (\pm 5, \pm 3), \qquad (\pm 3, \pm 5), \qquad (\pm 1 \ \pm 15),$$

$$pq = C = 6: \quad (\pm 6, \pm 1), \qquad (\pm 3, \pm 2), \qquad (\pm 2, \pm 3), \qquad (\pm 1, \pm 6).$$

To find a, b, p, q such that $aq + bp = 47$, we try various pairs. After a few tries we find that $(15, 1)$ and $(2, 3)$ will do, since $15 \cdot 3 + 1 \cdot 2 = 45 + 2 = 47$. Hence,

$$P_2 = 15x^2 + 47x + 6 = (15x+2)(x+3). \quad \blacktriangle$$

Sometimes embellishments are used to make the problem look harder.

EXAMPLE 3 $\qquad$ Factor $P_3 \equiv 15x^2 + 47xy^2 + 6y^4$.

Solution: Clearly the factorization must take the form

$$P_3 \equiv 15x^2 + 47xy^2 + 6y^4 = (ax+py^2)(bx+qy^2)$$
$$= abx^2 + (aq+bp)xy^2 + pqy^4.$$

This leads to the same system that we solved in Example 2. Thus $a = 15$, $b = 1$, $p = 2$, $q = 3$, and

$$15x^2 + 47xy^2 + 6y^4 = (15x+2y^2)(x+3y^2). \quad \blacktriangle$$

The perceptive reader will realize that Examples 2 and 3 are essentially the same.

In many cases a polynomial will be irreducible over the domain of the integers.

EXAMPLE 4   Try to factor $P_4 \equiv x^2 + 7x + 2$ using only polynomials with integer coefficients.

Solution: The set of integers (p, q) such that $pq = 2$ consists of $(\pm 1, \pm 2)$ and $(\pm 2, \pm 1)$. No matter what choice is made, $p + q \leq 1 + 2 = 3 < 7$. Hence P_4 is irreducible over the domain of integers. ▲

We will see later that this polynomial is reducible over the field of real numbers.

From Chapter 2, we recall the following formulas:

(19) $a^2 + 2ab + b^2 \qquad = (a+b)^2.$

(20) $a^3 + 3a^2b + 3ab^2 + b^3 = (a+b)^3.$

(21) $a^2 - b^2 \qquad\qquad = (a-b)(a+b).$

(22) $a^3 - b^3 \qquad\qquad = (a-b)(a^2 + ab + b^2).$

(23) $a^4 - b^4 \qquad\qquad = (a-b)(a+b)(a^2 + b^2).$

(24) $a^3 + b^3 \qquad\qquad = (a+b)(a^2 - ab + b^2).$

These formulas should be memorized because they are often useful in recognizing the factors of a polynomial.

EXAMPLE 5   Factor $P_5 \equiv x^2 - y^4 + 4xz + 4z^2.$

Solution: If we rearrange the terms in P_5 we have

$$P_5 = x^2 + 4xz + 4z^2 - y^4 = (x+2z)^2 - (y^2)^2.$$

We can use formula (21) with $a \equiv x + 2z$ and $b \equiv y^2$. Then

$$P_5 = (a-b)(a+b) = (x+2z-y^2)(x+2z+y^2)$$
$$= (x-y^2+2z)(x+y^2+2z). \blacktriangle$$

EXAMPLE 6   Factor $P_6 \equiv 8x^6 + 27y^9.$

Solution: If we set $a = 2x^2$ and $b = 3y^3$, then $a^3 + b^3 = 8x^6 + 27y^9$, and we can apply equation (24). This gives

$$8x^6 + 27y^9 = (2x^2 + 3y^3)(4x^4 - 6x^2y^3 + 9y^6). \blacktriangle$$

EXAMPLE 7 Factor $P_7 \equiv 2a^2bc + 6ab^2d + ac^2d + 3bcd^2$.

Solution: This polynomial does not seem to fit any pattern. We notice that the pair of coefficients $(2,6)$ is proportional to the pair $(1,3)$ so we group the first two terms and the last two terms. Then from each group we factor as much as possible. Thus

$$P_7 = (2a^2bc + 6ab^2d) + (ac^2d + 3bcd^2)$$
$$= 2ab(ac + 3bd) + cd(ac + 3bd)$$
$$= (2ab + cd)(ac + 3bd). \quad \blacktriangle$$

If the student will select a polynomial at random he will probably find that it is irreducible over the domain of the integers. It takes some care (but not too much) to select polynomials that can be factored over **Z**.

In certain situations we may wish to add a term to an expression so that the new expression is the square of a polynomial. Suppose, for example, that $E = Ax^2 + Bx$, where $A \neq 0$. If we add $B^2/4A$ to both sides, we have

$$E + \frac{B^2}{4A} = Ax^2 + Bx + \frac{B^2}{4A} = A\left(x^2 + \frac{B}{A}x + \frac{B^2}{4A^2}\right) = A\left(x + \frac{B}{2A}\right)^2$$

The right side is now the square of a polynomial. This operation is called *completing the square.*

EXAMPLE 8 Complete the square for $3x^2 - 7x$.

Solution: In this case $B^2/4A = (-7)^2/12$. Hence,

$$3x^2 - 7x + \frac{49}{12} = 3\left(x^2 - \frac{7}{3}x + \frac{49}{36}\right) = 3\left(x - \frac{7}{6}\right)^2. \quad \blacktriangle$$

In some cases the expression may have a nonzero constant. Further we may wish to preserve the value of the expression. These two items are illustrated in

EXAMPLE 9 If $E = 2x^2 + 5x - 3$, express E as the sum of a square and a constant.

Solution: Here $B^2/4A = 5^2/8$. Then

$$E = 2x^2 + 5x - 3 = 2x^2 + 5x + \frac{25}{8} - 3 - \frac{25}{8}$$
$$= 2\left(x^2 + \frac{5}{2}x + \frac{25}{16}\right) - \frac{24 + 25}{8}$$
$$= 2\left(x + \frac{5}{4}\right)^2 - \frac{49}{8}. \quad \blacktriangle$$

Exercise 2

In problems 1 through 45 factor the given polynomial into a product of polynomials that are irreducible over the domain of the integers. In some cases the given polynomial may already be irreducible.

1. $x^2 + 3x + 2$

2. $x^2 - 6x + 5$

3. $y^2 - 8y + 15$

4. $a^2 + 13ab + 12b^2$

5. $u^2 + 6u + 10$

6. $u^2 - 3u - 10$

7. $2a^3 - 9a^2 - 5a$

8. $3b^3 - b^2 - 10b$

9. $15x^2 - 2x - 24$

10. $y^2 - 9y + 12$

11. $x^3 + x^2 + 4x + 4$

12. $x^3 + x^2 - 9x - 9$

13. $x^7 - 8x^4y^3$

14. $27x^4 + xy^6$

15. $6ax + 8x + 3ay + 4y$

16. $15bx + 2y - 3by - 10x$

17. $12x^2 + 31xy - 30y^2$

18. $8y^2 - 26yz + 15z^2$

19. $2(x+y)^2 + 7xz + 7yz + 3z^2$

20. $x^2 + 4xy + 4y^2 - 5x - 10y + 4$

21. $4x^k - 3x^{2k} - 1$

22. $64y^7z^3 - 4y^3z^7$

23. $16y^2 + 27yz + 9z^2$

24. $16x^2 + 28xy + 9y^2$

25. $25x^2 - (3x+2)^2$

26. $(x+3)^2 - (2y+1)^2$

27. $(2a+3)^3 + a^3$

28. $(3b+1)^3 - (b+3)^3$

29. $5a^3bc - ab^2c + 10a^2bc^2 - 2b^2c^2$

30. $x^3y^2z^3 + 5x^2y^3z^3 - 14xy^4z^3$

31. $9u^2 - 6u(v-2w) + (v-2w)^2$

32. $81c^4 - (6c+1)^2$

33. $16x^4 - (x-y)^4$

34. $8x^4 + 26x^2y^2 - 99y^4$

35. $ax + 3bz + bx + 6cz + 2cx + 3az$

36. $a^2 + 4b^2 + 9c^2 + 4ab - 6ac - 12bc$

37. $(x^2+2x+2)^2 - 4(x^2+2x+2) + 3$

38. $(y^2+7y+8)^2 + 2(y^2+7y+8) - 8$

39. $2a^2 - 3b^2 - 5c^2 + 5ab + 3ac + 16bc$

40. $(2a^2+3a)^2 - (2a^2+3a) - 2$

41. $9(u-v)^2 + 6(u^2-v^2) + (u+v)^2$

42. $10(u-v)^2 - 7(u^2-v^2) + (u+v)^2$

43. $x^2 + 6xy + 9y^2 + 5x + 15y - 14$

44. $x^2 - y^2 + 2z^2 - 3xz + yz$

45. $2u^2 - 3v^2 - 5w^2 - uv + 3uw + 8vw$

46. Selecting integers for A, B, and C at random, (perhaps by rolling dice or drawing cards) compose five quadratic polynomials of the form (15). How many of these are irreducible over Z?

In problems 47 through 54 add a suitable constant to the given expression to complete the square.

47. $x^2 + 8x$

48. $y^2 - 6y$

49. $2x^2 - 10x$

50. $3z^4 + 6z^2$

51. $3y^2 + 2y$

52. $2x^2 - x$

53. $5x^4 - 3x^2$

54. $7x^2 + 2x$

In problems 55 through 60 write the given expression as the sum of a square and a constant.

55. $x^2 + 6x + 19$

56. $x^2 - 8x + 29$

57. $2y^2 + 3y + 5$

58. $3y^2 + 2y - 1$

59. $5x^2 + 15x + 13$

60. $3x^2 + 18x + 26$

4.4 RATIONAL EXPRESSIONS

Any expression which can be represented as the quotient of two polynomials is called a *rational expression*. For example,

$$(25) \qquad R_0 \equiv 1 + \cfrac{x}{1 + \cfrac{y}{1 + z}}$$

is a rational expression because (with some labor) the right side can be converted to

$$(26) \qquad R_0 = \frac{1 + x + y + z + xz}{1 + y + z},$$

and this is the quotient of two polynomials.

In this section we practice manipulations with rational expressions. As usual the symbols (variables) represent arbitrary real numbers, but if

$$(27) \qquad R = \frac{P}{Q}$$

where P and Q are polynomials, we must always demand that $Q \neq 0$. Hence, the domain of the variables must always be such that $Q \neq 0$ in (27). We make this agreement once and for all, so that it will be unnecessary to repeat it at every step when we are discussing a rational expression. However, we will on occasion mention it merely as a reminder. In the next example we will indicate clearly the importance of the condition $Q \neq 0$.

EXAMPLE 1   Simplify each of the following

$$(28) \qquad R_1 \equiv \frac{x^2 + 4x - 12}{x^2 - 4}, \qquad x \neq -2, \quad x \neq 2.$$

and

$$(29) \qquad R_1^\star \equiv \frac{x^2 + 8x - 20}{x^2 - 4}, \qquad x \neq -2, \quad x \neq 2.$$

Solution: In both of these expressions $Q = x^2 - 4 = (x+2)(x-2)$. Hence, the condition $Q \neq 0$ is just the condition $x \neq -2$ and $x \neq 2$. It is customary to place such conditions or limitations to the right of the equation as we have illustrated in (28) and (29).

To simplify these expressions we factor the numerator and denominator and after cancellation (see Theorem 18 in Appendix 2) we have

$$(30) \qquad R_1 = \frac{(x+6)(x-2)}{(x+2)(x-2)} = \frac{(x+6)\cancel{(x-2)}}{(x+2)\cancel{(x-2)}} = \frac{x+6}{x+2}$$

and

$$(31) \qquad R_1^\star = \frac{(x+10)(x-2)}{(x+2)(x-2)} = \frac{(x+10)\cancel{(x-2)}}{(x+2)\cancel{(x-2)}} = \frac{x+10}{x+2}.$$

The expressions on the extreme right in (30) and (31) are the simpler ones we are seeking.   ▲

Suppose that we use one of the forbidden values for x. If we put $x = 2$ in equation (30) a brief computation gives

$$(32) \qquad \frac{0}{0} = \frac{2+6}{2+2} = \frac{8}{4} = 2.$$

In equation (31) the same substitution $x = 2$, yields

$$(33) \qquad \frac{0}{0} = \frac{2+10}{2+2} = \frac{12}{4} = 3.$$

If we attempt to treat $0/0$ as a number then (32) and (33) together yield $3 = 2$. This is certainly one good reason for demanding that $Q \neq 0$ in (27). A rigorous treatment of the ratio P/Q when Q is near to 0 will be covered in calculus.

EXAMPLE 2   Simplify the expression

$$(34) \qquad R_2 \equiv \frac{x+4}{x-2} - \frac{x-4}{x+2}, \qquad x \neq \pm 2.$$

Solution: In equations (30) and (31) we cancelled common factors (see Theorem 18 in Appendix 2). Here the same theorem allows us to insert common factors so that each term in (34) has the same denominator. Inserting these factors we have

$$R_2 = \frac{x+4}{x-2} \cdot \frac{x+2}{x+2} - \frac{x-4}{x+2} \cdot \frac{x-2}{x-2} = \frac{x^2+6x+8}{x^2-4} - \frac{x^2-6x+8}{x^2-4}$$

$$= \frac{x^2+6x+8-(x^2-6x+8)}{x^2-4} = \frac{x^2+6x+8-x^2+6x-8}{x^2-4} = \frac{12x}{x^2-4}.$$

Clearly, this last expression is simpler than the one given in (34). Further, the two expressions are equal for all values of x except $x = \pm 2$. ▲

EXAMPLE 3   Simplify

$$R_3 \equiv \frac{a^2-b^2}{a^2+5ab+6b^2} \cdot \frac{a^2-5ab-14b^2}{a^2+8ab-33b^2} \cdot \frac{a^2+14ab+33b^2}{a^2-8ab+7b^2}.$$

Solution: Fortunately, each of the quadratic polynomials can be factored. By Theorem 17 in Appendix 2 we have

$$(35) \qquad R_3 = \frac{(a-b)(a+b)(a-7b)(a+2b)(a+3b)(a+11b)}{(a+2b)(a+3b)(a-3b)(a+11b)(a-7b)(a-b)}.$$

It is now clear that we can cancel certain factors that are in both the numerator and the denominator in (35). It may be well to introduce a system of marking items as they are canceled to avoid errors and to make the work easy to check. The reader may select his own system. One possible collection of marks is indicated below. From (35)

$$R_3 = \frac{(a-b)(a+b)(a-7b)(a+2b)(a+3b)(a+11b)}{(a+2b)(a+3b)(a-3b)(a+11b)(a-7b)(a-b)} = \frac{a+b}{a-3b}. ▲$$

EXAMPLE 4   Simplify

$$R_4 \equiv \frac{x^4-11x^2+18}{x^2-4x+4} \div \frac{x^2-8x+15}{x^2-11x+18}.$$

Solution: By Theorem 21 in Appendix 2 we invert the second fraction and replace division by multiplication. Then (factoring and canceling)

$$R_4 = \frac{(x^2-2)(x^2-9)}{(x-2)(x-2)} \cdot \frac{(x-2)(x-9)}{(x-3)(x-5)}$$

$$= \frac{(x^2-2)(x+3)(x-3)(x-2)(x-9)}{(x-2)(x-2)(x-3)(x-5)} = \frac{(x^2-2)(x+3)(x-9)}{(x-2)(x-5)}. \quad \blacktriangle$$

EXAMPLE 5   Simplify

(36)
$$R_5 \equiv \frac{x^2+2x+3}{x^2+1}.$$

Solution: Both the numerator and the denominator are irreducible over the field **R**, hence simplification (as we use the term here) is impossible. ▲

The desperate student might try to cancel x^2 obtaining

(37)
$$\frac{x^2+2x+3}{x^2+1} = \frac{x^2+2x+3}{x^2+1} = \frac{2x+3}{1}.$$

To see that this is wrong, put $x=1$ in (37). Then the left side is $6/2=3$, and the right side is $5/1=5$. Thus (37), if correct, gives $3=5$.

EXAMPLE 6   Simplify

(38)
$$R_6 \equiv \frac{\dfrac{x^2-x-2}{x^2+2x-3}}{\dfrac{x^2-4x-5}{x^2-x-12}}.$$

Solution: By Theorem 21 in Appendix 2, we invert the denominator in (38) and multiply. We combine several steps as indicated.

$$R_6 = \frac{x^2-x-2}{x^2+2x-3} \cdot \frac{x^2-x-12}{x^2-4x-5} = \frac{(x-2)(x+1)(x-4)(x+3)}{(x+3)(x-1)(x-5)(x+1)}.$$

$$= \frac{(x-2)(x-4)}{(x-1)(x-5)} = \frac{x^2-6x+8}{x^2-6x+5}. \quad \blacktriangle$$

Either of the last two forms may be accepted as simpler than (38), but the first of the two is more often preferred because it shows clearly that $R_6=0$ when $x=2$ or $x=4$, and is undefined when $x=1$ or $x=5$.

EXAMPLE 7   By combining the terms, simplify

(39) $$R_7 = \frac{2x-3}{x^2-1} - \frac{x+4}{x^2-5x+4} - \frac{x-5}{x^2-3x-4}.$$

Solution: We could use the formula

$$\frac{a}{b} + \frac{c}{d} + \frac{e}{f} = \frac{adf+bcf+bde}{bdf},$$

established in problem 12 of Exercise 1 in Chapter 2. But the computations would be rather complicated and the probability of making an error would be very high. Rather than use a common denominator that is the product of the three denominators in (39), we search for a simpler one that will do. To this end, we factor each denominator in (39):

$$b \equiv x^2 - 1 = (x-1)(x+1),$$
$$d \equiv x^2 - 5x + 4 = (x-1)(x-4),$$
$$f \equiv x^2 - 3x - 4 = (x+1)(x-4).$$

By inspection, it is clear that if $D \equiv (x-1)(x+1)(x-4)$, then each of b, d, and f will divide D. Further, D is the smallest polynomial with this property. Such an expression is called the *least common denominator* (denoted by *LCD*).

Now multiply each term in (39) by an appropriate factor to produce the *LCD*. We find

$$R_7 = \frac{(2x-3)}{x^2-1} \cdot \frac{x-4}{x-4} + \frac{-(x+4)}{x^2-5x+4} \cdot \frac{x+1}{x+1} + \frac{-(x-5)}{x^2-3x-4} \cdot \frac{x-1}{x-1}$$

$$= \frac{(2x^2-11x+12)-(x^2+5x+4)-(x^2-6x+5)}{(x-1)(x+1)(x-4)}$$

$$= \frac{(2-1-1)x^2+(-11-5+6)x+(12-4-5)}{(x-1)(x+1)(x-4)}$$

$$= \frac{-10x+3}{(x-1)(x+1)(x-4)}. \quad \blacktriangle$$

We now put in general terms the steps used in working Example 7. Suppose that we are to combine into a single fraction the sum

(40) $$R = \frac{n_1}{d_1} + \frac{n_2}{d_2} + \frac{n_3}{d_3} + \cdots + \frac{n_k}{d_k},$$

where each n_i and each d_i is a polynomial. If D is a polynomial such that each d_i is a factor of D (divides D) and if D has the smallest degree possible, then D is called the least common denominator (LCD) for the sum in (40). Suppose we have found such a D. We compute D/d_i and call this q_i. Then $d_i q_i = D$ and (40) can be put in the form

$$R = \frac{n_1}{d_1}\frac{q_1}{q_1} + \frac{n_2}{d_2}\frac{q_2}{q_2} + \frac{n_3}{d_3}\frac{q_3}{q_3} + \cdots + \frac{n_k}{d_k}\frac{q_k}{q_k}$$

$$= \frac{n_1 q_1}{D} + \frac{n_2 q_2}{D} + \frac{n_3 q_3}{D} + \cdots + \frac{n_k q_k}{D}$$

(41)
$$R = \frac{n_1 q_1 + n_2 q_2 + n_3 q_3 + \cdots + n_k q_k}{D}.$$

In going from (40) to (41) we have used the LCD to combine a sum of fractions into a single fraction. This is the usual situation, and (41) is regarded as a simpler form than (40). The reverse procedure (going from (41) to (40)) is called the *decomposition of a fraction into partial fractions*. This is considerably more difficult and is often needed in calculus. This topic can be deferred until the student meets it in calculus.

EXAMPLE 8

Simplify

(42)
$$R_8 \equiv \frac{\dfrac{2}{x-3} - \dfrac{3}{x-4}}{\dfrac{6}{x-5} - \dfrac{5}{x-4}}.$$

Solution: For the denominators $(x-3)$, $(x-4)$, $(x-5)$, and $(x-4)$ which appear in R_8, the product $(x-3)(x-4)(x-5)$ is the least common denominator. If we multiply the numerator and the denominator of R_8 by this quantity (see Theorem 18 in Appendix 2) we find that

$$R_8 = \frac{\dfrac{2}{x-3} - \dfrac{3}{x-4}}{\dfrac{6}{x-5} - \dfrac{5}{x-4}} \cdot \frac{(x-3)(x-4)(x-5)}{(x-3)(x-4)(x-5)}$$

$$= \frac{2(x-4)(x-5) - 3(x-3)(x-5)}{6(x-3)(x-4) - 5(x-3)(x-5)} = \frac{[2(x-4) - 3(x-3)](x-5)}{[6(x-4) - 5(x-5)](x-3)}$$

$$= \frac{(2x-8-3x+9)(x-5)}{(6x-24-5x+25)(x-3)} = \frac{(-x+1)(x-5)}{(x+1)(x-3)}$$

$$= -\frac{(x-1)(x-5)}{(x+1)(x-3)}. \quad \blacktriangle$$

30 JUNE all odd problems

Exercise 3

In problems 1 through 30 simplify the given expression.

1. $\dfrac{3x^2 + xy}{y^2 + 3xy}$

2. $\dfrac{5ab^3 + b^2}{x^2 + 5abx^2}$

3. $\dfrac{(a+b)^3}{a^2 - b^2}$

4. $\dfrac{x^2 - 8x + 15}{x^2 + 8x - 33}$

5. $\dfrac{3x^2 - 14x + 15}{6x^2 - x - 15}$

6. $\dfrac{7x^2 + 3x - 22}{5x^2 + 17x + 14}$

7. $\dfrac{ax - by + bx - ay}{4ax - 6by + 6bx - 4ay}$

8. $\dfrac{x^5 - y^2x^3 + a^3x^2 - a^3y^2}{-x^3 + yx^2 + a^2x - a^2y}$

9. $\dfrac{3y - 17x}{x^2 - y^2} + \dfrac{7}{x - y}$

10. $\dfrac{12xy}{x^3 + y^3} + \dfrac{4}{x + y}$

11. $2 - \dfrac{6}{x+2} + \dfrac{5}{x+3}$

12. $6 + \dfrac{3}{x+1} + \dfrac{22}{x-3}$

13. $\dfrac{x^2 + 4y^2}{x^2} \cdot \dfrac{(x-2y)^2}{x^4 - 16y^4} \cdot \dfrac{3x^2 + 6xy}{x - 2y}$

14. $\dfrac{3u + 6v}{6u + 15v} \cdot \dfrac{2u^2 + uv - 10v^2}{u^2 - 4uv + 4v^2} \cdot \dfrac{5u - 10v}{2u^2 - 8v^2}$

15. $\dfrac{a^4 - 81b^4}{a^2 - 6ab + 9b^2} \div \dfrac{a^2 + 9b^2}{4a^2 - 12ab}$

16. $\dfrac{ax + 5b^2x}{ab - 2b^2} \div \dfrac{3ay + 15b^2y}{ab^3 - 2b^4}$

17. $\left(\dfrac{1 - 2z}{7z} \div \dfrac{1 - 4z^2}{z^2} \right) \div \dfrac{2z - 1}{2z + 1}$

18. $\dfrac{1 - 2z}{7z} \div \left(\dfrac{1 - 4z^2}{z^2} \div \dfrac{2z - 1}{2z + 1} \right)$

19. $\dfrac{5}{x-1} - \dfrac{3}{x+2} + \dfrac{x^2 - 16}{x^2 + x - 2}$

20. $\dfrac{3}{x+3} - \dfrac{11}{x-5} + \dfrac{2(x^2 + 3)}{x^2 - 2x - 15}$

21. $\dfrac{8}{x^2 - 2x - 15} + \dfrac{11}{x^2 + 2x - 3} - \dfrac{4}{x^2 - 6x + 5}$

22. $\dfrac{7}{x^2-8x+12} - \dfrac{4}{x^2+5x-14} - \dfrac{3}{x^2+x-42}$

23. $\dfrac{x^2/y^2 - y^2/x^2}{x/y - y/x}$

24. $\dfrac{\dfrac{x^5}{y} - \dfrac{y^5}{x}}{\dfrac{1}{x^2} + \dfrac{1}{xy} + \dfrac{1}{y^2}}$

25. $\dfrac{\dfrac{2}{x+1} - \dfrac{1}{x-1}}{\dfrac{4}{x+1} + \dfrac{3}{x-1}}$

26. $\dfrac{\dfrac{x+y}{x-y} + \dfrac{x-y}{x+y}}{\dfrac{x+y}{x-y} - \dfrac{x-y}{x+y}}$

27. $\dfrac{1 + \dfrac{10xy}{x^2-5xy+6y^2}}{\dfrac{x+2y}{x-3y}\left(\dfrac{17y^2}{x^2-4y^2}+4\right)}$

28. $\dfrac{\dfrac{1+6x}{5-3x} + \dfrac{7-2x}{3-x}}{\dfrac{19-7x}{3x^2-14x+15}}$

29. $1 + \dfrac{x}{1 + \dfrac{x^2}{1 + \dfrac{x^3}{1+x^4}}}$

30. $1 + \dfrac{x}{2 + \dfrac{x}{3 + \dfrac{x}{4+x}}}$

4.5 RADICALS

If $x \neq 0$ and q is a positive integer, then there are precisely q numbers u, such that

(43) $$u^q = x.$$

Each of these numbers is called a qth root of x. If $q > 2$, then some of these qth roots must be complex numbers. Some examples of complex roots are given in the next exercise. Here we are concerned only with those qth roots that are real.

   If $x \in \mathbf{R}$ and q is an odd integer, then there is precisely one real qth root. This root is denoted by $\sqrt[q]{x}$ and is called the principal qth root of x. If q is even and $x < 0$, then there is no real qth root. If q is even and $x > 0$, then there are two qth roots that are real. The positive one is selected as the principal qth root. We summarize in

DEFINITION 2

Principal qth Root *Let q be a positive integer and let $x \in \mathbf{R}$. The principal qth root of x is denoted by $\sqrt[q]{x}$ and is determined thus:*

   If q is odd, $u^q = x$, and $u \in \mathbf{R}$, then $\sqrt[q]{x} = u$.

   If q is even, $u^q = x$, and $u \geq 0$, then $\sqrt[q]{x} = u$.

EXAMPLE 1   Find $\sqrt[4]{16}$, $\sqrt[3]{-729}$, and $\sqrt[5]{1/32}$.

Solution: From the definition

$$\sqrt[4]{16} = 2 \qquad \text{because } 2^4 = 16.$$

$$\sqrt[3]{-729} = -9 \quad \text{because } (-9)^3 = -729.$$

$$\sqrt[5]{\frac{1}{32}} = \frac{1}{2} \qquad \text{because } \left(\frac{1}{2}\right)^5 = \frac{1}{32}. \quad \blacktriangle$$

When $q = 2$, we write $\sqrt{x}$ in place of $\sqrt[2]{x}$. The symbol $\sqrt[q]{}$ is called a *radical sign*. The quantity under the sign is the *radicand*, and q is the *index* of the radical.

Theorem 3   *If q is a positive integer and x and y are real numbers such that $\sqrt[q]{x}$ and $\sqrt[q]{y}$ exist,‡ then*

(44) $$\left(\sqrt[q]{x}\right)^q = x,$$

(45) $$\sqrt[q]{x}\,\sqrt[q]{y} = \sqrt[q]{xy},$$

(46) $$\frac{\sqrt[q]{x}}{\sqrt[q]{y}} = \sqrt[q]{\frac{x}{y}}, \qquad y \neq 0.$$

If r is a positive integer and the roots exist, then

(47) $$\sqrt[q]{\sqrt[r]{x}} = \sqrt[qr]{x}$$

Proof: Equation (44) follows from the definition.

To prove (45), we suppose that $u = \sqrt[q]{x}$ and $v = \sqrt[q]{y}$. Then $\sqrt[q]{x}\,\sqrt[q]{y} = uv$ and hence $(uv)^q = u^q v^q = xy$. Consequently, uv is the principal qth root of xy. Equation (46) is proved in a similar fashion.

To prove (47), set $\sqrt[r]{x} = y$ and $\sqrt[q]{y} = u$. Then

$$\sqrt[q]{\sqrt[r]{x}} = \sqrt[q]{y} = u.$$

But $u^{qr} = (u^q)^r = y^r = x$. Hence, $u = \sqrt[q]{\sqrt[r]{x}} = \sqrt[qr]{x}.$ ∎

These laws of radicals can be used to simplify certain expressions. In sim-

‡ Here the term "exists" means that $\sqrt[q]{x}$ and $\sqrt[q]{y}$ are real numbers that satisfy the conditions specified in Definition 2.

plification the objective is to remove all radicals from the denominator (if there are any) and to make the powers in the radicand as small as possible. These criteria are illustrated in

EXAMPLE 2   Simplify each of the following (assuming that all of the variables represent positive numbers).

(a) $\sqrt{13,500}$, (b) $\sqrt{16a^3b^4c^{21}}$, (c) $\sqrt[3]{\dfrac{16y^8}{9x^5}}$, (d) $\sqrt{x^5} - 7\sqrt{x^3} + 12\sqrt{x}$.

Solution: We factor the radicand and use (45).

(a) $\sqrt{13,500} = \sqrt{135 \cdot 10^2} = \sqrt{9 \cdot 15 \cdot 10^2} = \sqrt{3^2}\sqrt{10^2}\sqrt{15}$

$\quad = 3 \cdot 10\sqrt{15} = 30\sqrt{15}.$

(b) $\sqrt{16a^3b^4c^{21}} = \sqrt{4^2a^2(b^2)^2(c^{10})^2\,ac} = 4ab^2c^{10}\sqrt{ac}.$

For (c) we introduce suitable factors in the radicand to make the denominator a perfect cube.

(c) $\sqrt[3]{\dfrac{16y^8}{9x^5} \cdot \dfrac{3x}{3x}} = \sqrt[3]{\dfrac{16 \cdot 3xy^8}{27x^6}} = \dfrac{\sqrt[3]{8y^6\,6xy^2}}{\sqrt[3]{(3x^2)^3}} = \dfrac{2y^2}{3x^2}\sqrt[3]{6xy^2}.$

(d) $\sqrt{x^5} - 7\sqrt{x^3} + 12\sqrt{x} = \sqrt{x^4}\sqrt{x} - 7\sqrt{x^2}\sqrt{x} + 12\sqrt{x}$

$\quad = x^2\sqrt{x} - 7x\sqrt{x} + 12\sqrt{x}$

$\quad = (x^2 - 7x + 12)\sqrt{x} = (x-3)(x-4)\sqrt{x}.$ ▲

EXAMPLE 3   Find $\sqrt{x^2}$ if $x < 0$.

Solution: The unsuspecting and trusting student will probably write $\sqrt{x^2} = x$, but this is an error. To see why, we observe that by Definition 2, we have $\sqrt{x^2} \geq 0$. On the other hand, by hypothesis $x < 0$. Hence $\sqrt{x^2} = x$ asserts that a nonnegative number equals a negative number, and this is impossible. The correct statement is

(48) $\sqrt{x^2} = |x|.$

This is true for all real x, positive, negative, or zero.   ▲

Any operation that removes radicals from the denominator of an expression is called *rationalizing the denominator*. One such operation has been illustrated in Example 2c. Another clever trick is illustrated in

EXAMPLE 4   Simplify the following: (a) $\dfrac{1}{\sqrt{13}-\sqrt{8}}$ (b) $\dfrac{\sqrt{x}-2\sqrt{y}}{\sqrt{x}+\sqrt{y}}$.

Solution: We observe that

$$\left(\sqrt{b}-\sqrt{a}\right)\left(\sqrt{b}+\sqrt{a}\right) = \left(\sqrt{b}\right)^2 - \left(\sqrt{a}\right)^2 = b - a.$$

This is the identity to use, if we want to rationalize the denominators in the given expressions.

(a) $\dfrac{1}{\sqrt{13}-\sqrt{8}}\cdot\dfrac{\sqrt{13}+\sqrt{8}}{\sqrt{13}+\sqrt{8}} = \dfrac{\sqrt{13}+\sqrt{8}}{13-8} = \dfrac{1}{5}\left(\sqrt{13}+\sqrt{8}\right).$

(b) $\dfrac{\sqrt{x}-2\sqrt{y}}{\left(\sqrt{x}+\sqrt{y}\right)}\cdot\dfrac{\left(\sqrt{x}-\sqrt{y}\right)}{\left(\sqrt{x}-\sqrt{y}\right)} = \dfrac{\sqrt{x}\sqrt{x}-2\sqrt{y}\sqrt{x}-\sqrt{x}\sqrt{y}+2\sqrt{y}\sqrt{y}}{x-y}$

$= \dfrac{x-2\sqrt{xy}-\sqrt{xy}+2y}{x-y}$

$= \dfrac{x-3\sqrt{xy}+2y}{x-y}.$ ▲

Exercise 4

In problems 1 through 32 simplify the given expressions. Assume that all variables represent positive numbers.

1. $\sqrt{8}$

2. $\sqrt{27}$

3. $\sqrt{3125}$

4. $\sqrt[3]{16/9}$

5. $\sqrt[3]{-81/16}$

6. $\sqrt[4]{1/125}$

7. $\sqrt[5]{2^6\cdot3^8}$

8. $\sqrt{1{,}000}$

9. $\sqrt[3]{-10{,}000}$

10. $\sqrt{8}+\sqrt{32}$

11. $\sqrt[3]{81}-\sqrt[3]{3}$

12. $\sqrt[4]{256}-\sqrt[4]{81}$

13. $\sqrt{x^3y^7}$

14. $\sqrt{x^5y^6z^9}$

15. $\sqrt{4a^5/b^7}$

16. $\sqrt{a^9b^7/9c^5}$

17. $\sqrt[3]{a^8}\sqrt[3]{a^5b^{13}}$

18. $\sqrt[3]{a^5b^3}/\sqrt[3]{a^2b^{11}}$

19. $\dfrac{\sqrt{a^3}+\sqrt{a^5}}{\sqrt{a}}$

20. $\dfrac{\sqrt{a^3}+a^5}{\sqrt{a}}$

21. $\sqrt[3]{\dfrac{3b^4}{49a^2}}$

22. $\sqrt[4]{\dfrac{5a^9}{8b^7}}$

23. $\dfrac{\sqrt[3]{ab^2}\sqrt[3]{8a^5b^7}}{\sqrt[3]{2a^2b^3}}$

24. $\dfrac{\sqrt[5]{a^9b^{11}}\sqrt[5]{a^{11}b^{13}}}{\sqrt[5]{81ab^2}}$

25. $\dfrac{\sqrt{15}+1}{\sqrt{5}-\sqrt{3}}$

26. $\dfrac{\sqrt{2}+\sqrt{8}}{\sqrt{11}-\sqrt{8}}$

27. $\dfrac{\sqrt{x}-2\sqrt{x^3}}{\sqrt{2x}+1}$

28. $\dfrac{\sqrt{(x+4)^3}+\sqrt{(x+1)^3}}{\sqrt{x+4}+\sqrt{x+1}}$

29. $3\sqrt{12}-5\sqrt{75}+7\sqrt{48}$

30. $7\sqrt{(8/5)}+3\sqrt{32/5}-\sqrt{242/5}$

31. $\sqrt[3]{27s^{11}t^7}-\sqrt[3]{-8s^{14}t^{10}}$

32. $\sqrt[4]{s^{26}t^{23}}+\sqrt[4]{16s^{22}t^{27}}$

In problems 33 through 36, let i be a number such that $i^2 = -1$ and assume that i obeys all of the usual algebraic laws (see Chapter 18 for details).

33. By computing the fourth power, show that each one of the numbers 1, -1, i, and $-i$ is a fourth root of 1. Thus 1 has at least four fourth roots; two of them are complex numbers (involve i).

★34. Show that each one of the four numbers

$$(1+i)/\sqrt{2}, \qquad (1-i)/\sqrt{2}, \qquad (-1+i)/\sqrt{2}, \quad \text{and} \quad (-1-i)/\sqrt{2}$$

is a fourth root of -1.

★35. Show that each one of the three numbers

$$1, \qquad \left(-1+\sqrt{3}i\right)/2, \quad \text{and} \quad \left(-1-\sqrt{3}i\right)/2$$

is a cube root of 1.

★36. Show that each one of the three numbers

$$-1, \qquad (1+\sqrt{3}i)/2, \quad \text{and} \quad (1-\sqrt{3}i)/2$$

is a cube root of -1.

★37. Prove that if x and y are positive, then $\sqrt{x+y} \neq \sqrt{x}+\sqrt{y}$.

4.6 RATIONAL EXPONENTS

We have defined x^n when n is a positive integer, and then we extended the definition so that Theorem 1 (page 52) is still true for any integer exponents. We now extend the definition so that rational numbers may also be used as exponents. Consider first the expression

$$E = x^{\frac{1}{q}} x^{\frac{1}{q}} x^{\frac{1}{q}} \cdots x^{\frac{1}{q}}.$$
$$\underbrace{\qquad\qquad\qquad}_{q \text{ factors}}$$

If the equation $x^m x^n = x^{m+n}$ is to be valid when m and n are rational numbers, then

$$E = x^{\frac{1}{q}+\frac{1}{q}+\frac{1}{q}+\cdots+\frac{1}{q}} = x^{\frac{q}{q}} = x^1 = x.$$

Consequently, the preservation of Theorem 1 forces us to adopt the definition

(49)
$$x^{\frac{1}{q}} = \sqrt[q]{x}.$$

Next, equation (3) of Theorem 1 forces us to agree that

(50)
$$\left(\sqrt[q]{x}\right)^p = (x^{1/q})^p = x^{p/q}.$$

Consequently, if we hope to preserve Theorem 1 when the exponents are rational numbers, we must make

DEFINITION 3   *Let p and q be integers where $q > 0$ and p and q have no common prime factors. If x is a real number for which $\sqrt[q]{x}$ exists, then*

(51)
$$x^{p/q} \equiv \left(\sqrt[q]{x}\right)^p$$

We pause to observe that we must have

$$\left(\sqrt[q]{x}\right)^p = \underbrace{\sqrt[q]{x}\,\sqrt[q]{x}\,\sqrt[q]{x}\cdots\sqrt[q]{x}}_{p \text{ factors}}$$

$$= \sqrt[q]{\underbrace{x\cdot x\cdot x\cdot\cdots\cdot x}_{p \text{ factors}}} = \sqrt[q]{x^p},$$

and hence in (51) we could also write $x^{p/q} \equiv \sqrt[q]{x^p}$.

Theorem 4   *Let m and n be arbitrary rational numbers and let x and y be positive numbers.*
PWO   *Then*

(2)
$$x^m x^n = x^{m+n},$$

(3)
$$(x^m)^n = x^{mn},$$

(4)
$$(xy)^n = x^n y^n,$$

(5)
$$\frac{x^m}{x^n} = x^{m-n},$$

and

(6)
$$\left(\frac{y}{x}\right)^n = \frac{y^n}{x^n}.$$

We have assumed that $x > 0$ and $y > 0$ for convenience. There are many cases in which Theorem 4 can be extended to include arbitrary real numbers x and y. Some care is needed, however, because (as we have seen in Example 3 of the preceding section) $\sqrt{(-1)^2} = 1$. Consequently, the equation

(3)
$$(x^m)^n = x^{mn}$$

fails to be true if $x = -1$, $m = 2$, and $n = 1/2$. Indeed with these numbers the left side of equation (3) is $\sqrt{(-1)^2} = \sqrt{1} = 1$, while the right side of the same equation is $(-1)^{2 \cdot (1/2)} = (-1)^1 = -1$.

The proof of Theorem 4 is not difficult, but it is tedious. As an example of the method to be used we prove (2) in the case that $m = p/q > 0$ and $n = r/s > 0$. We are to prove that

(52)
$$x^{\frac{p}{q}} x^{\frac{r}{s}} = x^{\frac{p}{q} + \frac{r}{s}} = x^{\frac{ps + qr}{qs}}.$$

Let L and R denote the left and right side of (52), respectively. Then by definition

$$R \equiv x^{\frac{ps+qr}{qs}} = \sqrt[qs]{x^{ps+qr}}.$$

In other words, R is the qsth root of x^{ps+qr}. On the other hand, the left side is the product of two positive numbers and if we raise it to the qs power we find

$$L^{qs} = \left(x^{\frac{p}{q}} x^{\frac{r}{s}}\right)^{qs} = \left(\sqrt[q]{x^p}\sqrt[s]{x^r}\right)^{qs}$$
$$= \left(\sqrt[q]{x^p}\right)^{qs}\left(\sqrt[s]{x^r}\right)^{qs} = (x^p)^s (x^r)^q$$
$$= x^{ps+qr}.$$

Hence L is also the qsth root of x^{ps+qr}. Thus, $L = R$.

EXAMPLE 1

Assuming that u, v, x, and y are positive, simplify

(a) $(-8)^{7/3} 9^{-3/2}$, (b) $\left(\dfrac{2u^{1/3}}{v^{1/2}}\right)^5 \left(\dfrac{u^{-1/6}}{v^{1/2}}\right)^7$,

(c) $(x^{1/3} + y^{1/3})(x^{2/3} - x^{1/3}y^{1/3} + y^{2/3})$.

Solution: We use Definition 3 and Theorem 4.

(a) $(-8)^{7/3} 9^{-3/2} = \dfrac{(\sqrt[3]{-8})^7}{(\sqrt{9})^3} = \dfrac{(-2)^7}{3^3} = \dfrac{-128}{27}$.

(b) $\left(\dfrac{2u^{1/3}}{v^{1/2}}\right)^5 \left(\dfrac{u^{-1/6}}{v^{1/2}}\right)^7 = \dfrac{32u^{5/3}}{v^{5/2}} \cdot \dfrac{u^{-7/6}}{v^{7/2}}$

$= 32 \dfrac{u^{5/3 - 7/6}}{v^{5/2 + 7/2}} = 32 \dfrac{u^{(10-7)/6}}{v^{12/2}} = 32 \dfrac{u^{1/2}}{v^6}$.

(c) $(x^{1/3} + y^{1/3})(x^{2/3} - x^{1/3}y^{1/3} + y^{2/3})$

$= x^{3/3} - \underline{x^{2/3}y^{1/3}} + \underwave{x^{1/3}y^{2/3}} + \underwave{x^{2/3}y^{1/3}} - \underline{x^{1/3}y^{2/3}} + y^{3/3}$

$= x + y$. ▲

We can define x^n when n is an irrational number and extend Theorem 1 so that it is true even for irrational exponents. This would include such expressions as $\sqrt{2^{\sqrt{3}}}$, 5^π, $(\sqrt{5} + \sqrt[3]{17})^{\sqrt{51} + \pi}$, etc. Such an extension is important for the theoretical aspects of mathematics but is unnecessary for our work.

4.7 ALGEBRAIC EXPRESSIONS

We start with a finite number of variables together with the real numbers. An expression is called an *algebraic expression* if it can be obtained from these elements by using a finite number of times the operations: addition, subtraction, multiplication, division, and taking the qth root (q an integer). Polynomials and rational expressions are examples of algebraic expressions. Further,

$$x^{1/3} + x^{4/3} + \sqrt{2}x^{7/3}, \qquad \dfrac{x + \sqrt{y}}{\sqrt[3]{z^2 + w}}, \qquad \sqrt{x^5 + \sqrt[3]{y^5 + \sqrt{z}}}$$

are also algebraic expressions, but these are not rational expressions. Some rather simple looking expressions are not algebraic. For example,

$$x^{\sqrt{2}}, \qquad 2^x, \qquad x^y, \qquad (x+y)^{\sqrt{17}}$$

are not algebraic expressions. However, $\pi^{\sqrt{2} + \sqrt{3}} x + \sqrt{7\pi}$ is a polynomial.

When an algebraic expression arises in connection with a specific problem, that problem will determine the domain of the variables involved. If we are given an expression with no knowledge of its source, then there is no way of knowing the domain of the variables. In this case we make a very simple agreement:

The domain of the variable (or variables) in an algebraic expression is the largest set for which the expression gives a real number.

EXAMPLE 1   Determine **D** the domain of x for each of the expressions:

$$E_1 = \frac{1}{x-3} + \frac{2}{x-5},$$

$$E_2 = \sqrt{x-2} + \sqrt{11-x}$$

Solution: Clearly E_1 is a real number for every x except $x = 3$ and $x = 5$. Hence $\mathbf{D}_1 = \mathbf{R} - \{3, 5\}$.

For the first term in E_2 to be real, we must have $x \geq 2$. For the second term to be real, we must have $x \leq 11$. For both terms to be real, we must have $2 \leq x \leq 11$. Consequently, $\mathbf{D}_2 = [2, 11]$. ▲

Exercise 5

In problems 1 through 33 simplify the given expression. In these problems assume that each variable represents a positive number.

1. $(-27)^{5/3}$ 2. $(-32)^{7/5}$ 3. $(27)^{-4/3}$

4. $(16)^{-5/4}$ 5. $(0.008)^{2/3}$ 6. $(0.0004)^{3/2}$

7. $\left(\dfrac{4}{9}\right)^{-3/2}$ 8. $\left(\dfrac{25}{36}\right)^{-3/2}$ 9. $\dfrac{4^{7/2}}{8^{5/3}}$

10. $x^{2/3}x^{3/4}$ 11. $y^{1/4}y^{-2/5}$ 12. $a^{5/7}a^{-1/2}a^{2/3}$

13. $\left(\dfrac{b^{1/2}b^{-2/3}}{b^{-3/4}}\right)^3$ 14. $\left(\dfrac{c^{1/5}c^{3/4}}{c^{-5/4}}\right)^5$ 15. $\dfrac{x^{3/5}y^{-3/4}}{x^{2/3}y^{-4/3}}$

16. $\left(\dfrac{x^{-3}y^{-2/3}}{x^{-3/4}y^{-4}}\right)^{12}$ 17. $\left(\dfrac{a^{1/5}b^{2/5}c^{3/5}}{a^{1/3}b^{2/3}c^{4/3}}\right)^{-1/2}$

18. $\left(\dfrac{x^{1/4}y^{-3/4}z^{-5/4}}{x^{-1/3}y^{-5/3}z^{-8/3}}\right)^{1/2}$ 19. $\left(\dfrac{49a^{-3}b^{2/3}}{81a^{2/3}b^{-2}}\right)^{-1/2}\left(\dfrac{a^3b^4}{a^{-2}b^{-3}}\right)^{1/3}$

20. $\left(\dfrac{3^{-2}x^{1/2}y^{-1/3}}{2^{-3}x^{1/4}y^{-2/3}}\right)^{3/2}\left(\dfrac{2^4x^{3/2}y^{1/3}}{3^5x^{2/3}y^{1/2}}\right)^{-1/2}$

21. $(x^{1/4}+y^{1/2})^2$ 22. $(x^{1/3}+y^{2/3})^3$

23. $(x^{4/5}+y^{3/2})(x^{4/5}-y^{3/2})$ 24. $(x^{3/2}-y^{1/2})(x^{1/2}+y^{1/2})^{-1}$

25. $\{[(a^{3/2})^{4/5}]^{8/3}\}^{5/16}$ 26. $\{[(b^{2/3}c)^{4/7}]^{3/8}\}^{7/2}$

27. $(a^2-x^2)^{1/2} + x^2(a^2-x^2)^{-1/2}$ 28. $(b^2-x^2)^{1/3} + 2x^2(b^2-x^2)^{-2/3}$

29. $(a^3-x^3)^{1/3}x^{-1/2} + x^{5/2}(a^3-x^3)^{-2/3}$

30. $x^{1/2}(5x^2 - 3x - 1) + x^{3/2}(2x + 5) - 7x^{5/2}$

31. $y^{3/2}(y - 5) - y^{1/2}(2y^2 - 7y + 3) + y^{-1/2}(y^3 - 2y^2 + 3y + 4)$

32. $x^{1/3}(x^3 + 3x + 7) - x^{4/3}(3 - 5x) + x^{7/3}(2x^2 - x - 5)$

33. $(4x - 3)^{4/3} - (4x - 3)^{1/3}(4x - 1) + (4x - 3)^{-2/3}(9x - 7)$

34. Explain in detail the classification of each expression in Section 4.7.

In problems 35 through 49 determine **D** the domain of the variable involved in accordance with the agreement stated in Section 4.7.

35. $\dfrac{15}{x-7} - \dfrac{19}{x-13}$ 36. $\dfrac{7}{x-15} + \dfrac{13}{x+19}$

37. $\sqrt{x+4}$ 38. $\sqrt{x+9}$

39. $\sqrt[3]{x+4}$ 40. $\sqrt[3]{x+(44)^3}$

41. $\sqrt{7-x^2}$ 42. $\sqrt{7-y^2}$

43. $\sqrt{(y-5)(y+3)}$ 44. $\sqrt{(z-17)(z+19)}$

45. $\sqrt{u+7} + \sqrt{5-u}$ 46. $\sqrt{v-5} + \sqrt{11-v}$

47. $\sqrt{x-10} + \sqrt{5-x}$ 48. $\sqrt{x-5} + \dfrac{1}{x-5} + \sqrt{5-x}$

49. $\sqrt{x+1} + \sqrt{x+2} + \sqrt{x+3} + \sqrt{x+4} + \sqrt{x+5}$

Review Questions

Try to answer the following questions as accurately as possible before consulting the text.

1. Give the definition of x^n: (a) when n is a natural number, (b) when n is a negative integer, and (c) when n is a rational number.

2. State the fundamental laws of exponents. (See Theorem 4.)

3. What constant term must be added to $Ax^2 + Bx$ to complete the square?

4. Explain the meaning of the term "least common denominator."

5. Explain why $\sqrt{(-1)^2}$ is not -1.

6. If m and n are rational numbers and x may be negative is the relation $(x^m)^n = x^{mn}$ always true?

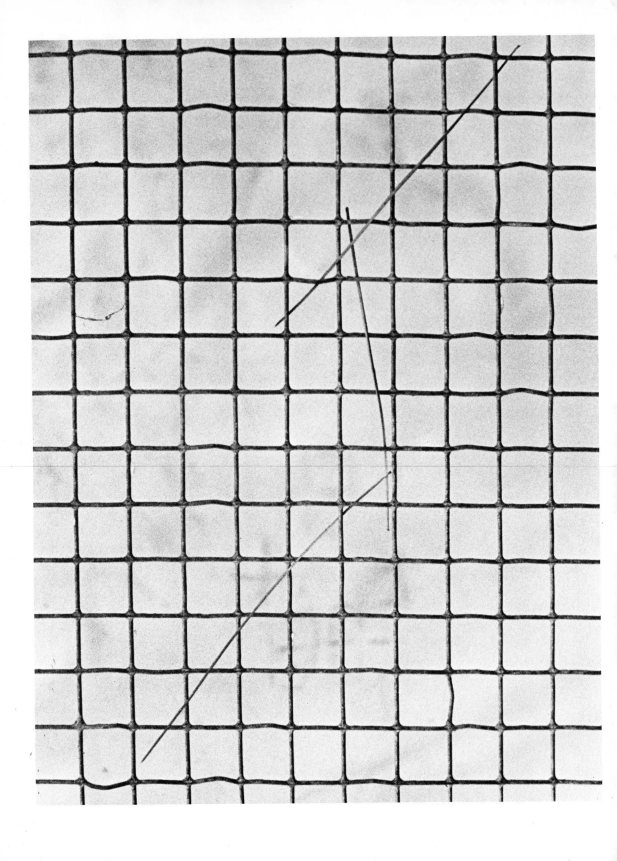

Equations in One Variable

This chapter is devoted to the solution of equations. However we only look at equations of a very simple type. We first must learn to solve these simple equations before we can approach the difficult ones that arise in every branch of science.

5.1 EQUIVALENT EQUATIONS

An equation is a statement that two expressions E_1 and E_2 are equal. Thus, in general, it has the form

$$(1) \qquad E_1 = E_2.$$

The expressions E_1 and E_2 may involve more than one variable and may be complicated. In this chapter we treat simple expressions in one variable.

If equation (1) is true for all values of the variable in its domain, then the equation is called an *identity*. For example, in the equation

$$(2) \qquad \sqrt{(2-x)(2+x)} = \sqrt{4-x^2}$$

the domain of x is the closed interval $[-2, 2]$, and equation (2) is true for every x in this domain. Hence equation (2) is an identity.

If we consider the equation

(3)
$$2 - x = \sqrt{4-x^2},$$

the common domain for x in the two expressions is again the closed interval $[-2, 2]$. However, equation (3) is not an identity (try $x = 1$ and observe that $2-1 \neq \sqrt{3}$).

If an equation is not an identity, it is called a *conditional equation*. A conditional equation is sometimes called an *open sentence*, because it is a statement that is neither true nor false, but becomes either true or false when a particular value is assigned to the variable.

Any value of a variable that makes an equation true is called a *solution* of the equation. It is also called a *root* of the equation. The set of all roots is called the *solution set* of the equation (or, for brevity, *the solution* of the equation). It is possible that a solution set may be the empty set. On the other hand, an equation may have infinitely many roots. To solve an equation means to find the solution set of the equation.

EXAMPLE 1   Solve equation (3).

Solution: We assume that equation (3) has at least one solution, and we let x represent one of its solutions. If (3) is indeed true, it remains true when we square both sides. This squaring operation gives

$$(2-x)^2 = 4 - x^2$$

(4)
$$4 - 4x + x^2 = 4 - x^2 \qquad \text{(expanding)}.$$

If (3) is true, then (4) is true. We next add $-4+x^2$ (the additive inverse of $4-x^2$) to both sides of equation (4). This operation gives

$$4 - 4x + x^2 + (-4+x^2) = 4 - x^2 + (-4+x^2)$$

$$(4-4) - 4x + (x^2 + x^2) = (4-4) + (-x^2 + x^2) \qquad \text{(grouping terms)}$$

$$-4x + 2x^2 = 0$$

(5)
$$-2x(2-x) = 0 \quad \text{(factoring)}.$$

If (3) is true, then (5) is true. If the product on the left side of (5) is zero, then by Theorem 3 in Chapter 2 one of the factors is zero. Hence either $-2x = 0$ or $2-x = 0$. Consequently, either $x = 0$ or $x = 2$.

We have found that if x is a solution of $2-x = \sqrt{4-x^2}$, then either $x = 0$ or $x = 2$. Symbolically, if $\mathbf{S}$ denotes the solution set, then $\mathbf{S} \subset \{0, 2\}$. There are two different ways of proving that $\mathbf{S} = \{0, 2\}$:

(A) If all of the steps in going from equation (3) to equation (5) are reversible, then indeed $\mathbf{S} = \{0, 2\}$.

(B) We can try each element in $\{0, 2\}$ in equation (3) to see if these values assigned to x make the statement true. We call this step *checking the solution*.

We will see in Section 5.6 that method **(A)** must be used with caution because when we square both sides of an equation (as we did in going from (3) to (4)) this step may not be reversible. We use method **(B)**.

If $x = 0$, equation (3) becomes $2 - 0 = \sqrt{4 - 0}$. This is true.

If $x = 2$, equation (3) becomes $2 - 2 = \sqrt{4 - 4}$. This is true.

We have proved that $\{0, 2\}$ is the solution set of $2 - x = \sqrt{4 - x^2}$. In other words, this equation is false for all x except $x = 0$ and $x = 2$. ▲

We leave this specific example and consider briefly the general theory of solving equations. Suppose that the expressions E_1 and E_2 involve only one variable which we denote by x (although any letter will do). Let $\mathbf{D}_1$ be the domain of x for E_1, and let $\mathbf{D}_2$ be the domain of x for E_2. Then the domain of the equation

$$(1) \qquad\qquad E_1 = E_2$$

is $\mathbf{D}_1 \cap \mathbf{D}_2 \equiv \mathbf{D}$. Frequently this will be $\mathbf{R}$, but not always. In Example 1, the domain of $E_1 \equiv 2 - x$ is $\mathbf{R}$, and the domain of $E_2 \equiv \sqrt{4 - x^2}$ is $[-2, 2]$. In this case $\mathbf{D}_1 \cap \mathbf{D}_2 = [-2, 2]$ and is not $\mathbf{R}$.

To keep matters simple we will usually omit a discussion of the domain. We can do this without harm if we agree (as in Chapter 4) that the domain is always the largest set for which all of the expressions involved give real numbers. With this agreement the domain of equation (3) is still the interval $[-2, 2]$.

DEFINITION 1

Equivalent Equations . *Two equations are equivalent if they both have the same solution set.*

To solve an equation, we replace the given equation by an equivalent equation that is easier to solve. In Example 1, equations (3), (4), and (5) are all equivalent, but equation (5) is easier to solve.

Theorem 1
PLE

Let A be any expression and let c be any nonzero constant. Then the equation

$$(1) \qquad\qquad E_1 = E_2$$

is equivalent to the equation

$$(6) \qquad\qquad E_1 + A = E_2 + A.$$

Further equation (1) *is equivalent to the equation*

(7) $$c E_1 = c E_2.$$

In particular, we can always select A to be $-E_2$ in (6), and when we do we find that equation (1) is equivalent to

(8) $$E_1 - E_2 = 0.$$

EXAMPLE 2

Solve the equation

(9) $$23x^2 - 41x - 300 = 4x^2 - 3x - 15.$$

Solution: We use Theorem 1 to put equation (9) in the form $E = 0$. In accordance with Theorem 1, we add to both sides the additive inverse of $4x^2 - 3x - 15$. Thus we find that

$$23x^2 - 41x - 300 - (4x^2 - 3x - 15) = 0$$

(10) $$23x^2 - 41x - 300 - 4x^2 + 3x + 15 = 0$$

$$(23 - 4)x^2 + (-41 + 3)x + (-300 + 15) = 0 \quad \text{(grouping terms)}$$

(11) $$19x^2 \qquad - 38x \qquad - 285 = 0.$$

In going from equation (9) to equation (10) we have merely transposed the terms from one side to the other with a change of sign in each term. Thus $4x^2$ on the right side of equation (9) disappears and reappears on the left side of equation (10) as $-4x^2$, etc.

Next we search for a common factor for the coefficients in equation (11), if one exists. Since 19 divides each coefficient in (11), we multiply both sides of (11) by $c = 1/19$ (see equation (7)) and obtain

$$x^2 - 2x - 15 = 0.$$

which by Theorem 1 is equivalent to the original equation. The left side can be factored easily

(12) $$(x - 5)(x + 3) = 0.$$

By Theorem 3 in Chapter 2 if a product is zero, then one of its factors must be zero. Therefore, if (12) is true, then either $x - 5 = 0$ or $x + 3 = 0$. Hence, either $x = 5$ or $x = -3$. Since each step is reversible, we have proved that the solution set (or solution) of equation (9) is $\{-3, 5\}$. The reader may check this solution by substituting $x = -3$ and $x = 5$ directly in equation (9) and showing that for each of these values of x the two sides are indeed equal. ▲

The reader may feel that the problem posed in Example 2 is difficult. Indeed the work involved in finding the solution set was long because we included a detailed explanation of each step. Once the steps are thoroughly understood, the work can be shortened. Rather than appeal to Theorem 1, we merely say "transposing" or "multiplying by c."

EXAMPLE 3   Solve the equation

(13) $4x^2 - 41x - 29 = 7x^2 - 11x + 46.$

Solution: Transposing the right side to the left side and changing signs we obtain

$$4x^2 - 41x - 29 - 7x^2 + 11x - 46 = 0$$

$$(4-7)x^2 + (-41+11)x + (-29-46) = 0 \quad \text{(grouping terms)}$$

(14) $-3x^2 \qquad\quad -30x \qquad\quad -75 = 0.$

We divide both sides of (14) by -3 and factor the result.

$$x^2 + 10x + 25 = 0$$
$$(x+5)^2 = 0$$
$$x + 5 = 0$$
$$x = -5.$$

Hence, the solution set is $\{-5\}$.   ▲

Although the factor $x+5$ occurs twice, we list the number -5 only once in the solution set. We will discuss this occurrence of a repeated factor in more detail in Chapter 19.

To check the solution we return to equation (13) and set $x = 5$. The left side gives

$$L \equiv 4(-5)^2 - 41(-5) - 29$$
$$= 100 + 205 - 29 = 276.$$

The right side gives

$$R \equiv 7(5)^2 - 11(-5) + 46$$
$$= 175 + 55 + 46 = 276.$$

We should observe that in Examples 2 and 3 we enjoyed "good luck." In Example 2 the given equation was equivalent to $x^2 - 2x - 15 = 0$ and this expression is easy to factor. In Example 3 the given equation is equivalent to $x^2 + 10x + 25 = 0$ and again factorization is easy. In Section 5.4 we will learn what to do with $E = 0$, when E is a 2nd degree polynomial that is difficult to factor.

Exercise 1

In problems 1 through 24 solve the given equation. Check that each element in the solution set is indeed a root of the given equation.

1. $(x+4)(x-7) = 0$

2. $5(x-8)(x+9) = 0$

3. $(x-1)(x+3)(x+9)^2 = 0$

4. $7(x+1)^2(x-3)^3(x+5)^7 = 0$

5. $x^2 - x - 30 = 0$

6. $2x^2 - 10x - 28 = 0$

7. $4x^2 - 10x - 24 = 0$

8. $6x^2 + 7x - 5 = 0$

9. $7x^2 - 791 = 19 - 3x^2$

10. $11x^2 - 999 = 4x^2 + 9$

11. $\dfrac{x}{2} + \dfrac{1}{3} = \dfrac{x}{4} + \dfrac{1}{5}$

12. $\dfrac{x}{3} - \dfrac{1}{5} = \dfrac{x}{7} - \dfrac{1}{9}$

13. $3(4x+5) = 5(4x+3)$

14. $3(5x+7) = 6(4x-2)$

15. $2x^2 + 3x + 4 = 2(x^2+3x+5)$

16. $3x^2 + 4x - 5 = 3(x^2-4x+5)$

17. $3x^2 - 6x + 11 = 3(7-2x+x^2)$

18. $3x^2 + 1 = 5(x^2+7) - 2(x^2+17)$

19. $2(2x^2-5x+13) = x^2 - 25x + 8$

20. $7(x^2+2x-5) = 5(x-1)(x+1) - 3(x-5)(x+2)$

21. $3(x+1)(x+3) = 5(x+2)(x-3) - 6(x+2)(x-4)$

22. $3(x-4)(x-5) = 5(x+5)(x+6) + 2(x+2)(x-4)$

23. $a(x+6) + b(2x+3) = x(a+2b+3)$

24. $2a(x+1) + 3b(x-2) = 2x(a+3) + 3x(b-4)$

In problems 25 through 28 find a value for b such that $x = 7$ is a solution of the given equation.

25. $x^2 + x + b = 0$

26. $x^2 + bx - 14 = 0$

27. $bx^2 - 11x - 21 = 0$

28. $bx^2 - 6x - 35b = 0$

29. What can be said about b and c if $x = 3$ is a solution of
$$bx^3 + cx^2 - 9cx - 9b = 0?$$

30. Repeat problem 29 if $x = 2$ is a solution of the equation.

5.2 LINEAR EQUATIONS IN ONE VARIABLE

The simplest type of equation is described in

DEFINITION 2

Linear Equation *If an equation in x can be put in the form*

(15)
$$ax + b = 0, \qquad a \neq 0,$$

using only transformations of the type described in Theorem 1, *then the equation is said to be a linear equation in x, or of first degree in x.*

Of course the letters are arbitrary. The equation $cy+d=0$ is linear in y if $c \neq 0$. Further, the equation may contain other variables. If such is the case, then a and b in (15) may represent expressions that involve variables other than x.

EXAMPLE 1

Discuss the linearity of the equation

(16)
$$abc^3xyz + bc^2yzw^2 + 5u^2w^3 = 0.$$

Solution: Since (16) can be put in the form

$$(abc^3yz)x + (bc^2yzw^2 + 5u^2w^3) = 0,$$

the equation is linear in x if $abc^3yz \neq 0$. Observe that it is also linear in y, z, a, and b, but the same equation is not linear in c, u, or w. ▲

Theorem 2
PLE

The linear equation (15) *always has a unique root,* $x = -b/a$.

EXAMPLE 2

Solve the linear equation in y

$$yx^5 + 3yx^2 - 5y + 17 = 0.$$

Solution: As an equation in x, this equation may be very difficult to solve. But it is equivalent to

$$(x^5 + 3x^2 - 5)y + 17 = 0.$$

If $x^5 + 3x^2 - 5 \neq 0$, then the equation is linear in y, and the solution is

$$y = \frac{-17}{x^5 + 3x^2 - 5}. \quad ▲$$

5.3 APPLICATIONS TO WORD PROBLEMS

Here we consider problems that are stated in words (no symbols, or only a very few). The object is to introduce symbols as needed and to transform the given information into an equation that we can solve. We must not expect to handle every word problem by introducing suitable symbols. For example, "How do we achieve peace in our time?" certainly poses an important word problem. But at the moment we have very little hope of solving this problem‡ either with mathematics or any other tool.

For those word problems that are amenable to a mathematical treatment, the following outline may be helpful.

(A) Search for those items in the problem that can be measured.

(B) Introduce letters (variables) as needed to represent the measures of those items.

(C) Read the problem carefully for information that gives an equation (or equations) relating these variables.

(D) If there are several equations and several variables, use some of the equations to transform one of the equations into an equation that involves only one variable.

(E) Solve the equation for the variable.

(F) Check the solution to see if it satisfies the conditions stated in the problem.

EXAMPLE 1   A Transvalley airplane leaves Wildroot Airport at 12:07 P.M. flying due north. Two hours later a Quick Serve plane departs from the same airport and overtakes the Transvalley plane at exactly 7:07 P.M. just when dinner is being served on both planes. If the Quick Serve plane flies 60 miles/hour faster than the Transvalley plane, what is the speed of each plane? How far are they from Wildroot airport when they meet?

Solution: We follow the suggested outline.

(A) Many items in the above problem can be measured, but the only ones of interest are those in the last sentence.

(B) Let x = speed (in miles/hour) of the Transvalley plane.
Let y = speed (in miles/hour) of the Quick Serve plane.
Let d_1 = distance traveled by the Transvalley plane.
Let d_2 = distance traveled by the Quick Serve plane.

‡ In October, 1938, Neville Chamberlain announced that he had solved this problem. No intelligent person really believed him at that time, and later events (e.g. World War II etc.) showed how wrong he was.

(C) From the statement of the problem,

(17)
$$y = x + 60.$$

Further, the distance traveled by the two planes is the same, when they meet. Thus $d_1 = d_2$. In addition to these equations we must use the relation between time, rate, and distance, when an object is traveling with a uniform motion ‡.

distance traveled = (rate of travel)(time),

(18)
$$d = rt.$$

For the Transvalley plane this gives $d_1 = 7x$, and for the Quick Serve plane, the same distance is given by $d_2 = 5y$. According to the problem $d_1 = d_2$ or

(19)
$$7x = 5y.$$

(D) Neither (17) nor (19) are sufficient to obtain a numerical answer. But we can use the expression $y = x + 16$ from (17) in (19). When we make this substitution we obtain

(20)
$$7x = 5(x + 60).$$

(E) Solving equation (20) we obtain

$$7x = 5x + 300 \qquad \text{(expanding)}$$
$$7x - 5x = 300 \qquad \text{(transposing)}$$
$$2x = 300 \qquad \text{(combining terms)}$$
$$x = 150 \text{ miles/hour} \quad \text{(dividing by 2).}$$

Once we have found x, we can use (17) to find y. Since $y = x + 60$, this gives $y = 150 + 60 = 210$ miles/hour. Finally $d_1 = 7x = 7(150) = 1,050$ miles.

(F) To check our solution, we read the problem carefully again and see if the values obtained really satisfy the given conditions. In this case we find that $210 = 150 + 60$, so the condition on the speed is satisfied. Further,

$$d_1 = 7(150) = 1,050 \text{ miles,}$$
$$d_2 = 5(210) = 1,050 \text{ miles.}$$

Hence, they have both traveled the same distance when they meet. ▲

‡ In any elementary treatment of problems of this type, we assume that the object is traveling with uniform motion (the speed is constant). If the speed changes, the question is meaningless. However, the work which follows will give the *average speed.*

EXAMPLE 2   An industrial chemist has 96 gallons of 70% sulphuric acid, and wishes to add enough water so that the resultant mixture is 40% sulphuric acid. How many gallons of water should be added?

Solution: Let $x \equiv$ number of gallons of water to be added. The amount of pure sulphuric acid is the same before and after the addition of water.

The amount of pure acid before addition = 0.70 (96).

The amount of pure acid after addition = 0.40 (96 + x).

Then,

$$0.70\,(96) = 0.40\,(96 + x)$$
$$7(96) = 4(96 + x) \qquad \text{(multiplying by 10)}$$
$$= 4(96) + 4x$$
$$7(96) - 4(96) = 4x$$
$$4x = (7 - 4)(96) = 3(96)$$
$$x = \frac{3(96)}{4} = 72 \text{ gallons.} \quad \blacktriangle$$

We let the reader check that the addition of 72 gallons of water does indeed yield a 40% sulphuric acid solution.

EXAMPLE 3   One pump working steadily will fill a reservoir in 30 days. A second pump will do the same job in 20 days. If both pumps are used together how long will it take to fill the reservoir?

Solution: This problem appears to be difficult because we do not have any information about the size of the reservoir. Thus it may seem difficult to select units. In other words, should we measure the volume of the reservoir in gallons or cubic yards, etc.? To illustrate that this is unimportant in this particular problem we will use a fictitious unit of volume, the zeek.

Let V = volume of the reservoir in zeeks.
Let r_1 = rate of the first pump in zeeks per day.
Let r_2 = rate of the second pump in zeeks per day.
Let x = number of days required when both pumps are operating.

Since volume pumped = (rate of pump)(time in operation) we have

$$\text{rate of pump} = \frac{\text{volume pumped}}{\text{time in operation}}$$

$$r_1 = \frac{V}{30} \text{ zeeks per day}, \qquad r_2 = \frac{V}{20} \text{ zeeks per day.}$$

When both pumps are operating the combined rate is $r_1 + r_2$. Hence, to fill the reservoir in x days we must have

(21) $\qquad V = (r_1 + r_2)x = \left(\dfrac{V}{20} + \dfrac{V}{30}\right)x.$

We first multiply both sides of (21) by $1/V$. Then

$$\left(\frac{1}{20} + \frac{1}{30}\right)x = 1$$

$$x = \frac{1}{\dfrac{1}{20} + \dfrac{1}{30}} = \frac{20(30)}{20 + 30} = \frac{600}{50} = 12 \text{ days.} \quad \blacktriangle$$

The reader is invited to work the same problem using gallons or cubic feet as a measure of the volume. Is it possible to compute the volume of the reservoir from the data?

Exercise 2

1. Suppose that the two airplanes in Example 1 fly in opposite directions and that at 8:07 P.M. they are 2,550 miles apart. Find the speed of each plane if the Quick Serve plane flies 40 miles/hour faster than the Transvalley plane.

2. A Transvalley plane leaves Sirloin City for Steakville 1,000 miles away at 7:43 A.M. At 8:43 A.M. a Quick Serve plane leaves Steakville for Sirloin City. If the Quick Serve plane travels 70 miles/hour faster than the Transvalley plane and passes over that plane at 10:43 A.M. find the speed of each plane.

3. A chemist has 26 gallons of 60% nitric acid. How much water should he add to obtain (a) a 40% nitric acid mixture, (b) a 20% nitric acid mixture? Is the answer to part (b) twice the answer to part (a)?

4. A tank car is partially full of 55% hydrochloric acid. After 9,000 gallons of water were added, a test showed that the mixture was 46% hydrochloric acid. How many gallons of 55% acid did the tank car contain initially?

5. Arvin can mow the lawn in 6 hours with an old-fashioned push type mower, and Bozo can do the same lawn in 2 hours with a power mower. How long will it take them to mow the lawn if they work together using both mowers?

6. Alice can paint her room in 5 hours 30 minutes. Jerry can paint the same room in 4 hours 30 minutes. If they work together how long will it take them to paint the room?

7. The sum of three consecutive integers is 249. Find the first of these three integers.

8. The sum of four consecutive integers is 274. Find the first of these four integers.

9. The sum of four consecutive odd integers is 1,096. Find the first one of these integers.

10. The sum of six consecutive odd integers in 492. Find the first one.

11. Find two numbers which differ by 4 and such that their squares differ by 200.

12. Find two numbers which differ by 6 and such that their squares differ by 267.

13. Two numbers have the sum 511, and one of the numbers is 6 times the other number. Find the numbers.

14. Three numbers have the sum 765. The second number is twice the first and the third number is three times the second number. Find the three numbers.

15. Zeno has found that if he increases his average speed from 50 miles/hour to 60 miles/hour he can save 15 minutes in driving from Wildwood to Lakeview. Find the distance from Wildwood to Lakeview.

16. One hour and 15 minutes after Arvin left home on his bicycle Bozo went after him in the car which averages 25 miles/hour faster than the bicycle. After driving 45 minutes Bozo caught up with Arvin. Find the speed of the car.

17. An airplane flying with a ground speed of 355 miles/hour finds that against the wind a certain trip requires 6 hours, but the return trip with the wind takes only 4 hours. Assuming that the wind is steady find the speed of the wind.

18. Going home from work, Zeno spends 15 minutes going steadily through town traffic, and when he hits route I-75 he increases his average speed by 30 miles/hour. If he spends 10 minutes on I-75 and the total distance from work to home (just at the I-75 exit) is 15 miles, find his speed on I-75.

19. Zeno has $10,000 invested, some at 6% and the rest at 4%. The total income gives him an average yield of 5.4%. How much was invested at 6%?

20. Nezo has invested a total of $8,000, some at 4% and the rest at 7%. If the 4% investment yields $155 more than the 7% investment how much was invested at 4%?

21. A rectangular lot is twice as long as it is wide. A road through the lot 20 feet wide and parallel to the long side will use 700 square feet of land more than a road 30 feet wide and parallel to the short side. Find the dimensions of the lot.

22. A retired professor sold his mansion. With the proceeds he bought a smaller house (for himself) at one-half the price of his mansion and a second house (to rent) at $2,000 more than one-third of the price of his mansion. He had $4,800 left which he spent on a new car. How much did he pay for the second house?

23. Arvin, Bozo, and Corky together earn $315 a week. Arvin's weekly wage is $60 less than twice Bozo's, while Corky earns $45 per week more than two-thirds Bozo's salary. Who has the biggest income, and how much is it?

24. Arvin, Bozo, and Corky bought a used car. Arvin put in $300 more than one-fifth of the cost of the car. Bozo contributed $90 more than Arvin. Corky's share was $200 less than one-fourth of the cost of the car. What did the car cost?

25. Bozo can paint a house in 36 hours. Corky can paint the same house in 24 hours. They started working together but after 12 hours (spread over 2 days) Bozo became discouraged and quit. How much time was necessary for Corky to finish the job?

26. Barney and Corrigan together can build a brick wall in 12 hours. If Corrigan works alone, he can do the job in 21 hours. How long would it take Barney to build the wall by himself?

27. My daughter is eleven years older than my son. In 3 years she will be twice as old as my son. How old are they now?

28. Bozo is five times as old as his son. In 18 years Bozo will only be twice as old as his son. How old is Bozo now?

29. Arvin, Bozo, and Corky working separately can pick all the oranges in a small grove in 36, 27, and 18 days, respectively. If Arvin and Bozo start together and Corky joins them 2 days later, how many days will it take to clear the grove of oranges?

30. When the boys from problem 29 moved to a second grove (same size) Arvin started working immediately. Bozo joined him 3 days later, and after another 3 days Corky started picking oranges too. How many days did Arvin work before the second grove was harvested?

31. Two cars racing on an oval track 4 miles long average 120 miles/hour and 105 miles/hour, respectively. If they start together, how long will it take for the first car to be one lap ahead of the second car?

5.4 QUADRATIC EQUATIONS

If an equation can be put in the form

$$(22) \qquad ax^2 + bx + c = 0, \qquad a \neq 0,$$

using only transformations of the type described in Theorem 1, then the equation is said to be a *quadratic equation* in x, or of *second degree* in x.

If we can factor the left side of (22), then the equation is easy to solve. We recall the technique in

EXAMPLE 1

Solve $x^2 - 3x - 10 = 0$.

Solution: By the methods developed in Section 5.1

$$x^2 - 3x - 10 = (x-5)(x+2) = 0$$

$$
\begin{array}{c|c}
x - 5 = 0 & x + 2 = 0 \\
x = 5, & x = -2. \quad \blacktriangle
\end{array}
$$

Suppose that in Example 1 we change -10 to -9, a very slight change. We still have a quadratic equation

$$(23) \qquad x^2 - 3x - 9 = 0,$$

but now the factors (if there are any) are not obvious. More powerful methods are needed. If we allow complex numbers to enter, then every quadratic equation has at least one root, but if we insist on the roots being real, then an equation such as $x^2 + 4 = 0$ has no root.

Theorem 3

The quadratic equation $(x-2)(x+2)$ $+2i$

$$(22) \qquad ax^2 + bx + c = 0, \qquad a \neq 0,$$

has at most two roots, r_1 and r_2. These are given by the formulas

$$(24) \qquad r_1 = \frac{-b + \sqrt{b^2 - 4ac}}{2a}, \qquad r_2 = \frac{-b - \sqrt{b^2 - 4ac}}{2a}.$$

Suppose that a, b, and c are real numbers.

If $b^2 - 4ac > 0$, then the equation has two real roots.

If $b^2 - 4ac = 0$, then the equation has one real root.

If $b^2 - 4ac < 0$, then the equation has no real root.

Proof: We let x represent a root of (22) so that indeed

$$(22) \qquad\qquad ax^2 + bx + c = 0.$$

Using a sequence of legitimate operations we will convert the left side of (22) into a perfect square so that we can take the square root of both sides. From (22) we have:

$$ax^2 + bx = -c \qquad \text{(transposing } c)$$

$$x^2 + \frac{b}{a}x \qquad = -\frac{c}{a} \qquad \text{(dividing by } a \neq 0)$$

$$x^2 + \frac{b}{a}x + \frac{b^2}{4a^2} = \frac{b^2}{4a^2} - \frac{c}{a} \qquad \left(\text{adding } \frac{b^2}{4a^2} \text{ to both sides}\right)$$

$$\left(x + \frac{b}{2a}\right)^2 = \frac{b^2 - 4ac}{4a^2} \qquad \text{(factoring the left side)}$$

$$x + \frac{b}{2a} = \pm\frac{\sqrt{b^2 - 4ac}}{2a} \qquad \text{(taking the square root of both sides)}$$

$$(25) \quad x = \frac{-b \pm \sqrt{b^2 - 4ac}}{2a} \qquad \left(\text{transposing } \frac{b}{2a}\right).$$

We have proved that if equation (22) has a root it is given by equation (25), But this is just the formula for the roots r_1 and r_2 given in (24). Conversely, suppose that x, given by (25), is indeed a real number. Then each step used in going from (22) to (25) is reversible. Hence if (25) is true so also is (22).
We examine $\sqrt{b^2 - 4ac}$. If $b^2 - 4ac > 0$ then $\sqrt{b^2 - 4ac}$ is a positive number, and equation (22) has two real roots. If $b^2 - 4ac = 0$, then the two formulas. in (24) give the same root $r = -b/2a$, and the equation has only one root. If $b^2 - 4ac < 0$, then $\sqrt{b^2 - 4ac}$ is not a real number. There are no real roots. ∎

In Chapter 18, we will introduce complex numbers. Then we will see that equation (22) has two complex roots, given by the very same formulas.

EXAMPLE 2   Solve $x^2 - 3x - 9 = 0$.

Solution: The formula given in Theorem 3,

$$(24) \qquad\qquad r_1, r_2 = \frac{-b \pm \sqrt{b^2 - 4ac}}{2a},$$

is called the *quadratic formula* and should be memorized. For the equation at hand, $a = 1$, $b = -3$, and $c = -9$. Then (24) gives for the roots

$$r_1, r_2 = \frac{-(-3) \pm \sqrt{(-3)^2 - 4(1)(-9)}}{2(1)}$$

$$= \frac{3 \pm \sqrt{9 + 36}}{2} = \frac{3 \pm \sqrt{45}}{2} = \frac{3 \pm 3\sqrt{5}}{2}.$$

Thus the solution set is $\left\{ \dfrac{3 + 3\sqrt{5}}{2}, \dfrac{3 - 3\sqrt{5}}{2} \right\}$. ▲

Suppose that we want to factor $x^2 - 3x - 9$. Does it help us to know the roots of the equation $x^2 - 3x - 9 = 0$?. This is covered in

Theorem 4

Let r_1 and r_2 be two distinct roots of

(22) $$ax^2 + bx + c = 0, \qquad a \neq 0.$$

Then

(26) $$ax^2 + bx + c = a(x - r_1)(x - r_2).$$

This theorem is a special case of a more general theorem that holds for an arbitrary polynomial of nth degree. Because this material is somewhat sophisticated, we defer the more general theorem and proof until Chapter 19.

Proof of Theorem 4: Since r_1 is a root, we have $ar_1^2 + br_1 + c = 0$. Subtracting zero in this special form we find

$$ax^2 + bx + c = ax^2 + bx + c - (ar_1^2 + br_1 + c)$$

$$= (ax^2 - ar_1^2) + (bx - br_1) + (c - c)$$

$$= a(x^2 - r_1^2) + b(x - r_1)$$

$$= a(x - r_1)(x + r_1) + b(x - r_1)$$

(27) $$ax^2 + bx + c = a(x - r_1)\left(x + r_1 + \frac{b}{a}\right).$$

From (27) it follows that $x = -(r_1 + b/a)$ is also a root, and hence it must be r_2. Then (27) gives (26). ∎

EXAMPLE 3   Factor $x^2 - 3x - 9$.

Solution: We already found the two roots of $x^2 - 3x - 9 = 0$ in Example 2. Using these roots and Theorem 4 we have

$$x^2 - 3x - 9 = \left(x - \frac{3+3\sqrt{5}}{2}\right)\left(x - \frac{3-3\sqrt{5}}{2}\right). \quad \blacktriangle$$

We encourage the student to expand the right side of this equation and prove that it does give the left side.

EXAMPLE 4   Find the roots of $5x^2 + 8x + 2 = 0$.

Solution: We use the quadratic formula with $a = 5$, $b = 8$, and $c = 2$. Then the roots are

$$r_1, r_2 = \frac{-8 \pm \sqrt{8^2 - 4(5)(2)}}{2(5)} = \frac{-8 \pm \sqrt{64 - 40}}{10}$$

$$= \frac{-8 \pm \sqrt{24}}{10} = \frac{-8 \pm 2\sqrt{6}}{10} = \frac{-4 \pm \sqrt{6}}{5}. \quad \blacktriangle$$

If we wish to factor $5x^2 + 8x + 2$, Theorem 4 gives

$$5x^2 + 8x + 2 = 5\left(x - \frac{-4+\sqrt{6}}{5}\right)\left(x - \frac{-4-\sqrt{6}}{5}\right)$$

$$= 5\left(x + \frac{4-\sqrt{6}}{5}\right)\left(x + \frac{4+\sqrt{6}}{5}\right).$$

EXAMPLE 5   Does $3x^2 - 9x + 7 = 0$ have real roots?

Solution: It is not necessary to compute r_1 and r_2. We only need to examine $b^2 - 4ac$. This quantity is called the *discriminant* of the quadratic (22). When $a = 3$, $b = -9$, and $c = 7$

$$b^2 - 4ac = (-9)^2 - 4(3)(7) = 81 - 84 = -3 < 0.$$

There are no real roots. $\quad \blacktriangle$

A word problem may lead to a quadratic equation with two real roots, one of which gives the solution to the word problem, while the other is physically meaningless. This is illustrated in

FIGURE 5.1

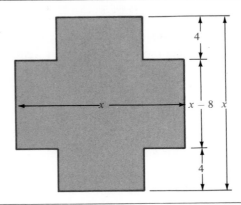

EXAMPLE 6

We wish to make a box by cutting 4-inch squares from each of the four corners of a square piece of tin. (See Figure 5.1.) Find the size of the piece of tin if the box has no top and the volume is 100 cubic inches.

Solution: Let x denote the length of one side of the piece of tin. Then the base of the box is $x - 2(4) = x - 8$ inches. For the volume we have

(28) $$V = \ell wh = (x-8)^2 4 = 100$$

$$x^2 - 16x + 64 = \frac{100}{4} = 25$$

$$x^2 - 16x + 39 = 0$$

$$(x-3)(x-13) = 0.$$

The solution set for the quadratic equation (28) is $\{3, 13\}$ but for the word problem $x = 3$ is physically meaningless. (Why?) The solution to the given problem is $x = 13$ inches. The piece of tin should be 13 inches by 13 inches, and not 3 inches by 3 inches. ▲

Exercise 3

In problems 1 through 16 find the real roots of the given quadratic equation either by factoring or by using the quadratic formula.

1. $x^2 - 6x + 3 = 0$ 2. $x^2 - 2x - 4 = 0$

3. $x^2 + 7x + 5 = 0$ 4. $x^2 + 5x + 3 = 0$

5. $x^2 - 5x + 4 = 0$ 6. $x^2 + 7x + 12 = 0$

7. $x^2 + x + 1 = 0$ 8. $x^2 + 2x + 3 = 0$

9. $4x^2 + 7x + 2 = 0$

10. $5x^2 - 9x + 3 = 0$

11. $4x^2 - 24x + 11 = 0$

12. $5x^2 - 8x + 3 = 0$

13. $6x^2 + 3x - 1 = 0$

14. $7x^2 - 6x - 5 = 0$

15. $6x^2 - 17x + 5 = 0$

16. $7x^2 - 9x - 10 = 0$

In problems 17 to 30 find all the real roots of the given fourth degree equation. *Hint:* In problems 17 through 24 let $y = x^2$.

17. $x^4 - 13x^2 + 36 = 0$

18. $x^4 - 6x^2 + 5 = 0$

19. $x^4 - 10x^2 + 5 = 0$

20. $x^4 - 6x^2 + 7 = 0$

21. $x^4 + 6x^2 - 9 = 0$

22. $x^4 + 5x^2 - 3 = 0$

23. $x^4 + 7x^2 + 6 = 0$

24. $x^4 + 8x^2 + 12 = 0$

25. $(x^2 + 4x)^2 + 4(x^2 + 4x) - 12 = 0$

26. $(x^2 - 3x)^2 - 2(x^2 - 3x) - 15 = 0$

27. $(x^2 - 5x)^2 - 10(x^2 - 5x) - 56 = 0$

28. $(x^2 - 2x)^2 - 5(x^2 - 2x) + 6 = 0$

29. $x^4 - 81 + 10(x^3 - 9x) = 0$

30. $x^4 - 16 + 6(x^3 + 4x) = 0$

31. Prove that if r_1 and r_2 are distinct roots of the quadratic equation $ax^2 + bx + c = 0$, then

$$r_1 + r_2 = -\frac{b}{a} \quad \text{and} \quad r_1 r_2 = \frac{c}{a}.$$

In problems 32 through 37 find a quadratic equation for which the sum and product of the roots have the given values. *Hint:* See problem 31.

32. The sum is 4, product is 2.

33. The sum is 7, product is 4.

34. The sum is 5/3, product is 1/2.

35. The sum is 7/4, product is 1/3.

36. The sum is -8, product is -7.

37. The sum is -5, product is -1.

In problems 38 through 43 find the value of k such that the given equation has only one real root.

38. $x^2 - 6x + k = 0$

39. $y^2 + 10y + k = 0$

40. $3z^2 + kz + 3 = 0$

41. $5u^2 + ku + 45 = 0$

42. $kv^2 + 5v + 6 = 0$

43. $kw^2 - 3w + 2 = 0$

44. A uniform border is added to two adjacent sides of a 9 feet by 12 feet rug. This adds 130 square feet to the area of the rug. Find the width of the border.

45. The sum of the squares of two adjacent odd integers is 290. Find the integers.

46. Two numbers differ by 5 and the sum of their squares is 13. Find the numbers.

47. Two numbers differ by 5 and the sum of their squares is 29. Find the numbers.

48. In a certain right triangle the sum of the lengths of the legs is 6 inches and the hypotenuse is 5 inches. Find the length of the shorter leg.

49. Do problem 48 if the sum of the lengths of the legs is 9 inches, and the hypotenuse is 7 inches long.

50. Do Example 6, if the volume of the box is to be 324 cubic inches.

51. Do Example 6, if we cut 3-inch squares from each corner and the volume is to be 150 cubic inches.

Occasionally a cubic equation can be solved by grouping terms in a suitable fashion. For example, the terms in $4x^3 - 4x^2 - 3x + 3 = 0$ can be grouped to give $4x^2(x-1) - 3(x-1) = 0$. Factoring, we have $(4x^2 - 3)(x-1) = 0$. Consequently, the roots are 1 and $\pm\sqrt{3}/2$. Use this technique in problems 52 through 55 to solve the given equation.

52. $x^3 - x^2 - 4x + 4 = 0$ 53. $4x^3 + 8x^2 - 3x - 6 = 0$

54. $9x^3 - 3x^2 - 3x + 1 = 0$ 55. $x^3 + 7x^2 - 4x - 28 = 0$

5.5 EQUATIONS INVOLVING RATIONAL EXPRESSIONS

If E is a rational expression, we can multiply both sides of $E = 0$ by a suitable polynomial P and obtain an equation $PE = 0$ that is usually easier to solve than $E = 0$. Strictly speaking, $PE = 0$ may not always be equivalent to $E = 0$ but if we are careful we can still use this process to solve the equation $E = 0$. Before going further with the theory we consider

EXAMPLE 1   Solve the equation

(29)
$$\frac{1}{x-5} + \frac{1}{x+13} = 0.$$

Solution: In its present form it would be difficult to guess at a solution. We multiply both sides of (29) by the LCD, $P = (x-5)(x+13)$. This gives

$$(x-5)(x+13)\frac{1}{x-5} + (x-5)(x+13)\frac{1}{x+13} = 0$$

(30)
$$x+13 \qquad +x-5 \qquad\qquad = 0$$

$$2x+8 = 0$$

$$x+4 = 0$$

$$x = -4.$$

Check Putting $x = -4$ in (29) we obtain

$$\frac{1}{-9} + \frac{1}{9} = 0. \quad \blacktriangle$$

Let us review this process. The polynomial $P = (x-5)(x+13)$ is zero when $x = 5$ and $x = -13$. When we replace equation (29), $E = 0$, by equation (30), $PE = 0$, we introduce the possibility that PE is zero when $x = 5$ and $x = -13$, because the first factor P is zero for these values of x. Consequently, there is a possibility that equation (30) is not equivalent to equation (29). However it only takes a few moments to check that the numbers $x = 5$ and $x = -13$ are not roots of (30). Thus, our fears are groundless; in this case equations (29) and (30) are equivalent. This situation is described precisely in

Theorem 5
PLE

Let P be a polynomial and let $P = 0$ have roots $s_1, s_2, \ldots, s_k$. If none of these roots are roots of the equation $PE = 0$, then

(31)
$$E = 0$$

and

(32)
$$PE = 0$$

are equivalent equations.

EXAMPLE 2

Solve the equation

(33)
$$\frac{1}{x-9} + \frac{2}{x-3} - \frac{3}{x-1} = 0,$$

and check each root.

Solution: The left side of (33) is the expression E in Theorem 5. We select for P the LCD of this expression. In other words, $P \equiv (x-9)(x-3)(x-1)$.

Multiplying both sides of $E = 0$ by P, we get

$$PE = (x-9)(x-3)(x-1)\left(\frac{1}{x-9} + \frac{2}{x-3} - \frac{3}{x-1}\right) = 0$$

$$(x-3)(x-1) + 2(x-9)(x-1) - 3(x-9)(x-3) = 0$$

$$x^2 - 4x + 3 + 2(x^2 - 10x + 9) - 3(x^2 - 12x + 27) = 0$$

$$(1+2-3)x^2 + (-4-20+36)x + (3+18-81) = 0$$

(34)
$$PE = 12x + (-60) = 0$$

$$x = 5.$$

Hence it appears that $x = 5$ is the only root of (33). What does Theorem 5 tell us in this situation? Since $P = (x-9)(x-3)(x-1)$, the roots of $P = 0$ are $s_1 = 9$, $s_2 = 3$, $s_3 = 1$. Since none of these are roots of (34), we conclude that (33) and (34) are indeed equivalent. Hence, $x = 5$ is the only root of (33).

Check   If we put $x = 5$ in equation (33) we find

$$\frac{1}{5-9} + \frac{2}{5-3} - \frac{3}{5-1} = 0$$

$$-\frac{1}{4} + 1 - \frac{3}{4} = 0. \quad \blacktriangle$$

We can still use Theorem 5 if the equation has the form $E_1 = E_2$. This is illustrated in

EXAMPLE 3   Solve the equation

(35)
$$\frac{3}{x+1} - \frac{2}{x} = \frac{1}{x+3}.$$

Solution: We multiply both sides of (35) by $P \equiv x(x+1)(x+3)$ which is the LCD for the set of denominators in (35). This gives

$$x(x+1)(x+3)\left(\frac{3}{x+1} + \frac{-2}{x}\right) = x(x+1)(x+3)\frac{1}{x+3}$$

$$3x(x+3) - 2(x+1)(x+3) = x(x+1)$$

$$3x^2 + 9x - 2(x^2 + 4x + 3) - (x^2 + x) = 0$$

$$(3-2-1)x^2 + (9-8-1)x - 6 = 0$$

(36)
$$-6 = 0.$$

The conclusion that $-6 = 0$ may at first seem shocking but if we review the reasoning behind the steps, then the shock vanishes. We assume that x represents a number for which (35) is true. If there is such a number, it follows that (36) is also true: $-6 = 0$. Since this latter statement is false, we conclude that (35) is never true for any real value of x. In other words (35) has no solution, or the solution set is $\varnothing$. ▲

Exercise 4

In problems 1 through 18 solve the given equation. Check each root you obtain to see that it satisfies the given equation

1. $\dfrac{1}{x+5} + \dfrac{1}{x+7} = 0$

2. $\dfrac{2}{x+5} + \dfrac{3}{x+10} = 0$

3. $\dfrac{2}{x-9} - \dfrac{3}{x-11} = 0$

4. $\dfrac{5}{x+1} - \dfrac{3}{x-7} = 0$

5. $\dfrac{x+1}{x+2} = \dfrac{x-3}{x+4}$

6. $\dfrac{x+5}{x-1} = \dfrac{x+3}{x-4}$

7. $\dfrac{x}{x^2+4} = \dfrac{x+1}{x^2+x+5}$

8. $\dfrac{x-1}{x^2-2} = \dfrac{x+2}{x^2+3x+4}$

9. $\dfrac{x-1}{x+3} = \dfrac{x-2}{x+2}$

10. $\dfrac{2x+3}{3x+4} = \dfrac{4x-5}{6x-11}$

11. $\dfrac{5}{x+2} = \dfrac{3}{x+1} + \dfrac{2}{x+3}$

12. $\dfrac{2}{x-1} - \dfrac{1}{x} = \dfrac{1}{x+1}$

13. $\dfrac{5}{y-3} = \dfrac{14}{y-1} - \dfrac{9}{y+4}$

14. $\dfrac{6}{y+1} = \dfrac{7}{y-1} - \dfrac{1}{y-5}$

15. $\dfrac{6}{x+1} = \dfrac{1}{x} + \dfrac{10}{x+4}$

16. $\dfrac{3}{x-2} + \dfrac{4}{x-3} = \dfrac{3}{x-4}$

17. $\dfrac{12}{x^2-4} + \dfrac{8}{x+2} = \dfrac{3}{x-2}$

18. $\dfrac{4}{x^2+4x+3} = \dfrac{x+8}{x+3} + \dfrac{1-x}{1+x}$

19. Find a number such that the number plus its reciprocal is (a) 4, (b) 10.

20. Find a number such that the number plus 10 times its reciprocal is (a) 7, (b) 14.

21. Find a number x such that the sum of the reciprocals of x and $x+2$ is the reciprocal of $x+4$. Check your answer.

22. Solve for x

$$\frac{1}{2-\dfrac{x}{3-\dfrac{x}{4}}} = \frac{1}{3}.$$

23. Solve for x in problem 22 when the right side is 1/4.

5.6 EQUATIONS INVOLVING RADICALS

If the expressions in the equation

(37) $$E_1 = E_2$$

involve radicals, it may be advantageous to raise both sides to a suitable power and try to solve

(38) $$E_1^n = E_2^n$$

rather than (37). If r is a root of (37), then r is also a root of (38). However, the converse is not always true. For example, if we square both sides of the equation $x = 1$ we obtain $x^2 = 1$. The original equation has only one root but the new equation $x^2 = 1$ has two roots, 1 and -1. Thus equation (38) may have a root s that is not a root of equation (37). Such a root is called an *extraneous* root. If n is odd, this extraneous root will be a complex number, but if n is even, then the extraneous root may be a real number. If we restrict ourselves to real roots, then equations (37) and (38) are equivalent when n is odd. We summarize these remarks in

Theorem 6   *Let S_n be the set of real roots of* (38) *and let* **S** *be the set of real roots of* (37).

If n is even, then $\mathbf{S}_n \supset \mathbf{S}$.

If n is odd, then $\mathbf{S}_n = \mathbf{S}$.

EXAMPLE 1   Solve the equation

(39) $$2\sqrt{x+6} - x = 3.$$

Solution: We transpose x and then square both sides obtaining

(40)
$$4(x+6) = (x+3)^2 = x^2 + 6x + 9$$
$$0 = x^2 + (6-4)x + 9 - 24 = x^2 + 2x - 15$$
$$0 = (x+5)(x-3)$$
$$x = 3, \quad \text{or} \quad x = -5.$$

Observe that if we squared both sides of equation (39) before transposing, the new equation would still involve $\sqrt{x+6}$ and would be harder to solve than the original equation.

Check If $x = 3$, equation (39) gives

$$2\sqrt{3+6} - 3 = 3$$
$$6 - 3 = 3.$$

If $x = -5$, equation (39) gives

$$2\sqrt{-5+6} - (-5) \overset{?}{=} 3$$
$$2 + 5 \neq 3.$$

In this example equations (39) and (40) are not equivalent. Equation (40) has the solution set $\{3, -5\}$ which contains $\{3\}$, the solution set of (39). ▲

If the expressions in $E_1 = E_2$ are very complicated we may need to apply Theorem 6 several times.

EXAMPLE 2   Solve the equation

(41) $$\sqrt{2x+4} + \sqrt{2x-3} = \sqrt{6x+13}.$$

Solution: We square both sides.

$$2x + 4 + 2x - 3 + 2\sqrt{2x+4}\sqrt{2x-3} = 6x + 13$$

$$2\sqrt{2x+4}\sqrt{2x-3} = 2x + 12 \qquad \text{(transposing)}$$

$$\sqrt{2x+4}\sqrt{2x-3} = x + 6 \qquad \text{(dividing by 2)}$$

$$4x^2 + 2x - 12 = x^2 + 12x + 36 \qquad \text{(squaring)}$$

(42) $$3x^2 - 10x - 48 = 0 \qquad \text{(combining terms)}$$

$$(x-6)(3x+8) = 0 \qquad \text{(factoring)}$$

$$x = 6 \qquad \text{or} \qquad x = -8/3.$$

Check When $x = 6$, equation (41) gives

$$\sqrt{12+4} + \sqrt{12-3} = \sqrt{36+13}$$
$$4 + 3 = 7.$$

When $x = -8/3$, the second term of (41) is $\sqrt{-16/3-3}$. Since this is not real, no further work is necessary. The solution set for equation (41) is $\{6\}$. The solution set for equation (42) is $\{6, -8/3\}$. ▲

Exercise 5

In problems 1 through 23 solve the given equation and check each root.

1. $\sqrt{x+6} + 2x = 9$

2. $\sqrt{x+7} + 5 = x$

3. $\sqrt{7x+14} - 2 = x$

4. $\sqrt{4x-2} + 5 = 2x$

5. $\sqrt{x+23} - \sqrt{x+16} = 1$

6. $\sqrt{x+13} - \sqrt{x+5} = 2$

7. $\sqrt{2x-7} - 1 = \sqrt{x-4}$

8. $\sqrt{5x+21} - 2 = \sqrt{x+13}$

9. $\sqrt{x+2} + \sqrt{x+16} = 7$

10. $\sqrt{x+19} + \sqrt{x+1} = 6$

11. $\sqrt{2x+3} + \sqrt{3x+4} = \sqrt{5x+9}$

12. $\sqrt{x-1} + \sqrt{3-x} = \sqrt{x+2}$

13. $\sqrt{6-x} = \sqrt{2x+5} - \sqrt{x-1}$

14. $\sqrt{2x+3} = \sqrt{5x+8} - \sqrt{3x+5}$

15. $3\sqrt{x} = \dfrac{20}{\sqrt{x}} - 11$

16. $6\sqrt{x} = \dfrac{5}{\sqrt{x}} - 13$

17. $\sqrt{\dfrac{x}{1-x}} + \sqrt{\dfrac{1-x}{x}} = \dfrac{5}{2}$

18. $\sqrt{\dfrac{2x}{1-3x}} + \sqrt{\dfrac{1-3x}{2x}} = \dfrac{13}{6}$

19. $\sqrt{2x^2-9x+4} - \sqrt{2x^2-7x+1} = 1$

20. $\sqrt{2x^2+5x-2} - \sqrt{2x^2+5x-9} = 1$

★21. $\dfrac{1-x^4}{x+x^3} = 2\sqrt{\dfrac{3}{x}} - x$

★22. $\dfrac{x^3+1}{x^2-1} = x + \sqrt{\dfrac{6}{x}}$

★23. $\dfrac{x+\sqrt{x^2-1}}{x-\sqrt{x^2-1}} - \dfrac{x-\sqrt{x^2-1}}{x+\sqrt{x^2-1}} = 8x\sqrt{x^2-3x+2}$

★5.7 MORE ABOUT INEQUALITIES

The material on inequalities covered in Section 3.2 was simple. Now that we have gained some skill at manipulation, we can consider a few inequalities that are slightly more involved.

According to Theorem 1 in Chapter 3, if a is any real number, then $a^2 \geq 0$. The equality sign occurs if and only if $a = 0$. This innocent sounding statement is the source of many remarkable results.

EXAMPLE 1

Prove that for any two real numbers

(43) $$2ab \leqq a^2 + b^2,$$

and that the equality sign occurs if and only if $a = b$.

Solution: By our remark above

(44) $$0 \leqq (a-b)^2.$$

Equality in (44) occurs if and only if $a = b$. Expanding (44) we have

$$0 \leqq a^2 - 2ab + b^2$$

Then by Theorem $\overline{13}$ in Chapter 3 (adding $2ab$ to both sides)

(43) $$2ab \leqq a^2 + b^2.$$

Since the equality sign occurs in (44) if and only if $a = b$, it also occurs in (43) under the same conditions. ▲

EXAMPLE 2

Prove that if a, b, c, and d are any four positive numbers, then

(45) $$ab + cd \leqq \sqrt{a^2 + c^2}\, \sqrt{b^2 + d^2}.$$

It is not easy to see the proper starting place for this problem, so we work backward. Thus, we start with the inequality (45) and see if we can deduce one that we know to be true. We call this operation the *analysis* of the problem.

Analysis If (45) is true, we can square both sides and obtain

(46) $$a^2b^2 + 2abcd + c^2d^2 \leqq (a^2 + c^2)(b^2 + d^2)$$

or

(47) $$a^2b^2 + 2abcd + c^2d^2 \leqq a^2b^2 + c^2b^2 + a^2d^2 + c^2d^2.$$

If we transpose (Theorem $\overline{13}$ in Chapter 3, add the same negative quantity to both sides) we obtain

(48) $$0 \leqq c^2b^2 - 2abcd + a^2d^2$$

or

(49) $$0 \leqq (cb - ad)^2.$$

But we know that this last inequality is always true. Hence, if we can reverse our steps, we can prove that the given inequality is also true.

Solution: We begin with the known inequality (49). On expanding we find that (48) is also true. Adding $a^2b^2 + 2abcd + c^2d^2$ to both sides of (48) we obtain (47). Factoring the right side of (47) and then taking the positive square root on both sides, we find (46) and (45). ▲

It is customary to do the analysis on scratch paper, and then write the solution in the proper order; i.e., in the reverse order of the analysis. The student should write out in detail the correct solution of this example, following the outline just given.

EXAMPLE 3   Without using tables, prove that

(50) $$\sqrt{2} + \sqrt{6} < \sqrt{3} + \sqrt{5}.$$

Solution: We give the analysis. Squaring both sides of (51) we find

(52) $$2 + 2\sqrt{2}\sqrt{6} + 6 < 3 + 2\sqrt{3}\sqrt{5} + 5.$$

On subtracting 8 from both sides and dividing by 2, we obtain

(53) $$\sqrt{2}\sqrt{6} < \sqrt{3}\sqrt{5}.$$

Finally we square both sides of (53) and obtain $12 < 15$. Since this last inequality is obviously true, we can start with this inequality, reverse our steps, and obtain (50). ▲

Exercise 6

In problems 1 through 6 determine which of the two given numbers is the larger.

1. $\sqrt{19} + \sqrt{21}$, $\sqrt{17} + \sqrt{23}$ 2. $\sqrt{11} - \sqrt{8}$, $\sqrt{17} - \sqrt{15}$

3. $\sqrt{17} + 4\sqrt{5}$, $5\sqrt{7}$ 4. $2\sqrt{2}$, $\sqrt[3]{23}$

5. $2\sqrt{3} + 3\sqrt{2}$, $5 + \sqrt{5}$ 6. $3\sqrt{2} - \sqrt{13}$, $\sqrt{17} - 2\sqrt{3}$

7. Prove that if $1 < k < n$, then $\sqrt{n-k} + \sqrt{n+k} < \sqrt{n-1} + \sqrt{n+1}$.

8. Prove that the inequality of Example 2 is true even if some or all of the numbers are negative.

In problems 9 through 20 prove the given inequality under the assumption that all of the quantities involved are positive. Determine the conditions under which the equality sign occurs.

9. $a + \dfrac{1}{a} \geq 2$

10. $\dfrac{a}{5b} + \dfrac{5b}{4a} \geq 1$

11. $\sqrt{\dfrac{c}{d}} + \sqrt{\dfrac{d}{c}} \geq 2$

12. $(c+d)^2 \geq 4cd$

13. $\dfrac{a+b}{2} \geq \sqrt{ab} \geq \dfrac{2ab}{a+b}$

14. $(a+5b)(a+2b) \geq 9b(a+b)$

15. $x^2 + 4y^2 \geq 4xy$

★ 16. $x^2 + y^2 + z^2 \geq xy + yz + zx$

★ 17. $\dfrac{c^2}{d^2} + \dfrac{d^2}{c^2} + 6 \geq \dfrac{4c}{d} + \dfrac{4d}{c}$

★ 18. $\dfrac{a+3b}{3b} \geq \dfrac{4a}{a+3b}$

★ 19. $cd(c+d) \leq c^3 + d^3$

★ 20. $4ABCD \leq (AB+CD)(AC+BD)$

21. Which of the above inequalities are still meaningful and true if the letters are permitted to represent negative numbers?

Review Questions

Try to answer the following questions as accurately as possible before consulting the text.

1. What is the difference between an identity and a conditional equation?

2. What is a solution set?

3. When are two equations equivalent?

4. What is a linear equation? How many different solutions are possible for a linear equation?

5. State the quadratic formula and the theorem that involves this formula (Theorem 3).

6. What is the discriminant of a quadratic equation, and why is it important?

Functions and Graphs

6

The reader has already met functions many times, but he may not have been aware of this during the encounter. The product of two numbers is a function of the two factors. The cube of a number is a function of the number. The number of distinct prime factors of a positive integer is a function of that integer.

In this chapter we sharpen this old intuitive idea of a function and make it precise. Further we review the rectangular coordinate system because it furnishes a good way of picturing certain kinds of functions.

To study functions with precision we must introduce some new terminology which may discourage the casual reader. With a little effort, the student will realize that the basic ideas are indeed simple and easy.

6.1 FUNCTIONS

A formula such as

(1)
$$y = 3x + 11$$

defines a function. A different formula such as

(2)
$$y = x^2$$

also defines a function (different from the first one). Further examples of functions defined by formulas are

(3)
$$z = 2\sqrt{25 - u^2},$$

and

(4)
$$w = \frac{1}{v+1} + \frac{3}{v+7} + \frac{5}{v+13}.$$

What is the essential character of these examples? The first formula gives a rule for finding a corresponding value of y for each value of x. For example, if $x = 4$, then equation (1) gives $3 \cdot 4 + 11 = 23$ for the corresponding value of y. Equation (2) also defines a function but a different one. If $x = 4$, then the corresponding value of y is $4^2 = 16$.

Formulas (3) and (4) also define functions, although different letters are used. If $u = 4$, then the associated or corresponding z is $2\sqrt{25 - 16} = 6$. If $v = -3$, the associated or corresponding w is $-\frac{1}{2} + \frac{3}{4} + \frac{1}{2} = \frac{3}{4}$. The use of the letters x and y in the discussion of a function is not a necessity but merely a convenience. The important item is the correspondence.

Although most of the functions considered in this book are given by formulas, the existence of a formula is not essential for the definition of a function. Further, the items related need not be numbers but can be elements of any set. For example, with every senator in the United States Senate, there is associated a unique state, the state that he represents in the Senate. With these examples in mind we arrive at

DEFINITION 1

Function *Let* **A** *and* **B** *be any two non-empty sets. If for each* $x \in$ **A** *there is a rule (method or procedure) which determines a unique* $y \in$ **B** *that corresponds to* x, *then the rule (method or procedure) determines a function from the set* **A** *to the set* **B**. *The letter used to denote an arbitrary element from* **A** *is called the independent variable. The letter used to indicate the corresponding element from* **B** *is called the dependent variable.*

A diagram will help us to understand this definition. In Figure 6.1, the sets **A** and **B** are represented schematically by the shaded regions, even though the sets may be sets of people, automobiles, or numbers. For each x in **A** there is a unique corresponding y in **B**. This correspondence is indicated by an arrow in Figure 6.1. Since y is unique, it is impossible for two arrows to start from the same point. However, it is quite possible for several different arrows to end at the same point. If y_1 corresponds to x_1 (or is associated with x_1), we say that y_1 is the *image* of x_1 under the function f. The element x_1 is called the *preimage* of y_1 or the *primitive* of y_1. We also say that the function f takes x_1 into y_1 or carries x_1 into y_1. Thus, the function diagrammed in Figure 6.1 takes x_3, x_4, and x_5 into the same element y_3 of **B**.

FIGURE 6.1

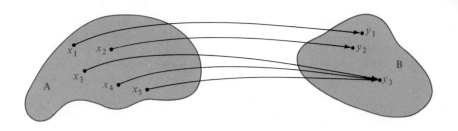

The function itself is the set of these correspondences. Consequently we say that the function f takes **A** to **B**, or f maps **A** to **B**. One frequently finds the notation

$$f: \mathbf{A} \to \mathbf{B} \qquad \text{(read, "}f\text{ takes }\mathbf{A}\text{ to }\mathbf{B}\text{")}$$

for a function. However, there is a much more convenient and useful notation

(5) $$y = f(x) \qquad \text{(read, "}y\text{ equals }f\text{ of }x\text{")}$$

which we will explore in detail in the next section. In this notation f does not multiply x, but f represents the rule (method or procedure) which we use to find y when x is given.

EXAMPLE 1   Find $f(2)$ if the function f is: (a) defined by equation (1), (b) defined by equation (2), (c) defined by equation (3), and (d) defined by equation (4).

Solution: Whenever the function f is given by a formula, we merely replace the independent variable by the particular value (in this case 2) and compute in accordance with the formula.

(a) Since $f(x) = 3x + 11$, then $f(2) = 3(2) + 11 = 17$.

(b) Since $f(x) = x^2$, then $f(2) = 2^2 = 4$.

(c) Since $f(u) = 2\sqrt{25 - u^2}$, then $f(2) = 2\sqrt{21}$.

(d) Since $f(v) = \dfrac{1}{v+1} + \dfrac{3}{v+7} + \dfrac{5}{v+13}$, then

$$f(2) = \frac{1}{3} + \frac{3}{9} + \frac{5}{15} = 1. \quad \blacktriangle$$

The concept of a function carries with it quite a few new terms.

Let f be a function from **A** *to* **B**. *The set* **A** *is called the domain of the function. Let* **C** *be the set of all y such that* $y = f(x)$ *for some x in* **A**. *The set* **C** *is called the range of the function. If* **C** = **B**, *then we say that f maps* **A** *onto* **B**. *If* **C** ≠ **B**, *then we say that f maps* **A** *into* **B**. *If each y in* **C** *is the image of just one x in* **A**, *then we say that f is a one-to-one mapping from* **A** *onto* **C**. *If* **A** *and* **C** *are sets of real numbers, then f is called a real-valued function of a real variable.*

The function pictured in Figure 6.1 does not give a one-to-one mapping. Neither does the function that associates with each senator, the state that he represents, since each state (normally) has two senators.

AGREEMENT

To have a suitable notation that reminds us of the concepts involved, we will use **D** for the domain of a function and **G** for the range of a function.

The more natural choice of **R** for the range of a function is forbidden because we are using **R** for the set of all real numbers.

EXAMPLE 2

Discuss the domain **D** and range **G** of each of the four functions defined by equations (1) through (4).

Solution: A function is not properly given unless the domain is stated in advance. Hence, the question about the domain is an improper one, since this information has not been supplied. In this text we are concerned with real-valued functions of a real variable. Hence, we make the following

AGREEMENT

Whenever a function is given by a formula (as in equations (1) through (4)), the domain is the largest set of real numbers for which the formula gives a real number ‡, unless otherwise specified.

With this agreement we have:

(a) If $f(x) = 3x + 11$, then the domain is **R**.

(b) If $f(x) = x^2$, then the domain is **R**.

(c) If $f(u) = 2\sqrt{25 - u^2}$, then the domain is the interval $[-5, 5]$.

(d) If $f(v) = \dfrac{1}{v+1} + \dfrac{3}{v+7} + \dfrac{5}{v+13}$, then the domain is $\mathbf{R} - \{-1, -7, -13\}$.

For the range **G** we have:

(a) For each real number y, the equation $y = 3x + 11$ can be solved for x. We find that $x = (y-11)/3$. Hence, if $f(x) = 3x + 11$, then the range **G** is **R**.

‡ See Chapter 4 Section 7 where we made a similar agreement.

(b) If y is any nonnegative number, then the equation $y = x^2$ can be solved. One solution is $x = \sqrt{y}$. Hence, if $f(x) = x^2$, then $\mathbf{G} = [0, \infty)$.

(c) If $f(u) = 2\sqrt{25 - u^2}$, then $-5 \leq u \leq 5$. If u is in this interval, then $0 \leq 2\sqrt{25 - u^2} \leq 10$. Clearly $\mathbf{G}$ is the closed interval $[0, 10]$.

(d) If $f(v) = \dfrac{1}{v+1} + \dfrac{3}{v+7} + \dfrac{5}{v+13}$,

then the equation $f(v) = c$ leads to a cubic equation. It can be proved that the equation always has at least one real root. If we grant this fact, then $\mathbf{G} = \mathbf{R}$.

Among these four functions only the first gives a one-to-one mapping from $\mathbf{D}$ to $\mathbf{G}$. ▲

EXAMPLE 3

Find the domain of

(6)
$$f(x) = \sqrt{x(x-6)(x-17)}.$$

Solution: If $x \geq 17$, then $x(x-6)(x-17) \geq 0$, and $\sqrt{x(x-6)(x-17)}$ is a real number. If $6 < x < 17$, then $x(x-6)(x-17) < 0$, and the formula does not give a real number for $f(x)$. The same is true if $x < 0$ because in this case all three factors are negative. If $0 \leq x \leq 6$, then $f(x) \geq 0$. Hence, the domain of this function is $[0, 6] \cup [17, \infty)$, the union of an interval and a ray. ▲

We are free to select any letter $x, y, z, u, v, \ldots$ for the independent variable. We can also select any (other) letter from the same set for the dependent variable. Thus, $u = f(v)$ or $x = f(w)$ are both suitable symbols for a function. We can also use different letters to represent different functions. The most popular ones for this purpose are f, g, h, F, G, H, and the Greek letters ϕ (phi) and ψ (psi). We may also use subscripts. For example, g_1, g_2, g_3, and g_4 can be used to represent four different functions.

There is a fine distinction between the symbol f and the symbol $y = f(x)$ for a function, but we prefer to avoid this distinction. We regard both symbols as representing a function, but the symbol $y = f(x)$ tells us which letters we have in mind for the dependent and independent variables.

Exercise 1

In problems 1 through 10 determine the domain of the given function in accordance with the agreement made on page 116. Also find the range of the function.

1. $f(x) = x + 5$

2. $f(x) = x^2 + 7$

3. $f(x) = \dfrac{1}{x+5}$

4. $f(x) = \dfrac{1}{x^2 + 7}$

5. $f(x) = 5\sqrt{x^2 - 9}$

6. $f(x) = 2 - \sqrt{(x+4)(x-7)}$

★ 7. $f(x) = \dfrac{6}{\sqrt{4-x^2}}$

★ 8. $f(x) = \dfrac{5}{1+x^2}$

★ 9. $f(x) = \sqrt{(x^2-1)(x^2-4)}$

★ 10. $f(x) = 3x + \sqrt{x-6}$

11. If $f(x) = x^2 - 3x + 2$, show that $f(0) = 2$, $f(1) = 0$, $f(2) = 0$, $f(3) = 2$, and $f(4) = 6$.

12. If $g(x) = x^2 - 5x + 3$, find $g(0)$, $g(1)$, $g(2)$, $g(3)$, and $g(4)$.

13. If $h(x) = x(x-1)(x-2)$, find $h(1)$, $h(2)$, $h(3)$, $h(4)$, $h(-1)$, and $h(-2)$.

14. If $u(x) = x^3 + 1$, find $u(-2)$, $u(-1)$, $u(0)$, $u(1)$, and $u(2)$.

15. If $v(x) = x^2 - 3$, find $v(1)$, $v(s)$, $v(t)$, and $v(y)$.

16. If $F(x) = 3x + 5$, find $F(-2)$, $F(s)$, $F(t)$, $F(y)$, and $F(x_1)$.

★ 17. If $f(x) = x^2 - x + 3$, find $f(x+1)$, $f(2x)$, $f(-x)$, $f(x+y)$, and $f(x-y)$.

In problems 18 through 22 a function is described in words. Find a formula for the function.

18. The surface area of a cube is a function of the length of an edge.

19. The volume of a cube is a function of the length of an edge.

20. The volume of a cube is a function of the surface area.

21. The area of an equilateral triangle is a function of the length of an edge.

22. The area of an equilateral triangle is a function of the length of the altitude.

23. Define the function $\pi(x)$ to be the number of primes less than or equal to x. By definition, the number 1 is not a prime. Find $\pi(6)$, $\pi(9)$, $\pi(26)$, $\pi(35)$, $\pi(\sqrt{98})$, and $\pi(\pi^2)$.

6.2 FUNCTION NOTATION

The notation $f(x)$ for a function permits us to indicate relations among functions in a very simple way. We illustrate the versatility of this notation with a few examples.

EXAMPLE 1   Let $f(x) = 4x - 7$. Is it true that

(7) $$f(x_1 + x_2) = f(x_1) + f(x_2)$$

for all pairs of real numbers x_1, x_2?

Solution: If the symbol $f(x)$ represents the product of f and x, then (7) would always be true (The Distributive Law). To show that (7) is false in general we only need to exhibit one pair for which it is false. We select (at random) $x_1 = 3$ and $x_2 = 4$. Computation with the given function yields

$$f(x_1) = f(3) = 4 \cdot 3 - 7 = 5, \qquad f(x_2) = f(4) = 4 \cdot 4 - 7 = 9,$$
$$f(x_1 + x_2) = f(3+4) = f(7) = 4 \cdot 7 - 7 = 21.$$

Since $21 \neq 5 + 9$ we see that in general $f(x_1 + x_2) \neq f(x_1) + f(x_2)$. ▲

EXAMPLE 2   Let $f(x) = 4x - 7$. Prove that no matter how we select x_1 and x_2, equation (7) is always false for this function.

Solution: For this particular function

$$f(x_1 + x_2) = 4(x_1 + x_2) - 7 = 4x_1 + 4x_2 - 7.$$
$$f(x_1) = 4x_1 - 7, \qquad f(x_2) = 4x_2 - 7.$$

If for some pair x_1, x_2, equation (7) were true, then we would have

$$4x_1 + 4x_2 - 7 = (4x_1 - 7) + (4x_2 - 7)$$
$$-7 = -14.$$

Hence, equation (7) is never true for the function $f(x) = 4x - 7$. ▲

EXAMPLE 3   Let $F(x) = x^2 - 5x + 17$. Prove that for all real x

$$(8) \qquad\qquad F(x+5) = F(-x).$$

Solution: For this particular function

$$F(x+5) = (x+5)^2 - 5(x+5) + 17 = x^2 + 10x + 25 - 5x - 25 + 17$$
$$= x^2 + 5x + 17.$$
$$F(-x) = (-x)^2 - 5(-x) + 17 = x^2 + 5x + 17 = F(x+5). \quad ▲$$

EXAMPLE 4   Let $g(x) = 3^x$. Prove that for all pairs of rational numbers x, y

$$(9) \qquad\qquad g(x+y) = g(x)g(y).$$

Solution: By the laws of exponents (Theorem 3 in Chapter 4),

$$g(x+y) = 3^{x+y} = 3^x \cdot 3^y = g(x)g(y). \quad ▲$$

EXAMPLE 5   If $F(x) = x^2 - 5x + 17$, find $F(x+1) - F(x)$ and $F(x+h) - F(x)$.

Solution: Direct computation with the given function yields

$$F(x+1) - F(x) = (x+1)^2 - 5(x+1) + 17 - (x^2 - 5x + 17)$$
$$= x^2 + 2x + 1 - 5x - 5 + 17 - x^2 + 5x - 17$$
$$= 2x - 4.$$

$$F(x+h) - F(x) = (x+h)^2 - 5(x+h) + 17 - (x^2 - 5x + 17)$$
$$= x^2 + 2xh + h^2 - 5x - 5h + 17 - x^2 + 5x - 17$$
$$= 2xh + h^2 - 5h. \quad \blacktriangle$$

6.3 COMPOSITE FUNCTIONS

We can construct complicated functions by composing simple ones. For example, if $u = x - 5$ and $y = \sqrt{u}$, then using the expression for u from the first function in the second one we obtain the composite function $y = \sqrt{x-5}$. A function that is obtained from simpler functions in this way is called a *composite function*.

The general situation is this. We have three sets **A**, **B**, and **C** (usually sets of real numbers) and two functions: $u = g(x)$ which maps **A** to **B** and $y = f(u)$ which maps **B** to **C** (see Figure 6.2). The composite function $F(x) \equiv f(g(x))$, obtained by using the expression for u from the first function in the second one, gives a mapping from **A** to **C**. If g maps x_1 to u_1 and f maps u_1 to y_1, then the composite function F maps x_1 to y_1.

EXAMPLE 1   Let $f(x) = x^2 + 2x$, and let $g(x) = 2x^2 - 1$. Find $f(g(x))$ and $g(f(x))$. Are these two functions the same (equal)?

Solution: In this example we have purposely dropped the letters y and u, which were helpful in the explanation of composite functions, but are not needed in this type problem. To compute $f(g(x))$ we merely replace x by $g(x)$ wherever x occurs in $f(x)$. Consequently,

$$f(g(x)) = (g(x))^2 + 2g(x) = (2x^2 - 1)^2 + 2(2x^2 - 1)$$
(10) $$f(g(x)) = 4x^4 - 4x^2 + 1 + 4x^2 - 2 = 4x^4 - 1.$$

FIGURE 6.2

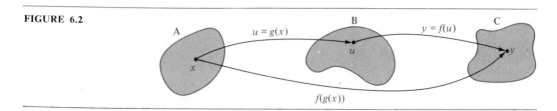

However, for $g(f(x))$ we have

$$g(f(x)) = 2(f(x))^2 - 1 = 2(x^2 + 2x)^2 - 1$$

(11) $$g(f(x)) = 2x^4 + 8x^3 + 8x^2 - 1.$$

Equations (10) and (11) show clearly that the two functions $f(g(x))$ and $g(f(x))$ are different. ▲

We can compose three or more functions if we wish.

EXAMPLE 2   Let $h(x) = x^3$, and let $g(x)$ and $f(x)$ be defined as in Example 1. Find $f(g(h(x)))$ in two ways.

Solution: We already know that $f(g(x))$ is given by

(10) $$f(g(x)) = 4x^4 - 1.$$

Therefore,

$$f(g(h(x))) = 4(h(x))^4 - 1 = 4(x^3)^4 - 1 = 4x^{12} - 1.$$

The reader is invited to compute $g(h(x))$ first obtaining $2x^6 - 1$ and then to find $f(g(h(x)))$ by computing $f(2x^6 - 1)$. This procedure should also yield $4x^{12} - 1$.

Example 1 shows that a Commutative Law for composite functions

(12) $$f(g(x)) = g(f(x))$$

is not true. Example 2 suggests that an Associative Law for composite functions may be true. To clarify the situation it is convenient to introduce the symbol $f \circ g$ to denote the composite function $F(x) \equiv f(g(x))$. With this notation the Associative Law takes the form

(13) $$f \circ (g \circ h) = (f \circ g) \circ h.$$

Theorem 1   *In any domain in which the functions in equation (13) are well defined, equation*
PWO   *(13) is true.*

Exercise 2

1. If $f(x) = 13x$, prove that $f(x + y) = f(x) + f(y)$ for all x, y.

2. If $g(x) = 3x + 2$ and $f(x) = 5x + 4$, prove that $f(g(x)) = g(f(x))$ for all x.

3. Suppose that $g(x) = ax + a - 1$ and $f(x) = bx + b - 1$. Prove that $f(g(x)) = g(f(x))$ for all x.

4. If $f(x) = 3x + 5$, prove that $f(4x) - 4f(x) + 15 = 0$ for all x.

5. If $f(x) = 7x - 3$, prove that $f(5x) - 5f(x) = 12$ for all x.

6. If $f(x) = x^2$, prove that $f(x+y) - f(x-y) = 4xy$ for all pairs x, y.

7. If $f(x) = x^2$, prove that $f(x+2) - 2f(x+1) + f(x) - 2 = 0$ for all x.

8. If $g(x) = 5a^x$, where $a > 0$, prove that $5g(x+y) = g(x)g(y)$ for all rational numbers x, y.

In problems 9 through 17 let $f(x) = x + 2$, $g(x) = 2x - 3$, and $h(x) = x^2 - x$.

9. Find $f\big(g(h(x))\big)$.

10. Find $g\big(h(f(x))\big)$.

11. Find $g\big(f(h(x))\big)$.

12. Find $h\big(g(f(x))\big)$.

13. For what values of x is $f\big(g(h(x))\big) = g\big(f(h(x))\big)$?

14. For what values of x is $h\big(g(f(x))\big) = g\big(h(f(x))\big)$?

15. Find $f^2\big(g^2(x)\big)$ and $g^2\big(f^2(x)\big)$.

16. Find $h\big(f(x) + g(x)\big)$.

17. Find $g\big(h(x) + 2f(x)\big)$.

6.4 THE RECTANGULAR COORDINATE SYSTEM

The reader has certainly met the rectangular coordinate system. However, this system of assigning coordinates to points in a plane is so important that we must present it once again. We begin with two directed lines that meet at right angles (see Figure 6.3). The point of intersection of these lines is called

FIGURE 6.3

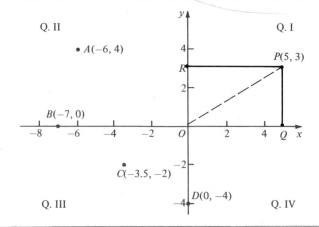

the *origin* and is usually lettered *O*. It is customary to take one of these lines to be horizontal and to take the direction to the right of *O* as the positive direction on this line. This horizontal line is called the *x-axis* or the *horizontal axis*. The other directed line, which is perpendicular to the *x*-axis, is called the *y-axis* or the *vertical axis*, and the positive direction is upward from *O*. These two axes divide the plane into four quadrants, which are labeled Q.I, Q.II, Q.III, and Q.IV as indicated in Figure 6.3.

Once a rectangular coordinate system has been chosen, any point in the plane can be located with respect to it. Suppose *P* is some point in the plane. Let *PQ* be a perpendicular to the *x*-axis, and let *PR* be a perpendicular to the *y*-axis. The directed distance $\overline{OQ}$ is called the *x*-coordinate of *P*, and the directed distance $\overline{OR}$ is called the *y*-coordinate of *P* (see Section 3.6, page 43).

For example, if *P* is the point shown in Figure 6.3, then $\overline{OQ} = 5$ and $\overline{OR} = 3$, so the *x*-coordinate is 5 and the *y*-coordinate is 3. When we wish to avoid the use of a particular letter such as *x* or *y*, we call $\overline{OQ}$ the *abscissa* of *P*, and we call $\overline{OR}$ the *ordinate* of *P*.

It is customary to write (*x*, *y*) to indicate the number pair $\overline{OQ}$ and $\overline{OR}$ that is associated with the point *P*. The numbers (*x*, *y*) are called the *coordinates* of the point *P*. Since the figure *OQPR* is a rectangle it is clear that $\overline{RP} = \overline{OQ}$ and $\overline{OR} = \overline{QP}$. Hence, a second definition is possible, namely:

The *x*-coordinate of *P* (the abscissa) is the directed distance of *P* from the *y*-axis.

The *y*-coordinate of *P* (the ordinate) is the directed distance of *P* from the *x*-axis.

A variety of other points together with their coordinates are shown in Figure 6.3. The reader should check each point carefully to see if the coordinates appear to be correct. This procedure can be reversed. Given the coordinates $A(-6, 4)$ for example, the point *A* can be located by moving 6 units to the left of *O* on the *x*-axis and then proceeding upward 4 units along a line parallel to the *y*-axis. Consequently, we have a one-to-one correspondance between the set of all points in the plane and the set of all ordered pairs of real numbers (*x*, *y*). This discussion gives

Theorem 2   *With a fixed pair of coordinates axes, each point P in the plane has a uniquely determined pair of coordinates (x, y) where x and y are real numbers. Conversely, for each ordered pair of real numbers (x, y) there is a uniquely determined point P in the plane that has the coordinates (x, y).*

Let *r* denote the distance of *P*(*x*, *y*) from the origin. This distance is the length of the diagonal of the rectangle *OQPR* (see Figure 6.3) with sides of length $|x|$ and $|y|$. By the Pythagorean Theorem $r = \sqrt{|x|^2 + |y|^2}$. Hence

(14) $$r = |OP| = \sqrt{x^2 + y^2}.$$

EXAMPLE 1   Find $|OA|$, $|OB|$, and $|OC|$ for the points A, B, and C shown in Figure 6.3.

Solution: By equation (14)

$$|OA| = \sqrt{(-6)^2 + 4^2} = \sqrt{36 + 16} = 2\sqrt{13},$$

$$|OB| = \sqrt{(-7)^2 + 0^2} = \sqrt{49} = 7,$$

$$|OC| = \sqrt{(-3.5)^2 + (-2)^2} = \sqrt{12.25 + 4} = \sqrt{16.25}. \quad \blacktriangle$$

The rectangular coordinate system is frequently called the Cartesian coordinate system in honor of its inventor René Descartes. A brief but highly entertaining account of his life can be found in *Men of Mathematics* by E. T. Bell (Simon and Schuster, 1937).

Exercise 3

1. On a rectangular coordinate system locate (plot) the points $(4, 3)$, $(3, -4)$, $(-2, -3)$, and $(-4, \sqrt{11})$. Find the distance of each of these points from the origin.

2. Find the distance from the origin for each of the points $(5, 12)$, $(-4, 7)$, $(-5, -4)$, and $(6, -3)$.

In problems 3 through 10 find the missing coordinate of P if P satisfies the given conditions.

3. $r = 5$, $x = 3$, P is in Q.IV.   4. $r = 13$, $y = -5$, P is in Q.III.

5. $r = \sqrt{29}$, $x = -5$, P is in Q.II.   6. $r = 13$, $x = 12$, P is in Q.I.

7. $r = 5$, $y = -4$, P is in Q.III.   8. $r = 2\sqrt{7}$, $y = 5$, P is in Q.II.

9. $r = \sqrt{11}$, $y = 2\sqrt{2}$, P is in Q.I.   10. $r = 10$, $x = 6$, P is in Q.IV.

In problems 11 through 19 a condition is given on the coordinates of P. In which quadrant must P lie if the coordinates satisfy the given condition? Ignore the possibility that the point may lie on a coordinate axis.

11. $x < 0$ 12. $y < 0$ 13. $1/x > 0$

14. $x/y > 0$ 15. $y/x < 0$ 16. $1/y < 0$

17. $y > 0$ 18. $x^3 > 0$ 19. $y^5 < 0$

In problems 20 through 34 a condition is given on the coordinates of P. What geometric figure is formed by the set of all points whose coordinates satisfy the given condition?

20. $x = 3$ 21. $y = 4$ 22. $y = -1$

23. $x = -5$ 24. $x^2 = 9$ 25. $y^2 = 4$

26. $y = -x$ 27. $x^2 = -4$ 28. $x^2 + y^2 > -4$

29. $x^2 + y^2 = 36$ 30. $x^2 = y^2$ 31. $x^2 = -y^2$

32. $|y| = 2$ 33. $|x| = 3$ 34. $|x| = |y|$

35. Is the distance given by equation (14) a directed distance?

★ 36. Let $P(x_1, y_1)$ and $Q(x_2, y_2)$ be two points in the plane. Discover and prove a formula for $|PQ|$, the distance between the two points.

6.5   THE GRAPH OF A FUNCTION

If we mark each point P in the plane for which the coordinates (x, y) are such that $y = f(x)$, then we obtain a picture of the function f. This picture (sketch, drawing) is extremely helpful because with it we can see at a glance important features of the function.

DEFINITION 3   ***Graph*** *Let* $y = f(x)$ *be a real-valued function of a real variable. The graph of f is the set of all points $P(x, y)$ on the plane for which x is in the domain of f, and y is the image of x under f.*

The graph of a function is *not* the picture or sketch with all of its inaccuracies and physical limitations. To the contrary, the graph is a collection of points. A sketch of a graph is an attempt at a physical picture of this collection of points. Despite the inevitable inaccuracies, a sketch of a graph is very useful and is often the first step in the study of a function.

EXAMPLE 1   Sketch the graph of

(15) $$y = \frac{1}{2}(x+3).$$

Solution: We have replaced the function $f(x) = (x+3)/2$ by $y = (x+3)/2$, because we intend to use y to represent the image of x under f. We will frequently make such a change without further explanation.

The graph of (15) consists of infinitely many points (x, y). We are content to find only a few, and for convenience we select small integers for x. For example, if $x = 1$, then (15) gives $y = (1+3)/2 = 2$. Hence the point $(1, 2)$ is a point of the graph of this function. To find more points of the graph we set $x = -5, -3, ..., 7$, compute the corresponding y, and arrange the pairs neatly in Table 1. The reader should check each entry. The points

TABLE 1	x	-5	-3	-1	1	3	5	7
	y	-1	0	1	2	3	4	5

FIGURE 6.4

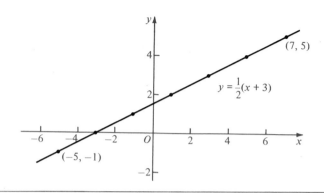

with these coordinates are indicated in Figure 6.4 by dots. A glance at these dots indicates clearly that they lie on a straight line. This is quite a coincidence, and we feel certain that if we computed more points for the graph of $y = (x+3)/2$, they would also lie on the same straight line.

The student should compute at least three more points and check that they also lie on the line shown in Figure 6.4. ▲

After we have acquired more mathematical machinery we will find that the graph of $y = (x+3)/2$ is a straight line, since this is a special case of the more general

Theorem 3
PWO

If either a or b is not zero, then the graph of

(16) $$ax + by + c = 0.$$

is a straight line.

If $b \neq 0$, we can solve (16) for y obtaining

(17) $$y = -\frac{a}{b}x - \frac{c}{b},$$

which gives (15) when $-a/b = 1/2$ and $-c/b = 3/2$.

EXAMPLE 2   Sketch the graph of

(18) $$y = x^2 - 4x + 7.$$

From this graph try to discover something of interest about the function.

Solution: Direct computation with the expression on the right side of (18) gives those points of the graph entered in Table 2.

x	-2	-1	0	1	2	3	4	5	6
y	19	12	7	4	3	4	7	12	19

When we prepare to sketch the graph of this function we observe that the y coordinates seem to be rather large, so we take the liberty of changing the size of the unit for the y-axis so that it is somewhat smaller than the unit for the x-axis. This is permitted as long as everyone concerned is aware of the change.‡ With this smaller unit the sketch is given in Figure 6.5.

What can we say that is interesting about this function? The experienced mathematician can derive all the information he needs from the formula $f(x) = x^2 - 4x + 7$. The student may take the liberty of leaning on the sketch. By inspection of Figure 6.5 he might conjecture (guess):

(A) The graph is a parabola.

(B) The least value (minimum) of $f(x)$ is 3, and this occurs when $x = 2$.

FIGURE 6.5

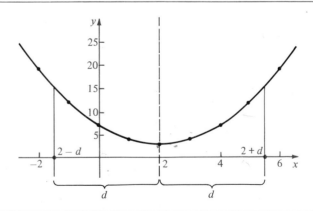

‡ This is a favorite trick of people who wish to confuse the audience with data. One superimposes two graphs, one for the cost of living as a function of time and the second for the average income of some group as a function of time. By an appropriate choice of the units, the plotter can show that the average income is increasing faster than the cost of living—or the cost of living is increasing faster than the average income—whichever way he wishes.

(C) The graph has some kind of symmetry about a vertical line through $x = 2$.

(D) As one moves from left to right the function is decreasing for $x \leqq 2$, and the function is increasing for $x \geqq 2$. ▲

★ EXAMPLE 3   Try to prove the conjectures (guesses) (A), (B), (C), and (D) formulated in the solution to Example 2.

Solution: (A) We need the definition of a parabola before we can prove that the graph of (18) is a parabola. This definition is usually covered in analytic geometry and so the student must postpone the proof until he studies that subject. It can be proved that the graph of $y = ax^2 + bx + c$ is always a parabola if $a \neq 0$.

(B) This conjecture is easy to settle. If we complete the square we have

$$(19) \qquad f(x) = x^2 - 4x + 7 = x^2 - 4x + 4 + 3 = (x-2)^2 + 3.$$

Then $f(x) = (x-2)^2 + 3 \geq 3$ because the term $(x-2)^2$ is never negative. Further, $f(x) = 3$ if and only if $(x-2)^2 = 0$, and this occurs if and only if $x = 2$.

(C) Symmetry about the vertical line through $x = 2$ means that the value of y when $x = 2+d$ is the same as the value of y when $x = 2-d$ (see Figure 6.5). In function notation this is expressed by the equation

$$(20) \qquad\qquad f(2+d) = f(2-d).$$

Is (20) really true for all d, when $f(x)$ is the function $x^2 - 4x + 7$? The picture certainly suggests that this is true. To settle the matter, we compute both sides of (20) explicitly. We find that

$$f(2+d) = (2+d)^2 - 4(2+d) + 7 = 4 + 4d + d^2 - 8 - 4d + 7$$
$$= d^2 + 3.$$
$$f(2-d) = (2-d)^2 - 4(2-d) + 7 = 4 - 4d + d^2 - 8 + 4d + 7$$
$$= d^2 + 3.$$

Consequently, equation (20) is indeed an identity, and the graph is symmetric with respect to the line $x = 2$.

(D) To investigate this conjecture we must first give the definition of an increasing (decreasing) function.

DEFINITION 4 *A function $f(x)$ is said to be increasing on an interval (or ray) **S** if for every pair x_1, x_2 in **S** with $x_1 < x_2$, we have*

(21) $$f(x_1) < f(x_2).$$

*A function $f(x)$ is said to be decreasing on **S** if for every pair x_1, x_2 in **S** with $x_1 < x_2$, we have*

(22) $$f(x_1) > f(x_2).$$

Suppose now that $x_1 < x_2$, and both are in the ray $[2, \infty)$. We compute $f(x_2) - f(x_1)$ for the function $f(x) = x^2 - 4x + 7$. We find that

$$
\begin{aligned}
f(x_2) - f(x_1) &= x_2^2 - 4x_2 + 7 - (x_1^2 - 4x_1 + 7) \\
&= x_2^2 - x_1^2 - 4(x_2 - x_1) \\
&= (x_2 - x_1)(x_2 + x_1 - 4) > 0,
\end{aligned}
$$

if $2 \leqslant x_1 < x_2$. Hence, $f(x_1) < f(x_2)$. We have proved that $f(x)$ is increasing on the ray $[2, \infty)$. ▲

EXAMPLE 4 Sketch the graph of

(23) $$y = |x - 2|.$$

Solution: If $x \geqslant 2$, then $|x-2| = x-2$, and the graph is a portion of a straight line. If $x < 2$, then $|x-2| = -(x-2) = 2-x$, and the graph is a portion of another straight line. A sketch of this graph is shown in Figure 6.6. The reader should use equation (23) to compute a few points of this graph and check that they do indeed lie on the indicated lines. ▲

FIGURE 6.6

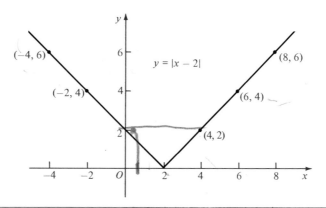

In each example considered thus far the sketch appeared to be an unbroken (continuous) curve. In the next example the graph consists of infinitely many pieces, called *components*. This is achieved by the introduction of a new function and a new symbol.

The symbol $[x]$ (read, "square bracket of x" or "the integer of x") denotes the greatest integer n such that $n \leq x$. As examples we have:

$$[7/2] = 3, \quad [\sqrt{111}] = 10, \quad [\pi] = 3, \quad [-7/2] = -4, \quad [79] = 79.$$

★ EXAMPLE 5   Sketch the graph of the function: (a) $y = [x]$, and (b) $y = [\sqrt{x}] - 1$.

Solution: Portions of these graphs are shown in Figures 6.7 and 6.8. We carry over the [, (notation from Section 4.7 to indicate whether the endpoint of a line segment is a part of the graph or not. (See Figure 3.2, page 40.) ▲

FIGURE 6.7   FIGURE 6.8

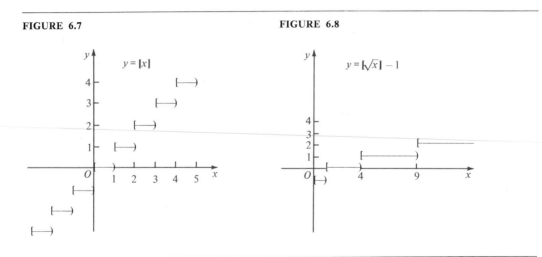

Exercise 4

In problems 1 through 6 the graph is a straight line. Determine the straight line by selecting two points at random. Then compute other points of the graph, and check that they lie on the line already drawn.

1. $y = 2x + 5$ 2. $y = 3x - 7$ 3. $y = \dfrac{1}{3}x - 2$

4. $y = \dfrac{1}{2}x + 4$ 5. $y = -x + 3$ 6. $y = -2x - 1$

In problems 7 through 32 sketch a graph of the given function for x in the indicated interval. In problems 7 through 14 find the maximum or minimum value of the function if there is one and the value of x at which it occurs.

7. $f(x) = x^2 - 4$, $[-3, 3]$

8. $f(x) = 6 - x^2$, $[-4, 4]$

9. $f(x) = 6x - x^2 + 5$, $[-4, 10]$

10. $f(x) = x^2 - 8x + 13$, $[0, 8]$

11. $f(x) = x^2 + 2x - 6$, $[-4, 4]$

12. $f(x) = x^2 + 4x - 1$, $[-4, 4]$

13. $f(x) = 3 - 3x - x^2$, $[-5, 3]$

14. $f(x) = 5 + 5x - x^2$, $[-1, 6]$

15. $f(x) = x^3/8$, $[-4, 4]$

16. $f(x) = x(x^2 - 4)/8$, $[-5, 5]$

17. $f(x) = \dfrac{10}{1 + x^2}$, $[-5, 5]$

18. $f(x) = \dfrac{4x}{1 + x^2}$, $[-5, 5]$

19. $f(x) = x^3 - 3x^2 - 9x + 13$, $[-5, 5]$

20. $f(x) = 7 + 36x + 3x^2 - 2x^3$, $[-5, 5]$

21. $f(x) = \dfrac{2x}{x - 3}$, $[-3, 8]$

22. $f(x) = \dfrac{3}{x + 2}$, $[-8, 3]$

23. $f(x) = \dfrac{x}{\sqrt{2 + x}\sqrt{3 - x}}$, $(-2, 3)$

24. $f(x) = \dfrac{2}{\sqrt{3 + x}\sqrt{4 - x}}$, $(-3, 4)$

25. $f(x) = -|x| + 1$, $[-4, 4]$

26. $f(x) = |-x + 1|$, $[-4, 4]$

27. $f(x) = 3 + x - |x|$, $[-4, 4]$

28. $f(x) = x + |x| - 2$, $[-4, 4]$

★ 29. $f(x) = x - [x]$, $[-4, 4]$

★ 30. $f(x) = \dfrac{12}{[x/2]}$, $[2, 10]$

★ 31. $f(x) = [4/x^2]$, $(0, 5)$

★ 32. $f(x) = 2x - [x]$, $[-4, 4]$

★ 33. Find a maximum set on which the function is increasing for the function given in problem (a) 1, (b) 2, (c) 5, (d) 6, (e) 11, (f) 12, (g) 13, (h) 14, (i) 15.

★ 34. If $f(x) = 4x/(1 + x^2)$, prove that if $x_2 > x_1$ and $x_1 x_2 < 1$, then $f(x_2) > f(x_1)$. (See problem 18.)

★ 35. Find a functional equation similar in form to equation (20) for $f(x) = 6x - x^2 + 5$. (See problem 9.)

★ 36. Do problem 35 for $f(x) = x^2 - 8x + 13$. (See problem 10.)

★ 37. Find two positive numbers such that their sum is 20 and their product is as large as possible.

★ 38. Prove that among all rectangles with the same perimeter 100, the one with the largest area is the square.

6.6 THE GRAPH OF AN EQUATION

An equation is a statement that two expressions are equal. These expressions may involve many variables, but we consider only equations which involve two variables, which we elect to name x and y.

DEFINITION 5

The graph of an equation is the set of all points $P(x, y)$ in the plane for which the equation is satisfied.

A sketch of the graph is a picture formed by marking enough of the points so that we obtain a clear idea of the graph.

Definition 5 is reminiscent of Definition 3 for the graph of a function. There is only a slight difference, but one that needs to be noted. Given a particular x, say x_1, there is at most one y_1 such that (x_1, y_1) is in the graph of a *function*, because in the definition of a function for each x_1 in the domain of f there is a unique y_1. By contrast, for an *equation* there may be several pairs $(x_1, y_1), (x_1, y_2), (x_1, y_3), \ldots$ for which the given equation is true.

EXAMPLE 1

Sketch the graph of

(24)
$$y^2 = \frac{1}{4}(x+3)^2$$

Solution: This equation is merely the square of equation (15) from the preceeding section. Thus, (24) is satisfied whenever $y = \pm(x+3)/2$. Consequently, the graph consists of the straight line shown in Figure 6.4 together with its reflection (mirror image) in the x-axis. These two lines are shown in Figure 6.9. The skeptical student should check several of the points on both straight lines to see that the coordinates satisfy equation (24). ▲

FIGURE 6.9

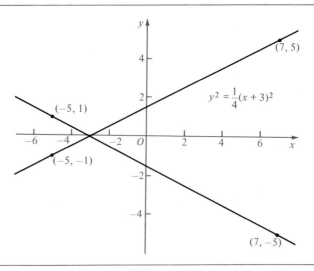

EXAMPLE 2   Sketch the graph of the equation

(25)
$$\frac{x^2}{16} + \frac{y^2}{4} = 1.$$

Solution: It is convenient to solve equation (25) for y. We find that

$$\frac{y^2}{4} = 1 - \frac{x^2}{16} = \frac{16 - x^2}{16}$$

(26)
$$y = \pm \sqrt{\frac{16 - x^2}{4}} = \pm \frac{1}{2} \sqrt{16 - x^2}.$$

Before finding specific points for the graph, we examine the right side of equation (26) because it will give us valuable information.

If $x^2 > 16$, the quantity under the radical is negative, and hence y is not real. Hence the only points of the graph are those for which $-4 \le x \le 4$.

Now we examine y. The maximum value of $|\sqrt{16 - x^2}|$ is 4. Hence, $-2 \le y \le 2$. Therefore the graph must lie in a box 8 units wide and 4 units high (see Figure 6.10).

A few of the points are listed in Table 3, and a sketch of the graph is shown in Figure 6.10.   ▲

TABLE 3	x	0	1	2	2.5	3	3.5	4
	y	± 2	$\pm \sqrt{15}/2$	$\pm \sqrt{3}$	$\pm \sqrt{39}/4$	$\pm \sqrt{7}/2$	$\pm \sqrt{15}/4$	0
	Approx.		± 1.936	± 1.732	± 1.561	± 1.323	± 0.968	

Observe that the graph (or curve‡) shown in Figure 6.10 has a certain amount of symmetry. Clearly the part below the x-axis is the mirror image of the part above the x-axis. The part to the left of the y-axis is the mirror image of the part to the right of the y-axis. In the first case we say that the curve (or graph) is *symmetric with respect to the x-axis* (*see* Example 1). In the second case we say that the *curve is symmetric with respect to the y-axis*. A curve is said to be *symmetric with respect to the origin* if for each point $P(x, y)$ on the curve, the point $P^\star(-x, -y)$ is also on the curve. It is clear that the curve shown in Figure 6.10 has all three types of symmetry.

‡ The distinction between a curve and a graph is somewhat complicated and must be deferred until the reader has learned more mathematics. In the meantime we will use the word curve on occasion, if the graph is also a curve.

FIGURE 6.10

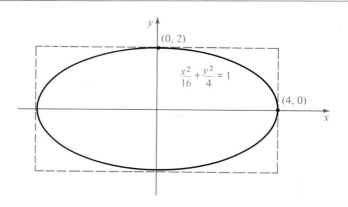

We will not pause for a rigorous discussion of symmetry and the various tests for it. Instead we will depend on the reader's mathematical intuition. We give below four tests for symmetry. Each test (except for the last one) is reasonably obvious once it receives careful thought.

In each test we make a change in the variables as indicated by the arrow in the second column of Table 4 and obtain a new equation from the original one. If the new equation is identical with the original one, or can be transformed into it by transposing terms or multiplying by a suitable constant, then the graph has the symmetry listed in the last column.

TABLE 4

	Change in the variables:	If the new equation is equivalent to the original one, then the graph is symmetric with respect to the:
(A)	$x \to -x$	y-axis
(B)	$y \to -y$	x-axis
(C)	$x \to -x$ and $y \to -y$	origin
(D)	$x \to y$ and $y \to x$	line $y = x$, the 45° line through the origin

EXAMPLE 3   Apply each of the tests in Table 4 to equation (25).

Solution: **(A)** Replacing x by $-x$ in equation (25) we obtain

(27) $$\frac{(-x)^2}{16} + \frac{y^2}{4} = 1.$$

Since $(-x)^2 = (-1)^2 x^2 = x^2$, equation (27) is identical with (25). Hence, the graph is symmetric with respect to the y-axis.

We leave tests (**B**) and (**C**) for the reader.

(**D**) If we replace x by y and y by x in equation (25) we obtain

(28)
$$\frac{y^2}{16} + \frac{x^2}{4} = 1 \quad \text{or} \quad \frac{x^2}{4} + \frac{y^2}{16} = 1.$$

Clearly (28) is essentially different from (25), so the graph in Figure 6.10 is not symmetric with respect to the line through the origin that bisects Quadrant I. (We knew this all along from the picture.) ▲

EXAMPLE 4

Determine the symmetries in the graph of : (a) $y = 1/x$ and (b) $x^2 + y^2 = 4xy$. Sketch the graph of each equation.

Solution: When the test (**D**) is applied to the equation $y = 1/x$, we obtain $x = 1/y$. This equation is equivalent to $y = 1/x$. Hence, the graph is symmetric with respect to the 45° line through the origin. Test (**C**) also reveals that the graph is symmetric with respect to the origin. The graph is shown in Figure 6.11.

When the tests are applied to the equation $x^2 + y^2 = 4xy$, we find that the graph has the same symmetries as the graph of $y = 1/x$. To sketch the graph we note that we can put the equation in the form $y^2 - (4x)y + x^2 = 0$ and use the quadratic formula to solve for y in terms of x. With a little labor, we find that $y = (2 \pm \sqrt{3})x$. Hence, the graph consists of the two straight lines shown in Figure 6.12. ▲

FIGURE 6.11   FIGURE 6.12

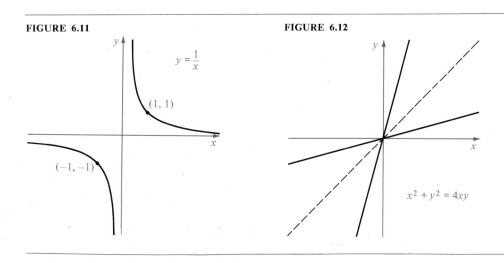

$$y = \frac{1}{x}$$

$(1, 1)$

$(-1, -1)$

$x^2 + y^2 = 4xy$

Exercise 5

1. Select a pair of numbers at random for a and b and plot the points $P(a, b)$, $Q(-a, b)$, $R(a, -b)$, $S(-a, -b)$, and $T(b, a)$. What can you say about the symmetry of the last four points and P? This problem is the basis for proving the statements in Table 4.

2. What can you say about the graph of

   (a) $x^3 + 3x^2 y + 3xy^2 + y^3 = (x+y)^3$,   (b) $x^2 + y^2 = -17$,

   (c) $x^4 + y^4 = 0$,   (d) $(y^2+1)[1/x] = 0$,

   (e) $y^2 - x^2 = 0$?

   Would you refer to any of these graphs as curves?

In problems 3 through 22 sketch the graph of the given equation and determine any symmetries that the graph may possess.

3. $y^2 = (2x+5)^2$ 4. $y^2 = 2x + 5$

5. $x^2 + y^2 = 25$ 6. $x^2 y^2 = 1$

7. $(x-3)(y-2) = 0$ 8. $xy + 4y - 3x - 12 = 0$

9. $\dfrac{x^2}{16} + \dfrac{y^2}{25} = 1$ 10. $\dfrac{x^2}{16} - \dfrac{y^2}{25} = 1$

11. $y^2 = 8 + x^3$ 12. $y^2 = x^2(1-x^2)$

13. $y^2 = x(1-x^2)$ 14. $y^2 = x^2(x-1)$

15. $y^2 = \dfrac{x(x-6)}{x-3}$ 16. $y^2 = \dfrac{x-1}{(x-4)(x-7)}$

17. $x^2 + xy - 6y^2 = 0$ 18. $y^2 + yx^2 - 2x^4 = 0$

19. $|x| + |y| = 1$ 20. $|x+y| = 1$

21. $|y| = x^2$ 22. $|x| + 3|y| = 5$

23. Prove that the graph of $x^2 + y^2 = r^2$ is a circle if $r \neq 0$. Where is the center of the circle?

6.7 VARIATION

A number of functions have special names or titles which are often used in place of the symbols, or along with the symbols. For simplicity, we have avoided these special titles, but the interested reader will find them in Table 5.

TABLE 5		
The identity function: $f(x) = x$.		The square root function: $f(x) = \sqrt{x}$.
The constant function: $f(x) = a$.		The absolute value function: $f(x) = |x|$.
The squaring function: $f(x) = x^2$.		The greatest integer function: $f(x) = [x]$.

There are still more functions with special names that we will soon meet: the exponential function, the logarithm function, the sine function, and five other related functions in trigonometry.

We have inherited from the Middle-Ages (or earlier) certain special phrases to describe functions. These phrases are still in use today and in fact, they are rather convenient, particularly in the physical sciences, so we now discuss them.

DEFINITION 6

The statement "u varies as v" means that there is a constant k such that

(29)
$$u = kv.$$

The constant k is called the constant of proportionality.

Other phrases are also in use. Thus, "*u* is directly proportional to *v*" and "*u* varies directly as *v*" both mean that for some constant *k* we have $u = kv$. It should be noted that *u* and *v* may be functions. For example, if $v = x^2$, $u = y + 10$, and "*u* varies as *v*", then $y + 10 = kx^2$.

EXAMPLE 1

Put each of the following statements into equation form:

(a) The area of a circle varies directly as the square of the radius.

(b) Hooke's Law. The elongation of an elastic rod varies directly as the force applied.

Solution: (a) With the usual meaning for the symbols we have $A = kr^2$. In this case it can be proved that $k = \pi$.

(b) We are free to introduce symbols whenever they are needed. Let *s* denote the elongation (change in length) of the rod, and let *F* denote the force applied (to both ends of the rod). Then the statement gives

(30)
$$s = kF. \quad \blacktriangle$$

In this example the role played by *k*, the constant of a proportionality, stands out more clearly. The elongation depends on the radius of the rod and the material used. Thus, the constant *k* must change as we change these items. Suppose that we agree on the material and the radius of the rod. Is *k* now a constant? Certainly the amount of elongation given by a thousand pound force is fixed, but the value of *k* depends on whether we are measuring *s* in inches, feet, or centimeters. Thus, in the physical sciences the constant *k* absorbs the units used. In other words, if we change the units of measurement, the value of *k* will change.

DEFINITION 7   *The statement "u varies inversely as v" means that there is a constant k such that*

(31)
$$u = \frac{k}{v}.$$

Suppose that the force is held fixed in Example 1(b), but the radius of the rod is regarded as variable. Then the elongation varies inversely as the square of the radius, and (31) gives

(32)
$$s = k\frac{1}{r^2}.$$

Notice that the k in (32) is not the same constant as the one in (30). In (32) k depends on many items including the force applied, but not the radius of the rod. In (30) k depends on many items including the radius of the rod, but not the force applied. In fact s varies directly as F and inversely as the square of r, so that there is a constant k (different from the other two) such that

(33)
$$s = k\frac{F}{r^2}.$$

EXAMPLE 2   Suppose that y varies directly as u and the square of v, and varies inversely as w and the cube of x. If $y = 12$ when $u = 3$, $v = 4$, $w = 5$, and $x = 2$, find y when $u = 4$, $v = 9$, $w = 25$, and $x = 3$.

Solution: By our understanding of the use of words, the first sentence gives

(34)
$$y = k\frac{uv^2}{wx^3}.$$

To determine k we use the fact that $y = 12$ when $u = 3$, $v = 4$, $w = 5$ and $x = 2$. Then (34) gives

(35)
$$12 = k\frac{3(4^2)}{5(2^3)} = \frac{6}{5}.$$

Whence, $k = 12 \cdot (5/6) = 10$. Consequently, (34) yields

(36)
$$y = 10\frac{uv^2}{wx^3}.$$

Now we can answer the question posed. If we substitute in (36) we find

$$y = 10\frac{4(9^2)}{25(3^3)} = \frac{10(12)}{25} = \frac{24}{5}. \quad \blacktriangle$$

Exercise 6

1. If y varies directly as x, and $y = 2$ when $x = 15$, find y when x is 75.

2. If y varies directly as x, and $y = 11$ when $x = 44$, find y when $x = 18$.

3. Do problem 1 if y varies inversely as x.

4. Do problem 2 if y varies inversely as x.

5. If y varies directly as x and t^2, and inversely as r, and $y = 1$ when $x = 2$, $t = 3$, and $r = 4$, find y when $x = 3$, $t = 4$, and $r = 8$.

6. If y varies directly as x^2 and inversely as t and r^3, and $y = 1/9$ when $x = 4$, $t = 5$, and $r = 6$, find y when $x = 6$, $t = 10$, and $r = 2$.

7. Do problem 5 if y varies directly as t^2 and inversely as x^3 and r.

8. Do problem 6 if y varies directly as x and r and inversely as t^2.

9. Do problem 5 if y varies directly as the sum of x and t and inversely as the sum of x and r.

10. Do problem 6 if y varies directly as the sum of x and r and inversely as the sum of x and t.

11. Neglecting air resistance, the distance a body falls when starting from rest varies directly as the square of the time. If a body falls 64 feet in 2 seconds how far will it fall (a) in 3 seconds, (b) in t seconds?

12. The intensity of a light on a plane surface varies inversely as the square of the distance of the surface from the source of the light. If we double the distance of the source from the plane, what happens to the intensity?

13. Boyle's Law. The volume of a gas at constant temperature varies inversely as the pressure. If a certain mass of gas at 70° F. had a volume of 270 cubic inches, find the volume if the pressure is tripled while the temperature is held constant. If you try this experimentally the gas will get hotter, and it will be necessary to cool it back to 70° F. before measuring the pressure.

14. The strength S of a beam varies directly as the width and the square of the depth and inversely as the length of the beam. If a beam 2 inches wide, 4 inches deep, and 8 feet long will support 600 pounds, what weight will it support if we increase the depth to 6 inches and the length to 12 feet?

15. If the beam in problem 14 is turned sideways so that the width is 4 inches, the depth is 2 inches, and the length is 8 feet, find the weight that it can support.

16. Find the percentage loss in strength of a wooden beam if in order to smooth its surface, the width is decreased from 2 inches to 1.6 inches, and the depth is decreased from 4 inches to 3.6 inches.

17. The resistance of a wire to an electrical current varies directly as its length and inversely as the square of its diameter. If a wire 90 feet long with a diameter 0.015 inch has a resistance of 10 ohms, find the resistance of a 160 foot long wire with a diameter 0.01 inch if it is made of the same material.

18. A wire (made of the same material as the one used in problem 17) is 200 feet long and has a diameter 0.02 inch. Find the resistance.

19. The crushing load of a circular pillar varies directly as the fourth power of its diameter and inversely as the square of its height. If the crushing load of a pillar 15 feet high and 5 inches in diamter is 50 tons, find the crushing load if the diameter is increased to 8 inches and the length to 20 feet.

20. What must be the diameter of a pillar (made of the same material used in problem 19) if the height is 12 feet and the crushing load is 1,250 tons.

21. Suppose that y varies directly as x and r^2 and inversely as s^4. If x, r, and s each varies inversely as t, how does y vary in relation to t?

★22. If x varies directly as y and $k \neq \pm 1$, prove that $x^2 + y^2$ varies directly as $x^2 - y^2$.

★23. Let a and b be two constants, and suppose that x varies directly as y. Prove that if $ak^3 - b \neq 0$, then $ax^3 + y^3$ varies directly as $ax^3 - by^3$.

★24. Assume for simplicity that all of the letters represent positive numbers. Prove that if

$$\frac{x_1}{y_1} = \frac{x_2}{y_2} = \frac{x_3}{y_3},$$

then

$$\frac{x_1}{y_1} = \frac{x_2}{y_2} = \frac{x_3}{y_3} = \frac{ax_1 + bx_2 + cx_3}{ay_1 + by_2 + cy_3}.$$

★25. Under the assumptions of problem 24 prove that

$$\frac{x_1}{y_1} = \frac{x_2}{y_2} = \frac{x_3}{y_3} = \sqrt[n]{\frac{ax_1^n + bx_2^n + cx_3^n}{ay_1^n + by_2^n + cy_3^n}}.$$

★**6.8 ALTERNATE DEFINITION OF A FUNCTION**

No matter what the subject may be, the student should consult at least one book other than his textbook. In algebra and trigonometry the textbooks all cover the same material with only minor additions or omissions. However, there is one difference that may cause the reader some discomfort so we take the liberty of discussing it here.

We recall from Definition 1 (page 114) that a function is determined if there is a rule (method or procedure) that associates with each element of set **D** a uniquely determined element from a set **G**. This definition seems to be natural because in each example studied so far the function was presented by a rule. Thus, the function $y = x^2$ (or $f(x) = x^2$) is defined by the rule: Select a real number x, then its associate (or image) y is obtained by squaring x.

There is however, some logical basis for objecting to this definition. After all, a function is not a rule or a procedure. These only serve to determine the function. The modern attitude is that the function stands apart from the mechanism by which it is constructed. We may create a function in any way we wish (rule, method, etc.), but the definition of a function should stand by itself.

What is a function? We can form an ordered pair (x, y) where x is an element of the domain of the function f, and y is its image $f(x)$. Since the image is unique, it is clear that for any two distinct ordered pairs (x, y) and $(x^\star, y^\star)$ we must have $x \neq x^\star$. The modern point of view is that the function is just this set of ordered pairs.

DEFINITION 8   ***Function***   *A function f is a set* **S** *of ordered pairs (x, y), such that if (x, y) and $(x^\star, y^\star)$ are distinct elements of* **S**, *then $x \neq x^\star$. The set of all x such that (x, y) is in* **S** *is called the domain of f. The set of all y such that (x, y) is in* **S** *is called the range of f.*

The condition $x \neq x^\star$ merely states that each x in the domain of f has a unique image. All other concepts relating to functions can be defined by means of ordered pairs. One may argue that Definition 8 is better than Definition 1. However, it is a trivial matter to prove that these definitions are equivalent. Any f that can be defined as a set of ordered pairs can also be defined by a rule and conversely.

Since the rule definition is natural, easy to understand, and is the one that is always used when building a function, we have selected Definition 1 in preference to Definition 7. Further, the notation $y = f(x)$ has many advantages. The reader needs only to compare the notation $y = x^2 + 5x - 7$ (or $f(x) = x^2 + 5x - 7$) with the notation

$$\mathbf{S} = \{(x, y) | y = x^2 + 5x - 7\},$$

which gives the same function.

There is one source of confusion that is present with either notation. As we have mentioned earlier, the letters x and y are merely convenient symbols that represent elements from **D** and **G**. Any letters will do. For example, $y = x^2$, $f(t) = t^2$, $r = s^2$, $\{(u, u^2)\}$, and $\{(v, v^2)\}$ all represent the same

FIGURE 6.13

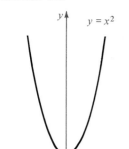

$y = x^2$

FIGURE 6.14

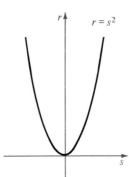

$r = s^2$

FIGURE 6.15

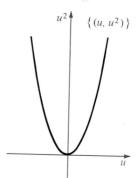

$\{(u, u^2)\}$

function. (See Figures 6.13, 6.14, and 6.15.) If we want to emphasize that we are selecting a particular element from the domain of f, we may use $x_0, x_1, a, \ldots$ and write $y_0 = f(x_0)$, $y_1 = f(x_1)$, $b = f(a), \ldots$ for the images of $x_0, x_1, a, \ldots$, respectively.

★6.9 CARTESIAN PRODUCTS AND RELATIONS

We have previously used the word "relation" in a loose intuitive manner. We are now in a position to give the word a precise meaning. First we need

DEFINITION 9

Cartesian Product *Let* **A** *and* **B** *be two sets. The set of all pairs* (x, y) *where* $x \in$ **A** *and* $y \in$ **B** *is called the Cartesian product of* **A** *and* **B** *and is denoted by* **A** × **B**.

For example, if **A** $= \{1, 2\}$ and **B** $= \{a, b, c\}$, then

$$\mathbf{A} \times \mathbf{B} = \{(1, a), (1, b), (1, c), (2, a), (2, b), (2, c)\}.$$

DEFINITION 10

Relation *Any subset* **S** *of* **A** × **B** *is called a relation. If* **A** *and* **B** *are both the same set* **U**, *any subset of* **U** × **U** *is called a relation in* **U**.

A simple example of a relation is an inequality. We may write the relation as $2x > y$ but from a purist point of view this statement defines a set **S** of pairs (x, y) for which the inequality is true, and **S** is a subset of **R** × **R**. In this case the set **S** defined by $2x > y$ will contain such pairs as $(1, 1)$, $(2, 3)$, $(150, 299), \ldots$, but the pair $(150, 301)$ will not be in **S**.

From this definition it follows that every real-valued function of a real variable is also a relation in **R** × **R**. However, a relation need not be a function, because it may violate the uniqueness condition required of a function. For example, the relation **S** obtained from $2x > y$ contains $(6, 11)$, $(6, 10), (6, 9), \ldots (6, -1), \cdots$ infinitely many pairs with the same first element 6, whereas if f is a function the set can contain at most one pair $(6, y)$.

This abstract concept of a relation is not really needed in elementary algebra or trigonometry but does occupy an important place in many of the branches of modern mathematics. When considering a relation we often want to know if it is reflexive, symmetric, transitive, or any combination of these.

DEFINITION 11   **Reflexive**   *A relation* **S** *in* **U** × **U** *is said to be reflective if for every* x *in* **U** *the pair* (x, x) *is in* **S**.

DEFINITION 12   **Symmetric**   *A relation* **S** *in* **U** × **U** *is said to be symmetric if whenever* (x, y) *is in* **S**, *the pair* (y, x) *is also in* **S**.

DEFINITION 13   **Transitive**   *A relation* **S** *in* **U** × **U** *is said to be transitive if whenever* (x, y) *is in* **S** *and* (y, z) *is in* **S** *we also have* (x, z) *in* **S**.

The student should regard these items as abstractions of the relation "equals" and this may help him to keep the definitions in mind. The "equals" relation is reflexive, symmetric, and transitive. To see this, let **E** be the set of pairs (x, y) for which $x = y$.

Equality is reflexive because $x = x$. Thus, $(x, x) \in$ **E**.

Equality is symmetric because if $x = y$, then $y = x$. Thus, if $(x, y) \in$ **E**, then $(y, x) \in$ **E**.

Equality is transitive because if $x = y$ and $y = z$, then $x = z$. Thus ,if $(x, y) \in$ **E** and $(y, z) \in$ **E**, then $(x, z) \in$ **E**.

Equality is not the only relation that is reflexive, symmetric, and transitive. For example, if **U** is the set of all straight lines in the plane and we relate those pairs that are either parallel or identical, then this relation is also reflexive, symmetric, and transitive.

DEFINITION 14   **Equivalence Relation**   *A relation in* **U** *is called an equivalence relation if it is reflexive, symmetric, and transitive.*

This concept of an equivalence relation is important for a proper presentation of fractions. Let **S** be the set of all fractions a/b where a and b are integers and $b \neq 0$. We relate two such fractions a/b and c/d by writing

(37) $$\frac{a}{b} = \frac{c}{d} \quad \text{if and only if} \quad ad = bc.$$

For example,

$$\frac{3}{7} = \frac{39}{91} \qquad \text{because} \qquad 3 \times 91 = 273 = 7 \times 39.$$

Theorem 4
PLE

The relation defined by the condition $ad = bc$ is an equivalence relation in the set of fractions.

The interested reader should use the condition (37) to prove that this relation among fractions is:

Reflexive: $\qquad \dfrac{a}{b} = \dfrac{a}{b}.$

Symmetric: If $\dfrac{a}{b} = \dfrac{c}{d},$ then $\dfrac{c}{d} = \dfrac{a}{b}.$

Transitive: If $\dfrac{a}{b} = \dfrac{c}{d}$ and $\dfrac{c}{d} = \dfrac{e}{f},$ then $\dfrac{a}{b} = \dfrac{e}{f}.$

Exercise 7

In problems 1 through 16 we give a statement which when true defines a relation in some suitable set. In each case determine if the relation is reflexive (R), symmetric (S), transitive (T), or none of these (N). Is the relation a function?

1. y is the brother or the sister of x.

2. y is the son or daughter of x.

3. y is the mother of x.

4. x and y are integers and $x+y$ is even.

5. x and y are integers and $x+y$ is odd.

6. x and y are straight lines and y is perpendicular to x.

7. $y > x$

8. $y \geqq x$

9. $y \geqq x+5$

10. $y = x+5$

11. x and y are triangles and y is congruent to x.

12. x and y are triangles and y is similar to x.

13. x and y are quadrilaterals and y has the same area as x.

14. x and y are triangles and at least one angle of y is equal to one angle of x.

15. x and y are subsets of some universal set and $y \subset x$.

16. x and y are subsets of some universal set and $y = x$.

17. Suppose that **A** and **B** are finite sets with m and n elements, respectively. How many elements are in $\mathbf{A} \times \mathbf{B}$?

18. If **A** has n elements, how many different subsets are contained in **A**?

Review Questions

Try to answer the following questions as accurately as possible before consulting the text.

1. Give the definition of a function (a) as a rule, (b) as a set of ordered pairs.

2. What is the distinction between the three statements (a) f maps **A** to **B**, (b) f maps **A** into **B**, and (c) f maps **A** onto **B**?

3. What are the standard letters (in this text) for the domain and range of a function?

4. Explain what is meant by a composite function.

5. What is the graph of a function?

6. What is the graph of an equation? How does it differ from the graph of a function?

7. What can you say about the graph of $ax + by + c = 0$?

8. What can you say about the graph of $y = ax^2 + bx + c$?

9. Give the definition of (a) an increasing function, (b) a decreasing function.

10. Explain the meaning of the symbol $[x]$.

11. State the tests for symmetry of a graph (see Table 4).

12. Define the Cartesian product of two sets.

13. If **A** and **B** are finite sets, how many elements are in the set $\mathbf{A} \times \mathbf{B}$.?

14. Define (a) a relation, (b) a reflexive relation, (c) a symmetric relation, (d) a transitive relation.

15. State the definition of an equivalence relation.

The Trigonometric Functions

7

The trigonometric functions originated with man's need to solve triangles. Such problems arise frequently in surveying, navigation, and astronomy, and these problems were so important in ancient and medieval times that the techniques for solving triangles were perfected several hundred years ago. But research on the trigonometric functions is still carried on today. New applications have been found, far removed from the geometry of the triangle. Further, the trigonometric functions have been the source of some of the deepest and most beautiful results in modern mathematics. To give each item its proper emphasis, we introduce the trigonometric functions in a purely theoretical manner, and we push the more mundane aspects, such as solving triangles, to the background. (See Chapter 12.)

7.1 ANGLES

In geometry an angle is defined as the figure formed by two rays (or line segments) issuing from a common point O, called the *vertex*. (See Figure 7.1.) The geometric angle is said to be in *standard position* if it is placed on a rectangular coordinate system in such a way that the vertex is at the origin, one ray lies on the positive x-axis, and the second ray lies in the upper half-plane defined by the condition $y > 0$ (see Figure 7.2). The angle can be measured either by the amount of rotation necessary to turn OA into OB

FIGURE 7.1

FIGURE 7.2

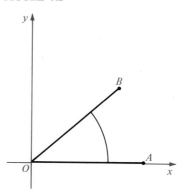

or by the length of the arc of a circle with center at O that is intercepted by the two rays. In geometry the measure of the angle is always regarded as positive.

In trigonometry we extend this concept of an angle. In order to make the distinction clear, we will call the new quantity a *trigonometric angle* and refer to the older quantity as a *geometric angle*. Later, when the distinction is well understood, we can drop the adjectives and refer to the figure AOB as an angle.

It is also important to distinguish between an angle and its measure. When we need to make such a distinction we will use θ (Greek theta) for the angle and t for the measure of the angle. This careful distinction will also be dropped in due time, after we have become familiar enough with the two items to use one symbol for both the angle and its measure. For the measure we have in mind we need

DEFINITION 1   ***Unit Circle***   *The circle* $x^2 + y^2 = 1$, *which has center at the origin and radius* 1 *is called the unit circle. The circumference* ‡ *of this circle is* 2π.

The trigonometric angle is a directed angle. One of the rays OA is called the *initial side* (or initial ray) of the angle, and the other ray OB is called the *terminal side*. We may think of the angle as being generated by rotating the initial ray OA into the terminal ray OB. The direction of rotation is indicated by an arrow. The angle is regarded as positive if the rotation is in a counterclockwise direction (see Figure 7.3). The angle is regarded as negative if the rotation is in a clockwise direction (see Figure 7.4). Further, the amount of

‡ It may come as a surprise to the student that this item is a part of the definition but in fact this is the definition of π.

FIGURE 7.3

FIGURE 7.4

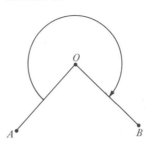

rotation for a trigonometric angle is not restricted. One complete revolution will bring the initial side back to its starting place, so that the terminal side will coincide with the initial side (see Figure 7.5). An angle of $3\frac{1}{8}$ revolutions is shown in Figure 7.6. Just as for the geometric angle, we say that the trigonometric angle is in *standard position* if the vertex is at the origin and the initial side is along the positive x-axis.

There are two important methods of assigning a measure to an angle which we now discuss. The most common unit is the degree obtained by assigning the number 360° to the angle generated by one complete revolution. Thus, we obtain an angle of one degree if we divide the unit circle into 360 equal arcs and draw rays from the center of the circle to two successive points of the subdivision. An angle of one minute (1′) is defined as 1/60 of a degree. In Figure 7.7 we show a few angles with their measure in degrees. We use this unit only because we have inherited it from the Babylonians who selected it more than 2,500 years ago. It is possible that they chose 360 because they thought that a year consisted of 360 days, or perhaps they used 360 because it has so many divisors (24). Whatever may be the reason for their choice, the degree is used almost universally in the elementary applications of trigonometry and for better or worse we are stuck with it.

FIGURE 7.5

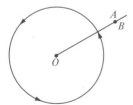

FIGURE 7.6

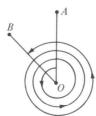

FIGURE 7.7

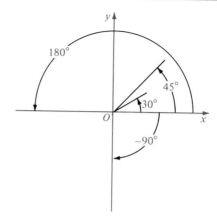

The radian is the "natural" unit for measuring angles. First, an angle of one radian cuts off from the unit circle an arc of unit length. Second, when radian measure is used, certain formulas in calculus become simpler.

DEFINITION 2   **Radian Measure** *Let t be a real number. Starting at the point $A(1,0)$ on the unit circle, mark a second point B on the unit circle such that*

  (1) *The length of the arc $\overset{\frown}{AB}$ is $|t|$.*

  (2) *The arc $\overset{\frown}{AB}$ has a counterclockwise direction if $t > 0$ and has a clockwise direction if $t < 0$.*

Then the rays OA and OB and the directed arc $\overset{\frown}{AB}$ form a trigonometric angle θ and the radian measure of this angle is t.

Observe that $|t|$ may exceed 2π. If it does, then the arc $\overset{\frown}{AB}$ will cover some parts of the unit circle more than once ($\overset{\frown}{AB}$ wraps all the way around the circle. See Figure 7.8). Further if $|t|$ is very large, the arc $\overset{\frown}{AB}$ will wrap around the unit circle many times.

We now look at the relation between the radian measure and the degree measure of an angle. By definition, one complete revolution is 360° and it is also 2π radians (since the circumference of the unit circle is 2π). Consequently, we have the equation

(1) $$360° = 2\pi \text{ radians.}$$

This equation may look a little strange because as numbers $360 \neq 2\pi$, but, we can manipulate with equation (1) as long as we keep in mind that it asserts that two angles are equal. If we divide both sides of equation (1) by 2,

FIGURE 7.8

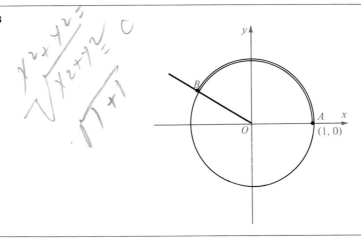

we obtain

(2) $$180° = \pi \text{ radians.}$$

We regard equation (2) as the fundamental one for converting from radians to degree or from degrees to radians.

EXAMPLE 1   What is the radian measure of a 1° angle?

Solution: Divide both sides of (2) by 180. This gives

$$1° = \frac{\pi}{180} \text{radian} = \frac{3.14159 \cdots}{180} = 0.017453 \cdots \text{radian.} \quad ▲$$

EXAMPLE 2   How many degrees are there in a 1-radian angle?

Solution: Divide both sides of (2) by π. This gives

$$1 \text{ radian} = \frac{180°}{3.14159 \cdots} = 57.296 \cdots° = 57° \, 17.75 \cdots'. \quad ▲$$

EXAMPLE 3   What is the radian measure of an 85° angle?

Solution: For 1° we divide both sides of (2) by 180. To obtain 85° we multiply the resulting equation on both sides by 85. Thus,

$$85° = 85 \times 1° = 85 \times \frac{\pi}{180} \text{ radians} = \frac{17\pi}{36} \text{ radians.} \quad ▲$$

We will often write $\theta = 85°$ in place of θ is an angle whose measure is $85°$ We may use such a shorthand as long as we are aware that the angle θ is not the same thing as its measure ‡.

It is sometimes convenient to write $\theta = 3^{(r)}$ instead of $\theta = 3$ radians. In this text we will go even further and drop the superscript. The statement "$\theta = 3$" means that θ is an angle whose measure is 3 radians, while $\theta = 3°$ means an angle of 3 degrees. Similarly, $\theta = \pi/4$ means that θ is an angle whose measure is $\pi/4$ radians.

Exercise 1

1. The number 360 has more divisors than any other positive integer $n < 360$. Show that 360 has 24 divisors by listing them and counting them.

2. Among the positive integers less than 360, the number 240 has the greatest number of divisors. Show that this number has 20 divisors by listing them,

3. Find the radian measure for each of the following angles: (a) $60°$, (b) $240°$. (c) $720°$, (d) $-135°$, (e) $12°$, (f) $132°$, (g) $36°$, (h) $-150°$.

4. Find the radian measure for each of the following angles: (a) $90°$, (b) $450°$, (c) $-1080°$, (d) $45°$, (e) $225°$, (f) $3°$, (g) $21°$, (h) $-210°$.

5. Find the degree measure of each of the following angles: (a) $\pi/4$, (b) $-9\pi/4$, (c) $5\pi/2$, (d) $11\pi/120$, (e) $-2\pi/9$, (f) $19\pi/36$, (g) 7π, (h) -8π.

6. Find the degree measure of each of the following angles: (a) $\pi/12$, (b) 3π, (c) $7\pi/6$, (d) $-11\pi/4$, (e) $13\pi/45$, (f) $-17\pi/30$, (g) $5\pi/8$, (h) -4π.

7. Draw a picture showing the following angles: (a) $\theta = 1$ (radian), (b) $\theta = 2$, (c) $\theta = 1/2$, (d) $\theta = \pi/4$, (e) $\theta = \pi$, (f) $\theta = 5\pi$, (g) $\theta = -6\pi$.

8. The terminal side of an angle in standard position lies on the positive x-axis. Give all the possible values for the radian measure of this angle.

9. Do problem 8 if the terminal side lies (a) on the negative x-axis, (b) on the positive y-axis, (c) on the negative y-axis.

10. Do problem 8 if the terminal side lies in the first quadrant and makes a geometric angle of $30°$ with the initial side.

★11. How would you attempt to define the length of an arc (a piece of a curve)?

★12. Let t be the radian measure of an angle in standard position. Let C be a circle of radius r with center at the origin. Finally, let s be the length of the arc of C intercepted by the angle. Sketch a proof that

(3)
$$s = rt.$$

Note: This equation is often used as the definition of radian measure.

‡ The correct form is $m(\theta) = 85°$ (read, "m of θ is $85°$") where m represents the measure. However, there is no need for such a sophisticated notation.

In problems 13 through 19 use 22/7 for π in order to simply computation. Give the answer to three significant figures.

13. Find the length of arc intercepted on a circle of radius 9 inches by a central angle of 44°.

14. How far does the tip of the minute hand of a clock move in 35 minutes if the hand is 6 inches long?

15. If a car travels 11 feet without skidding and the radius of the wheels is 1 foot, through how many radians does each wheel turn? How many degrees?

16. If a truck travels 21 feet without skidding and the radius of the wheels is 2 feet 6 inches, through how many radians does each wheel turn? How many degrees?

17. Assuming that the radius of the earth is 3960 miles, find the length of arc on the earth's surface of one degree latitude on a meridian. What is the length of an arc of one minute of latitude?

18. Washington, D.C. is located at 38° 52′ north latitude. How far is it from the equator?

19. Seattle, Washington is located at 47° 42′ north latitude. How far is it from the North Pole?

★ 20. Eratosthenes (275 B.C., 194 B.C.) conjectured that the earth was round and computed the radius of the earth in the following manner. The Egyptian city of Alexandria (in which he worked as a librarian) lies directly north of Syene (his birthplace), so that the sun attains its maximum angle of elevation at the same time for both cities. On a certain day the sun was directly overhead at Syene, while on the same day at Alexandria the maximum angle of elevation for the sun was 82° 48′. Assuming that Alexandria is 498 miles north of Syene, compute the radius of the earth.

21. A flywheel of diameter 12.0 feet makes 500 revolutions in an hour. (a) How far does a point on the perimeter travel in one hour? (b) If a bug clinging to the flywheel travels one mile during the hour how far is the bug from the center of the wheel?

7.2 THE TRIGONOMETRIC FUNCTIONS

These six new functions, together with their abbreviations (symbols) are:

$\sin \theta$ (sine of θ),	$\csc \theta$ (cosecant of θ),
$\cos \theta$ (cosine of θ),	$\sec \theta$ (secant of θ),
$\tan \theta$ (tangent of θ),	$\cot \theta$ (cotangent of θ).

FIGURE 7.9

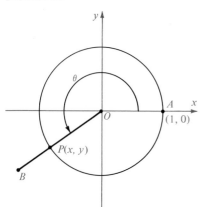

FIGURE 7.10

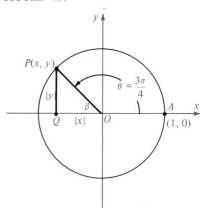

Originally these functions were defined as ratios of sides in a right triangle (see Chapter 13, Theorem 1). A definition which is meaningful for any angle is presented in

DEFINITION 3

Let θ be a trigonometric angle. Place θ in standard position (see Figure 7.9) and let P be the point of intersection of the terminal side and the unit circle. If the coordinates of P are (x, y), then

(4) $\sin \theta \equiv y$,

(5) $\cos \theta \equiv x$,

(6) $\tan \theta \equiv \dfrac{y}{x}, \quad x \neq 0$,

(7) $\csc \theta \equiv \dfrac{1}{y}, \quad y \neq 0$,

(8) $\sec \theta \equiv \dfrac{1}{x}, \quad x \neq 0$,

(9) $\cot \theta \equiv \dfrac{x}{y}, \quad y \neq 0$.

FIGURE 7.11

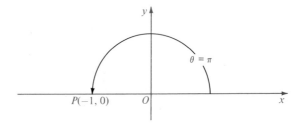

EXAMPLE 1   Find each of the trigonometric functions of $3\pi/4$.

Solution: This angle (also 135°) is shown in Figure 7.10. Since $\beta = \pi/4$ (or 45°), the triangle OPQ is isosceles; therefore $|x| = |y|$. Further, P is on the unit circle, $x^2 + y^2 = 1$. Hence $2x^2 = 1$ and $|x| = 1/\sqrt{2}$. For this particular point $x < 0$ and $y > 0$. Consequently, $x = -1/\sqrt{2}$ and $y = 1/\sqrt{2}$. Therefore

$$\sin\frac{3\pi}{4} = y = \frac{1}{\sqrt{2}} = \frac{\sqrt{2}}{2}, \qquad \csc\frac{3\pi}{4} = \frac{1}{y} = \frac{1}{1/\sqrt{2}} = \sqrt{2},$$

$$\cos\frac{3\pi}{4} = x = -\frac{1}{\sqrt{2}} = -\frac{\sqrt{2}}{2}, \qquad \sec\frac{3\pi}{4} = \frac{1}{x} = \frac{1}{-1/\sqrt{2}} = -\sqrt{2},$$

$$\tan\frac{3\pi}{4} = \frac{y}{x} = \frac{1/\sqrt{2}}{-1/\sqrt{2}} = -1, \qquad \cot\frac{3\pi}{4} = \frac{x}{y} = \frac{-1/\sqrt{2}}{1/\sqrt{2}} = -1. \quad ▲$$

Using these techniques we can compute the trigonometric functions of the angles 0, $\pi/6$, $\pi/4$, $\pi/3$, and $\pi/2$. This important task is left for the student as problem 2 of the next exercise.

EXAMPLE 2   Find each of the trigonometric functions of π.

Solution: This angle (also 180°) is shown in Figure 7.11. The point P has coordinates $(-1,0)$. Whenever a division by 0 occurs, we write that the function is undefined. We use $\theta = 180°$ instead of $\theta = \pi$, so that the student has an opportunity to become accustomed to both systems of measure. Then

$$\sin 180° = y = 0, \qquad\qquad \csc 180° \text{ is undefined,}$$

$$\cos 180° = x = -1, \qquad\qquad \sec 180° = \frac{1}{x} = \frac{1}{-1} = -1,$$

$$\tan 180° = \frac{y}{x} = \frac{0}{-1} = 0, \qquad \cot 180° \text{ is undefined.} \quad ▲$$

At first it may seem that there are too many trigonometric functions and that the burden of learning the definitions, theorems, properties, etc., may be too much for the student. However, many of the relations among the trigonometric functions show patterns which lighten the burden. For example, the six functions form three pairs of reciprocals as described in

Theorem 1   *If θ is any angle for which the functions are defined, then:*

(a) *$\sin\theta$ and $\csc\theta$ are reciprocals,*

(b) *$\cos\theta$ and $\sec\theta$ are reciprocals,*

(c) *$\tan\theta$ and $\cot\theta$ are reciprocals.*

In equation form these relations are:

(10) $\quad \sin\theta\csc\theta = 1,$ or $\sin\theta = \dfrac{1}{\csc\theta},$ or $\csc\theta = \dfrac{1}{\sin\theta},$

(11) $\quad \cos\theta\sec\theta = 1,$ or $\cos\theta = \dfrac{1}{\sec\theta},$ or $\sec\theta = \dfrac{1}{\cos\theta},$

(12) $\quad \tan\theta\cot\theta = 1,$ or $\tan\theta = \dfrac{1}{\cot\theta},$ or $\cot\theta = \dfrac{1}{\tan\theta}.$

Proof: Equation (10) follows directly from equations (4) and (7) because

$$\sin\theta\csc\theta = y \cdot \frac{1}{y} = 1.$$

Similarly equation (11) follows directly from equations (5) and (8). Equation (12) follows directly from equations (6) and (9). ▮

The reader is urged to check this theorem numerically using the values in Examples 1 and 2. Note that in Example 2, $\cot 180°$ and $\csc 180°$ are undefined. Consequently, no claim is made about equations (12) and (10) for this angle.

EXAMPLE 3

Find each of the trigonometric functions of $\alpha = 19\pi/4$.

Solution: Observe that $19\pi/4 = 4\pi + 3\pi/4$. Since 4π is twice the circumference of the unit circle, it follows that when the angle α is in standard position, the terminal side of α will coincide with the terminal side of the angle $\theta = 3\pi/4$ (see Figure 7.10). Consequently, the point P on the terminal side of α and on the unit circle will coincide with the point for $\theta = 3\pi/4$. Then x and y are the same for both angles. Therefore, the trigonometric functions of $19\pi/4$ are identical with those of $3\pi/4$ given in the solution to Example 1. ▲

This example suggests the next definition and Theorems 2 and 3 that follow.

DEFINITION 4

Coterminal Angles *Two angles α and θ are said to be coterminal if the terminal side of α coincides with the terminal side of θ when both angles are in standard position.*

Theorem 2
PLE

Two angles α and θ are coterminal if and only if there is an integer n such that

(13) $$\alpha = \theta + 2n\pi.$$

Theorem 3
PLE

If α and θ are coterminal angles then any trigonometric function of α is equal to the same trigonometric function of θ.

In equation form, Theorem 3 states that if $\alpha = \theta + 2n\pi$, where $n \in \mathbf{Z}$, then

(14) $\sin \alpha = \sin \theta$,

(15) $\cos \alpha = \cos \theta$,

(16) $\tan \alpha = \tan \theta$,

(17) $\csc \alpha = \csc \theta$,

(18) $\sec \alpha = \sec \theta$,

(19) $\cot \alpha = \cot \theta$.

Exercise 2

1. Prove Theorems 2 and 3.

2. From plane geometry it is easy to determine the ratios of the sides of a right triangle when one angle is $\pi/6 = 30°$ or $\pi/4 = 45°$. With this information one can obtain all of the trigonometric functions of these angles. In the following table check the entries given, and fill in the remaining blanks with the proper values. The entries in Table 1 are important and should be memorized.

TABLE 1

θ	Sin θ	Cos θ	Tan θ	Cot θ	Sec θ	Csc θ
0	0	1	0	undefined		
$\pi/6$ 30°	$1/2$					
$\pi/4$ 45°	$\dfrac{\sqrt{2}}{2}$					
$\pi/3$ 60°	$\dfrac{\sqrt{3}}{2}$			$\dfrac{\sqrt{3}}{3}$		
$\pi/2$ 90°	1					1

3. Find all of the trigonometric functions for each of the following angles: (a) $2\pi/3$, (b) $-\pi/4$, (c) $7\pi/6$, (d) 4π, (e) $-3\pi/2$.

4. Find all of the trigonometric functions for each of the following angles: (a) $4\pi/3$, (b) $-\pi/6$, (c) $11\pi/3$, (d) $-5\pi/4$, (e) $3\pi/2$.

5. Make a table showing the algebraic sign of each of the trigonometric functions when the terminal side lies in each of the four quadrants. There should be twenty-four entries in the table.

6. *Alternate definition of the trigonometric functions.* Place the angle θ in standard position. Select a point $P(x, y)$ on the terminal side, but not

FIGURE 7.12

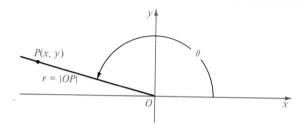

necessarily on the unit circle (see Figure 7.12). Let $r > 0$ be the distance from P to the origin. Then

$$\sin \theta \equiv \frac{y}{r}, \qquad \csc \theta \equiv \frac{r}{y},$$

$$\cos \theta \equiv \frac{x}{r}, \qquad \sec \theta \equiv \frac{r}{x},$$

$$\tan \theta \equiv \frac{y}{x}, \qquad \cot \theta \equiv \frac{x}{y},$$

provided of course that none of the denominators are zero. Prove that this definition is equivalent to Definition 3. *Hint:* Use similar triangles.

7. When θ is in standard position, the point $P(x, y)$ is on the terminal side. Make a sketch showing P and θ $(0 < \theta < 2\pi)$. Compute $\sin \theta$ and $\cos \theta$ for (a) $P(12, 5)$, (b) $P(-4, 3)$, (c) $P(-4, -5)$, (d) $P(1, -2)$.

8. Do problem 7 for the points (a) $P(3, 4)$, (b) $P(5, -12)$, (c) $P(-2, -3)$, (d) $P(-5, 3)$.

7.3 THE TRIGONOMETRIC FUNCTIONS OF A REAL NUMBER

Definition 3 (page 154) gives the trigonometric functions of an angle. But we cannot always regard the independent variable as an angle. Indeed, suppose that we wish to consider the composite function $y = \sin u$, where $u = t^2$. The composition gives $y = \sin t^2$. Now, if t represents an angle, then what is the meaning of t^2 (the square of an angle)? On the other hand, if we agree that u is an angle so that $\sin u$ is well defined, then we have $t = \sqrt{u}$, the square root of an angle. If we hold stubbornly to Definition 3, we will have trouble in forming composite functions (although this is not the only difficulty).

The way out of our dilemma is quite simple, and the reader may have guessed it already. It is contained in

DEFINITION 5   *Let t be a real number, and let θ be the angle whose radian measure is t. Then*

(20) $\sin t \equiv \sin \theta,$ (23) $\csc t \equiv \csc \theta,$

(21) $\cos t \equiv \cos \theta,$ (24) $\sec t \equiv \sec \theta,$

(22) $\tan t \equiv \tan \theta,$ (25) $\cot t \equiv \cot \theta.$

Now, the composite function $\sin t^2$ is well defined for all real t. For example, if $t = 3\pi$, then $\sin t^2 = \sin 9\pi^2$. Of course we do not have at hand a numerical value for $\sin 9\pi^2$; and, in fact, this is an irrational number. But we will learn how to find a rational number that is close to $\sin 9\pi^2$ when we discuss tables of the trigonometric functions in Section 6.

Once the distinction between $\sin \theta$ and $\sin t$ is clearly understood we can put the matter aside. We can use the symbol θ to represent either an angle, a real number, or the radian measure of an angle (which is a real number). One student may think of $\sin \theta$ as the sine of an angle, while another may regard $\sin \theta$ as the sine of a real number. As far as this text is concerned, the answers to the exercises will be the same for both students.

Exercise 3

1. As functions of a real variable, give the domain **D** and the range **G** of each of the trigonometric functions.

2. In the interval $(0, \pi/2)$ each of the trigonometric functions is either increasing or decreasing. Name those that are increasing.

3. Do problem 2 for the interval $(\pi/2, \pi)$.

4. Is there any interval in which four of the trigonometric functions are increasing?

7.4 ALGEBRAIC SIGNS OF THE TRIGONOMETRIC FUNCTIONS

We call θ a third quadrant angle (or a Q.III angle) if its terminal side lies in the third quadrant when it is in standard position. We use similar terms for the other quadrants. Since $y > 0$ in Q.I and Q.II and $\sin \theta = y$, we see that

$\sin \theta$ is positive if θ is a Q.I or a Q.II angle. If θ is a Q.III or Q.IV angle, then $y < 0$ and $\sin \theta$ is negative. A similar argument gives the algebraic sign for $\cos \theta$, $\sec \theta$, and $\csc \theta$. For $\tan \theta$ and $\cot \theta$ we must examine the ratio y/x, which is positive if x and y are both positive (Q.I) and also if x and y are both negative (Q.III). The signs of the trigonometric functions for the various quadrants are given in Table 2. The student should check each entry.

TABLE 2	Quadrant	Sin	Cos	Tan	Cot	Sec	Csc
	I	+	+	+	+	+	+
	II	+	−	−	−	−	+
	III	−	−	+	+	−	−
	IV	−	+	−	−	+	−

7.5 REDUCTION TO FUNCTIONS OF AN ACUTE ANGLE

If we know the trigonometric functions for each θ in the interval $[0, \pi/2]$ then it is an easy matter to obtain the trigonometric functions for any other θ. Perhaps the reader has already conjectured this from his experiences with the earlier examples and problems. We now formalize this result.

DEFINITION 6   **Reference Angle**   *Put θ in standard position. The terminal side of θ and the x-axis form two geometric angles ρ_1 and ρ_2 (Greek rho). The smaller of these two angles is called the reference angle and will be denoted by ρ.*

FIGURE 7.13   **FIGURE 7.14**

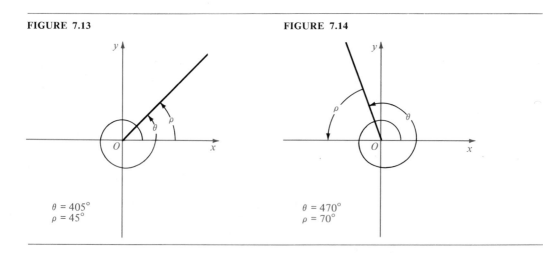

$\theta = 405°$ $\theta = 470°$
$\rho = 45°$ $\rho = 70°$

FIGURE 7.15

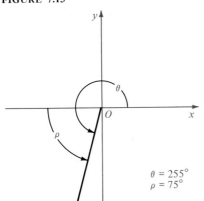

$\theta = 255°$
$\rho = 75°$

FIGURE 7.16

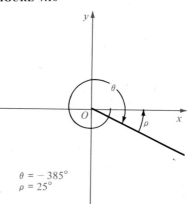

$\theta = -385°$
$\rho = 25°$

Four different possibilities for θ are shown in Figures 7.13, 7.14, 7.15, 7.16, and the reference angle ρ is noted in each case. According to the definition, we always have $0 \leq \rho \leq \pi/2$ (or $0° \leq \rho \leq 90°$). Suppose that θ is a number and not an angle. In this case we use the angle for which θ is the radian measure, find its reference angle, and let ρ be the radian measure of the reference angle. Then ρ is called the *reference number* for the number θ.

Whether θ is an angle or number, we have

Theorem 4

Let θ be an arbitrary angle (or number) and let ρ be the reference angle (or number). Then

(26) $|\sin\theta| = \sin\rho$,

(27) $|\cos\theta| = \cos\rho$,

(28) $|\tan\theta| = \tan\rho$,

(29) $|\csc\theta| = \csc\rho$,

(30) $|\sec\theta| = \sec\rho$,

(31) $|\cot\theta| = \cot\rho$.

Proof: The terminal side of θ may lie in any one of the four quadrants (or may coincide with one of the axes). It is sufficient to give the proof if θ is a Q.III angle, because the other possible cases are handled similarly.

If P is the point (x, y), let $|x| = a$, $|y| = b$, and let B be the point (a, b). Then, as indicated in Figure 7.17, the triangle OPQ is congruent to the triangle OBC. Hence, the angle BOC is equal to ρ. Consequently,

$$|\sin\theta| = |y| = b = \sin\rho,$$
$$|\cos\theta| = |x| = a = \cos\rho,$$
$$|\tan\theta| = \left|\frac{y}{x}\right| = \frac{b}{a} = \tan\rho,$$

and similarly for the other three relations. ▌

FIGURE 7.17

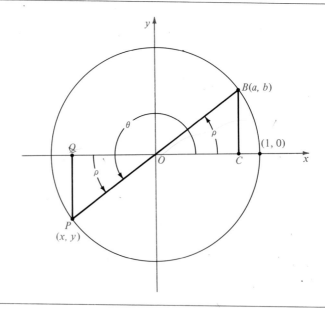

EXAMPLE 1   Find tan 2205°.

Solution: We subtract integer multiples of 180°, adjusted so that the difference (in absolute value) is as close as possible to 0. Clearly,

$$(32) \qquad\qquad 2205° - 12(180°) = 2205° - 2160° = 45°.$$

Hence $\rho = 45°$. But we know from Exercise 2, problem 2, that $\tan 45° = 1$.
Therefore, $|\tan 2205°| = 1$. But $2205° = 6(360°) + 45°$, so $2205°$ is a Q.I
angle, and $\tan \theta > 0$. Consequently, $\tan 2205° = 1$. ▲

EXAMPLE 2   Find $\sin \dfrac{10\pi}{3}$ and $\tan \dfrac{10\pi}{3}$.

Solution: This time we subtract integer multiples of π to find ρ. From

$$(33) \qquad\qquad \frac{10}{3}\pi - 3\pi = \frac{10\pi - 9\pi}{3} = \frac{1}{3}\pi,$$

we see that $\rho = \pi/3$. But we already know that $\sin(\pi/3) = \sqrt{3}/2$, and

$\tan(\pi/3) = \sqrt{3}$ (see Exercise 2, problem 2). Finally $10\pi/3$ is a Q.III angle (or number). Hence (see Table 2)

$$\sin\frac{10\pi}{3} = -\frac{\sqrt{3}}{2}, \qquad \tan\frac{10\pi}{3} = \sqrt{3}. \quad \blacktriangle$$

If θ is a second or third quadrant angle, we must subtract (or add) an odd multiple of π to find $\pm\rho$. If θ is a first or fourth quadrant angle, we must subtract (or add) an even multiple of π to find $\pm\rho$.

Exercise 4

In problems 1 through 12 find the quadrant for the given θ, and find ρ.

1. $85\pi/7$	2. $81\pi/11$	3. $93\pi/5$	4. 90.1π
5. $13\pi/8$	6. $-67\pi/6$	7. $-1000°$	8. $3758°$
9. $-1999°$	10. $975°$	11. $1234°$	12. $-2199°$

In problems 13 through 24, find sin, cos, and tan for each θ listed. It is easy to find csc, sec, and cot by finding the reciprocals of these numbers.

13. $25\pi/6$	14. $13\pi/4$	15. $-20\pi/3$	16. $23\pi/2$
17. $9\pi/4$	18. $-15\pi/2$	19. $540°$	20. $870°$
21. $-240°$	22. $-780°$	23. $-1110°$	24. $-2160°$

7.6 TABLES OF THE TRIGONOMETRIC FUNCTIONS

The reader has certainly noticed that in the theory, θ may be any real number, but that in the problems, we cling closely to the values 0, $\pi/6$, $\pi/4$, $\pi/3$, $\pi/2$, and to numbers for which these are the reference numbers. As we have seen, the trigonometric functions for these values of θ are relatively easy to compute, but for any other rational multiple of π, the computation is more difficult. Fortunately for us, these difficult computations were done many years ago; the results have been arranged in tables for convenience. Consequently trigonometry is an easy subject because the hard labor of compiling the tables has already been done. A table of values of the trigonometric functions is given as Table B in the appendix. A brief portion of that table is reproduced here so that we can discuss the arrangement of the items.

TABLE 3	Radians	Degrees	Sin	Tan	Cot	Cos		
	.3142	**18° 00'**	.3090	.3249	3.078	.9511	**72° 00'**	1.2566
	.3171	10	.3118	.3281	3.047	.9502	71° 50'	1.2537
	.3200	20	.3145	.3314	3.018	.9492	40	1.2508
	.3229	30	.3173	.3346	2.989	.9483	30	1.2479
	.3258	40	.3201	.3378	2.960	.9474	20	1.2450
	.3287	50	.3228	.3411	2.932	.9465	10	1.2421
	.3316	**19° 00'**	.3256	.3443	2.904	.9455	**71° 00'**	1.2392
	.3345	10	.3283	.3476	2.877	.9446	70° 50'	1.2363
	.3374	20	.3311	.3508	2.850	.9436	40	1.2334
	.3403	30	.3338	.3541	2.824	.9426	30	1.2305
	.3432	40	.3365	.3574	2.798	.9417	20	1.2275
	.3462	50	.3393	.3607	2.773	.9407	10	1.2246
	.3491	**20° 00'**	.3420	.3640	2.747	.9397	**70° 00'**	1.2217

Suppose that we want sin 18°. We look in the column headed degrees and find 18° 00' in the first row below the heading. In the same row, but in the next column, we find 0.3090. Since this column is headed Sin, we infer that

$$\sin 18° 0' = 0.3090.$$

In the same manner, looking under the headings Tan, Cot, and Cos, and in the proper rows we find that:

$$\tan 18° 30' = 0.3346 \qquad \cot 19° 10' = 2.877 \qquad \cos 19° 50' = 0.9407.$$

The values of sec and csc are not tabulated here because they are never really needed in the applications to computation. For example, multiplication by $\sec \theta$ is the same as dividing by $\cos \theta$, and the values of $\cos \theta$ are in the table.

This table can be used for the inverse process. Suppose that we want to find the angle θ such that $\tan \theta = 0.3378$. We look in the column headed tan, until we find this number and then observe that it is in the same row with 18° 40'. We infer that $0.3378 = \tan 18° 40'$ or $\theta = 18° 40'$.

If θ is a real number, or the radian measure of an angle, then we use the first column. For example,

$$\sin 0.3200 = 0.3145, \qquad \cos 0.3432 = 0.9417.$$

It should be noted that the vast majority of the entries in Table B are not correct, but are only approximations. Consequently, all of the $=$ symbols used in this section are erroneous. To be precise we should use a new symbol $\approx$ (read, "is approximately equal to") rather than the $=$ sign. We will not slavishly adhere to the demands of perfection, but will continue to use (and abuse) the = sign whenever this practice seems to be harmless. However, we will use ≈ whenever there is some reason to emphasize that we are using an approximate value.

Exercise 5

In problems 1 through 16 find the value of the trigonometric function.

1. $\sin \dfrac{\pi}{60}$ 2. $\cos \dfrac{\pi}{15}$ 3. $\tan \dfrac{\pi}{10}$ 4. $\cot \dfrac{\pi}{5}$

5. $\cos 0.7359$ 6. $\tan 0.5992$ 7. $\cot 0.4654$ 8. $\sin 0.0145$

9. $\cot 4° 50'$ 10. $\tan 9° 10'$ 11. $\cos 21° 40'$ 12. $\sin 24° 50'$

13. $\sin 44°50'$ 14. $\cos 39° 10'$ 15. $\tan 31° 0'$ 16. $\cot 0° 20'$

In problems 17 through 31 find the value of the independent variable as follows: For a Roman letter give the answer in degrees between 0° and 90°. For a Greek letter give the answer in radians between 0 and $\pi/2$.

17. $\sin A = 0.0029$ 18. $\cos B = 0.9995$ 19. $\tan C = 0.1405$

20. $\cot D = 343.8$ 21. $\sin E = 0.2334$ 22. $\cos F = 0.9520$

23. $\cot G = 2.300$ 24. $\tan \alpha = 0.9490$ 25. $\cos \beta = 0.9171$

26. $\sin \gamma = 0.4848$ 27. $\tan \theta = 0.7400$ 28. $\cot \phi = 1.054$

29. $\cot \alpha = 11.06$ 30. $\sin \beta = 0.6604$ 31. $\tan \gamma = 0.3121$

7.7 THE COFUNCTIONS

The reader probably has noticed that in Table B, the angle θ listed in the second column runs only to 45° ($\pi/4$ radians). What do we do if $45° < \theta \leqslant 90°$? At first glance it seems as though we should double the length of the tables to include this missing interval. However, there is a set of trigonometric identities that makes such an extension of the tables unnecessary. We need two definitions and one theorem.

DEFINITION 7   *Complement*   *If A and B are angles and*

(34) $$A + B = 90°,$$

then A and B are said to be complementary angles. Each of A and B is said to be the complement of the other. If α and β are real numbers and

(35) $$\alpha + \beta = \frac{\pi}{2},$$

then α and β are said to be complementary numbers. Each of α and β is said to be the complement of the other.

For example, 68° and 22° are complementary angles. We will see in a moment that if we want sin 68°, it is the same as cos 22° and this latter item is in Table B. Similarly, we will see that tan 68° = cot 22°.

DEFINITION 8   ***Cofunction***   *The cosine function is called the cofunction of the sine function. The cotangent function is called the cofunction of the tangent function. The cosecant function is called the cofunction of the secant function. Further, in each of the above, the second named function is called the cofunction of the first.*

Theorem 5   *If A and B are complementary angles, then any trigonometric function of A is equal to the cofunction of B. If α and β are complementary numbers, then any trigonometric function of α is equal to the cofunction of β.*

In equation form Theorem 5 states that:

(36) $\qquad \sin A = \cos(90° - A), \qquad \sin \alpha = \cos\left(\dfrac{\pi}{2} - \alpha\right),$

(37) $\qquad \cos A = \sin(90° - A), \qquad \cos \alpha = \sin\left(\dfrac{\pi}{2} - \alpha\right),$

(38) $\qquad \tan A = \cot(90° - A), \qquad \tan \alpha = \cot\left(\dfrac{\pi}{2} - \alpha\right),$

(39) $\qquad \cot A = \tan(90° - A), \qquad \cot \alpha = \tan\left(\dfrac{\pi}{2} - \alpha\right),$

and two more identities for the secant and cosecant, which are not really needed. This theorem accounts for the prefix co in the words cosine, cotangent, and cosecant. All of these equations can be condensed into one equation:

$$f(\alpha) = \text{cof}\left(\frac{\pi}{2} - \alpha\right).$$

We will see in due time that this theorem is true without any restriction on the range of the variables. We will be content to prove this theorem for $0° < A, B < 90°$, because we intend to use this theorem for A and B in this interval.

FIGURE 7.18

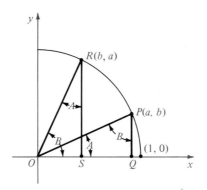

Proof: We place the angle A in standard position, and let $P(a, b)$ be the point of intersection of the terminal side with the unit circle. From P we draw a line segment PQ perpendicular to the x-axis at Q to form the right triangle OPQ shown in Figure 7.18. The angle OPQ is the complement of A and is labeled B in the figure. We next locate the point $R(b, a)$ on the unit circle, and again draw RS perpendicular to the x-axis at S to form the right triangle ORS. Since both triangles have sides of length $a, b, 1$, they are congruent. Hence, the angle SOR is also B, and in this triangle B is in standard position. Then, by the definition of the trigonometric functions,

$$\sin A = b = \cos B = \cos(90° - A),$$

$$\cos A = a = \sin B = \sin(90° - A),$$

$$\tan A = \frac{b}{a} = \cot B = \cot(90° - A),$$

$$\cot A = \frac{a}{b} = \tan B = \tan(90° - A). \quad \blacksquare$$

EXAMPLE 1   Find $\sin 68°$ and $\tan 68°$.

Solution: Combining equations (36), (38), and Table B we obtain

$$\sin 68° = \cos 22° = 0.9272, \qquad \tan 68° = \cot 22° = 2.475. \quad \blacktriangle$$

Let us now take another look at Table B. At the bottom of each column headed by a trigonometric function, is the appropriate cofunction. Thus, Cos is at the bottom of the column headed Sin, and Sin is at the bottom of the column headed Cos. In the right-hand column each angle is the complement of its mate standing in the same row in the left-hand column. For example, 20° and 70° stand in the same row; 38° 50′ and 51° 10′ stand in the same row, etc. The table is arranged so that Theorem 5 and Definition 8 can be completely ignored if the following very simple rule is used.

RULE

If $0° \leq A \leq 45°$, use the left-hand column for the angle and the headings at the top. If $45° \leq A \leq 90°$, use the right-hand column for the angle and the headings (footings?) at the bottom.

A similar rule holds for radian measure.

Exercise 6

1. In Figure 7.18, the angle A is drawn so that $0° < A < 45°$. Make a figure showing an angle A such that $45° < A < 90°$. Is the proof given in the text still valid in this case?

In problems 2 through 13 find the value of the trigonometric function.

2. $\sin 63° 10'$ 3. $\tan 52° 30'$ 4. $\cot 45° 50'$ 5. $\cos 49° 40'$

6. $\sin 74° 10'$ 7. $\tan 78° 40'$ 8. $\cot 79° 30'$ 9. $\cos 66° 00'$

10. $\sin 84° 20'$ 11. $\tan 80° 10'$ 12. $\cot 85° 00'$ 13. $\cos 76° 50'$

In problems 14 through 25 find the value of the independent variable. For a Roman letter give the answer in degrees between $0°$ and $90°$. For a Greek letter give the answer in radians between 0 and $\pi/2$.

14. $\sin A = 0.9511$ 15. $\cos B = 0.1822$ 16. $\tan C = 3.450$

17. $\cot D = 0.3249$ 18. $\cos E = 0.3256$ 19. $\tan F = 1.492$

20. $\cot G = 0.8002$ 21. $\sin \alpha = 0.9799$ 22. $\cos \beta = 0.7009$

23. $\sin \gamma = 0.7771$ 24. $\tan \theta = 1.949$ 25. $\cot \phi = 0.3443$

7.8 GRAPHS OF THE TRIGONOMETRIC FUNCTIONS

Although the tables and the various theorems proved so far tell us much about the trigonometric functions, the reader may still have a hazy idea of the behavior of these functions. To clarify the matter we will use what we now know (tables and theorems) to sketch the graphs of these functions. In this section we give these graphs with only a brief discussion. We will take up the subject again and with more details in Chapter 10.

It is easy to sketch the graph of the sine function using values from Table B. However, it is worthwhile to get a dynamic view of the graph. We return to the unit circle and the definition of the sine function. Suppose, as shown in Figure 7.19, the point P moves on the unit circle in a counterclockwise manner starting from the point $A(1,0)$. If θ is a real number (or the radian measure of the angle), then by definition θ is also the length of the arc $\overset{\frown}{AP}$. Further, $\sin \theta = y$, the directed distance $\overline{QP}$ (indicated by the arrows in the

FIGURE 7.19

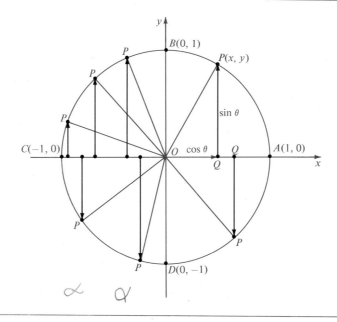

figure). With this picture in mind it is easy to trace the behavior of $\sin\theta$ as a function of θ. As θ increases from 0 to $\pi/2$ (point B in Figure 7.19), $\sin\theta$ increases from 0 to its maximum value 1. As θ increases from $\pi/2$ to π (point C in Figure 7.19), $\sin\theta$ decreases from 1 to 0. The reader should continue to follow the fortunes of $\sin\theta$ as P completes the circuit. The graph of $y = \sin\theta$ is shown in Figure 7.20. The student should check some of the points (or better still, sketch his own graph). The directed line segment $\overline{QP}$ is often called the *line representation* of the sine function.

FIGURE 7.20

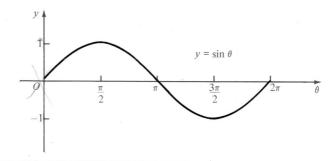

FIGURE 7.21

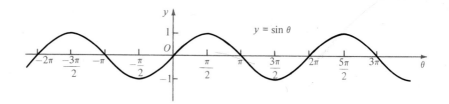

FIGURE 7.22

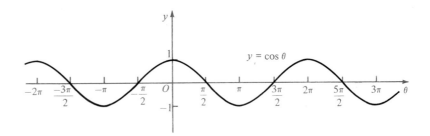

FIGURE 7.23

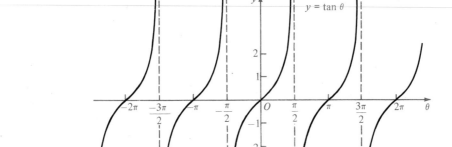

If P continues to travel around the unit circle, the pattern shown in Figure 7.20 is repeated, and we have a *periodic graph*. A portion of the graph is shown in Figure 7.21. The sine function is a *periodic function* with period 2π and is the simplest example (and our first example) of a periodic function. We will return to this topic in Chapter 10.

Since $\cos\theta = x$, the directed line segment $\overline{OQ}$ in Figure 7.19 is a line representation of the cosine function. Using this segment, or values from Table B, we obtain the sketch shown in Figure 7.22. The student will notice at once that the graph of $\cos\theta$ seems to be congruent to the graph of $\sin\theta$. If we move the graph of $y = \cos\theta$ to the right $\pi/2$ units, we obtain the graph of $y = \sin\theta$ shown in Figure 7.21. This fact depends on the identity $\sin(\theta + \pi/2) = \cos\theta$, and this identity will be proved in Chapter 9.

There are line representations for the other trigonometric functions, but we omit them. Sketches of the graphs of $\tan\theta$, $\cot\theta$, $\sec\theta$, and $\csc\theta$ are shown in Figures 7.23, 7.24, 7.25, and 7.26, respectively. The reader should check a few points on each graph, or better still, sketch his own first, and then compare his with the ones given here.

FIGURE 7.24

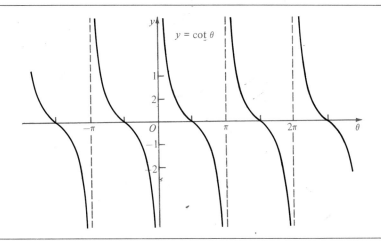

FIGURE 7.25

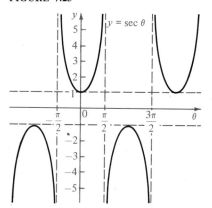

FIGURE 7.26

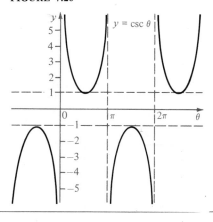

7.9 ODD AND EVEN FUNCTIONS

The graphs sketched in the preceding section have a certain symmetry. If we use the tests for symmetry developed in Chapter 6, Section 5 we find:

(I) If $f(-x) = f(x)$, then the graph of $y = f(x)$ is symmetric with respect to the y-axis.

(II) If $f(-x) = -f(x)$, then the graph of $y = f(x)$ is symmetric with respect to the origin.

We give special names to functions of this type in

DEFINITION 9   *A function f is said to be an even function if for every x in the domain of f*

$$(40) \qquad\qquad f(-x) = f(x).$$

The function is said to be an odd function if for every x in the domain of f

$$(41) \qquad\qquad f(-x) = -f(x).$$

EXAMPLE 1   Let $f(x) \equiv x^n$ where n is an integer. Is this function odd, even, or neither?

Solution: Let x be arbitrary ($x \neq 0$, if $n < 0$). Then

$$(42) \qquad f(-x) = (-x)^n = (-1)^n x^n = (-1)^n f(x).$$

If n is an even integer, then $(-1)^n = 1$, and (42) gives $f(-x) = f(x)$. This is equation (40). Hence, if n is even, the function x^n is an even function.
If n is an odd integer, then $(-1)^n = -1$, and (42) gives $f(-x) = -f(x)$. This is equation (41). Hence, if n is odd, the function x^n is an odd function.   ▲

This example also reveals the source of the names odd and even in Definition 9. However, these functions are not the only functions that are odd or even.

Theorem 6   *Among the trigonometric functions, sine, tangent, cotangent, and cosecant are odd, and cosine and secant are even.*

The reader should examine Figures 7.21 through 7.26 to see if the graphs are consistent with Theorem 6.
In equation form Theorem 6 states that for all θ in the domain of the function

(43) $\sin(-\theta) = -\sin\theta,$ (44) $\csc(-\theta) = -\csc\theta,$

(45) $\tan(-\theta) = -\tan\theta,$ (46) $\cot(-\theta) = -\cot\theta,$

(47) $\cos(-\theta) = \cos\theta,$ (48) $\sec(-\theta) = \sec\theta.$

FIGURE 7.27   FIGURE 7.28

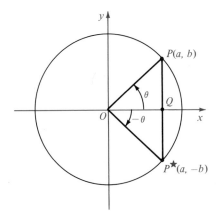

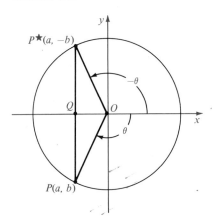

The first four of these equations have the form (41), required of an odd function. The last two have the form of (40), required of an even function.

Proof of Theorem 6: We put θ in standard position and let P be the intersection point of the terminal side with the unit circle. However, we use the notation (a, b) for the coordinates of P rather than the customary (x, y). Now consider the point $P^\star$ whose coordinates are $(a, -b)$. $P^\star$ will also be on the unit circle. In addition, P and $P^\star$ will be on the same vertical line. These points are shown in Figures 7.27 and 7.28 which illustrate two of the many possible cases for θ. In particular, Figure 7.28 shows the case in which θ and b are negative and in this event $-\theta$ and $-b$ are positive. In every case the triangles OQP and $OQP^\star$ will be congruent. Hence, $P^\star$ is on the terminal side of $-\theta$. Consequently, by the definitions of the trigonometric functions,

$$\sin(-\theta) = -b = -\sin\theta,$$

$$\cos(-\theta) = a = \cos\theta,$$

$$\tan(-\theta) = \frac{a}{-b} = -\frac{a}{b} = -\tan\theta.$$

The other three are proved in a similar manner. ∎

Exercise 7

1. Prove that the maximum value for $\sin\theta$ is 1, and the minimum value is -1.

2. Prove that the assertion of problem 1 is also true for $\cos\theta$.

3. Prove that $\tan \theta$ does not have a maximum.

4. In what portion of the interval $[0, 2\pi]$ is $\sin \theta$ decreasing?

5. In what portion of the interval $[0, 2\pi]$ is $\cos \theta$ decreasing?

6. Is there any interval in which $\tan \theta$ is decreasing?

7. Find all values of θ for which $\sin \theta = 0$. These are called the zeros of the function.

8. Find all the zeros of (a) $\cos \theta$, (b) $\tan \theta$, (c) $\cot \theta$, (d) $\sec \theta$, (e) $\csc \theta$.

9. Without looking at the graphs or Table B, complete the following table giving the behavior of the function (increasing or decreasing) in the given interval.

TABLE 3

Interval	Sin θ	Cos θ	Tan θ	Cot θ	Sec θ	Csc θ
$0 < \theta < \pi/2$, Q.I	Inc.	Dec.	Inc.	Dec.	Inc.	Dec.
$\pi/2 < \theta < \pi$, Q.II						
$\pi < \theta < 3\pi/2$, Q.III						
$3\pi/2 < \theta < 2\pi$, Q.IV						

10. Prepare a table similar to the one in problem 9 showing the range of each trigonometric function in each interval.

11. Draw a figure for the proof of Theorem 6 when (a) $\theta < 0$ and θ is in Q.II, (b) $\theta > 0$ and θ is in Q.IV.

★12. Prove that if $f(x)$ is an odd function and $g(x)$ is an even function, then the product $p(x) = f(x)g(x)$ is an odd function, and the quotient $q(x) = f(x)/g(x)$ is an odd function.

★13. Prove that if both $f(x)$ and $g(x)$ are odd functions, then both $p(x)$ and $q(x)$ in problem 12 are even functions.

★14. Prove that if $f(x)$ is an odd function, then the power $P(x) = [f(x)]^n$ is an odd function when n is an odd integer and an even function when n is an even integer. What happens if $f(x)$ is an even function? The function $[f(x)]^n$ is frequently written $f^n(x)$, for brevity.

★15. Prove that the sum (or difference) of two odd functions is an odd function, and the sum (or difference) of two even functions is an even function.

★ 16. Is there a function that is both odd and even?

17. If $f(x)$ is an odd function and $g(x)$ is an even function, what can be said about $f(x)+g(x)$?

Review Questions

Try to answer the following questions as accurately as possible before consulting the text.

1. Explain the difference between a geometric angle and a trigonometric angle.

2. State the definition of the six trigonometric functions.

3. Give all of the trigonometric functions for 0°, 30°, 45°, 60°, and 90°.

4. Give the definition of the reference angle. Why is the reference angle important?

5. Name the cofunction for each of the six trigonometric functions. State the basic theorem about cofunctions. (See Theorem 5.)

6. Sketch the graph of each of the six trigonometric functions.

7. Give the definition of (a) an odd function and (b) an even function.

8. Which of the trigonometric functions are odd functions and which are even functions?

Trigonometric Identities and Equations

8

The distinction between an identity and an equation has been discussed in Chapters 1 and 5 (see pages 4 and 83) and the reader should review that material. In this chapter we concentrate on proving identities that involve the trigonometric functions. The subject matter is extremely important for later mathematical work, but for the present, the student should regard it merely as a good practice field in which he can display his talent for algebraic manipulation, using trigonometric functions.

Proving trigonometric identities should be fun for a person who has a normal, healthy, inquisitive mind. The converse, however, is not necessarily true.

8.1 REVIEW OF ALGEBRAIC IDENTITIES

We consider an equation $E_1 = E_2$ involving two expressions. These expressions may involve many variables, but in most of our examples and problems E_1 and E_2 will involve only one variable. To simplify the language, we state the definitions for the case of one variable. Whenever it is convenient to do so, we may transpose E_2 to the left side and put the equation in the form $E = 0$. We recall our agreement that, unless otherwise specifically stated, the domain **D** of an equation is the largest set of values of the variable for which the expressions in the equation are well defined.

DEFINITION 1   *Identity*   *An equation is called an identity if it is true for all values of the variable in **D**, the domain of the equation.*

DEFINITION 2   *Conditional Equation*   *An equation is called a conditional equation if there is at least one value of the variable in **D** for which the equation is not true. The subset **S** of **D** for which the equation is true is called the solution set of the equation.*

EXAMPLE 1   Prove the identity

$$(1) \qquad \frac{3+x}{1-x} - \frac{3-x}{1+x} = \frac{8x}{1-x^2}$$

Solution: The expressions are not well defined if $x = 1$ or $x = -1$. Thus $\mathbf{D} = \mathbf{R} - \{-1, 1\}$. In proving an identity we prefer to work only with one side because it is neater and encourages orderly and logical thinking.‡ In this case we work on the left side and use the rules of algebra for adding fractions. If $x \in \mathbf{D}$, then for the left side we have

$$\frac{3+x}{1-x} - \frac{3-x}{1+x} \qquad \Big| \qquad = \frac{8x}{1-x^2}$$

$$= \frac{(3+x)(1+x) - (3-x)(1-x)}{(1-x)(1+x)}$$

$$= \frac{3 + 4x + x^2 - (3 - 4x + x^2)}{1 - x^2}$$

$$= \frac{3 + 4x + x^2 - 3 + 4x - x^2}{1 - x^2}$$

$$= \frac{8x}{1-x^2} \qquad \Big| \qquad = \frac{8x}{1-x^2}. \quad \blacktriangle$$

Since the equal sign is reflexive, $E_1 = E_2$ if and only if $E_2 = E_1$. This means that in any given identity we may select the side on which we prefer to work. Quite naturally we choose to simplify the complicated side rather than complicate the simple side.

An equation may involve several variables. This is the point of

EXAMPLE 2   Solve the conditional equation

$$(2) \qquad\qquad (x - 2y)^2 = x^2 - xy + y^2.$$

‡ There are different opinions on this point. The student should consult the local authorities.

Solution: Since $(x-2y)^2 = x^2 - 4xy + 4y^2$ for all values of x and y (an identity), we can write the conditional equation (2) as

$$(3) \qquad\qquad x^2 - 4xy + 4y^2 = x^2 - xy + y^2.$$

Hence, $3y^2 - 3xy = 0$ or $3y(y - x) = 0$. Therefore, (2) is satisfied (true) if $y = 0$ or if $y = x$ and only in these cases. ▲

Why not always use $\equiv$ for identities, and $=$ for conditional equations? We observe that we used an identity in solving equation (2). If we insisted on the distinction between $\equiv$ and $=$ then both symbols would appear in the solution of this problem and the result might be confusing. Further, the work required to make such a distinction might be more trouble than the gain. In what follows, we will often use the equal sign for an identity. The reader should have no trouble distinguishing between the two.

EXAMPLE 3

In the following set of eight equations, four are identities and four are conditional equations. Find the conditional equations and solve them.

$$(4) \qquad\qquad (x-4)(x+4) = x^2 - 16,$$

$$(5) \qquad\qquad (y+2)^3 = y^3 + 6y^2 + 12y + 8,$$

$$(6) \qquad\qquad \frac{A}{B} + \frac{1}{2} = \frac{2A + B}{2B},$$

$$(7) \qquad\qquad 10^{m+n} = 10^m 10^n,$$

$$(8) \qquad\qquad (x-4)(x+4) = 6x,$$

$$(9) \qquad\qquad (y+2)^3 = y^3 + 5y^2 + 8y + 5,$$

$$(10) \qquad\qquad \frac{A}{B} + \frac{1}{2} = \frac{A+1}{B+2},$$

$$(11) \qquad\qquad 10^{m^2+n} = 10^m 10^{n^2}.$$

Solution: Clearly, the first four equations are identities. Each is useful in solving one of the next four conditional equations.

Using (4) we find that (8) is equivalent to

$$(12) \qquad\qquad x^2 - 16 = 6x$$

or $x^2 - 6x - 16 = 0$. But $x^2 - 6x - 16 = (x-8)(x+2)$. Hence, $x = 8$ or $x = -2$.

Using (5) we find that (9) is equivalent to

$$(13) \qquad y^3 + 6y^2 + 12y + 8 = y^3 + 5y^2 + 8y + 5$$

or $y^2 + 4y + 3 = 0$. But $y^2 + 4y + 3 = (y+3)(y+1)$. Hence, $y = -1$ or $y = -3$.

Using (6) we find that (10) is equivalent to

$$(14) \qquad \frac{2A+B}{2B} = \frac{A+1}{B+2}$$

or $(2A+B)(B+2) = 2B(A+1)$. This gives $B^2 + 2AB + 4A + 2B = 2AB + 2B$. Hence, $B^2 + 4A = 0$ or $A = -B^2/4$.

Using (7) we find that (11) is equivalent to

$$(15) \qquad 10^{m^2+n} = 10^{m+n^2}$$

or $m^2 + n = m + n^2$. Hence $m^2 - n^2 - (m-n) = 0$ or $(m-n)(m+n-1) = 0$. Thus (11) is satisfied if $m = n$ or $m = 1 - n$ and only in those cases. ▲

Exercise 1

1. Check that the identity (1) is true for $x = 0$, $x = 2$, $x = 3$, and $x = 1/2$ by showing that both sides of this equation give, respectively, 0, $-16/3$, -3, and $16/3$.

2. Check that the identity (4) is true for $x = 1, 3, 5$, and 7 by showing that both sides of the equation give, respectively, -15, -7, 9, and 33.

In problems 3 through 9 prove each of the identities. Work only on one side.

3. $(a+b)^2 - (a-b)^2 = 4ab$ 4. $(x+y)^3 - (x-y)^3 = 2y(3x^2+y^2)$

5. $(A+B)^4 - (A-B)^4 = 8AB(A^2+B^2)$

6. $\dfrac{1+x}{1-x} + \dfrac{1-x}{1+x} = 2 + \dfrac{4x^2}{1-x^2}$ 7. $\dfrac{3+x}{1-x} + \dfrac{3-x}{1+x} = 6 + \dfrac{8x^2}{1-x^2}$

8. $\dfrac{\dfrac{3}{2} + \dfrac{x-5}{x+4}}{\dfrac{11}{4} - \dfrac{3x+21}{2x+8}} = 2$ 9. $\dfrac{\dfrac{5}{2} - \dfrac{9x-8}{4x-3}}{\dfrac{1}{4} + \dfrac{x+3}{8x-6}} = \dfrac{2}{3}$

★ 10. Prove that if n is an integer greater than 4, then $n^2 - 16$ is not a prime.

★ 11. Prove that if n is a positive integer, then $n^2 + 6n + 5$ is not a prime.

★ 12. Here is an old mind-reading trick. Joe asks Jim to (a) Think of a number, but keep it secret. (b) Double the number. (c) Add 36. (d) Divide the result by 2. (e) Subtract the original number. Then Joe proudly announces that the number Jim has in his mind at the end of these steps is 18.

   This trick depends on a simple algebraic identity. Find and prove this identity, and, thus, expose the trick.

★ 13. What is the converse of the last statement in the introduction to this chapter?

8.2 ELEMENTARY TRIGONOMETRIC IDENTITIES

We begin with a theorem that gives the fundamental trigonometric identities. All other trigonometric identities in this chapter are proved by manipulation with these eight basic ones. The student should memorize these equations, (16) through (23).

Theorem 1　　*For any θ for which the denominator is not zero*

$$(16) \quad \tan \theta = \frac{1}{\cot \theta}, \qquad\qquad (19) \quad \tan \theta = \frac{\sin \theta}{\cos \theta},$$

$$(17) \quad \sec \theta = \frac{1}{\cos \theta}, \qquad\qquad (20) \quad \cot \theta = \frac{\cos \theta}{\sin \theta},$$

$$(18) \quad \csc \theta = \frac{1}{\sin \theta},$$

$$(21) \qquad\qquad \sin^2 \theta + \cos^2 \theta = 1,$$

$$(22) \qquad\qquad 1 + \tan^2 \theta = \sec^2 \theta,$$

$$(23) \qquad\qquad 1 + \cot^2 \theta = \csc^2 \theta.$$

We observe that the notation $\sin^2 \theta$ means $(\sin \theta)^2$, the square of the sine of θ. If we wanted to indicate the sine of the square of θ, then we would write $\sin(\theta^2)$ or $\sin \theta^2$.

Proof: The identities (16), (17), and (18) have already been proved in Theorem 1 in Chapter 7. They are listed here again for ready reference. We also recall that these identities can be put in various equivalent forms:

(24) $$\tan\theta\cot\theta = 1, \qquad \cot\theta = \frac{1}{\tan\theta},$$

(25) $$\cos\theta\sec\theta = 1, \qquad \cos\theta = \frac{1}{\sec\theta},$$

(26) $$\sin\theta\csc\theta = 1, \qquad \sin\theta = \frac{1}{\csc\theta}.$$

For the remaining identities we place the angle whose measure is θ (in degrees or radians) in standard position and let $P(x, y)$ be the point of intersection of the terminal side with the unit circle. Since P is on the unit circle this means that for every θ,

(27) $$x^2 + y^2 = 1.$$

By definition $\tan\theta = y/x$, $y = \sin\theta$, and $x = \cos\theta$. Consequently,

$$\tan\theta \equiv \frac{y}{x} = \frac{\sin\theta}{\cos\theta},$$

and this is (19). In the same way

$$\cot\theta \equiv \frac{x}{y} = \frac{\cos\theta}{\sin\theta},$$

and this is (20). From (27) and the definitions $x = \cos\theta$ and $y = \sin\theta$ we have

$$\cos^2\theta + \sin^2\theta = 1,$$

and, except for commuting the terms, this is (21). If $x \neq 0$ we can divide both sides of (27) by x^2 and obtain

$$1 + \left(\frac{y}{x}\right)^2 = \left(\frac{1}{x}\right)^2$$

or

(22) $$1 + \tan^2\theta = \sec^2\theta.$$

Similarly, if we divide both sides of (27) by $y^2 \neq 0$, we get $(x/y)^2 + 1 = (1/y)^2$

or

(23) $$1 + \cot^2 \theta = \csc^2 \theta. \quad \blacksquare$$

These identities can be put in various forms as we have already indicated for the first three. For example, (21) can be written as

$$\sin^2 \theta = 1 - \cos^2 \theta, \quad \cos^2 \theta = 1 - \sin^2 \theta = (1 - \sin \theta)(1 + \sin \theta).$$

We can also write

(28) $$\sin \theta = \pm \sqrt{1 - \cos^2 \theta},$$

where the $\pm$ sign is needed because in Q.III and Q.IV $\sin \theta < 0$, and $\sqrt{1 - \cos^2 \theta} \geq 0$ by the definition of the symbol $\sqrt{}$.

It would be ridiculous to memorize all of the various forms for the identities of Theorem 1. It is much wiser to memorize the identities (16) through (23), and then to derive the various other forms by suitable manipulations whenever they are needed.

EXAMPLE 1   Prove that $\sec A \sin A = \tan A$.

Solution:

$$\sec A \sin A = \tan A$$

$$= \frac{1}{\cos A} \sin A$$

$$= \frac{\sin A}{\cos A}$$

$$= \tan A \qquad = \tan A. \quad \blacktriangle$$

EXAMPLE 2   Prove that $5 \sin^2 B + 3 \cos^2 B - 2 = 2 \sin^2 B + 1$.

Solution:

$$5 \sin^2 B + 3 \cos^2 B - 2 = 2 \sin^2 B + 1$$

$$= 2 \sin^2 B + 3 \sin^2 B + 3 \cos^2 B - 2$$

$$= 2 \sin^2 B + 3 (\sin^2 B + \cos^2 B) - 2$$

$$= 2 \sin^2 B + 3 - 2 \qquad = 2 \sin^2 B + 1. \quad \blacktriangle$$

EXAMPLE 3   Prove that $\dfrac{2 \sin^2 C - 1}{\sin C \cos C} = \tan C - \cot C$.

Solution:

$$\frac{2\sin^2 C - 1}{\sin C \cos C} = \tan C - \cot C$$

$$= \frac{2\sin^2 C - (\sin^2 C + \cos^2 C)}{\sin C \cos C}$$

$$= \frac{\sin^2 C - \cos^2 C}{\sin C \cos C}$$

$$= \frac{\sin^2 C}{\sin C \cos C} - \frac{\cos^2 C}{\sin C \cos C}$$

$$= \frac{\sin C}{\cos C} - \frac{\cos C}{\sin C} \qquad = \tan C - \cot C. \quad \blacktriangle$$

EXAMPLE 4   Prove that $(\sin D + \cos D)(\sec D - \csc D) = \tan D - \cot D$.

Solution: $(\sin D + \cos D)(\sec D - \csc D) = \tan D - \cot D$

$$= \sin D \sec D + \cos D \sec D - \sin D \csc D - \cos D \csc D$$

$$= \frac{\sin D}{\cos D} \quad + \quad 1 \quad - \quad 1 \quad - \frac{\cos D}{\sin D}$$

$$= \tan D - \cot D \qquad\qquad\qquad\qquad = \tan D - \cot D. \quad \blacktriangle$$

Note that in this identity each one of the six trigonometric functions appears exactly once.

Exercise 2

Prove each of the following identities. The identities in this list are simple and no one of the first 25 should require more than three minutes to prove.

1. $\sin A \cot A = \cos A$

2. $\sin B \sec B \cot B = 1$

3. $\csc C \tan C = \sec C$

4. $\cos D \csc D = \cot D$

5. $\cos^2 E \csc^2 E = \csc^2 E - 1$

6. $\sec^2 F - \sec^2 F \sin^2 F = 1$

7. $\dfrac{\sin^2 A}{\csc^2 A} = \sin^4 A$

8. $\dfrac{\cos^3 B}{\cot^3 B} = \sin^3 B$

9. $\dfrac{\cos^3 C}{\cot^2 C} = \cos C \sin^2 C$

10. $\dfrac{\cos^3 D}{\tan^2 D} = \cos^5 D \csc^2 D$

11. $\dfrac{\cot E}{\csc E} = \cos E$

12. $\dfrac{\tan^2 F}{\sec^2 F} = \sin^2 F$

13. $\csc^2 H \sec^5 H = \tan^5 H \csc^7 H$

14. $\sin \alpha \cos \alpha (\tan \alpha + \cot \alpha) = 1$

15. $(1 - \sec^2 \gamma)(1 - \csc^2 \gamma) = 1$

16. $1 + \tan^4 \alpha + 2 \tan^2 \alpha = \sec^4 \alpha$

17. $1 - 2 \sin^2 \theta = 2 \cos^2 \theta - 1$

18. $\dfrac{\sin^4 B - \cos^4 B}{\sin^2 B - \cos^2 B} = 1$

19. $\dfrac{\cos^2 C}{1 + \sin C} = 1 - \sin C$

20. $\dfrac{\sec^2 D - \tan^2 D}{\csc^2 D} = \sin^2 D$

21. $\dfrac{1 - \tan^2 E}{1 + \tan^2 E} = \cos^2 E - \sin^2 E$

22. $\sin^3 G \cos^3 G = \tan^6 G \cos^9 G \csc^3 G$

23. $\sec \beta \csc \beta (\sin \beta + \cos \beta) = \sec \beta + \csc \beta$

24. $\dfrac{3 \cos \phi + 5 \sin \phi \cot \phi}{2 \sin \phi} = 4 \cot \phi$

25. $\dfrac{1 + \cot A + \sec A}{\csc A} = \sin A + \cos A + \tan A.$

The following identities are a little more difficult than the first 25, but no one of them should require more than six minutes to prove.

26. $2 \csc \alpha - \cot \alpha \cos \alpha = \sin \alpha + \csc \alpha$

27. $2 \sin^2 \beta (\tan^2 \beta + 1) + 1 = \sec^2 \beta + \tan^2 \beta$

28. $(1 + \tan \gamma)(1 - \cot \gamma) = \tan \gamma - \cot \gamma$

29. $\sec \theta \csc \theta - 2 \cos \theta \csc \theta = \tan \theta - \cot \theta$

30. $\tan A (1 + \csc A) = \sec A (1 + \sin A)$

31. $(1 - \sec B)(\cos^2 B + \cos B) = -\sin^2 B$

32. $(1 + \sin C)(\sec C - \tan C) = \cos C$

33. $\cos^2 D (1 + \tan D)^2 = \sin^2 D (1 + \cot D)^2$

34. $(\sin E + 2 \cos E)^2 + (2 \sin E - \cos E)^2 = 5$

35. $(\sec F + 3 \tan F)^2 - (3 \sec F + \tan F)^2 = -8$

36. $\sec G \csc G (\tan G + \cot G) = \sec^2 G + \csc^2 G$

37. $\cos^3 A \sin A + \sin^3 A \cos A = \sin A \cos A$

38. $1 + \tan^2 B - \sin^2 B - \sin^2 B \tan^2 B = 1$

39. $\sin^3 C (1 + \cot C) + \cos^3 C (1 + \tan C) = \sin C + \cos C$

40. $D \sec^2 D + \csc D - D \tan^2 D - \cot D \cos D = D + \sin D$

41. $\dfrac{\cos^3 E - \sin^3 E}{\cos E - \sin E} = 1 + \sin E \cos E.$ 42. $\dfrac{\tan^3 F + 1}{\tan F + 1} = \sec^2 F - \tan F$

43. $1 + \tan^6 \alpha = \sec^2 \alpha (3 - 3 \sec^2 \alpha + \sec^4 \alpha)$

44. $\sec^6 \beta - 1 = \tan^2 \beta (3 + 3 \tan^2 \beta + \tan^4 \beta)$

45. $\sin^6 \gamma + \cos^6 \gamma = 1 - 3 \sin^2 \gamma + 3 \sin^4 \gamma$

46. $\dfrac{\cos^2 \theta}{\sin^2 \theta - 7 \sin \theta + 6} = \dfrac{1 + \sin \theta}{6 - \sin \theta}$

47. $\dfrac{\tan^2 \theta - 8}{\tan^2 \theta + 2 \sec \theta - 14} = \dfrac{3 + \sec \theta}{5 + \sec \theta}$

★48. $(\sec \phi + \tan \phi - 1)(\sec \phi - \tan \phi + 1) = 2 \tan \phi$

★49. $(1 + \tan \alpha + \cot \alpha)^2 = 1 + \sec^2 \alpha + \csc^2 \alpha + 2 \sec \alpha \csc \alpha$

★50. $(1 + \cot \beta + \csc \beta)^2 = 2(1 + \csc \beta)(\cot \beta + \csc \beta)$

8.3 MORE COMPLICATED TRIGONOMETRIC IDENTITIES

We recall three theorems about fractions. These theorems can be used in proving trigonometric identities because for fixed θ, each trigonometric function of θ is a real number.

The Addition of Fractions From Chapter 2 we recall that if $b \neq 0$ and $d \neq 0$, then

(29)
$$\frac{a}{b} + \frac{c}{d} = \frac{ad + bc}{bd}.$$

EXAMPLE 1   Prove that $\dfrac{\tan \theta}{\csc \theta - \cot \theta} - \dfrac{\sin \theta}{\csc \theta + \cot \theta} = \sec \theta + \cos \theta.$

Solution: Working only on the left side we have

$$\frac{\tan \theta}{\csc \theta - \cot \theta} + \frac{- \sin \theta}{\csc \theta + \cot \theta}$$

$$= \frac{\tan \theta (\csc \theta + \cot \theta) - \sin \theta (\csc \theta - \cot \theta)}{(\csc \theta - \cot \theta)(\csc \theta + \cot \theta)} \qquad \text{[by (29)]}$$

$$= \frac{\tan\theta\csc\theta + 1 - 1 + \sin\theta\cot\theta}{\csc^2\theta - \cot^2\theta}$$

$$= \frac{\dfrac{\sin\theta}{\cos\theta}\dfrac{1}{\sin\theta} + \dfrac{\sin\theta}{1}\dfrac{\cos\theta}{\sin\theta}}{1}$$

$$= \frac{1}{\cos\theta} + \cos\theta,$$

$$= \sec\theta + \cos\theta. \quad \blacktriangle$$

The Reduction of Compound Fractions   If $b \neq 0$, $c \neq 0$, and $d \neq 0$, then

(30)
$$\frac{\dfrac{a}{b}}{\dfrac{c}{d}} = \frac{a}{b}\cdot\frac{d}{c} = \frac{ad}{bc}.$$

When there are only three terms on the left side of (30), a fourth term "1" must be inserted in a convenient spot. Thus, the same theorem also gives

(31)
$$\frac{\dfrac{a}{b}}{c} = \frac{a/1}{\dfrac{b}{c}} = \frac{a}{1}\cdot\frac{c}{b} = \frac{ac}{b},$$

and

(32)
$$\frac{\dfrac{a}{b}}{c} = \frac{\dfrac{a}{b}}{c/1} = \frac{a}{b}\cdot\frac{1}{c} = \frac{a}{bc}.$$

Notice that the two results in (31) and (32) are quite different.

EXAMPLE 2   Prove that
$$\frac{\csc\theta}{\cot\theta + \tan\theta} = \cos\theta.$$

Solution: Again we work only on the left side. This time, for convenience, we arrange the work in a line rather than in a column.

$$\frac{\csc\theta}{\cot\theta + \tan\theta} = \frac{\dfrac{1}{\sin\theta}}{\dfrac{\cos\theta}{\sin\theta} + \dfrac{\sin\theta}{\cos\theta}} = \frac{\dfrac{1}{\sin\theta}}{\dfrac{\cos^2\theta + \sin^2\theta}{\sin\theta\cos\theta}}$$

$$= \frac{1}{\sin\theta}\frac{\sin\theta\cos\theta}{\cos^2\theta + \sin^2\theta} = \frac{\cos\theta}{1} = \cos\theta. \quad \blacktriangle$$

Cancellation and Its Inverse If $b \neq 0$ and $d \neq 0$, then

(33)
$$\frac{ad}{bd} = \frac{a}{b}.$$

Since the equality sign is symmetric, equation (33) (like any equation) can be read from left to right and also from right to left. Thus, the fraction a/b can always be altered in form by introducing any factor d both in the numerator and in the denominator. When d is suitably selected, this operation can be very helpful.

EXAMPLE 3   Prove that

(34)
$$\frac{1}{\sec X(1 - \sin X)} = \sec X + \tan X.$$

Solution: Working on the left side, we use (33) with $d \equiv 1 + \sin X$. This gives

$$\frac{1}{\sec X(1 - \sin X)} = \frac{1 + \sin X}{\sec X(1 - \sin X)(1 + \sin X)} = \frac{1 + \sin X}{\sec X(1 - \sin^2 X)}.$$

We now see why the factor d was chosen to be $1 + \sin X$. This selection gives $1 - \sin^2 X = \cos^2 X$ in the denominator. Then the left side of (34) is

$$\frac{1 + \sin X}{\sec X \cos^2 X} = \frac{1 + \sin X}{(\sec X \cos X)\cos X} = \frac{1 + \sin X}{\cos X} = \sec X + \tan X. \quad \blacktriangle$$

Exercise 3

Prove each of the following identities.

1. $\dfrac{\tan B \sin B}{\tan B - \sin B} = \dfrac{\sin B}{1 - \cos B}$

2. $\dfrac{\cos C - \sin C}{\cos C + \sin C} = \dfrac{\cot C - 1}{\cot C + 1}$

3. $\csc E = \dfrac{1 + \cot E}{\sin E + \cos E}$

4. $\sec G \csc G = \tan G + \cot G$

5. $\dfrac{1 + \sin \alpha}{1 - \sin \alpha} = -\dfrac{1 + \csc \alpha}{1 - \csc \alpha}$

6. $\dfrac{\tan \theta - \cot \theta}{\sin \theta - \cos \theta} = \sec \theta + \csc \theta$

7. $(\sec \theta + \tan \theta)^2 = \dfrac{1 + \sin \theta}{1 - \sin \theta}$

8. $\dfrac{\sin^2 \theta}{\tan \theta - \sin \theta} = \dfrac{\sin \theta \cos \theta}{1 - \cos \theta}$

9. $\dfrac{\tan \theta - 1}{1 - \cot \theta} = \tan \theta$

10. $\dfrac{1 - \sin \theta}{\cos \theta} = \dfrac{\cos \theta}{1 + \sin \theta}$

11. $\dfrac{\sin\theta+\cos\theta}{\sec\theta+\csc\theta}=\sin\theta\cos\theta$

12. $\dfrac{\sin\theta}{\csc\theta-\cot\theta}=1+\cos\theta$

13. $\dfrac{3}{1+2\cos^2\theta}=\dfrac{3\sec^2\theta}{3+\tan^2\theta}$

14. $\dfrac{\tan A\sin A}{1+\cos A}=\sec A-1$

15. $\tan^2 B-\sin^2 B=\tan^2 B\sin^2 B$

16. $\dfrac{\sec^4 D+\tan^4 D}{\sec^2 D\tan^2 D}-2=\dfrac{\cos^4 D}{\sin^2 D}$

17. $\dfrac{1}{\cos E+\sin E\tan E}=\cos E$

18. $\dfrac{\sec^2 F-\tan^2 F}{\sec F-\cos F}=\cot F\csc F$

19. $\dfrac{2-\cos^2\alpha}{1-\sin^2\alpha}=\sec^4\alpha-\tan^4\alpha$

20. $\dfrac{\csc\beta-\cot\beta}{\csc\beta+\cot\beta}=\dfrac{1-\cos\beta}{1+\cos\beta}$

21. $\dfrac{1}{1-2\sin\gamma}=\dfrac{2\sec\gamma+\tan\gamma}{2\cos\gamma-3\tan\gamma}$

22. $(\cot^2\gamma-\cos^2\gamma)^2=\dfrac{\cos^8\gamma}{\sin^4\gamma}$

23. $\dfrac{\cos A\cot A}{1+\sin A}=\csc A-1$

24. $\dfrac{1}{1+\cos^2 B}=\dfrac{\sec^2 B}{\tan^2 B+2}$

25. $\dfrac{\tan C-\sin C}{\sin^3 C}=\dfrac{\sec C}{1+\cos C}$

26. $\dfrac{\tan D-\cot D}{\tan D+\cot D}=2\sin^2 D-1$

27. $\dfrac{1}{\sin E+\cos E\cot E}=\sin E$

28. $\dfrac{\cos^2\alpha\cot^2\alpha}{\cot\alpha-\cos\alpha}=\cos\alpha+\cot\alpha$

29. $\sec A=\dfrac{1+\tan A}{\sin A+\cos A}$

30. $\dfrac{\cos B}{1-\sin B}=\dfrac{\cot B\cos B}{\cot B-\cos B}$

★31. $\dfrac{\csc^4 D+\cot^4 D}{\csc^2 D\cot^2 D}=\dfrac{\sin^4 D}{\cos^2 D}+2$

★32. $\dfrac{\cot\theta\cos\theta}{\cot\theta+\cos\theta}=\dfrac{\cot\theta-\cos\theta}{\cot\theta\cos\theta}$

★33. $\dfrac{1+\sin\theta+\cos\theta}{1+\sin\theta-\cos\theta}=\dfrac{\sin\theta}{1-\cos\theta}$

★34. $\dfrac{\sec\alpha+\tan\alpha}{\cos\alpha-\tan\alpha-\sec\alpha}=-\csc\alpha$

35. $\dfrac{1}{1+\sin A}+\dfrac{1}{1-\sin A}=2\sec^2 A$

36. $\dfrac{\sin D-\cos D}{\cos D}+\dfrac{\cos D+\sin D}{\sin D}=\sec D\csc D$

37. $\dfrac{1}{\sec F-1}-\dfrac{1}{\sec F+1}=2\cot^2 F.$

38. $\dfrac{1+\sin\beta}{\cos\beta}+\dfrac{\cos\beta}{1+\sin\beta}=2\sec\beta$

39. $\dfrac{1-\cos\gamma}{\sin\gamma} - \dfrac{\sin\gamma}{1-\cos\gamma} = -2\cot\gamma$

40. $\dfrac{\cos C}{\tan C+\sec C} - \dfrac{\cos C}{\tan C-\sec C} = 2$

41. $(\tan G+\cot G)^2 = \sec^2 G\csc^2 G$

42. $\dfrac{\tan\alpha}{\cos\alpha-1} - \dfrac{\tan\alpha}{\cos\alpha+1} = \dfrac{-2}{\sin\alpha\cos\alpha}$

43. $\dfrac{\tan\beta}{1-\cot\beta} + \dfrac{\cot\beta}{1-\tan\beta} = 1 + \tan\beta + \cot\beta$

44. $\cos^2\beta + 2\sin^2\beta + \sin^2\beta\tan^2\beta = \sec^2\beta$

45. $\sin\gamma\cos\gamma(\sec\gamma+\csc\gamma)^2 - 2 = \sec\gamma\csc\gamma$

46. $\dfrac{\sin^2\theta}{\sin\theta-\cos\theta} - \dfrac{\sin\theta+\cos\theta}{\tan^2\theta-1} = \sin\theta + \cos\theta$

47. $\dfrac{\cos\theta}{\sin\theta+\cos\theta} - \dfrac{\sin\theta}{\sin\theta-\cos\theta} = \dfrac{\cot^2\theta+1}{\cot^2\theta-1}$

48. $\dfrac{\sin C}{\cot C+\csc C} - \dfrac{\sin C}{\cot C-\csc C} = 2$

★49. $\left(\dfrac{\cos\theta}{\tan\theta} + \dfrac{\sin\theta}{\cot\theta}\right) \Big/ (\tan\theta+\cot\theta-1) = \sin\theta + \cos\theta$

★50. $\cos\beta + \cos\beta\cot\beta + \sin\beta\tan\beta + \sin\beta = \dfrac{1}{\cos\beta} + \dfrac{1}{\sin\beta}$

8.4 SOME ELEMENTARY
TRIGONOMETRIC EQUATIONS

A trigonometric equation is an equation in which trigonometric functions occur. Such equations arise frequently in mathematics, physics, and in all types of engineering. To solve a trigonometric equation completely, means to find all values of the variable for which the equation is a true statement. Frequently there are infinitely many such values. In order to restrict the number of solutions, we usually ask that the variable lie in some suitable interval. For example, we may seek all solutions θ such that $0° \leq \theta < 360°$ (or, in radians, $0 \leq \theta < 2\pi$). For simplicity we use only the degree measure for θ in this section. All of the rules used in solving algebraic equations can be applied directly to trigonometric equations.

EXAMPLE 1 Solve the equation

(35) $$8 \sin \theta - 6 = 2 \sin \theta - 3$$

(a) completely, (b) for θ in the interval $0° \leqq \theta < 360°$.

Solution: (a) Transposition gives

$$8 \sin \theta - 2 \sin \theta = 6 - 3$$
$$6 \sin \theta = 3$$
$$\sin \theta = 1/2.$$

Thus θ must be in Q.I or in Q.II. The reference angle is 30°, so two of the solutions are $\theta = 30°$ and $\theta = 180° - 30° = 150°$. All angles coterminal with these angles are also solutions. But these are all of the solutions, and they can be listed by writing

$$\theta = 30° + n360°, \qquad \theta = 150° + n360°, \qquad n \in \mathbf{Z}.$$

(b) If we ask that $0° \leqq \theta < 360°$, then only two of the infinitely many solutions for (35) will lie in this interval. These are $\theta = 30°, 150°$. ▲

EXAMPLE 2 Find all solutions of the equation

(36) $$2 \cos^2 \theta - 5 \cos \theta - 3 = 0$$

for θ in the interval $0° < \theta < 720°$.

Solution: We turn this into an algebraic equation by setting $x = \cos \theta$. Then (36) becomes $2x^2 - 5x - 3 = 0$. Factoring, we obtain

$$(2x + 1)(x - 3) = 0.$$

Either $2x + 1 = 0,$ $x = -1/2.$ or $x - 3 = 0,$ $x = 3.$

Thus $\cos \theta = x = -1/2.$ $\cos \theta = x = 3.$

Either $\theta = 120° + n360°.$ No real θ.

or $\theta = 240° + n360°.$

To obtain the solutions in the given interval set $n = 0$ or 1. The solutions are $\theta = 120°, 240°, 480°, 600°$. ▲

EXAMPLE 3 Solve the equation

(37) $$\tan \theta + 3 \sec^2 \theta = 8$$

for θ in the interval $0° < \theta < 180°$.

Solution: Using $\sec^2\theta = \tan^2\theta + 1$ in (37) we obtain

$$\tan\theta + 3(\tan^2\theta + 1) = 8.$$

If now we let $\tan\theta = u$, then the corresponding algebraic equation is

$$3u^2 + u - 5 = 0.$$

By the quadratic formula, $u = \left(-b \pm \sqrt{b^2 - 4ac}\right)/2a$, and hence

$$u = \frac{-1 \pm \sqrt{1 + 60}}{6} \approx \frac{-1 \pm 7.810}{6} = 1.135 \qquad \text{or} \qquad -1.468.$$

Consequently, $\tan\theta \approx 1.135$ or -1.468. When we turn to Table B, we do not find any angle listed for which $\tan\theta = 1.135$ or $|\tan\theta| = 1.468$. This difficulty will disappear after we have covered interpolation (the science of reading between the lines) in Section 11.7 and Section 12.4. In the meantime, we will use that number in Table B, that is nearest to the one we are seeking. In other words, we will give our solution (or solutions) to the nearest $10'$. With this agreement we find that $\theta = 48°\,40'$ or $\theta = 124°\,20'$.   ▲

EXAMPLE 4   Solve the equation

(38) $$\sin\theta \tan^2\theta - 3 + \tan^2\theta - 3\sin\theta = 0$$

for θ in the interval $-180° < \theta < 180°$.

Solution: In Example 3 it was possible to make a substitution so that only one trigonometric function appeared. In (38) the appropriate substitution is not obvious. Fortunately, we can group the terms and factor. We have

$$\sin\theta \tan^2\theta + \tan^2\theta - 3 - 3\sin\theta = 0,$$

$$\tan^2\theta(\sin\theta + 1) - 3(\sin\theta + 1) = 0,$$

$$(\tan^2\theta - 3)(\sin\theta + 1) = 0.$$

Either $\tan^2\theta - 3 = 0$,   or $\sin\theta + 1 = 0$,

$\tan^2\theta = 3$, $\sin\theta = -1$,

$\tan\theta = \pm\sqrt{3}$, $\theta = -90°$.

$\theta = 60°,\ 120°,\ -60°,\ -120°.$

Arranging the answers in increasing order, we have $\theta = -120°,\ -90°,\ -60°,$ $60°,\ 120°.$ ▲

EXAMPLE 5

Solve

(39) $$\sin \theta + \cos \theta = 1$$

for θ in the interval $0° \leqq \theta < 360°$.

Solution: If we substitute $\cos \theta = \pm \sqrt{1 - \sin^2 \theta}$ and transpose we find

(40) $$\sin \theta - 1 = \pm \sqrt{1 - \sin^2 \theta}.$$

On squaring both sides of this equation we obtain

$$\sin^2 \theta - 2\sin \theta + 1 = 1 - \sin^2 \theta,$$
$$2\sin^2 \theta - 2\sin \theta = 0,$$
$$2\sin \theta(\sin\theta - 1) = 0.$$

Either $2\sin \theta = 0,$ | or $\sin \theta - 1 = 0,$

$\sin \theta = 0,$ | $\sin \theta = 1,$

$\theta = 0°, 180°.$ | $\theta = 90°.$

In the first four examples, each of the steps was reversible, so that logically no check was necessary for the answers. In this example the step in which we squared both sides of (40) is not reversible. Every solution of (40) is in the list $\theta = 0°, 90°, 180°$, but it is possible that the list contains entries that are not solutions (see Section 5.6). Hence, it is necessary to check each of the values of θ found above.

Check If $\theta = 0°,$ then $\sin \theta + \cos \theta = 0 + 1 = 1.$

If $\theta = 90°,$ then $\sin \theta + \cos \theta = 1 + 0 = 1.$

If $\theta = 180°,$ then $\sin \theta + \cos \theta = 0 + (-1) \neq 1.$

Hence, there are only *two* solutions for equation (40), $\theta = 0°$ or $\theta = 90°.$ ▲

EXAMPLE 6

Solve

(41) $$2\cos(3\theta + 24°) = \sqrt{2}$$

for θ in the interval $0° \leqq \theta < 360°$.

Solution: This equation is equivalent to $\cos(3\theta + 24°) = \sqrt{2}/2$. Thus the angles in question are $45°, 315°$, etc. But these are the values for $3\theta + 24°$.

Hence we must write:

either $\qquad\qquad 3\theta + 24° = 45° + n360°, \qquad n \in \mathbf{Z}$

or $\qquad\qquad 3\theta + 24° = 315° + n360°, \qquad n \in \mathbf{Z}.$

Subtracting 24° from each side and dividing by 3 we find that

either $\qquad\qquad \theta = \dfrac{45° - 24° + n360°}{3} = 7° + n120°$

or $\qquad\qquad \theta = \dfrac{315° - 24° + n360°}{3} = 97° + n120°.$

Taking $n = 0$, 1, and 2 successively, we find that $\theta = 7°$, 97°, 127°, 217°, 247°, 337°. Any other integer value for n will give a θ that lies outside of the prescribed interval. ▲

Exercise 4

Solve each of the following equations for θ in the interval $0° \leqq \theta < 360°$. Give answers to the nearest 10′.

1. $\sin(\theta + 10°) = \sqrt{3}/2$
2. $\cos(\theta - 15°) = 1/2$
3. $\tan(\theta - 70°) = -1$
4. $\cot(\theta + 230°) = -\sqrt{3}$
5. $2\sin^2\theta - 7\sin\theta + 3 = 0$
6. $2\cos^2\theta + 11\cos\theta + 5 = 0$
7. $\sin^2\theta - \cos\theta + 1 = 0$
8. $6(1 + \sin\theta) = \cos^2\theta$
9. $\sin\theta + 2\csc\theta = 4$
10. $5\cos\theta + \sec\theta = 5$
★11. $\tan 2\theta = 2\cot 2\theta$
★12. $\sin 3\theta = 2\cos 3\theta$
13. $8\tan\theta + 3 + 2\sec^2\theta = 0$
14. $7\csc^2\theta = 6(1 + \cot\theta)$
15. $3\sin\theta - \cos\theta = -3$
16. $2\sin\theta + \cos\theta = 2$
★17. $8\sin\theta + \cos\theta = 4$
★18. $11\sin\theta + 3\cos\theta = 9$
19. $2\sin 2\theta - \tan 2\theta = 0$
20. $\csc\theta + 3\cot\theta = 0$
21. $\sin^2\theta + \sin\theta\cos\theta - 6\cos^2\theta = 0$
22. $2\sin^2\theta + 7\sin\theta\cos\theta + 3\cos^2\theta = 0$
23. $1 - 2\sin\theta = \cot\theta - 2\cos\theta$
24. $3 + \cos\theta = 3\tan\theta + \sin\theta$
25. $4\cos^3\theta + 3 = 4\cos^2\theta + 3\cos\theta$

26. $4\sin^3\theta + 8\sin^2\theta = 3\sin\theta + 6$

★★27. Prove that for a root of the equation $a\sin\theta + b\cos\theta = c$, where a, b, and c are integers, $\sin\theta$ is rational if and only if $a^2 + b^2 - c^2$ is the square of an integer.

Review Questions

Try to answer the following questions as accurately as possible before consulting the text.

1. What is the definition of an identity in one variable?

2. Give an example of a conditional equation in two variables.

3. Name those pairs of trigonometric functions that are reciprocals.

4. Complete the following trigonometric identities

   (a) $\cos^2\theta = 1 -$ (b) $\tan^2\theta = \sec^2\theta -$

   (c) $\csc^2\theta = 1 -$ (d) $\sin\theta = \tan\theta(\underline{\quad\quad})$

   (e) $\cos\theta = \cot\theta(\underline{\quad\quad})$ (f) $1 = \sin\theta\cot\theta(\underline{\quad\quad})$

5. Complete the following algebraic identities

   (a) $\dfrac{A}{B} - \dfrac{C}{D} = \dfrac{}{BD}$ (b) $\dfrac{A/B}{C/D} = \dfrac{A}{}$

   (c) $\dfrac{A}{B/C} = \dfrac{A}{}$ (d) $\dfrac{A/B}{C} = \dfrac{A}{}$

The Addition Formulas and Related Topics

9

If A, B, and C are any numbers, then

$$C(A-B) = CA - CB.$$

Hence we might expect, by analogy, that

(1) $$\cos(A-B) = \cos A - \cos B.$$

However, formula (1) is *false*. To prove that (1) is not an identity, we need only one counterexample; that is, one pair A, B for which the two sides of this equation are not equal. Suppose we set $A = 0$ and $B = 0$. Since $\cos 0 = 1$, the left side of (1) gives

$$\cos(0-0) = \cos 0 = 1.$$

The right side of (1) gives

$$\cos 0 - \cos 0 = 1 - 1 = 0.$$

Since $1 \neq 0$, equation (1) is not an identity. Our first objective is to replace the false identity (1) by the correct one:

$$\cos(A-B) = \cos A \cos B + \sin A \sin B.$$

197

From this identity it is easy to derive a large number of other important identities. The rest of this chapter is devoted to proving these identities and examining some of their applications.

The student is warned that this chapter contains an abundance of new and important formulas. These formulas must be memorized as they are proved or the student will soon become hopelessly confused.

9.1 A DISTANCE FORMULA

Let $P(2, 1)$ and $Q(6, 4)$ be two points in the plane. Suppose that we wish to compute the distance between P and Q. If lines are drawn parallel to the coordinates axes through P and Q as indicated in Figure 9.1, then the triangle PQR is a right triangle. Further, the sides have lengths

$$a = |PR| = 6 - 2 = 4$$

and

$$b = |RQ| = 4 - 1 = 3.$$

Therefore the distance between P and Q is the hypotenuse of the right triangle PQR and this distance is given by

$$d = \sqrt{a^2 + b^2} = \sqrt{(6-2)^2 + (4-1)^2} = \sqrt{16+9} = 5.$$

Now, we replace the numbers with letters, and obtain a general formula that is correct for any pair of points P and Q.

FIGURE 9.1

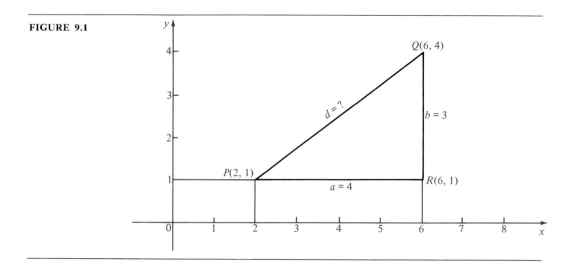

Theorem 1   *If $P(x_1, y_1)$ and $Q(x_2, y_2)$ are arbitrary points in the plane, then the distance between P and Q is given by*

(2) $$d = \sqrt{(x_2 - x_1)^2 + (y_2 - y_1)^2}.$$

Proof: Let L_1 be the horizontal line $y = y_1$ that passes through $P(x_1, y_1)$ and let L_2 be the vertical line $x = x_2$ that passes through $Q(x_2, y_2)$ (see Figure 9.2). We let R be the point of intersection of L_1 and L_2. Consequently,

FIGURE 9.2

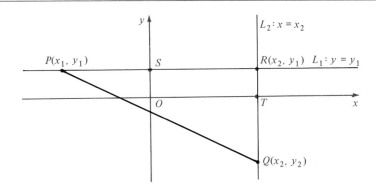

R has the coordinates (x_2, y_1), and PQR is a right triangle. Let S be the point of intersection of L_1 with the y-axis, and let T be the point of intersection of L_2 with the x-axis. Now the coordinates of a point are the directed distances from the two axes. Hence, by Theorem 26 of Chapter 3 on directed distances

(3) $$\overline{PR} = \overline{PS} + \overline{SR} = \overline{SR} - \overline{SP} = x_2 - x_1,$$

(4) $$\overline{QR} = \overline{QT} + \overline{TR} = \overline{TR} - \overline{TQ} = y_1 - y_2.$$

Consequently, the lengths of the sides of the right triangle are $a = |x_2 - x_1|$ and $b = |y_1 - y_2|$. By the Pythagorean Theorem

(5) $$d = |PQ| = \sqrt{a^2 + b^2} = \sqrt{(x_2 - x_1)^2 + (y_1 - y_2)^2}.$$

Since $(y_1 - y_2)^2 = (y_2 - y_1)^2$, equation (5) gives equation (2). ∎

Formula (2) is easy to remember because it is just a disguised form of the Pythagorean Theorem.

Exercise 1

In problems 1 through 14 find the distance between the two given points *without* making a drawing. Then make a careful drawing to scale, and check your answers by measuring the distance with a ruler.

1. $(1, 1)$ and $(6, 13)$ 2. $(7, 6)$ and $(4, 2)$

3. $(2, -4)$ and $(-1, 0)$ 4. $(-5, 1)$ and $(-1, 4)$

5. $(-2, -2)$ and $(4, 6)$ 6. $(10, 1)$ and $(-8, 1)$

7. $(-3, -1)$ and $(-9, -1)$ 8. $(5, 5)$ and $(-6, -6)$

9. $(8, 10)$ and $(1, 5)$ 10. $(-2, -3)$ and $(-3, -2)$

11. $(4, -3)$ and $(-7, -8)$ 12. $(9, 11)$ and $(2, 6)$

13. $(-4, -5)$ and $(-5, -4)$ 14. $(2, -3)$ and $(-3, -5)$

15. Is formula (2) still true, if the subscripts 1 and 2 are interchanged?

16. Is it always possible to select the letters P and Q so that $x_2 \geqq x_1$ and $y_2 \geqq y_1$?

17. How does formula (2) simplify if the point P is at the origin?

★18. Does the proof of Theorem 1 depend on the drawing or is the argument valid no matter where the points are in the plane?

9.2   THE ADDITION FORMULAS
FOR THE COSINE

The first of these formulas is given in

Theorem 2

For all real numbers‡ A and B

(6) $$\cos(A - B) = \cos A \cos B + \sin A \sin B.$$

Proof: For simplicity we suppose first that $0 < B < A < 2\pi$, so that $A - B > 0$. These conditions on A and B assure us that the figure (Figure 9.3) which represents the data will have the appearance that a normal person might expect. In Figure 9.3 we draw the angles A and B in standard position and we let $P(x_1, y_1)$ and $Q(x_2, y_2)$ be the coordinates of the intersection points of the terminal lines of A and B with the unit circle. By the definition of the trigonometric functions

(7) $$\begin{cases} x_1 = \cos A, \quad y_1 = \sin A, \\ x_2 = \cos B, \quad y_2 = \sin B. \end{cases}$$

---

‡ Recall that the trigonometric functions are defined for real numbers as well as for angles. When we use $A, B, \ldots$ as labels for angles in the figures, it is understood that these are the angles whose radian measures are $A, B, \ldots$. Of course, Theorem 2 (and all the theorems that flow from it) are also true if A and B represent angles measured in degrees.

FIGURE 9.3

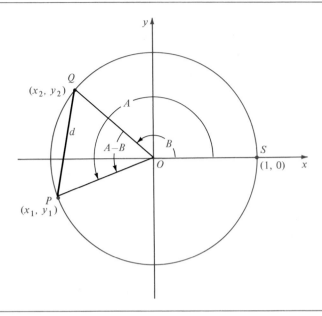

Let d be the distance between P and Q. Our plan is to compute d^2 in two different ways and to equate the results of the two computations. From Figure 9.3, Theorem 1, and equation (7) we have

$$d^2 = (x_2 - x_1)^2 + (y_2 - y_1)^2 = (\cos B - \cos A)^2 + (\sin B - \sin A)^2$$

$$= \cos^2 B - 2 \cos B \cos A + \cos^2 A + \sin^2 B - 2 \sin B \sin A + \sin^2 A$$

$$= (\sin^2 B + \cos^2 B) + (\sin^2 A + \cos^2 A) - 2(\cos A \cos B + \sin A \sin B).$$

But by Theorem 1 of Chapter 8 we have $\sin^2 \theta + \cos^2 \theta = 1$ for any θ. Hence,

(8) $$d^2 = 2 - 2(\cos A \cos B + \sin A \sin B).$$

We next rotate everything in Figure 9.3 (except the axes), in a clockwise direction (backwards) through an angle B. The result of this rotation is shown in Figure. 9.4. It carries the point Q into the point $S(1,0)$ and carries the point P into a point $R(x_3, y_3)$. Since the distance d between P and Q does not change during the rotation, d is also the distance between S and R. Now OR is the terminal line of the angle $A - B$. Hence, for the coordinates of R we have

(9) $$x_3 = \cos(A - B), \qquad y_3 = \sin(A - B).$$

FIGURE 9.4

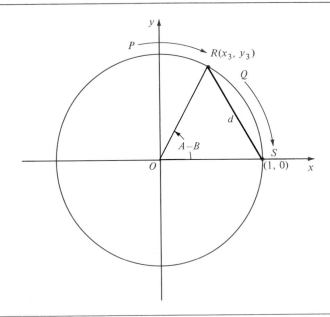

We now compute d^2 using the points R and S. Theorem 1 gives

$$d^2 = (x_3 - 1)^2 + (y_3 - 0)^2 = (\cos(A - B) - 1)^2 + (\sin(A - B) - 0)^2$$

$$= \cos^2(A - B) - 2\cos(A - B) + 1 + \sin^2(A - B)$$

(10) $$d^2 = 2 - 2\cos(A - B).$$

Finally we compare (10) and (8). This yields

(11) $$2 - 2\cos(A - B) = 2 - 2(\cos A \cos B + \sin A \sin B).$$

If we subtract 2 from both sides and then multiply by $-1/2$, we obtain (6). Thus, (6) is true whenever $0 < B < A < 2\pi$.

The trusting student may assume that the argument can be shown to be valid for arbitrary A and B. The more conscientious student should check the following argument.

I. It is easy to show that (6) is still true if $B = 0$, or $A = B$, or $A = 2\pi$. Thus (6) is true if $0 \leq B \leq A \leq 2\pi$.

II. Suppose that $0 \leq A \leq B \leq 2\pi$. We can apply (6) to $\cos(B - A)$. Since cosine is an even function (see Theorem 6 in Chapter 7) we have

$$\cos(A-B) = \cos[(-1)(A-B)] = \cos(-A+B) = \cos(B-A)$$
$$= \cos B \cos A + \sin B \sin A \qquad \text{[by (6)]}$$
$$= \cos A \cos B + \sin A \sin B.$$

Hence, (6) is true for all A, B in the interval $[0, 2\pi]$.

III. Suppose that A and B are arbitrary real numbers. Let $C = A - 2m\pi$ and let $D = B - 2n\pi$, where m and n are integers (which may be negative) which are selected‡ so that C and D are in $[0, 2\pi]$. By part **II** we can apply equation (6) to $\cos(C-D)$. We obtain

$$\cos(C-D) = \cos C \cos D + \sin C \sin D$$

(12) $\cos[(A-2m\pi) - (B-2n\pi)]$
$$= \cos(A-2m\pi)\cos(B-2n\pi) + \sin(A-2m\pi)\sin(B-2n\pi).$$

We now use Theorem 3 in Chapter 7 (page 157) on coterminal angles. Since $A - 2m\pi$ is coterminal with A, and $B - 2n\pi$ is coterminal with B, and since

$$(A-2m\pi) - (B-2n\pi) = A - B - 2(m-n)\pi$$

is coterminal with $A - B$, equation (12) gives

(6) $$\cos(A-B) = \cos A \cos B + \sin A \sin B. \quad \blacksquare$$

Theorem 3   *For all real A and B*

(13) $$\cos(A+B) = \cos A \cos B - \sin A \sin B.$$

Proof: We can always write that $B = -(-B)$ and apply Theorem 2 to $\cos[A-(-B)]$. By Theorem 6 in Chapter 7, cosine is an even function and sine is an odd function. Hence

$$\cos(A+B) = \cos[A-(-B)]$$
$$= \cos A \cos(-B) + \sin A \sin(-B)$$
$$= \cos A \cos B + \sin A (-\sin B)$$
$$= \cos A \cos B - \sin A \sin B. \quad \blacksquare$$

Formulas (6) and (13) are called *the addition formulas* for the cosine.

‡ For example, if $A = 34.75\pi$ and $B = 49.23\pi$, then we set $m = 17$ and $n = 24$. Then $C = 34.75\pi - 34\pi = 0.75\pi$ and $D = 49.23\pi - 48\pi = 1.23\pi$. Both C and D lie in the interval $[0, 2\pi]$.

Theorem 4   *For all real θ*

(14)
$$\cos\left(\frac{\pi}{2} - \theta\right) = \sin\theta$$

and

(15)
$$\sin\left(\frac{\pi}{2} - \theta\right) = \cos\theta.$$

Of course if θ is the degree measure of an angle, these formulas are replaced by

(14⋆)
$$\cos(90° - \theta) = \sin\theta,$$

(15⋆)
$$\sin(90° - \theta) = \cos\theta.$$

Proof: In equation (6) we set $A = \frac{\pi}{2}$. We obtain

$$\cos\left(\frac{\pi}{2} - B\right) = \cos\frac{\pi}{2}\cos B + \sin\frac{\pi}{2}\sin B$$
$$= 0 \cdot \cos B + 1 \cdot \sin B = \sin B$$

(16)
$$\cos\left(\frac{\pi}{2} - B\right) = \sin B.$$

When B is replaced by θ, equation (16) gives (14). Next in (16) set $\pi/2 - B = \theta$ so that $B = \pi/2 - \theta$. Then (16) gives (15).   ∎

Theorem 5
PLE   *For all real θ*

(17)
$$\tan\left(\frac{\pi}{2} - \theta\right) = \cot\theta, \qquad \cot\left(\frac{\pi}{2} - \theta\right) = \tan\theta,$$

(18)
$$\sec\left(\frac{\pi}{2} - \theta\right) = \csc\theta, \qquad \csc\left(\frac{\pi}{2} - \theta\right) = \sec\theta.$$

Theorems 4 and 5 can be summarized as "A trigonometric function of θ is the cofunction of the complement of θ."

EXAMPLE 1   Compute $\cos 15°$ using the angles $45°$ and $30°$.

Solution: By Theorem 2

$$\cos 15° = \cos(45° - 30°) = \cos 45° \cos 30° + \sin 45° \sin 30°$$

$$= \frac{\sqrt{2}}{2}\frac{\sqrt{3}}{2} + \frac{\sqrt{2}}{2}\frac{1}{2} = \frac{\sqrt{6} + \sqrt{2}}{4} \approx \frac{2.4495 + 1.4142}{4} \approx 0.9659.$$

This coincides with the value given in Table B.   ▲

EXAMPLE 2

Prove that $\cos(-20°) = \sin 110°$.

Solution: In equation (15★) of Theorem 4, set $\theta = -20°$. This gives

$$\cos(-20°) = \sin[90° - (-20°)] = \sin(90° + 20°) = \sin 110°. \quad \blacktriangle$$

EXAMPLE 3

Simplify the expression $E = \cos\dfrac{3\theta}{5}\cos\dfrac{\theta}{3} - \sin\dfrac{3\theta}{5}\sin\dfrac{\theta}{3}$.

Solution: We recognize that this expression has the same form as the right side of (13) with $A = 3\theta/5$ and $B = \theta/3$. Whence, E is the cosine of the sum of these angles; that is, $E = \cos(14\theta/15)$. $\quad \blacktriangle$

EXAMPLE 4

Given that A and B are acute angles with $\tan A = 3/5$ and $\cos B = 4/7$, compute $\cos(A+B)$. Is $A+B$ in Q.I or Q.II?

Solution: Here it is convenient to use the alternate definition of the trigonometric functions given in Chapter 7, Exercise 2, problem 6. With the data we can construct angles A and B in standard position as indicated in Figure 9.5. From the figure we can find the trigonometric functions that we need.

FIGURE 9.5

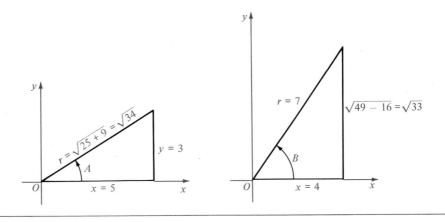

Thus,

$$\cos A = \frac{5}{\sqrt{34}}, \qquad \sin A = \frac{3}{\sqrt{34}},$$

$$\cos B = \frac{4}{7}, \qquad \sin B = \frac{\sqrt{33}}{7},$$

$$\cos(A+B) = \cos A \cos B - \sin A \sin B = \frac{5}{\sqrt{34}}\frac{4}{7} - \frac{3}{\sqrt{34}}\frac{\sqrt{33}}{7} = \frac{20 - 3\sqrt{33}}{7\sqrt{34}}.$$

Since A and B are acute, $A+B$ must lie either in Q.I or Q.II. To determine the quadrant, it is sufficient to find the sign of $\cos(A+B)$. Now $20^2 = 400$ and $(3\sqrt{33})^2 = 9 \times 33 = 297$. Since $400 > 297$, then $20 > 3\sqrt{33}$. Therefore $\cos(A+B) > 0$. Hence $A+B$ is in Q.I. ▲

Exercise 2

In problems 1 through 8 simplify each of the expressions by reducing it to a single term. Wherever possible, obtain a numerical value without using the tables.

1. $\cos 15° \cos 5° + \sin 15° \sin 5°$

2. $\cos 1500° \cos 500° + \sin 1500° \sin 500°$

3. $\cos 950° \cos 550° - \sin 950° \sin 550°$

4. $\cos 102° \cos 48° - \sin 48° \sin 102°$

5. $\cos 25° \cos 185° - \sin 25° \sin 185°$

6. $\cos 140° \cos 5° + \sin 140° \sin 5°$

7. $\cos 3A \cos 2A - \sin 3A \sin 2A$

8. $\sin 4A \sin 6A + \cos 6A \cos 4A$

9. Find $\cos 165°$ using the angles $135°$ and $30°$

10. Find $\cos 195°$ using the angles $150°$ and $45°$

11. Find $\cos 105°$ using the angles $315°$ and $210°$

12. Find $\cos 165°$ using the angles $300°$ and $135°$

In problems 13 through 24 prove that the given equations are identities.

13. $\cos(180° + \theta) = -\cos\theta$

14. $\cos(180° - \theta) = -\cos\theta$

15. $\cos(30° + A) + \cos(150° - A) = 0$

16. $\cos(60° + B) - \cos(300° - B) = 0$

17. $\dfrac{\cos 5B}{\sin 2B} - \dfrac{\sin 5B}{\cos 2B} = \cos 7B \sec 2B \csc 2B$

18. $\dfrac{\cos 4B}{\sin 3B} + \dfrac{\sin 4B}{\cos 3B} = \sec 3B \cos B \csc 3B$

19. $\cos(C+D)\cos D + \sin(C+D)\sin D = \cos C$

20. $\cos(3F+4G)\cos(2F+5G) + \sin(3F+4G)\sin(2F+5G) = \cos(F-G)$

21. $-2\cos(135° - H) = \sqrt{2}(\cos H - \sin H)$

22. $\cos(A+B) + \cos(A-B) = 2\cos A \cos B$

23. $\cos(A+B) - \cos(A-B) = -2\sin A \sin B$

24. $\dfrac{\cos(A+B)}{\cos(A-B)} = \dfrac{1-\tan A \tan B}{1+\tan A \tan B}$

25. Complete the proof of Theorem 2 by proving part **I** on page 202.

26. Prove Theorem 5.

27. If A and B are complementary angles, prove that $\cos(7A+5B) = -\sin 2B$.

28. If A and B are complementary angles, prove that $\cos(5A-2B) = \sin 7B$.

29. If A and B are acute angles with $\sin A = 2/3$ and $\sin B = 3/4$, find $\cos(A+B)$. Is $A+B$ in Q.I or is it in Q.II? Do not use tables.

30. If A and B are Q.I angles and if $\sin A = 2/5$ and $\sin B = 10/11$, find $\cos(A+B)$. Is $A+B$ in Q.I or in Q.II? Do not use tables.

31. If A and B are Q.III angles and if $\tan A = 2/3$ and if $\tan B = 4/3$, find $\cos(A+B)$ and determine the quadrant for the angle $A+B$.

★ 32. Let n be a positive integer and suppose that A and B are acute angles with $\tan A = (n-1)/n$ and $\tan B = (n+1)/n$. Find $\cos(A+B)$ and determine the quadrant for the angle $A+B$.

★ 33. If C and D are acute angles with $\tan C = 3\sqrt{3}$ and $\tan D = \sqrt{3}/2$, prove that $C+D = 120°$.

★ 34. Given that E, F, and G are acute angles with $\tan E = 2/3$, $\tan F = 3/4$, and $\tan G = 11/23$, prove that $E+F = G+45°$.

35. Is Theorem 4 of this chapter already contained in Theorem 5 of Chapter 7 (page 166)? If so, it was unnecessary to prove Theorem 4. Explain.

9.3 THE ADDITION FORMULAS FOR THE SINE AND TANGENT

These formulas are given in

Theorem 6   *For all real A and B*

(19) $$\sin(A+B) = \sin A \cos B + \cos A \sin B,$$

(20) $$\sin(A-B) = \sin A \cos B - \cos A \sin B,$$

(21) $$\tan(A+B) = \frac{\tan A + \tan B}{1 - \tan A \tan B},$$

$$(22) \qquad \tan(A-B) = \frac{\tan A - \tan B}{1 + \tan A \tan B},$$

There are similar formulas for the reciprocal functions: cotangent, secant, and cosecant, but they are not very useful and can be ignored.

Proof: By Theorem 4, equation (14)

$$(23) \qquad \sin(A+B) = \cos\left[\frac{\pi}{2} - (A+B)\right] = \cos\left[\left(\frac{\pi}{2} - A\right) - B\right].$$

If we apply Theorem 2 to the right side of (23), we obtain

$$\sin(A+B) = \cos\left(\frac{\pi}{2} - A\right)\cos B + \sin\left(\frac{\pi}{2} - A\right)\sin B$$

$$(19) \qquad \sin(A+B) = \sin A \cos B + \cos A \sin B.$$

For equation (20) we can use (19). We find that

$$\sin(A-B) = \sin[A + (-B)] = \sin A \cos(-B) + \cos A \sin(-B)$$
$$= \sin A \cos B + \cos A(-\sin B)$$

$$(20) \quad \sin(A-B) = \sin A \cos B - \cos A \sin B.$$

From equations (19) and (20) we have

$$\tan(A+B) = \frac{\sin(A+B)}{\cos(A+B)} = \frac{\sin A \cos B + \cos A \sin B}{\cos A \cos B - \sin A \sin B}.$$

Next, we divide both the numerator and denominator of this fraction by $\cos A \cos B$ in order to convert to tangent functions. This gives

$$\tan(A+B) = \frac{\dfrac{\sin A \cos B + \cos A \sin B}{\cos A \cos B}}{\dfrac{\cos A \cos B - \sin A \sin B}{\cos A \cos B}}$$

$$(21) \quad \tan(A+B) = \frac{\dfrac{\sin A \cos B}{\cos A \cos B} + \dfrac{\cos A \sin B}{\cos A \cos B}}{\dfrac{\cos A \cos B}{\cos A \cos B} - \dfrac{\sin A \sin B}{\cos A \cos B}} = \frac{\tan A + \tan B}{1 - \tan A \tan B}.$$

The technique used in going from (19) to (20) can be used to derive (22) from (21). ∎

EXAMPLE 1

Compute $\tan 75°$.

Solution: $\tan 75° = \tan(30° + 45°) = \dfrac{\tan 30° + \tan 45°}{1 - \tan 30° \tan 45°}$

$$= \frac{\sqrt{3}/3 + 1}{1 - 1\sqrt{3}/3} = \frac{\sqrt{3} + 3}{3 - \sqrt{3}} = \frac{\sqrt{3} + 3}{3 - \sqrt{3}}\frac{3 + \sqrt{3}}{3 + \sqrt{3}}$$

$$= \frac{9 + 6\sqrt{3} + 3}{9 - 3} = 2 + \sqrt{3} \approx 3.732. \quad \blacktriangle$$

EXAMPLE 2

Simplify the expression $\dfrac{\sin(\alpha + \beta) - \sin(\alpha - \beta)}{\sin(\alpha + \beta) + \sin(\alpha - \beta)}$.

Solution: For the numerator N we have

$$N = \sin\alpha\cos\beta + \cos\alpha\sin\beta - (\sin\alpha\cos\beta - \cos\alpha\sin\beta)$$

$$= 2\cos\alpha\sin\beta.$$

For the denominator D we add the terms in the parentheses rather than subtract. This gives $D = 2\sin\alpha\cos\beta$. Therefore,

$$\frac{\sin(\alpha + \beta) - \sin(\alpha - \beta)}{\sin(\alpha + \beta) + \sin(\alpha - \beta)} = \frac{N}{D} = \frac{2\cos\alpha\sin\beta}{2\sin\alpha\cos\beta} = \cot\alpha\tan\beta. \quad \blacktriangle$$

Exercise 3

In problems 1 through 8 simplify each of the expressions by reducing it to a single term. Wherever possible, obtain a numerical value without using the tables.

1. $\sin 335°\cos 305° - \cos 335°\sin 305°$

2. $\sin 215°\cos 170° - \cos 215°\sin 170°$

3. $\sin 535°\cos 545° + \cos 535°\sin 545°$

4. $\sin 405°\cos 505° + \sin 505°\cos 405°$

5. $\dfrac{\tan 62° - \tan 17°}{1 + \tan 62°\tan 17°}$

6. $\dfrac{\tan 435° - \tan 300°}{1 + \tan 435°\tan 300°}$

7. $\dfrac{\tan(C + D) + \tan(C - D)}{1 - \tan(C + D)\tan(C - D)}$

8. $\dfrac{\tan(2E + F) + \tan 2(F - E)}{1 - \tan(2E + F)\tan 2(F - E)}$

9. Find $\sin 105°$ and $\tan 105°$, using the angles $45°$ and $60°$.

10. Find $\sin 15°$ and $\tan 15°$ using the angles $45°$ and $60°$.

11. Find $\sin 75°$ and $\tan 75°$ using the angles $315°$ and $240°$.

12. Find $\sin 15°$ and $\tan 15°$ using the angles $150°$ and $225°$.

13. Find $\tan(A+B)$ and $\tan(A-B)$ if $\tan A = 3/7$ and $\tan B = 4/7$.

14. Find $\tan(A+B)$ and $\tan(A-B)$ if $\tan A = 2$ and $\tan B = 3$.

15. Find $\sin(A+B)$ and $\sin(A-B)$ for the angles of problem 13, assuming that A and B are acute angles.

16. Find $\sin(A+B)$ and $\sin(A-B)$ for the angles of problem 14, assuming that A and B are acute angles.

★17. Given that $\sin 1° = 0.017452$ and $\cos 1° = 0.999848$, compute $\sin 2°$ and $\cos 2°$ to six decimal places.

★18. Using the data and results of problem 17, compute $\sin 3°$ and $\cos 3°$ to six decimal places.

Prove the following identities.

19. $\sin(90°+\alpha) = \cos\alpha$

20. $\sin(270°+\beta) = -\cos\beta$

21. $\tan(45°+\gamma) = \dfrac{1+\tan\gamma}{1-\tan\gamma}$

22. $\tan(135°-\theta) = -\dfrac{1+\tan\theta}{1-\tan\theta}$

23. $\sin(\alpha+30°) + \sin(\alpha-30°) = \sqrt{3}\sin\alpha$

24. $\sin(\alpha+45°) - \sin(\alpha-135°) = \sqrt{2}(\sin\alpha+\cos\alpha)$

25. $\dfrac{\cos(C+D) - \cos(C-D)}{\cos(C+D) + \cos(C-D)} = -\tan C\tan D$

26. $\dfrac{\sin 3A}{\sin 2A} + \dfrac{\cos 3A}{\cos 2A} = \sin 5A \sec 2A \csc 2A$

27. $\dfrac{\sin 3B}{\sin 2B} - \dfrac{\cos 3B}{\cos 2B} = \sec 2B \sin B \csc 2B$

★28. $\cos(A+B+C) = \cos A \cos B \cos C - \sin A \sin B \cos C$
$\qquad - \sin A \cos B \sin C - \cos A \sin B \sin C$

★29. $\sin(A+B+C) = \sin A \cos B \cos C + \cos A \sin B \cos C$
$\qquad + \cos A \cos B \sin C - \sin A \sin B \sin C$

★30. $\tan(A+B+C) = \dfrac{\tan A + \tan B + \tan C - \tan A \tan B \tan C}{1 - \tan A \tan B - \tan B \tan C - \tan C \tan A}$

9.4 THE DOUBLE-ANGLE FORMULAS

These formulas are given in

Theorem 7   *For any real number* A

(24) $\sin 2A = 2\sin A \cos A,$

(25) $\cos 2A = \cos^2 A - \sin^2 A,$

(26) $\tan 2A = \dfrac{2\tan A}{1 - \tan^2 A}.$

Each of these formulas is proved by setting $B = A$ in the appropriate addition formula. If we set $B = A$ in (19), we find

$$\sin(A + A) = \sin A \cos A + \cos A \sin A$$

(24) $\sin 2A = \sin A \cos A + \sin A \cos A = 2\sin A \cos A.$

Similarly, if we set $B = A$ in (13) we deduce (25). If we set $B = A$ in (21), we obtain (26). ∎

EXAMPLE 1   Prove the identity $\cos 2A = 1 - 2\sin^2 A$.

Solution: Since $\cos^2 A = 1 - \sin^2 A$, we can use this in (25). Indeed

$$\cos 2A = \cos^2 A - \sin^2 A = (1 - \sin^2 A) - \sin^2 A = 1 - 2\sin^2 A. \quad \blacktriangle$$

EXAMPLE 2   Given that $\sin 15° = (\sqrt{6} - \sqrt{2})/4$ and $\cos 15° = (\sqrt{6} + \sqrt{2})/4$, compute $\sin 30°$.

Solution: From equation (24)

$$\sin 30° = 2\sin 15° \cos 15° = 2\frac{\sqrt{6} - \sqrt{2}}{4}\frac{\sqrt{6} + \sqrt{2}}{4}$$

$$= 2\frac{\left(\sqrt{6}\right)^2 - \left(\sqrt{2}\right)^2}{16} = \frac{6 - 2}{8} = \frac{1}{2}. \quad \blacktriangle$$

EXAMPLE 3   Express $\sin 3A$ in terms of the trigonometric functions of A.

Solution: From Theorem 6, equation (19)

$$\sin 3A = \sin(2A + A) = \sin 2A \cos A + \cos 2A \sin A$$

$$= 2\sin A \cos A \cos A + (\cos^2 A - \sin^2 A)\sin A$$

$$= 3\sin A \cos^2 A - \sin^3 A. \quad \blacktriangle$$

9.5 THE HALF-ANGLE FORMULAS

These formulas are given in

Theorem 8   *For any real number θ*

(27)
$$\sin\frac{\theta}{2} = \pm\sqrt{\frac{1-\cos\theta}{2}},$$

(28)
$$\cos\frac{\theta}{2} = \pm\sqrt{\frac{1+\cos\theta}{2}},$$

(29)
$$\tan\frac{\theta}{2} = \frac{\sin\theta}{1+\cos\theta}.$$

Proof: If we add the two identities

(30) $$\cos^2 A + \sin^2 A = 1$$

(25) $$\cos^2 A - \sin^2 A = \cos 2A$$

we obtain

(31) $$2\cos^2 A \qquad = 1 + \cos 2A.$$

If we subtract (25) from (30), we obtain

(32) $$2\sin^2 A = 1 - \cos 2A.$$

If in each of these last two equations we divide by 2 and take the square root on both sides, we find from (31):

(33)
$$\cos A = \pm\sqrt{\frac{1+\cos 2A}{2}},$$

and from (32)

(34)
$$\sin A = \pm\sqrt{\frac{1-\cos 2A}{2}}.$$

Finally if we set $2A = \theta$ and, hence, $A = \theta/2$ in (33) and (34), we obtain (28) and (27), respectively.

To prove formula (29) notice that

$$\frac{\sin 2A}{1+\cos 2A} = \frac{2\sin A \cos A}{\sin^2 A + \cos^2 A + \cos^2 A - \sin^2 A}$$

$$= \frac{2\sin A \cos A}{2\cos^2 A} = \tan A.$$

The same substitutions, $2A = \theta$ and $A = \theta/2$, will transform this identity into (29). ∎

The $\pm$ sign in (27) is unavoidable, because the radical sign always means the positive (or zero) square root, while $\sin(\theta/2)$ may be negative. A similar remark applies to equation (28). By taking the quotient of (27) and (28) we find

(35)
$$\tan\frac{\theta}{2} = \pm\sqrt{\frac{1-\cos\theta}{1+\cos\theta}}.$$

However this formula is not as nice as (29) because in (29) the nuisance $\pm$ sign does not appear.

EXAMPLE 1

(a) Compute $\cos 120°$ from $\cos 240°$. (b) Compute $\cos 60°$ from $\cos 120°$.

Solution: (a) We assume as known that $\cos 240° = -1/2$. Then from equation (28) we find

$$\cos 120° = \pm\sqrt{\frac{1+\cos 240°}{2}} = \pm\sqrt{\frac{1-1/2}{2}} = \pm\sqrt{\frac{1}{4}} = \pm\frac{1}{2}.$$

But we select the answer $-1/2$ because in Q.II the cosine is negative.

(b) We assume that $\cos 120° = -1/2$. Then from (28)

$$\cos 60° = \pm\sqrt{\frac{1+\cos 120°}{2}} = \pm\sqrt{\frac{1-1/2}{2}} = \pm\sqrt{\frac{1}{4}} = \pm\frac{1}{2}.$$

But we select the answer $+1/2$ because in Q.I the cosine is positive. ▲

Observe that in (a) and (b) the computation is exactly the same. Hence, we can not determine the sign in (28) unless we know the quadrant for $\theta/2$

EXAMPLE 2

Compute $\cos 7° 30'$

Solution: By Example 1 of Section 9.2, we have $\cos 15° = (\sqrt{6}+\sqrt{2})/4$ Hence by (28)

$$\cos 7° 30' = \sqrt{\frac{1+(\sqrt{6}+\sqrt{2})/4}{2}} = \frac{1}{2}\sqrt{\frac{4+\sqrt{6}+\sqrt{2}}{2}} \approx 0.9914. \ ▲$$

EXAMPLE 3

Simplify the expression $\sqrt{1-\cos 6A}$.

Solution: We can write

$$\sqrt{1-\cos 6A} = \sqrt{2}\sqrt{\frac{1-\cos 6A}{2}} = \sqrt{2}\sin 3A,$$

if both sides are positive (or zero). However the radical on the left side is never negative (by definition), so the right side must be adjusted in an appropriate way. This can be done using the absolute value sign. Indeed, for all angles A, we have $\sqrt{1-\cos 6A} = \sqrt{2}|\sin 3A|$. ▲

Exercise 4

In problems 1 through 8 simplify each expression.

1. $\dfrac{2\tan 20°}{1-\tan^2 20°}$

2. $\cos^2 5° - \sin^2 5°$

3. $2\sin 76° \cos 76°$

4. $\sqrt{\dfrac{1-\cos 426°}{2}}$

5. $\sqrt{\dfrac{1+\cos 200°}{2}}$

6. $\dfrac{\sin 284°}{1+\cos 284°}$

7. $\dfrac{\sin A \cos A}{\cos^2 A - \sin^2 A}$

8. $\dfrac{\tan 8B}{1-\tan^2 8B}$

9. Find $\sin 15°$ and $\cos 15°$ using $\cos 30° = \sqrt{3}/2$.

10. If A is an acute angle and $\cos A = 3/5$, find $\sin(A/2)$ and $\cos(A/2)$.

11. If B is an acute angle and $\cos B = 4/5$, find $\sin(B/2)$ and $\cos(B/2)$.

12. Find $\sin 2A$ and $\cos 2A$ for the angle A of problem 10.

13. Find $\sin 2B$ and $\cos 2B$ for the angle B of problem 11.

14. Given $\tan(A/2) = 1/3$, find $\sin A$ and $\cos A$.

15. If $180° < A < 270°$ and $\tan A = 3/4$, find $\sin(A/2)$ and $\cos(A/2)$.

16. If $270° < B < 360°$ and $\cos B = 5/13$, find $\sin(B/2)$ and $\cos(B/2)$.

★ 17. If $n > 1$, $\cos C = 2n/(1+n^2)$, and C is an acute angle, find $\sin(C/2)$ and $\cos(C/2)$.

★ 18. Find $\sin 2C$ and $\cos 2C$ for the angle C of problem 17.

★ 19. Prove that if $0 < A < 90°$, then $\sin 2A < 2\sin A$.

★ 20. Prove that if $0 < B < 45°$, then $\tan 2B > 2\tan B$. What happens to this inequality if $45° < B < 90°$?

21. Express $\cos 3C$ in terms of $\cos C$.

22. Express $\tan 4D$ in terms of $\tan D$.

In problems 23 through 36 prove the given identities.

23. $2\cos^2 E - \cos 2E = 1$

24. $\cos^2 2F - 4\sin^2 F \cos^2 F = \cos 4F$

25. $\dfrac{1 - 2\sin^2 G}{\sin G \cos G} = 2\cot 2G$

26. $\dfrac{\sin 4H}{2\sin 2H} = \cos^4 H - \sin^4 H$

27. $\dfrac{\cos \alpha}{\cos \alpha - \sin \alpha} - \dfrac{\sin \alpha}{\cos \alpha + \sin \alpha} = \sec 2\alpha$

28. $\dfrac{\cos \beta}{\cos \beta - \sin \beta} + \dfrac{\sin \beta}{\cos \beta + \sin \beta} = 1 + \tan 2\beta$

29. $\dfrac{1 - \tan \gamma}{1 + \tan \gamma} = \dfrac{1 - \sin 2\gamma}{\cos 2\gamma}$

30. $\dfrac{1 - \tan^2 \theta}{1 + \tan^2 \theta} = \cos 2\theta$

31. $\dfrac{1 + \sin 2A + \cos 2A}{1 + \sin 2A - \cos 2A} = \cot A$

32. $\dfrac{\sin B + \sin 2B}{1 + \cos B + \cos 2B} = \tan B$

33. $\dfrac{\tan(C/2) + \cot(C/2)}{\cot(C/2) - \tan(C/2)} = \sec C$

34. $\dfrac{2}{\tan(D/2) + \cot(D/2)} = \sin D$

35. $\sin \theta = \dfrac{2\tan(\theta/2)}{1 + \tan^2(\theta/2)}$

36. $\cos \theta = \dfrac{1 - \tan^2(\theta/2)}{1 + \tan^2(\theta/2)}$

★37. Prove that $\sin \theta$ and $\cos \theta$ are both rational if and only if $\tan(\theta/2)$ is either rational or does not exist.

★9.6 **PRODUCT TO SUM FORMULAS; SUM TO PRODUCT FORMULAS**

We compare the two addition formulas for the sine function, namely

(19) $$\sin(A + B) = \sin A \cos B + \cos A \sin B$$

and

(20) $$\sin(A - B) = \sin A \cos B - \cos A \sin B.$$

We observe that the right sides of these two equations are the same except for the connecting sign. This suggests that adding the two equations or taking their difference should lead to something interesting. Indeed, if we add these two equations, we find

(36) $$\sin(A + B) + \sin(A - B) = 2\sin A \cos B.$$

If we subtract equation (20) from (19) we find

(37) $$\sin(A+B) - \sin(A-B) = 2\cos A \sin B.$$

If now we divide each of the equations (36) and (37) by 2 and then put the right side first, we have the formulas of

Theorem 9   *For all real numbers A and B*

(38) $$\sin A \cos B = \frac{1}{2}[\sin(A+B) + \sin(A-B)],$$

and

(39) $$\cos A \sin B = \frac{1}{2}[\sin(A+B) - \sin(A-B)].$$

This theorem is important because it gives a formula for replacing a product of two trigonometric functions by a sum of two trigonometric functions.

EXAMPLE 1   Express $\sin 5\theta \cos 3\theta$ as a sum of trigonometric functions.

Solution: By equation (38)

$$\sin 5\theta \cos 3\theta = \frac{1}{2}[\sin(5\theta + 3\theta) + \sin(5\theta - 3\theta)]$$

$$= \frac{1}{2}[\sin 8\theta + \sin 2\theta]. \quad \blacktriangle$$

A similar manipulation can be performed with the addition formulas for the cosine function:

(6) $$\cos(A-B) = \cos A \cos B + \sin A \sin B$$

and

(13) $$\cos(A+B) = \cos A \cos B - \sin A \sin B.$$

If we add these two equations, we find

(40) $$\cos(A+B) + \cos(A-B) = 2\cos A \cos B.$$

If we subtract equation (13) from (6), we find

(41) $$\cos(A-B) - \cos(A+B) = 2\sin A \sin B.$$

Again we divide each of these equations by 2 and put the right side first. We then have

Theorem 10   *For all real numbers A and B*

(42)
$$\cos A \cos B = \frac{1}{2}[\cos(A+B) + \cos(A-B)],$$

and

(43)
$$\sin A \sin B = -\frac{1}{2}[\cos(A+B) - \cos(A-B)].$$

This theorem gives two more formulas for replacing a product of two trigonometric functions by a sum of two trigonometric functions.

If we want formulas that replace a sum (or difference) of two trigonometric functions by a product, it is sufficient to take (36), (37), (40), and (41). However, for convenience, we would like to replace the sum $A+B$ by C and the difference $A-B$ by D. Suppose we do this. If we add the two equations

$$A + B = C,$$
$$A - B = D,$$

we obtain
$$2A = C + D.$$
Subtraction gives
$$2B = C - D.$$

Hence $A = (C+D)/2$ and $B = (C-D)/2$. Making these substitutions in (36), (37), (40), and (41) we find

Theorem 11   *For all real numbers C and D*

(44)
$$\sin C + \sin D = 2 \sin \frac{C+D}{2} \cos \frac{C-D}{2},$$

(45)
$$\sin C - \sin D = 2 \cos \frac{C+D}{2} \sin \frac{C-D}{2},$$

(46)
$$\cos C + \cos D = 2 \cos \frac{C+D}{2} \cos \frac{C-D}{2},$$

(47)
$$\cos C - \cos D = -2 \sin \frac{C+D}{2} \sin \frac{C-D}{2}.$$

These formulas can be used to convert sums of two trigonometric functions into products of two trigonometric functions.

EXAMPLE 2   Convert $\cos 5\alpha - \cos 3\alpha$ into a product of trigonometric functions.

Solution: By (47) with $C = 5\alpha$ and $D = 3\alpha$,

$$\cos 5\alpha - \cos 3\alpha = -2\sin\frac{5\alpha + 3\alpha}{2}\sin\frac{5\alpha - 3\alpha}{2} = -2\sin 4\alpha \sin\alpha. \quad \blacktriangle$$

EXAMPLE 3   Simplify the expression $(\cos 7\beta + \cos\beta)/(\sin 7\beta + \sin\beta)$.

Solution: We apply (46) to the numerator and (43) to the denominator. This gives

$$\frac{\cos 7\beta + \cos\beta}{\sin 7\beta + \sin\beta} = \frac{2\cos\dfrac{7\beta + \beta}{2}\cos\dfrac{7\beta - \beta}{2}}{2\sin\dfrac{7\beta + \beta}{2}\cos\dfrac{7\beta - \beta}{2}} = \frac{2\cos 4\beta\cos 3\beta}{2\sin 4\beta\cos 3\beta} = \cot 4\beta. \quad \blacktriangle$$

Exercise 5

Convert the following products into sums of trigonometric functions.

1. $\sin 15°\cos 5°$
2. $\cos 44°\cos 20°$
3. $\cos A\cos 3A$
4. $\cos 3B\sin B$
5. $\cos 2C\cos 3C\cos 4C$
6. $\cos 2D\sin 3D\cos 4D$

Convert the following expressions into products of trigonometric functions.

7. $\sin\theta + \sin 3\theta$
8. $\cos\theta + \cos 5\theta$
9. $\cos 2\theta + \cos 6\theta$
10. $\sin 3\theta - \sin 9\theta$

In problems 11 through 22 prove the given identities.

11. $\cos(\alpha + \beta)\cos(\alpha - \beta) = \cos^2\alpha - \sin^2\beta$

12. $\dfrac{\cos 3\gamma - \cos\gamma}{\sin 3\gamma - \sin\gamma} = -\tan 2\gamma$

13. $\dfrac{\sin 5A + \sin A}{\cos 5A + \cos A} = \tan 3A$

14. $\dfrac{\sin 6A + \sin 4A}{\sin 4A + \sin 2A} = \cos 2A + \sin 2A\cot 3A$

15. $\sin B + \sin 2B + \sin 3B = \sin 2B(1 + 2\cos B)$

16. $\cos 2C + \cos 4C + \cos 6C = \cos 4C(1 + 2\cos 2C)$

★ 17. $\sin D - \sin 2D + \sin 3D - \sin 4D = -4\sin\dfrac{D}{2}\cos D\cos\dfrac{5D}{2}$

★ 18. $\cos E - \cos 2E + \cos 3E - \cos 4E = 4\sin\dfrac{E}{2}\cos E\sin\dfrac{5E}{2}$

19. $\dfrac{\sin\alpha + \sin 2\alpha + \sin 3\alpha}{\cos\alpha + \cos 2\alpha + \cos 3\alpha} = \tan 2\alpha$

20. $\dfrac{\sin\alpha - \sin 2\alpha + \sin 3\alpha}{\cos\alpha - \cos 2\alpha + \cos 3\alpha} = \tan 2\alpha$

21. $\dfrac{\sin(A+B)\cos(A-B) + \cos(A+B)\sin(A-B)}{\cos(A+B)\cos(A-B) - \sin(A+B)\sin(A-B)} = \tan 2A$

★ 22. $\dfrac{\sin^2(C+D) + \sin^2(C-D)}{2\cos^2 C\cos^2 D} = \tan^2 C + \tan^2 D$

★★ 23. If $\cos(A+B) = 1/3$ and $\cos(A-B) = 1/2$, find $\cos A$ and $\cos B$. Assume that $0° < A < B < 90°$.

★★ 24. If $\cos(A+B) = 1/5$ and $\cos(A-B) = 4/5$, find $\cos A$ and $\cos B$. Assume that $0° < A < B < 90°$.

★★ 25. A generalization of the two preceding problems. Let $\cos(A+B) = x$ and $\cos(A-B) = y$, where $0 < x < y < 1$. If $0° < A < B < 90°$, find $\cos A$ and $\cos B$ in terms of x and y. Also find $\sin A$ and $\sin B$ in terms of x and y.

★★ 26. Check the answers in problem 25 by showing that $\sin^2 A + \cos^2 A = 1$ and $\sin^2 B + \cos^2 B = 1$ for the values obtained in problem 25.

**9.7 MORE ABOUT
TRIGONOMETRIC EQUATIONS**

Now that we have more information about trigonometric functions (addition formulas, sum to product formulas, etc.) we are in a position to solve trigonometric equations that were difficult or untouchable with our more primitive tools.

EXAMPLE 1

Solve the equation

(48) $$5\sin 2\theta = 3\cos\theta$$

for θ in the interval $0° \leqq \theta < 360°$.

Solution: If we use the double-angle formula, equation (24), on the left side of (48), we obtain

$$10 \sin \theta \cos \theta = 3 \cos \theta$$

$$10 \sin \theta \cos \theta - 3 \cos \theta = 0$$

$$\cos \theta (10 \sin \theta - 3) = 0.$$

Either $\cos \theta = 0$ or $10 \sin \theta - 3 = 0$

$$\theta = 90°, 270°$$

$$\sin \theta = \frac{3}{10} = 0.3000$$

$$\theta = 17° 30', 162° 30'.$$

Hence, the solutions are $\theta = 17° 30', 90°, 162° 30', 270°$. ▲

EXAMPLE 2   Solve the equation

(49) $$12 \sin \theta + 5 \cos \theta = 8$$

for θ in the interval $0° \leq \theta \leq 360°$.

Solution: We will use equation (19) in the form

(50) $$\sin(\theta + \alpha) = \sin \theta \cos \alpha + \cos \theta \sin \alpha.$$

Our objective is to identify the left side of (49) with the right side of (50) by selecting α appropriately We must have $\sin^2 \alpha + \cos^2 \alpha = 1$, and to obtain this we divide both sides of (49) by $\sqrt{(12)^2 + 5^2} = 13$ Thus (49) is equivalent to

(51) $$\frac{12}{13} \sin \theta + \frac{5}{13} \cos \theta = \frac{8}{13}.$$

In selecting α a diagram may be helpful. As indicated in Figure 9.6, we select a point on the unit circle with coordinates $(12/13, 5/13)$ to obtain α. Then $\cos \alpha = 12/13$, $\sin \alpha = 5/13$, and equation (51) can be put in the form

$$\sin \theta \cos \alpha + \cos \theta \sin \alpha = \frac{8}{13}$$

$$\sin(\theta + \alpha) \approx 0.6154$$

(52) $$\theta + \alpha = 38° 0' + n \, 360°, \qquad 142° 0' + n \, 360°.$$

FIGURE 9.6

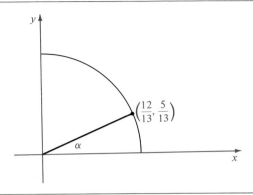

Since $\cos\alpha = 12/13 \approx 0.9231$, Table B gives $\alpha \approx 22°\,40'$. Subtracting α from both sides in (52) we have

$$\theta = 15°\,20' + n\,360°, \qquad 119°\,20' + n\,360°.$$

To obtain the solutions in the required interval set $n = 0$. ▲

EXAMPLE 3   Solve the equation

(53) $\sin 7\theta + \sin \theta = 0$

for θ in the interval $0° \leqq \theta < 360°$.

Solution: By equation (44) this sum can be converted into a product, and we obtain the equivalent equation

$$2\sin\frac{7\theta+\theta}{2}\cos\frac{7\theta-\theta}{2} = 0.$$

Either $\sin 4\theta = 0$ | or $\cos 3\theta = 0$

$4\theta = n\,180°$ | $3\theta = 90° + n\,180°$

$\theta = n\,45°$ | $\theta = 30° + n\,60°.$

Using $n = 0, 1, 2, 3, 4, 5, 6, 7$ in the first solution and $n = 0, 1, 2, 3, 4, 5$, in the second solution, gives those values of θ in the interval $0° \leqq \theta < 360°$. Hence $\theta = 0°, 30°, 45°, 90°, 135°, 150°, 180°, 210°, 225°, 270°, 315°, 330°.$ ▲

★**9.8   A GENERAL PRINCIPLE**

A fundamental theorem in algebra states that a polynomial equation of nth degree

$$a_0 x^n + a_1 x^{n-1} + a_2 x^{n-2} + \cdots + a_{n-1} x + a_n = 0, \qquad a_0 \neq 0,$$

cannot have more than n roots. For example, a quadratic equation has at most two roots. Is there a similar principle for trigonometric equations? There is, but it is somewhat more complicated.

For convenience we consider polynomials that involve only the sine and cosine functions. Naturally a term such as $\sin^3 \theta$ counts as a term of 3rd degree. But $\sin 3\theta$ will also be counted as a term of 3rd degree. To find the degree of the product of two terms, we add the degrees of each term. For example, $\sin 4\theta \cos 5\theta$ counts as a term of 9th degree. Then we have

Theorem 12
PWO

If n is the largest degree of the terms of a polynomial equation in $\sin \theta$ and $\cos \theta$, then this equation has at most $2n$ roots in the interval $0° \leqq \theta < 360°$.

The proof of this theorem is difficult and lies outside the scope of this book. However, we can observe this theorem in operation in our examples. Equation (49) is obviously of 1st degree, and there are two solutions in $0° \leqq \theta < 360°$. Equation (48) is of 2nd degree and has four roots in this interval. Finally equation (53) is of 7th degree, so it may have fourteen roots in this interval. Actually the final list contains only twelve entries, but this deficiency can be explained by pointing out that $\theta = 90°$ and $\theta = 270°$ are obtained from both factors, and, hence, each should be listed twice and counted as multiple roots.

Finally, we observe that an equation may not have any solutions. For example, the equation $\sin^2 \theta - 7 \sin \theta + 12 = 0$ is equivalent to the equation $(\sin \theta - 3)(\sin \theta - 4) = 0$, so that either $\sin \theta = 3$ or $\sin \theta = 4$. Since neither of these possibilities can actually occur for real θ, this equation has no real solutions. It is possible to give a satisfactory definition of the trigonometric functions for complex values of θ. When this is done this equation will have complex solutions. The details of this fascinating story must be reserved for a course in complex variables.

Exercise 6

In problems 1 through 18 find all the solutions in the interval $0° \leqq \theta < 360°$. Give answers to the nearest ten minutes. Check that the number of solutions satisfies the requirements of Theorem 12.

1. $\tan 2\theta = 2 \tan \theta$ 2. $\sin 2\theta = \sin \theta$

3. $\cos 2\theta = \sin \theta$ 4. $\sin 4\theta = \cos 2\theta$

5. $2 \sin 5\theta \cos 3\theta - 2 \cos 5\theta \sin 3\theta = \sqrt{3}$

6. $\cos 6\theta \cos 2\theta + \sin 6\theta \sin 2\theta = 1$

7. $\tan 4\theta - \tan 2\theta = 1 + \tan 2\theta \tan 4\theta$

8. $1 + \tan \theta + \tan 2\theta = \tan \theta \tan 2\theta$

9. $\tan(3\theta - 10°) = \cot(2\theta + 15°)$

10. $\sin \theta \csc(2\theta + 15°) = \cos \theta \sec(2\theta + 15°)$

11. $\cos 5\theta + \cos 3\theta = 0$ 12. $\sin 5\theta + \sin \theta = 0$

★13. $\sin \theta + \cos \theta = \sin \theta \tan \theta + \cos \theta \cot \theta$

★14. $\sin \theta + \cos \theta = \sqrt{3} \sin \theta \cos \theta$

★15. $\sin 2\theta + \cos 3\theta = \sin 4\theta$ ★16. $\cos \theta - \cos 5\theta = \sin 3\theta$

★★17. $\sin 6\theta \cos \theta = \sin 8\theta \cos 3\theta$ ★★18. $\tan \theta + \tan 2\theta = \tan 3\theta$

★19. In a certain right triangle the sum of the lengths of the two sides is 7/5 of the length of the hypotenuse. Find the angles of this triangle.

★20. The area of a right triangle is 1/8 of the square of the length of the hypotenuse. Find the angles of this triangle.

★21. In an isosceles triangle the sum of the two equal altitudes is 1/2 of the longer altitude. Find the base angles of this triangle.

★22. The sides of a right triangle form a geometric progression. Find the smallest angle of this triangle.

★★23. The sides of a right triangle form an arithmetic progression. Find the smallest angle of this triangle.

★24. A homogeneous straight wire AB, 34.0 inches long is bent to form a right angle at a point C, 14.0 inches from the end A. The wire is then supported at C by a horizontal pin. If the wire is free to rotate about the pin at C, find the angle that the arm CB makes with the horizontal plane when the wire is in equilibrium under the forces of gravity.

★25. A square is inscribed in a second square in such a way that the ratio of their areas is 8/13. If θ is the smaller of the two angles that one side of the square makes with a side of the other square, find θ.

Review Questions

Try to answer the following questions as accurately as possible before consulting the text.

1. State the formula for the distance between two points $P(x_1, y_1)$ and $Q(x_2, y_2)$.

2. Give the right side of each of the following formulas.

$\sin(A+B) =$

$\sin(A-B) =$

$\cos(A+B) =$

$\cos(A-B) =$

$\tan(A+B) =$

$\tan(A-B) =$

$\sin 2A =$

$\cos 2A =$

$\tan 2A =$

3. Give the right side of the following half-angle formulas.

$\sin \dfrac{\theta}{2} =$

$\cos \dfrac{\theta}{2} =$

$\tan \dfrac{\theta}{2} =$

4. Give the right side of the following product to sum formulas.

$\sin A \cos B =$

$\cos A \sin B =$

$\cos A \cos B =$

$\sin A \sin B =$

5. Give the right side of the following sum to product formulas.

$\sin C + \sin D =$

$\sin C - \sin D =$

$\cos C + \cos D =$

$\cos C - \cos D =$

6. What is the maximum number of roots that an nth degree trigonometric equation can have in the interval $[0, 360°)$?

7. What is the maximum number of roots that the equation

$$\cos 9\theta \cos \theta + \cos 7\theta \cos 3\theta + 2\cos^2 5\theta = 0$$

can have in the interval $[0, 360°)$?

Graphs and Inverse Functions

We are already familiar with the concept of the graph of a function. In this chapter we go a little deeper in our study of graphs. Further, we look carefully at the graph of the inverse of a function.

The inverse of a function is really a very simple concept, but it may seem confusing because the role of the independent and dependent variables are interchanged. Since x is traditionally used for the independent variable and y is traditionally used for the dependent variable, some confusion is almost certain to occur when the letters are interchanged. To avoid the difficulty we will use the pair (u, v) rather than (x, y) in Section 10.1, but we will return to the standard (x, y) in Section 10.2.

10.1 THE INVERSE OF A FUNCTION

Consider the function f defined by the formula

$$(1) \qquad v = f(u) = 2u + 7.$$

For each u in the domain of f, this formula gives a unique v.

Suppose that we start with a fixed v and search for the u that is the pre-image of v. We do not have to search long, because in this simple example the equation can be solved for u in terms of v. We find that

(2)
$$u = \frac{1}{2}(v-7) \equiv g(v).$$

This equation defines a new function g that is called the *inverse* of the function f. Thus, if u and v are any two real numbers such that

$$v = f(u),$$

then for this same pair of numbers

$$u = g(v).$$

If we wish to use the concept of ordered pairs for a function, then we say that the pair (u, v) belongs to the function f if and only if the pair (v, u) belongs to the inverse function g.

The particular example treated above is easy because the mapping defined by equation (1) is one-to-one from the domain of f to the range of f. Let us carry through the analysis for a more complicated example and note the difficulties that arise. Indeed, we now let f be the function defined by

(3)
$$v = u^2 - 2u + 3 \equiv f(u).$$

For each fixed v we can solve equation (3) for u, using the quadratic formula. We find that

$$0 = u^2 - 2u + (3-v)$$

$$u = \frac{-(-2) \pm \sqrt{4 - 4(3-v)}}{2} = 1 \pm \sqrt{1 - (3-v)}$$

(4)
$$u = 1 \pm \sqrt{v-2}.$$

We would like to announce that equation (4) defines a new function that is the inverse function of f defined by equation (3). There is one slight hitch. If (4) defines u as a function of v, then for each v in the domain of this function there must be a unique u that corresponds to v. Equation (4) shows that if $v > 2$, then there are two corresponding numbers: $u_1 = 1 + \sqrt{v-2}$ and $u_2 = 1 - \sqrt{v-2}$. Hence, we avoid the word function, and instead we speak of equation (4) as giving the *inverse relation* of the function f defined by (3). The graph of $v = u^2 - 2u + 3$ (shown in Figure 10.1) reveals the source of our difficulty. From this figure it is clear that for each $v_0 > 2$, the horizontal line $v = v_0$ meets the graph in two points, and v_0 is the image of two distinct values of u given by equation (4). Suppose that v' (read, "v prime") is a number such

FIGURE 10.1

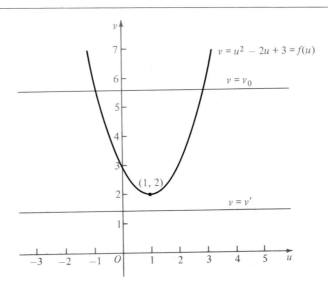

that $v' < 2$. Figure 10.1 shows that the line $v = v'$ does not meet the graph, and hence v' does not have a preimage under the given f. This explains why the quantity under the radical in equation (4) is negative when $v < 2$. We infer from the graph, or equation (4), that the range of f is the ray $[2, \infty)$.

To obtain an inverse function for f, we reduce the domain of f to some smaller set $\mathbf{D}_1 \subset \mathbf{D}$ such that f is one-to-one on $\mathbf{D}_1$. Thus, if we take $\mathbf{D}_1$ to be the ray $[1, \infty)$, then $f(u) \equiv u^2 - 2u + 3$ is increasing in $\mathbf{D}_1$, and the function

$$(5) \qquad\qquad u = 1 + \sqrt{v-2} \equiv g_1(v)$$

is the inverse function of $f(u)$ in $\mathbf{D}_1$.

On the other hand, if we take $\mathbf{D}_2$ to be the ray $(-\infty, 1]$, then the function $f(u) \equiv u^2 - 2u + 3$ is decreasing in $\mathbf{D}_2$, and the function

$$(6) \qquad\qquad u = 1 - \sqrt{v-2} \equiv g_2(v)$$

is the inverse function of $f(u)$ in $\mathbf{D}_2$. With this example in mind we are prepared for the general definition.

DEFINITION 1   ***Inverse Function***   *Let f be a function whose domain is* $\mathbf{D}$ *and whose range is* $\mathbf{G}$. *Let g be a function whose domain is* $\mathbf{G}$ *and whose range is contained in* $\mathbf{D}$. *If for every v in* $\mathbf{G}$

$$(7) \qquad\qquad f(g(v)) = v,$$

then g is called an *inverse function of f*. The union of all such inverse functions gives the *inverse relation* of f.

EXAMPLE 1

Use Definition 1 to prove that $g_1(v) \equiv 1 + \sqrt{v-2}$ is an inverse function for $f(u) = u^2 - 2u + 3$.

Solution: The range of f is the ray $[2, \infty)$. This same ray is the domain of g_1. For any v in this ray we must compute $f(g(v))$. Consequently, we must replace u in $f(u) = u^2 - 2u + 3$ by $u = g_1(v) = 1 + \sqrt{v-2}$. We find that

$$f(g_1(v)) = \left(1 + \sqrt{v-2}\right)^2 - 2\left(1 + \sqrt{v-2}\right) + 3$$

$$= 1 + 2\sqrt{v-2} + v - 2 - 2 - 2\sqrt{v-2} + 3 = v. \quad \blacktriangle$$

There are many ways of selecting an inverse function for $v = u^2 - 2u + 3$, but the only reasonable ones are the two given by equations (5) and (6). Whenever there is more than one way to select an inverse function, it is customary to select one particular inverse function and call it *the inverse function* or the *principal branch* of the inverse function. It would be more appropriate to call the inverse function the principal branch of the inverse relation.

In selecting the principal branch of the inverse relation, we usually settle on some simple function or on one that has some nice property. A mathematician who prefers increasing functions would naturally select the function $g_1(v) = 1 + \sqrt{v-2}$ as the inverse function of $f(u) = u^2 - 2u + 3$, while his more pessimistic colleague who is intrigued by decreasing functions might select $g_2(v) = 1 - \sqrt{v-2}$. This particular example is not very important, so it is not necessary to make a definite choice. In Section 10.3 we will discuss the inverse trigonometric functions. These functions are important, and here there is universal agreement on their definitions.

Exercise 1

In problems 1 through 12 find an inverse function by solving the given equation for the independent variable in terms of the dependent variable. In each case, state the domain **D** and the range **G** of the inverse function you have selected. Note that in some cases the solution may not be unique.

1. $v = 3u - 5$

2. $v = -4u + 11$

3. $v = -\dfrac{1}{2}u + \dfrac{5}{2}$

4. $v = \dfrac{11}{3}u - \dfrac{7}{3}$

5. $v = \dfrac{2u+3}{u-1}$

6. $v = \dfrac{-u+5}{2u+7}$

7. $v = u^2 + 2u - 15$

8. $v = u^2 - 4u + 3$

9. $y = -x^2 + 4x + 5$ 10. $s = -r^2 - 6r + 10$

11. $z = t^3$ 12. $w = z^4$

13. Prove that the function $f(x) = \dfrac{-x+2}{3x+1}$ is its own inverse.

14. Select several of the inverse functions found in problems 1 through 12, and prove that they satisfy equation (7).

15. Find an inverse function for

$$f(x) = x + \frac{1}{x},$$

where f is defined on the domain $x > 0$.

16. Find an inverse function for

$$f(x) = x - \frac{16}{x} + 3,$$

where $f(x)$ is defined on the domain $x > 0$.

17. Let a, b, c, and d be constants such that $ad - bc \neq 0$. Find a formula for the inverse of the function $y = (ax+b)/(cx+d)$. Use your formula to check the answers to problems 5 and 6.

18. For the functions given in problems 1 through 13, find those for which the inverse relation is also a function.

10.2 THE GRAPH OF AN INVERSE FUNCTION

The concept of an inverse function is independent of the letters used. To examine the graph of an inverse function, we return to the customary letters x and y.

Let $y = f(x)$ carry the particular number a into b. In other words, let

(8) $b = f(a)$.

Then under the inverse relation b will have one or more image points, one of which must be a. If we use g to denote the appropriate branch of the inverse relation, then

(9) $a = g(b)$.

Thus, if the graph of $y = f(x)$ contains the point (a, b), then the graph of the inverse relation contains the point (b, a). We recall from Section 6.6 that if P is the point (a, b) and T is the point (b, a), then the points P and T are symmetric with respect to the line $y = x$. From this fact we obtain immediately

Theorem 1
PLE

Let F be the graph of $y = f(x)$, and let G be the graph of the inverse relation. Then G is the reflection of F in the line $y = x$.

If G_1 is the graph of an inverse function of f, then G_1 is contained in G. We usually indicate an inverse function by writing $x = g(y)$. However, at this moment we interchange the letters x and y and write $y = g(x)$ because we wish to continue with x as the independent variable and y as the dependent variable. If $y = g(x)$ is an inverse function of $y = f(x)$, then the defining equation (7), which states that $f(g(v)) = v$, now has the form

(10) $$f(g(x)) = x$$

for every x in the domain of g.

EXAMPLE 1

If $y = x^2 - 2x + 3$, sketch the graph of the inverse relation. Find two possible inverse functions, and indicate the graph of each.

Solution: Except for the change of letters, this is the same function discussed in Section 10.1 (see equation (3)), and the graph of f is indicated in Figure 10.1. To obtain the graph G of the inverse relation we merely reproduce this graph (see Figure 10.2) and then reflect it about the line $y = x$.

FIGURE 10.2

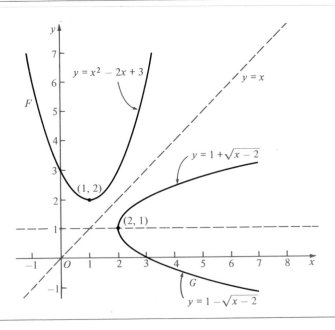

Two possible inverse functions have been discussed in Section 10.1. After the required change in the letters, equations (5) and (6) give

$$y = g_1(x) = 1 + \sqrt{x-2}$$

and

$$y = g_2(x) = 1 - \sqrt{x-2}.$$

The portion of G that lies on or above the line $y = 1$ is the graph of $y = g_1(x)$. The portion of G that lies on or below the line $y = 1$ is the graph of $y = g_2(x)$. ▲

Exercise 2

1. Prove that the points $P(a, b)$ and $T(b, a)$ are symmetric with respect to the line $y = x$.

2. Sketch the graph for each of the functions given in problems 1, 3, 5, 7, 9, 15, and 16 in Exercise 1. In each case use Theorem 1 to sketch the graph of the inverse relation. Then indicate that portion of the graph that corresponds to your selection of an inverse function.

3. In the next section we will discuss the inverse trigonometric functions. Try to anticipate this material in the following way. Without looking at the text, sketch the graph of the inverse relations of (a) $y = \sin x$, (b) $y = \cos x$, (c) $y = \tan x$, (d) $y = \cot x$, (e) $y = \sec x$, and (f) $y = \csc x$. For the first four functions try to decide how you would restrict the domain of f in order to obtain the inverse function (principal value of the inverse relation).

10.3 THE INVERSE TRIGONOMETRIC FUNCTIONS

We recall the graph of $y = \sin x$ shown in Figure 10.3. By Theorem 1, the graph of the inverse relation is obtained by reflecting the graph of $y = \sin x$ in the line $y = x$. The result of this operation is shown in Figure 10.4. In place of the "neutral" symbol $g(x)$ for the inverse relation, we use special symbols for the inverse trigonometric relations. For $y = \sin x$, the inverse relation is denoted by the collection of symbols

$$(11) \qquad\qquad y = \arcsin x$$

(read, "y is an arc whose sine is x" or "y is an angle whose sine is x"). Some authors use the symbol

$$(12) \qquad\qquad y = \sin^{-1} x$$

FIGURE 10.3   FIGURE 10.4

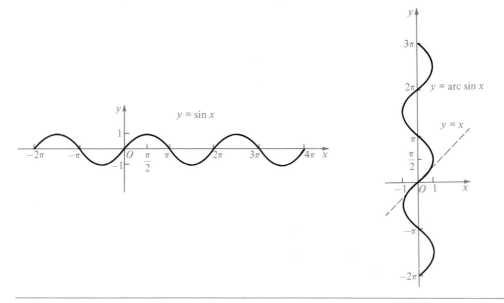

(read, "y is an inverse sine of x") to indicate the same relation, but the notation in (12) is falling from favor because it might be confused with $y = (\sin x)^{-1} = \csc x$. Whatever notation is used, the graph of (11) or (12) is shown in Figure 10.4.

EXAMPLE 1   Find $y = \arcsin(1/2)$.

Solution: We must find all y such that $\sin y = 1/2$. One such y is $\pi/6$, and a second one is $5\pi/6$. All other y are coterminal with one of these. If S denotes the set of all y for which $\sin y = 1/2$, then

(13) $S = \{y \mid y = \pi/6 + 2n\pi \quad \text{or} \quad y = 5\pi/6 + 2n\pi, \quad n \in Z\}$

Thus, $y = \arcsin(1/2)$ is the set S. ▲

   Because we obtain more than one number for $y = \arcsin(1/2)$, the relation $y = \arcsin x$ is frequently called a *multiple-valued function*.
   To obtain the principal branch of the inverse relation, we select a suitable piece of the graph of $y = \arcsin x$. In this particular case the universal choice is that part of Figure 10.4 that lies between the horizontal lines $y = -\pi/2$ and $y = \pi/2$. This amounts to restricting the domain of the primitive function $y = \sin x$ to be the interval $[-\pi/2, \pi/2]$. It is customary to use a capital letter to indicate the function obtained in this way.

DEFINITION 2   *The function $y = \operatorname{Sin} x$ is the function whose domain is $[-\pi/2, \pi/2]$ and in this interval*

(14) $\operatorname{Sin} x \equiv \sin x.$

The function

(15) $y = \operatorname{Arc\,sin} x$

(read, "y is the arc whose sine is x") is the inverse function of $y = \operatorname{Sin} x$.

   The graphs of these two new functions are indicated in Figure 10.5 and 10.6. The function $y = \operatorname{Arc\,sin} x$ is called *the inverse of the sine function* or *the principal branch of the inverse sine function*. The essential fact is that

(15) $y = \operatorname{Arc\,sin} x$

if and only if

(16) $x = \sin y$ and $-\pi/2 \leq y \leq \pi/2.$

EXAMPLE 2   Find $y = \operatorname{Arc\,sin}(1/2)$.

Solution: This example is similar to Example 1, but now we are asking for $\operatorname{Arc\,sin}(1/2)$ and not $\operatorname{arc\,sin}(1/2)$. Thus we must select from **S**, given by (13), the one element in the interval $[-\pi/2, \pi/2]$. Clearly this is $\pi/6$. Hence, $\operatorname{Arc\,sin} 1/2 = \pi/6$.   ▲

FIGURE 10.5   FIGURE 10.6

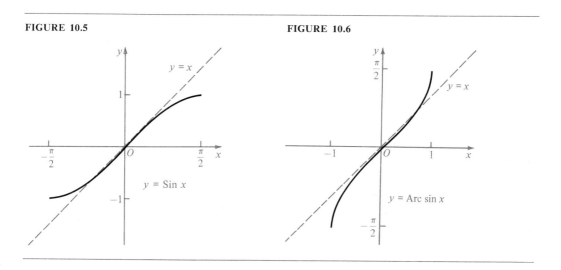

We now give a parallel treatment for the remaining trigonometric functions. The graphs of $y = \cos x$, $y = \tan x$, and $y = \cot x$ are given in Figures 10.7, 10.9, and 10.11, respectively. The graphs of the corresponding inverse relations are given in Figures 10.8, 10.10, and 10.12, respectively. In these graphs the "cap" functions and their inverses are shown in black.

DEFINITION 3   *The domain of $y = \text{Cos } x$ is $[0, \pi]$ and in this interval*

(17) $\text{Cos } x \equiv \cos x.$

The domain of $y = \text{Tan } x$ is $(-\pi/2, \pi/2)$ and in this interval

(18) $\text{Tan } x \equiv \tan x.$

The domain of $y = \text{Cot } x$ is $(0, \pi)$ and in this interval

(19) $\text{Cot } x \equiv \cot x.$

DEFINITION 4   *The function $y = \text{Arc cos } x$ is the inverse function of $y = \text{Cos } x$. The function $y = \text{Arc tan } x$ is the inverse function of $y = \text{Tan } x$. The function $y = \text{Arc cot } x$ is the inverse function of $y = \text{Cot } x$.*

FIGURE 10.7

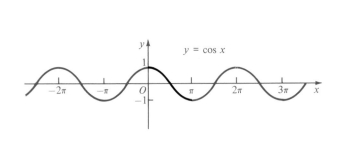

$y = \cos x$

FIGURE 10.8

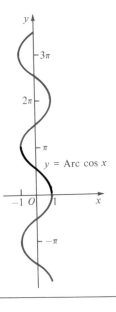

$y = \text{Arc cos } x$

FIGURE 10.9

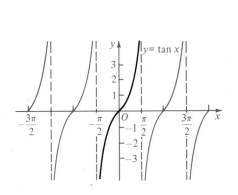

FIGURE 10.10

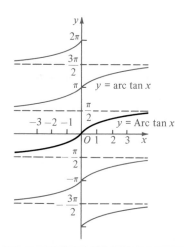

FIGURE 10.11

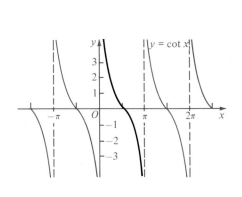

FIGURE 10.12

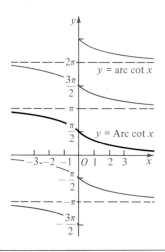

From Definitions 3 and 4 we see that

(20) $y = \operatorname{Arc\,cos} x$

if and only if

(21) $x = \cos y$ and $0 \leqq y \leqq \pi.$

Further we have

(22) $$y = \text{Arc}\tan x$$

if and only if

(23) $$x = \tan y \quad \text{and} \quad -\pi/2 < y < \pi/2.$$

For the cotangent function,

(24) $$y = \text{Arc}\cot x$$

if and only if

(25) $$x = \cot y \quad \text{and} \quad 0 < y < \pi.$$

We can proceed to define $\text{Arc}\sec x$ and $\text{Arc}\csc x$, but these functions are not really needed and hence we may omit them.

EXAMPLE 3   Find (a) $\text{Arc}\cos(\sqrt{2}/2)$, (b) $\text{Arc}\tan(-1)$, and (c) $\text{Arc}\cot\sqrt{3}$.

Solution: In each case we recognize the correct θ from experience. Thus,

$$\text{(a)} \ \cos\frac{\pi}{4} = \frac{\sqrt{2}}{2}, \qquad \text{(b)} \ \tan\left(-\frac{\pi}{4}\right) = -1, \qquad \text{(c)} \ \cot\frac{\pi}{6} = \sqrt{3}.$$

All that remains is to check that each θ lies in the proper interval. This gives

$$\text{(a)} \ \frac{\pi}{4} = \text{Arc}\cos\frac{\sqrt{2}}{2}, \quad \text{(b)} \ -\frac{\pi}{4} = \text{Arc}\tan(-1), \quad \text{(c)} \ \frac{\pi}{6} = \text{Arc}\cot\sqrt{3}. \ \blacktriangle$$

EXAMPLE 4   Find $\sec(\text{Arc}\tan(-4/5))$.

Solution: It is helpful to sketch the angle $\theta \equiv \text{Arc}\tan(-4/5)$ in standard position. Since $\text{Tan}\,\theta = -4/5$ and $-\pi/2 < \theta < \pi/2$ by definition, θ must be in Q.IV. Consequently, we can construct θ by drawing a line from 0 to the point $P(5, -4)$, as indicated in Figure 10.13. Then $r = \sqrt{25+16} = \sqrt{41}$. Therefore, $\sec(\text{Arc}\tan(-4/5)) = \sec\theta = \sqrt{41}/5$. $\blacktriangle$

EXAMPLE 5   Find $\cos(\text{Arc}\sin v)$.

Solution: A sketch of $\theta \equiv \text{Arc}\sin v$ will be helpful. We do not know whether v is positive, zero, or negative. Since we cannot picture all three cases at once, we select one case (see Figure 10.14), but we keep in mind that

FIGURE 10.13

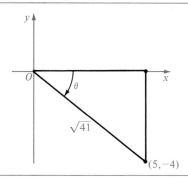

FIGURE 10.14

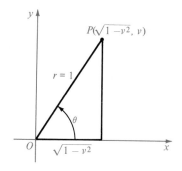

the remaining cases may need separate consideration. From the figure, we see that $\cos\theta = \cos(\text{Arc}\sin v) = \sqrt{1-v^2}$. If $v < 0$, then $\text{Arc}\sin v < 0$, and θ is in Q.IV. But $\cos\theta \geqq 0$ for θ in Q.IV. Hence $\cos(\text{Arc}\sin v) = \sqrt{1-v^2}$, just as in the first case. We leave it for the student to show that when $v = 0$, the formula

(26) $$\cos(\text{Arc}\sin v) = \sqrt{1-v^2}$$

is still true.

Exercise 3

In problems 1 through 10 find the angle in degrees to the nearest ten minutes. In problems 11 through 18 find the angle in radians.

1. $\text{Arc}\cos 0$ 2. $\text{Arc}\tan 0$ 3. $\text{Arc}\sin(-1/2)$

4. $\text{Arc}\cos(-\sqrt{2}/2)$ 5. $\text{Arc}\cot(-1)$ 6. $\text{Arc}\cos(-1)$

7. $\text{Arc} \tan 0.2401$ 8. $\text{Arc} \sin(-0.3007)$ 9. $\text{Arc} \cos(-0.7642)$

10. $\text{Arc} \cot 1.220$ 11. $\text{Arc} \sin(-1)$ 12. $\text{Arc} \cos(-1)$

13. $\text{Arc} \tan \sqrt{3}$ 14. $\text{Arc} \sin(-\sqrt{2}/2)$ 15. $\text{Arc} \cot 1$

16. $\text{Arc} \sin(\sqrt{3}/2)$ 17. $\text{Arc} \cos(-1/2)$ 18. $\text{Arc} \tan 1$

In problems 19 through 40 simplify the given expression. Find a numerical value whenever possible.

19. $\cos(\text{Arc} \tan(3/4))$ 20. $\tan(\text{Arc} \cos(3/5))$

21. $\sin(\text{Arc} \cos(5/13))$ 22. $\cos(\text{Arc} \sin(7/25))$

23. $\tan[\text{Arc} \sin(-0.7009)]$ 24. $\sin[\text{Arc} \tan(-0.4245)]$

25. $\cos(\text{Arc} \cos a)$ 26. $\tan(\text{Arc} \tan b)$

27. $\cot(\text{Arc} \cot c)$ 28. $\sin(\text{Arc} \sin d)$

29. $\sin(\text{Arc} \tan e)$ 30. $\csc(\text{Arc} \cos f)$

31. $\sec(\text{Arc} \sin g)$ 32. $\tan(\text{Arc} \cos h)$

33. $\sin(2 \text{Arc} \cos u)$ 34. $\cos(2 \text{Arc} \sin u)$

35. $\cos(2 \text{Arc} \tan u)$ 36. $\sin(2 \text{Arc} \cot u)$

37. $\sin\left(\dfrac{1}{2} \text{Arc} \sin u\right)$ 38. $\cos\left(\dfrac{1}{2} \text{Arc} \cos u\right)$

39. $\tan\left(\dfrac{1}{2} \text{Arc} \cos u\right)$ 40. $\sec\left(\dfrac{1}{2} \text{Arc} \cos u\right).$

In problems 41 through 51 identify the statement as true or false and give your reasons.

41. $2 \text{Arc} \sin u = \text{Arc} \sin 2u$

42. $2 \text{Arc} \sin u = \text{Arc} \cos(1 - 2u^2)$, for $u \geq 0$

43. $\text{Arc} \tan u = \dfrac{\text{Arc} \sin u}{\text{Arc} \cos u}$ 44. $\cos\left(\dfrac{\pi}{2} - \text{Arc} \sin \dfrac{2}{3}\right) = \dfrac{2}{3}$

45. $\tan\left(\pi - \text{Arc} \tan \dfrac{7}{8}\right) = -\dfrac{7}{8}$ 46. $\text{Arc} \cot v = \dfrac{1}{\text{Arc} \tan v}$

47. $\text{Arc} \cot v = \text{Arc} \tan \dfrac{1}{v}$ 48. $\text{Arc} \cos(-u) = \text{Arc} \cos u$

49. $\text{Arc} \sin(-w) = -\text{Arc} \sin w$ 50. $\text{Arc} \cos(-u) = \pi - \text{Arc} \cos u.$

★ **10.4 COMBINATIONS OF THE**
INVERSE TRIGONOMETRIC
FUNCTIONS

By using combinations of the inverse trigonometric functions, we can create problems that are somewhat more involved than those of the preceding section.

EXAMPLE 1

If $C = \text{Arc}\tan 3 + \text{Arc}\sin(5/13)$, find the quadrant for C without using tables.

Solution: Let $A = \text{Arc}\tan 3$, and let $B = \text{Arc}\sin(5/13)$. Since A and B are both in Q.I, then $C = A + B$ is either in Q.I or Q.II. Hence, it will suffice to determine the sign of $\cos C$. By constructing suitable triangles we find that

$$\sin A = \frac{3}{\sqrt{10}} = \frac{3\sqrt{10}}{10}, \qquad \sin B = \frac{5}{13},$$

$$\cos A = \frac{1}{\sqrt{10}} = \frac{\sqrt{10}}{10}, \qquad \cos B = \frac{12}{13}.$$

Hence

$$\cos C = \cos A \cos B - \sin A \sin B = \frac{\sqrt{10}}{10}\frac{12}{13} - \frac{3\sqrt{10}}{10}\frac{5}{13}$$

$$= \frac{(12-15)\sqrt{10}}{130} = \frac{-3\sqrt{10}}{130} < 0.$$

Therefore, C is in Q.II. ▲

EXAMPLE 2

Simplify the expression $\sin(\text{Arc}\sin x + \text{Arc}\sin y)$.

Solution: Let $A = \text{Arc}\sin x$, and $B = \text{Arc}\sin y$. Then

$$\sin A = x, \qquad \cos A = \sqrt{1-x^2}, \qquad \sin B = y, \qquad \cos B = \sqrt{1-y^2}.$$

Consequently, the given expression is equal to

$$\sin(A+B) = \sin A \cos B + \cos A \sin B = x\sqrt{1-y^2} + y\sqrt{1-x^2},$$

which is certainly simpler than the original expression. ▲

EXAMPLE 3

Prove that $\qquad \text{Arc}\cot(1/9) + \text{Arc}\cot(4/5) = 3\pi/4$.

Solution: Let $A = \text{Arc}\cot(1/9)$, and let $B = \text{Arc}\cot(4/5)$. Then, $\tan A = 9$ and $\tan B = 5/4$. Therefore,

$$\tan(A+B) = \frac{\tan A + \tan B}{1 - \tan A \tan B} = \frac{9 + 5/4}{1 - 9(5/4)}$$

$$= \frac{36+5}{4-45} = \frac{41}{-41} = -1.$$

Whence $A + B = 3\pi/4 + n\pi$, where n is an integer. But A and B are both in Q.I. Therefore n must be zero. ▲

Exercise 4

In problems 1 through 4 determine the sine and cosine of the given angle without using tables. Determine the quadrant of the angle.

1. $\text{Arc}\cos(3/5) + \text{Arc}\tan(5/12)$ 2. $\text{Arc}\tan 2 + \text{Arc}\sin(3/5)$

3. $\text{Arc}\tan 3 + \text{Arc}\sin(1/3)$ 4. $\text{Arc}\cot(2/5) + \text{Arc}\tan(3/7)$

5. Find $\sin(\text{Arc}\sin x + 2\,\text{Arc}\tan x)$.   6. Find $\cos(2\,\text{Arc}\cos y + \text{Arc}\tan y)$.

7. Prove that $\text{Arc}\tan(2/3) + \text{Arc}\tan(1/5) = \pi/4$.

★ 8. Prove that if $u > 1$, then $2\,\text{Arc}\tan u + \text{Arc}\sin(2u/(1 + u^2)) = \pi$.

★ 9. Graph the functions for the indicated intervals:

    (a) $y = \text{Arc}\cos x^2$, $-1 \leqq x \leqq 1$,

    (b) $y = \text{Arc}\cos(\cos x)$, $-4\pi \leqq x \leqq 4\pi$,

    (c) $y = (\text{Arc}\cos x)^2$, $-1 \leqq x \leqq 1$.

★ 10. Graph the functions for the indicated intervals:

    (a) $y = \text{Arc}\sin x^2$, $-1 \leqq x \leqq 1$,

    (b) $y = (\text{Arc}\sin x)^2$, $-1 \leqq x \leqq 1$,

    (c) $y = \text{Arc}\sin(\sin x)$, $-4\pi \leqq x \leqq 4\pi$.

★ 11. Which of the following functions defined for the interval $[-1, 1]$ are odd functions, even functions, or neither:

    (a) $\text{Arc}\tan x$,   (b) $(\text{Arc}\sin x)(\text{Arc}\cos x)$,

    (c) $(\text{Arc}\sin x)(\text{Arc}\tan x)$,   (d) $x^2 + \text{Arc}\cos x$,

    (e) $x^3 - \text{Arc}\cos x + \pi/2$,   (f) $x\,\text{Arc}\sin x$?

10.5 PERIODIC FUNCTIONS

In Section 7.8 we stated that $\sin x$ is a periodic function. Geometrically this means that the graph consists of a basic pattern that is repeated infinitely to form the full graph (see Figure 10.3 for the graph of $\sin x$). This repetition of a basic pattern is expressed symbolically by the equation

(27) $$\sin(x + 2\pi) = \sin x,$$

which is true for every x. The number 2π is called the period of $\sin x$. For an arbitrary function f, we have

DEFINITION 5   ***Period***   *A nonconstant function $f(x)$ is said to be periodic if there is a number $q \neq 0$ such that*

(28) $$f(x+q) = f(x)$$

for every x. The number q is called a period of the function f. If p is the smallest positive number in the set of all periods of f, then p is called the period of f.

EXAMPLE 1   Find the period of (a) $\sin 3x$ and (b) $\tan(x/7)$.

Solution: (a) We test $\sin 3x$ using equation (28). When we replace x by $x+q$ in $\sin 3x$, we obtain

(29) $$\sin 3(x+q) = \sin(3x+3q).$$

If $3q = 2n\pi$, then the right side of (29) is $\sin 3x$ for every x. Consequently, q is a period if $3q = 2n\pi$. Since 2π is *the* period for $\sin\theta$ (smallest positive period), we obtain the smallest q when $3q = 2\pi$. In other words, $p = 2\pi/3$ is *the* period of $\sin 3x$. A portion of the graph of $y = \sin 3x$ is shown in Figure 10.15. In the same figure we give a portion of the graph of $y = \sin x$ in color. Note that the multiplier 3 tends to "squash" the curve. This "squashing" effect is described accurately in Theorem 3.

   (b) We apply the same argument to $\tan(x/7)$. The essential difference is that *the* period of $\tan\theta$ is π, not 2π. Again, we replace x by $x+q$. Then,

(30) $$\tan\frac{x+q}{7} = \tan\left(\frac{x}{7} + \frac{q}{7}\right).$$

If $q/7 = n\pi$, then the right side of equation (30) is $\tan(x/7)$ for every x. To obtain the period we set $n = 1$. Then $p = 7\pi$ is *the* period of $\tan(x/7)$. ▲

FIGURE 10.15

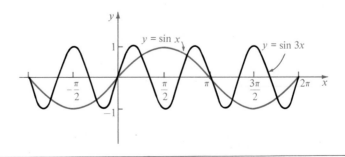

FIGURE 10.16

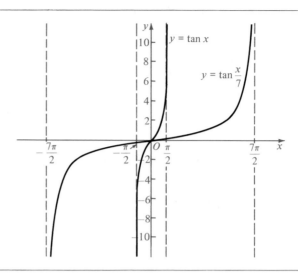

A portion of the graph of $y = \tan(x/7)$ is shown in Figure 10.16.

The same type of argument will prove the following general theorem, which applies to an arbitrary periodic function. Notice that in this theorem, b may be positive or negative.

Theorem 2
PLE

Let $f(x)$ be a periodic function, and let p be the period of $f(x)$. If $a \neq 0$, and $b \neq 0$, then the function

(31) $$g(x) \equiv af(bx+c).$$

has the period $p/|b|$.

Since we already know the period for each of the trigonometric functions, Theorem 2 gives immediately

Theorem 3
PLE

If $a \neq 0$, and $b \neq 0$, then $2\pi/|b|$ is the period for each of the functions:

$$a\sin(bx+c), \qquad a\cos(bx+c), \qquad a\sec(bx+c), \qquad a\csc(bx+c).$$

However, the period is $\pi/|b|$ for the two functions

$$a\tan(bx+c) \qquad and \qquad a\cot(bx+c).$$

We next look at the effect of the multiplier a in equation (31). If the maximum value of $f(x)$ is M and $a > 0$, then clearly the maximum value of

$af(bx+c)$ is aM. The same relationship holds for the minimum (least) value of $f(x)$. For example, the maximum value of $3\sin(5x+7)$ is obviously 3, and the minimum value of this function is -3. If the graph of $f(x)$ is "centered" about the x-axis, then the maximum value of $f(x)$ is called the *amplitude of $f(x)$*. For example, the graph of $a\sin(bx+c)$ is "centered" about the x-axis and the amplitude is $|a|$. In the general case, the amplitude is "the distance from the crest of the wave to the center of the wave." This is stated precisely in

DEFINITION 6   *Amplitude   Let $f(x)$ be a periodic function. Suppose that M is the maximum value of $f(x)$, and m is the minimum value of $f(x)$. Then the quantity $(M-m)/2$ is called the amplitude of $f(x)$.*

Finally we examine the effect of c in equation (31), but only for the particular function $f(x) = \sin x$. We note that $\sin 0 = 0$, so the "wave" starts at the origin (see Figure 10.3).

Now consider the altered function

(32) $y = a\sin(bx+c)$.

For this function $y = 0$ when $bx+c = n\pi$. We select the particular case $n = 0$ and find that $bx+c = 0$ or

(33) $$x = -\frac{c}{b}.$$

Thus the "wave" can be thought of as starting at $-c/b$. More precisely, the graph of $y = a\sin(bx+c)$ crosses the x-axis at $x = -c/b$. This quantity is called the *phase shift* of the sine wave or the phase shift of the function. For simplicity, we assume that $b > 0$. Then the phase shift is to the left a distance c/b if $c > 0$. The phase shift is to the right a distance $|c/b|$ if $c < 0$. We illustrate these concepts in

EXAMPLE 2   For the function

(34) $$y = 3\sin\left(2x + \frac{\pi}{3}\right) + 1,$$

find (a) the period, (b) the amplitude, and (c) the phase shift. Use this data to sketch the graph.

Solution: (a) By Theorem 3, the period $p = 2\pi/2 = \pi$. Hence, we can sketch the graph for $0 \leq x \leq \pi$ (or any other suitable interval of length π) and obtain the full graph by repeating the pattern an infinite number of times.

FIGURE 10.17

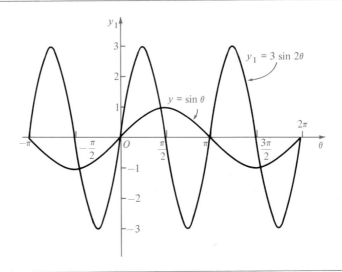

(b) Since the maximum and minimum values of $\sin \theta$ are 1 and -1, respectively, the maximum and minimum values of $3\sin(2x+\pi/3)+1$ are $3+1=4$ and $-3+1=-2$, respectively. By Definition 6, the amplitude is

$$\frac{M-m}{2} = \frac{4-(-2)}{2} = \frac{6}{2} = 3,$$

a number that we could easily guess at the outset.

(c) The phase shift is

$$-\frac{c}{b} = -\frac{\pi/3}{2} = -\frac{\pi}{6}.$$

To sketch the graph, we first consider the simpler function $y_1 = 3\sin 2\theta$, which has the same amplitude and period, but is "centered" about the x-axis and has zero phase shift. This graph is a sine wave and is shown in Figure 10.17.

To obtain the graph of $y = 3\sin(2x+\pi/3)+1$, we replace 2θ in y_1 by $2x+\pi/3$. This amounts to a phase shift of $\pi/6$ to the left. This can be accomplished by moving the y_1-axis in Figure 10.17, $\pi/6$ units to the right. Finally, the term $+1$ in equation (34) lifts the entire curve in Figure 10.17 up one unit. This can be accomplished by moving the θ-axis down one unit. The result of these two shifts is the graph of $y = 3\sin(2x+\pi/3)+1$ shown in Figure 10.18. ▲

FIGURE 10.18

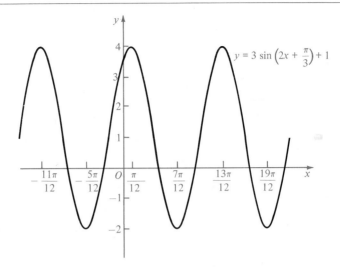

$$y = 3 \sin \left(2x + \frac{\pi}{3}\right) + 1$$

EXAMPLE 3   Prove that if at least one of the coefficients A, B is not zero, then the graph of

(35) $$y = A \sin bx + B \cos bx$$

is always a sine wave. Find the period, amplitude, and phase shift of this function.

Solution: We replace the coefficients in (35) by more suitable ones. The trick is to multiply and divide by $\sqrt{A^2 + B^2}$. Thus, (35) is equivalent to

(36) $$y = \sqrt{A^2 + B^2} \left(\frac{A}{\sqrt{A^2 + B^2}} \sin bx + \frac{B}{\sqrt{A^2 + B^2}} \cos bx \right).$$

Now let α be such that

(37) $$\cos \alpha = \frac{A}{\sqrt{A^2 + B^2}} \quad \text{and} \quad \sin \alpha = \frac{B}{\sqrt{A^2 + B^2}},$$

where $0 \leqq \alpha \leqq 2\pi$. Then (36) becomes

$$y = \sqrt{A^2 + B^2} (\sin bx \cos \alpha + \cos bx \sin \alpha)$$

(38) $$y = \sqrt{A^2 + B^2} \sin (bx + \alpha).$$

From (38) it is clear that the graph of (35) has period $2\pi/|b|$, amplitude $\sqrt{A^2 + B^2}$, and phase shift $-\alpha/b$, where α is defined by the equation set (37). ▲

EXAMPLE 4   Find the period, amplitude, and phase shift of

(39) $y = 3 \sin 4x + 2 \cos 4x$

Solution: Following the method of Example 3

$$y = \sqrt{3^2 + 2^2} \left(\frac{3}{\sqrt{3^2 + 2^2}} \sin 4x + \frac{2}{\sqrt{3^2 + 2^2}} \cos 4x \right)$$

$$= \sqrt{13} \sin (4x + \alpha),$$

where $\alpha = \operatorname{Arc} \cos (3/\sqrt{13}) \approx 0.5881$. The period is $2\pi/4 = \pi/2$, the amplitude is $\sqrt{13}$, and the phase shift is $\alpha/4 \approx 0.1470$ units to the left. ▲

Exercise 5

1. Let p be the period of $f(x)$. What can you say about the numbers $2p$, $3p$, $4p$, $-p$, $-2p$, and $-3p$?

2. Suppose $f(x)$ has one period q and a second period r. Prove that if m and n are integers, then $mq + nr$ is also a period of $f(x)$.

In problems 3 through 9 determine whether or not the given function is periodic. Supply a proof of your assertion.

3. $y = \operatorname{Arc} \sin x$ 4. $y = \operatorname{Arc} \tan x$

★ 5. $y = x - [x]$ ★ 6. $y = x^2 - [x^2]$

★ 7. $y = \sin x + \sin 3x$ ★ 8. $y = \sin 2x + \cos^2 4x$

★ 9. $y = f(\sin 5x)$, where $f(x)$ is a function whose domain contains the interval $[-1, 1]$. Assume that $f(x)$ is neither a constant nor periodic.

★ 10. Let $f(x)$ and $g(x)$ be functions that have the same period p. Let A and B be arbitrary constants, and let m and n be positive integers. Prove that each of the following functions is periodic or a constant.

   (a) $y = Af(x) + Bg(x)$ (b) $y = f^m(x) g^n(x)$

   (c) $y = Af^m(x) + Bg^n(x)$ (d) $y = \dfrac{1}{Af^2(x) + Bg^2(x)}.$

In problems 11 through 20 state the period of the given function. Sketch the graph over some interval whose length is twice the period. Give the amplitude, if the function has one.

11. $y = 3 \sin 4x$ 12. $y = -4 \cos 3x$

13. $y = \frac{1}{2}\tan\frac{x}{3}$ 14. $y = 2\sin\frac{x}{5}$

15. $y = \sin x + 2\cos x$ 16. $y = 4\cos 2x + 5\sin 2x$

17. $y = \sin 2x - 3\cos 2x$ 18. $y = 2\sin x - 3\cos x$

19. $y = \cos^2 3x$ 20. $y = 2 - 3\sin^2 x$

21. Give the phase shift for the functions in problems (a) 15, (b) 16, (c) 17, and (d) 18.

22. Show that the formula $(M - m)/2$ for the amplitude, stated in Definition 6, gives the distance from the crest of the wave to the center of the wave.

23. If C and D are arbitrary real numbers, then it is not always possible to find α such that $\cos\alpha = C$ and $\sin\alpha = D$. Prove that if $C^2 + D^2 = 1$, then there is an α such that $\cos\alpha = C$ and $\sin\alpha = D$.

24. Use the result of problem 23 to show that the equation set (37) always has a solution.

10.6 GRAPHING BY THE ADDITION OF ORDINATES

In sketching the graph of an equation such as

$$(40) \qquad\qquad y = \frac{1}{2}x + \cos x,$$

the computation of points on the graph can be tedious. If we are interested only in the general shape of the graph and do not demand a truly accurate drawing, then we can avoid much of the labor. The key idea is to decompose (40) into two parts.

$$(41) \qquad\qquad y_1 = \frac{1}{2}x$$

and

$$(42) \qquad\qquad y_2 = \cos x,$$

and then obtain the graph of (40) by adding y_1 and y_2. This can usually be done "optically" without computation. The method applies to the sum of any number of functions, although it may be unwieldy and of little use if there are more than three functions. The method is called the *addition of ordinates* or the *composition of ordinates*.

FIGURE 10.19

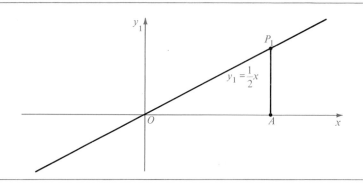

EXAMPLE 1

Use the method described above to sketch the graph of equation (40).

Solution: The graph of $y = x/2$ is shown in Figure 10.19. Since the graph is a straight line, only two simple computations are required.

The graph of $y = \cos x$ is shown in Figure 10.20. Since we have already made a careful sketch of this graph (see Figure 7.22), it takes only a moment to recall its appearance and to reproduce it.

Finally, to obtain the graph of $y = x/2 + \cos x$ we "compose" the two graphs in Figures 10.19 and 10.20. The result is shown in Figure 10.21. Note that we obtain a cosine wave "riding" a straight line. Geometrically, the ordinate y in Figure 10.21 is the directed line segment $\overline{AP}$ and this is $\overline{AP_1} + \overline{AP_2}$, the sum of the corresponding directed line segments in Figures 10.19 and 10.20.

EXAMPLE 2

Sketch the graph of $y = \sin x + \dfrac{1}{4}\sin 3x$.

Solution: The graphs of the two pieces: $y_1 = \sin x$ and $y_2 = \frac{1}{4}\sin 3x$ are shown together in Figure 10.22. The composite graph is shown in Figure 10.23. Note that the period of $\sin x$ is 2π, and the period of $\sin 3x$ is $2\pi/3$. Hence, the period of the composite function is 2π. It suffices to sketch the graph for the interval $[0, 2\pi]$. ▲

FIGURE 10.20

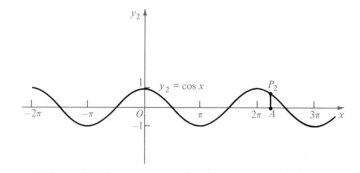

FIGURE 10.21

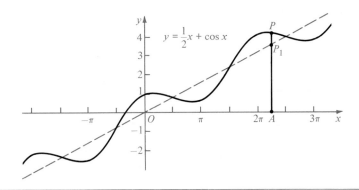

FIGURE 10.22

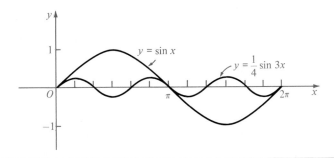

FIGURE 10.23

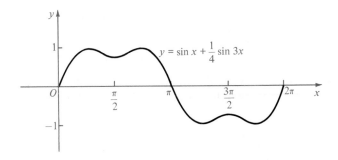

Exercise 6

In problems 1 through 11 sketch the graph of the given function for x in the indicated interval.

1. $y = |\sin x|, \quad 0 \leq x \leq 2\pi$

2. $y = \sin^2 x + 2\cos^2 x, \quad 0 \leq x \leq 2\pi$

3. $y = \sin x - \dfrac{1}{3}\sin 3x, \quad 0 \leq x \leq 4\pi$

4. $y = \cos x + \cos 2x, \quad 0 \leq x \leq 4\pi$

5. $y = \sin x + \cos 2x, \quad 0 \leq x \leq 4\pi$

6. $y = x + \sin x, \quad -2\pi \leq x \leq 4\pi$

★ 7. $y = \dfrac{1}{4}x(1 + \sin x), \quad 0 \leq x \leq 5\pi$

8. $y = \cos x + \dfrac{1}{3}\cos 3x, \quad 0 \leq x \leq 4\pi$

★ 9. $y = \sin x + \dfrac{1}{3}\sin 3x + \dfrac{1}{5}\sin 5x, \quad 0 \leq x \leq 2\pi$

★10. $y = \sin x - \dfrac{1}{2}\sin 2x + \dfrac{1}{3}\sin 3x, \quad 0 \leq x \leq 2\pi$

Review Questions

Try to answer the following questions as accurately as possible before consulting the text.

1. Give the definition of an inverse function.

2. How does an inverse relation differ from an inverse function?

3. State the geometric theorem that involves the graph of a function and the graph of its inverse relation.

4. Give the definitions of the four inverse trigonometric functions.

5. Sketch the graphs of the inverse trigonometric functions.

6. Give the definition of (a) a periodic function, (b) the period, and (c) the amplitude.

7. What can you say about the graph of $y = A \sin bx + B \cos bx$?

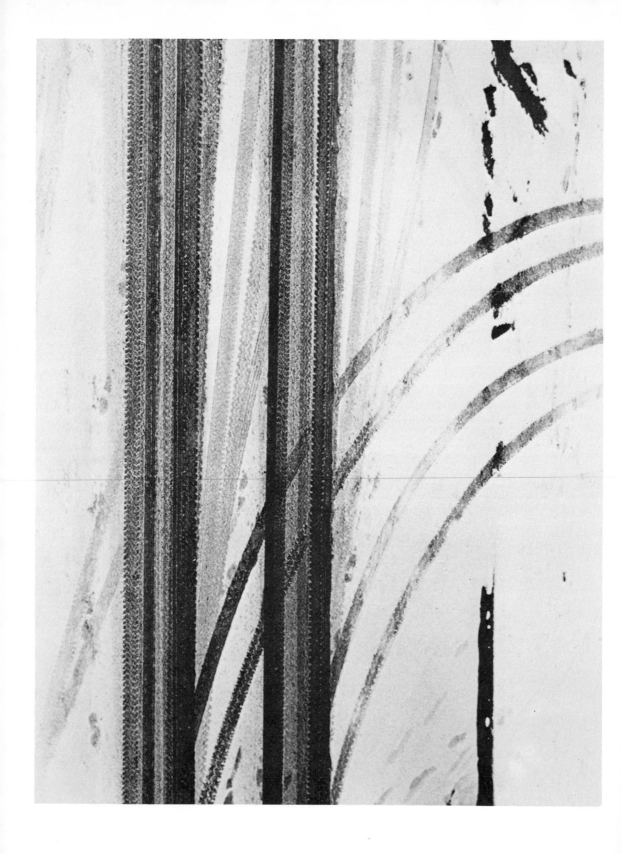

Logarithms

11

Logarithms were invented to simplify computation. With the advent of modern calculators, the importance of logarithms began to wane. As the desk calculators became more efficient, less expensive, and more readily available, the practical need for studying logarithms steadily decreased. There are still good reasons for studying logarithms. First, the logarithmic function has an important place in engineering, applied mathematics, and pure mathematics, so that the function would be worthy of study even if it were never used as an aide to computation. Second, it is of interest to place ourselves for a moment in the 17th century when there was a tremendous amount of computation to be done, and no buttons to press (in fact no electricity available to run the calculators if there had been any calculators to run). It is certainly exciting to watch the birth and development of an idea which at that time was a truly remarkable advance in man's knowledge. Finally, logarithms may still be useful as a computation aid (a) when a calculator is not readily available or (b) when computing a quantity such as $(3.456)^{2.753}$, which is beyond the capability of the usual calculator. Although John Napier (1550–1617), the inventor of logarithms, is justly famous for this work, he thought that his many other activities were far more important. In fact Napier believed that his reputation would rest on a religious book, *A Plaine Discovery of the Whole Revelation of St. John*, in which, among other things, he tried to prove that the world would end sometime between 1688 and 1700.

11.1 THE EXPONENTIAL FUNCTION

We recall the laws of exponents covered in Chapter 5. However, for our present purpose we now use different letters. If $b > 0$, and u and v are arbitrary real numbers, then

(1) $b^u b^v = b^{u+v}$,

(2) $\dfrac{b^u}{b^v} = b^{u-v}$,

(3) $(b^u)^v = b^{uv}$,

(4) $b^0 = 1$,

(5) $b^{-u} = \dfrac{1}{b^u}$,

(6) $b^{u/v} = \sqrt[v]{b^u} = \left(\sqrt[v]{b}\right)^u$.

EXAMPLE 1   Illustrate each of the above identities with a numerical example.

Solution: We give one possible selection of numbers. Of course, the reader may choose other numbers.

(1e) $2^3\, 2^5 = 2^8 = 256$,

(2e) $\dfrac{5^7}{5^4} = 5^3 = 125$,

(3e) $(3^{1/4})^8 = 3^2 = 9$,

(4e) $1{,}000{,}000^0 = 1$,

(5e) $(0.1)^{-2} = \dfrac{1}{(0.1)^2} = 100$,

(6e) $27^{4/3} = \left(\sqrt[3]{27}\right)^4 = 3^4 = 81$. ▲

The study of logarithms requires some knowledge of the laws of exponents, because the logarithmic function is the inverse of a particular exponential function.

DEFINITION 1   *Let b be a fixed positive number. Then, the function defined by*

(7)
$$y = b^x$$

for all real x, is called an exponential function. The number b is called the base of the function.

When we turn to the definition and the use of logarithms, there are only two values for b that are of interest. However, in developing the theory we prefer to let b be an arbitrary fixed positive number, but it is convenient to select b so that $b > 1$.

To obtain a feeling for the behavior of the function $f(x) = b^x$, we sketch the graph for the base $b = 2$. It is easy to locate points of the graph when x

FIGURE 11.1

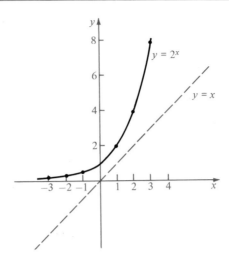

is an integer and to approximate points of the graph when x is not an integer. It can be proved that the function 2^x is continuous and increasing, but the proof is long and requires careful and special preparation; so we omit it. If we plot a few of the points when x is an integer and use the fact that 2^x is continuous and increasing, we arrive at the sketch shown in Figure 11.1.

Exercise 1

1. Sketch the graph of each of the following exponential functions: (a) $y = 3^x$, (b) $y = (3/2)^x$, (c) $y = 1^x$, (d) $y = (2/3)^x$, (e) $y = (1/2)^x$.

2. Select numbers b, u, and v, and use these numbers to illustrate equations (1) through (6).

11.2 THE LOGARITHMIC FUNCTION

The logarithmic function is the inverse of the exponential function. Of course, the inverse function depends on the base b. Hence, in the notation $y = \log x$ for the logarithmic function, we adjoin b as a subscript and we write $y = \log_b x$ (read, "y is the logarithm of x to the base b") to denote this new function.

DEFINITION 2

Suppose that $b > 0$ and $b \neq 1$. The inverse function of the exponential function

(7)
$$y = b^x$$

FIGURE 11.2

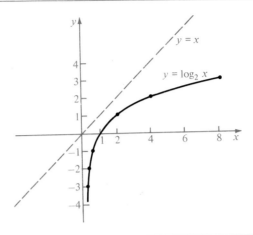

is called the logarithmic function to the base b and is denoted by

(8) $$y = \log_b x.$$

We obtain the graph of $y = \log_b x$ by reflecting the graph of $y = b^x$ across the line $y = x$. Using this technique on the graph shown in Figure 11.1, we obtain the graph of $y = \log_2 x$ shown in Figure 11.2.

Observe that in going from equation (7) to equation (8), we have interchanged x and y, so that x is the independent variable in both equations.

We now use the rather neutral symbols N and u (in place of x and y) and let N and u have the same meaning as we pass from the exponential function to its inverse. Then the definition can be put in the form

(9) $$u = \log_b N \quad \text{if and only if} \quad b^u = N.$$

This latter form is often more useful. Restated in words, we have the alternative

DEFINITION 3   *The logarithm of a number N to the base b is the number u that must be used as an exponent on b to give N.*

The following examples illustrate the use of the criterion given in (9):

$$\log_2 8 \quad = 3 \quad \text{because} \quad 2^3 \quad = 8,$$
$$\log_2 16 \quad = 4 \quad \text{because} \quad 2^4 \quad = 16,$$
$$\log_{16} 16 \quad = 1 \quad \text{because} \quad 16^1 \quad = 16,$$
$$\log_5 25 \quad = 2 \quad \text{because} \quad 5^2 \quad = 25,$$

$$\log_5 1 \quad = 0 \qquad \text{because} \qquad 5^0 \quad = 1,$$

$$\log_4 2 \quad = 0.5 \qquad \text{because} \qquad 4^{0.5} = \sqrt{4} = 2,$$

$$\log_4 0.5 \quad = -0.5 \qquad \text{because} \qquad 4^{-0.5} = 1/\sqrt{4} = 0.5,$$

$$\log_4 0.0625 = -2 \qquad \text{because} \qquad 4^{-2} \quad = 1/16 = 0.0625,$$

$$\log_{10} 100 \quad = 2 \qquad \text{because} \qquad 10^2 \quad = 100,$$

$$\log_{64} 4 \quad = \frac{1}{3} \qquad \text{because} \qquad 64^{1/3} = \sqrt[3]{64} = 4.$$

Exercise 2

Find each of the following logarithms.

1. $\log_{10} 10$ 2. $\log_{10} 1000$ 3. $\log_{10} 10{,}000$ 4. $\log_{10} 0.1$

5. $\log_{10} 0.01$ 6. $\log_{10} 0.001$ 7. $\log_2 32$ 8. $\log_3 81$

9. $\log_4 256$ 10. $\log_4 64$ 11. $\log_8 2$ 12. $\log_8 4$

13. $\log_8 32$ 14. $\log_8 16$ 15. $\log_2 \dfrac{1}{16}$ 16. $\log_3 \dfrac{1}{81}$

17. $\log_9 \dfrac{1}{3}$ 18. $\log_9 \dfrac{1}{27}$ 19. $\log_{1/2} 4$ 20. $\log_{1/3} 27$

Find the unknown N, b, or u in each of the following equations.

21. $\log_b 32 = 5$ 22. $\log_b 81 = 4$ 23. $\log_9 N = \dfrac{1}{2}$

24. $\log_8 N = \dfrac{2}{3}$ 25. $\log_4 32 = u$ 26. $\log_8 \sqrt{2} = u$

27. $\log_{\sqrt{2}} 8 = u$ 28. $\log_{32} 8 = u$ 29. $\log_3 N = 4$

30. $\log_4 N = 3$ 31. $\log_b \dfrac{1}{9} = -\dfrac{1}{2}$ 32. $\log_5 N = -2$

33. $\log_4 N = -3$ 34. $\log_b \dfrac{1}{7} = -\dfrac{1}{3}$ 35. $\log_{\sqrt{3}} 81 = u$

★ 36. $\log_4 (\log_4 16) = u$ ★ 37. $\log_5 (\log_3 243) = u$

★ 38. $\log_b (\log_b 27^3) = 2$ ★ 39. $\log_2 (\log_2 N) = 3$

40. What is the domain and range of the function (a) 2^x, (b) 10^x?

41. What is the domain and range of the function (a) $\log_2 x$, (b) $\log_{10} x$?

42. Explain why $b^{\log_b N} = N$ for every pair of positive numbers N, b, with $b \neq 1$.

11.3 THE THREE LAWS OF LOGARITHMS

Using only the definition of a logarithm and the laws of exponents, equations (1), (2), and (3), we can prove three important theorems about logarithms. In these theorems the base b is assumed to be the same throughout, and, hence, we may omit the subscript b from $\log_b$

Theorem 1   *The logarithm of a product‡ is equal to the sum of the logarithms of its factors,*

(10) $$\log MN = \log M + \log N.$$

Proof: Let

(11) $$u = \log M \quad \text{and} \quad v = \log N.$$

Then by (9), the definition of a logarithm,

(12) $$M = b^u \quad \text{and} \quad N = b^v.$$

From the laws of exponents, equation (1), we find that

(13) $$MN = b^u b^v = b^{u+v}.$$

We observe that $u+v$ is the exponent on b that gives MN. Hence, by the definition of a logarithm, $u+v$ is the logarithm of MN to the base b. Using this fact and equation set (11), we obtain

(14) $$\log MN = u + v = \log M + \log N. \quad \blacksquare$$

Theorem 2   *The logarithm of a fraction is equal to the logarithm of the numerator minus the logarithm of the denominator*

(15) $$\log \frac{M}{N} = \log M - \log N.$$

Proof: Again we let $u = \log M$ and $v = \log N$. Then we have equation set (12). This time we use equation (2) of the laws of exponents, and we find that

(16) $$\frac{M}{N} = \frac{b^u}{b^v} = b^{u-v}.$$

---

‡ Since the domain of the logarithm function is the set of positive numbers, we must assume that M and N are positive. Henceforth, whenever we write $\log x$, it is understood that $x > 0$.

Since $u - v$ is the exponent on b that gives M/N, we have

(17)
$$\log \frac{M}{N} = u - v = \log M - \log N. \quad \blacksquare$$

Theorem 3

The logarithm of the kth power of a number is equal to k times the logarithm of the number,

(18)
$$\log M^k = k \log M.$$

Proof: Again we let $u = \log M$ and then $M = b^u$. We raise both sides of this equation to the kth power and use equation (3) of the laws of exponents. Then we obtain

(19)
$$M^k = (b^u)^k = b^{ku}.$$

Since ku is the exponent on b that gives M^k, we have

(20)
$$\log M^k = ku = k \log M. \quad \blacksquare$$

Since each of the first three laws of exponents gave a corresponding theorem for logarithms, our curiosity should be aroused about the other three laws of exponents given by equations (4), (5), and (6). Equation (4) gives

(21)
$$\log 1 = 0,$$

equation (5) gives

(22)
$$\log \frac{1}{N} = -\log N,$$

and equation (6) gives

(23)
$$M^{\log N} = N^{\log M}.$$

Equation (23) is never needed in elementary work and can safely be ignored by the student. The other two equations are important. Their proofs are posed as problems 29 and 30 in Exercise 3.

Exercise 3

In problems 1 through 20 all logarithms are to the base 10. Given that $\log 2 = 0.301$, $\log 3 = 0.477$, and $\log 7 = 0.845$, compute each of the following logarithms.

Hint: Remember that $\log 10 = 1$ and $5 = 10/2$.

1. $\log 4$ 2. $\log 5$ 3. $\log 6$ 4. $\log 8$

5. $\log 9$ 6. $\log 16$ 7. $\log 81$ 8. $\log \dfrac{1}{3}$

9. $\log \dfrac{1}{2}$ 10. $\log \sqrt{3}$ 11. $\log \sqrt[4]{7}$ 12. $\log \sqrt[6]{14}$

13. $\log 70$ 14. $\log 700$ 15. $\log \dfrac{14}{3}$ 16. $\log \dfrac{21}{2}$

17. $\log \dfrac{3}{14}$ 18. $\log \dfrac{2}{21}$ 19. $\log \sqrt{3000}$ 20. $\log \sqrt[3]{2000}$

In problems 21 through 28 identify each assertion (equation) as true or false. If it is false, indicate how to correct the assertion to make it true.

21. $(\log A)^4 = 4 \log A$ 22. $\log ABC^2 = \log A + \log B + 2 \log C$

23. $\log X - \log Y = \dfrac{X}{Y}$ 24. $\log \dfrac{A+B}{C} = \log A + \log B - \log C$

25. $\log B = 2 \log \sqrt{B}$ 26. $\log \dfrac{D}{EF} = \dfrac{\log D}{\log E + \log F}$

27. $3 \log(\log B) = \log(\log B^3)$ 28. $2 \log A^3 = \log A^8 - \log A^2$

29. Prove that $\log 1 = 0$ by setting $N = 1$ in equation (9).

30. Prove that equation (22) is true by setting $M = 1$ in equation (15).

31. Starting with equation (10), prove that $\log MNP = \log M + \log N + \log P$.

★ 32. Starting with equation (6), prove that equation (23) is true. *Hint:* Set $b^u = M$ and $b^{1/v} = N$.

11.4 VARIOUS SYSTEMS OF LOGARITHMS

Before logarithms can be useful as an aid to computation, extensive tables must be prepared. Before these tables can be prepared, a suitable base must be selected.

If the base is $e = 2.7182 \cdots$, the logarithms are known as *natural logarithms*, or Naperian logarithms, after the inventor of logarithms.

If the base is 10, the logarithms are known as *common logarithms*, or Briggsian logarithms, after Henry Briggs (1561–1630), who first suggested the use of 10 as a base.

Since e is an irrational number, the student will raise his eyebrows at the name "natural" for logarithms with such a horrible base. After all, what number is more natural than 10? A course in differential calculus will help to explain this nomenclature. As the student goes further in mathematics he will see why e, along with π, is one of the most important irrational numbers in natural science and why logarithms with this base are called natural logarithms. For the rest of this text, however, the base will be 10, and the notation $\log N$ will mean the logarithm to the base 10.

In Exercise 3, we used $\log 2 = 0.301$. This means that

$$2 = 10^{0.301} = 10^{301/1000}$$

or if 10 is raised to the power indicated by the numerator (301) and if the root is extracted as indicated by the denominator (1000), the result should be 2. Strictly speaking, the result will not be 2, but will be an irrational number very close to 2. It can be proved that in order to obtain 2 exactly, the exponent must be an irrational number and 0.301 represents just the first three digits of this number. This proof and the methods used to obtain the approximations, such as $\log 2 = 0.301$, are too difficult to be given in this book. The student should feel quite pleased that all of the logarithms he will need have already been computed, and the results are recorded to four decimal places in Table C of the Appendix. More extensive tables to a greater number of decimal places are available if the need should ever arise. Thus, all of the difficult work connected with logarithms has already been done.

11.5 A TABLE OF LOGARITHMS

Table C of the Appendix gives (to four decimal places) the logarithm of every three digit number between 1 and 10. We will see shortly that with this table we can find (to four decimal places) the logarithm of any number.

Let us first examine Table C in detail. The *border* of the table, which consists of the first column and the top row, gives the number N. The first column gives the first two digits of N. The top row gives the third digit of N. The logarithm of N is found in the *body* of the table, the part to the right of the first column and below the first row.

Suppose that we want to find $\log 8.76$. The first two digits 8.7 are located in the first column. The third digit 6 is found in the top row. The entry in the body of the table under 6, and in the same row with 8.7 is 0.9425. Hence, we infer (correctly) that

(24) $$\log 8.76 = 0.9425.$$

Table C can also be used in the reverse direction. Thus if we are given $\log N$, we look in the body of the table for $\log N$ and then determine N from the corresponding entry on the border. For example, if $\log N = 0.0899$, we locate this logarithm in the space under 3 in the same row with 1.2. Hence, if $\log N = 0.0899$, then $N = 1.23$. We speak of N as the *antilogarithm* of 0.0899, and we write

(25)   antilog $0.0899 = 1.23.$

Suppose that $\log N = 0.4807$, and we wish to find N. If we look in Table C, we do not find 0.4807 but numbers on either side

$$\log 3.02 = 0.4800$$
$$\log N \quad = 0.4807$$
$$\log 3.03 = 0.4814$$

from which we might guess (correctly) that $N = 3.025$. This process of reading between the lines of a table is called *interpolation*. We postpone a serious study of interpolation until Section 11.7.

EXAMPLE 1   Find: (a) $\log 4.79$,   (b) $\log 5.00$,   (c) $\log 9.70$.

Solution: Using Table C we find:

(a) $\log 4.79 = 0.6803$,   (b) $\log 5.00 = 0.6990$,   (c) $\log 9.70 = 0.9868$. ▲

EXAMPLE 2   Find N if:

(a) $\log N = 0.4440$,   (b) $\log N = 0.7767$,   (c) $\log N = 0.9090$.

Solution: We find these numbers in the body of Table C

$$\log 2.78 = 0.4440 \quad \text{or} \quad \text{antilog } 0.4440 = 2.78,$$
$$\log 5.98 = 0.7767 \quad \text{or} \quad \text{antilog } 0.7767 = 5.98,$$
$$\log 8.11 = 0.9090 \quad \text{or} \quad \text{antilog } 0.9090 = 8.11.$$

Hence, (a) $N = 2.78$, (b) $N = 5.98$, (c) $N = 8.11$. ▲

We will now show how to find the logarithm of any positive number using Table C. First observe that when the base is 10,

$$\log 1000 \;=\; 3 \quad \text{because} \quad 10^3 = 1000,$$
$$\log \;\;100 \;=\; 2 \quad \text{because} \quad 10^2 = \;\;100,$$
$$\log \;\;\;10 \;=\; 1 \quad \text{because} \quad 10^1 = \;\;\;10,$$

$$\log \quad 1 \quad = \quad 0 \quad \text{because} \quad 10^0 \quad = \quad 1,$$

$$\log \quad 0.1 \quad = \quad -1 \quad \text{because} \quad 10^{-1} = \frac{1}{10} = 0.1,$$

$$\log \quad 0.01 = -2 \quad \text{because} \quad 10^{-2} = \frac{1}{100} = 0.01.$$

Suppose now that we wish to find $\log 876{,}000$. Using powers of 10 we can write

$$876{,}000 = 8.76 \times 10^5.$$

Then by Theorem 1 for the logarithm of a product

$$\log 876{,}000 = \log(8.76 \times 10^5) = \log 8.76 + \log 10^5 = 5 + \log 8.76.$$

From Table C, $\log 8.76 = 0.9425$. Hence, $\log 876{,}000 = 5.9425$.

This process is general. Before stating the rule for finding the logarithm of any number, we need

DEFINITION 4   *Scientific Notation* ‡   *A number N is said to be written in scientific notation if it has the form*

(26) $N = A \times 10^c$

where c is an integer and $1 \leqq A < 10$.

If N is in scientific notation, then by Theorem 1

$$\log N = \log A + \log 10^c$$

(27) $\log N = c + \log A.$

Now c is an integer that is easy to compute, and $\log A$ can be found directly from Table C. Hence, (27) gives the procedure for finding the logarithm of any positive number. It is convenient to give special names to the parts in equation (27).

DEFINITION 5   *When $N = A \times 10^c$ is in scientific notation, then $\log A$ is called the mantissa of the logarithm, and c is called the characteristic.*

Table C is a table of mantissas, and the mantissa is always between zero and one. The characteristic is always an integer.

‡ See Appendix 6 for more detailed information on scientific notation.

RULE 1

To find the logarithm of a number N, use Table C to find the fractional part of the logarithm (the mantissa) using only the digits of N and ignoring the decimal point. Then add to this fraction, the integer c (the characteristic) determined by equation (26).

If we move the decimal point q places to convert N to a number A such that $1 \leqq A < 10$, then

$$(28) \qquad \begin{cases} c = q > 0 & \text{if} & N \geqq 10, \\ c = 0 & \text{if} & 1 \leqq N < 10, \\ c = -q < 0 & \text{if} & N < 1. \end{cases}$$

If we are given $\log N$ and wish to find N, then the above procedure is reversed.

EXAMPLE 3

Find: (a) $\log 35.7$, (b) $\log 259{,}000{,}000$, (c) $\log 0.000952$.

Solution: In (a) we move the decimal point one place to convert $N = 35.7$ to $A = 3.57$. Hence, $q = 1$ and $c = 1$. In (b) $q = 8$ and $c = 8$. In (c) $q = 4$ and $c = -4$. From Table C we have

(a) $\qquad \log 35.7 = \log(3.57 \times 10^1) = \quad 1 + 0.5527 = 1.5527,$

(b) $\quad \log 259{,}000{,}000 = \log(2.59 \times 10^8) = \quad 8 + 0.4133 = 8.4133,$

(c) $\quad \log 0.000952 = \log(9.52 \times 10^{-4}) = -4 + 0.9786 = 0.9786 - 4.$ ▲

EXAMPLE 4

Find N if: (a) $\log N = 4.3201$ and (b) $\log N = 0.8887 - 6$.

Solution: We use Table C to determine the digits from the mantissa. Then we locate the decimal point from the characteristic. We find that

(a) $\qquad N = \text{antilog} \, 4.3201 = 2.09 \times 10^4 = 20{,}900,$

(b) $\qquad N = \text{antilog} \, 0.8887 - 6 = 7.74 \times 10^{-6} = 0.00000774.$ ▲

Exercise 4

In problems 1 through 16 use Table C to find the logarithm of the given number.

1. 7.65	2. 8.52	3. 9.00	4. 3.00
5. 1.35	6. 2.46	7. 6.42	8. 9.63
9. 230	10. 3570	11. 191,000	12. 4870×10^6
13. 0.0230	14. 0.00357	15. 0.000084	16. 0.070×10^{-3}

In problems 17 through 32 use Table C to find N to three significant figures.

17. $\log N = 0.1553$ 18. $\log N = 0.2430$

19. $\log N = 0.3139$ 20. $\log N = 0.4425$

21. $\log N = 0.4900$ 22. $\log N = 0.7177$

23. $\log N = 0.8808$ 24. $\log N = 0.9877$

25. $\log N = 7.5315$ 26. $\log N = 1.7076$

27. $\log N = 5.5775$ 28. $\log N = 4.9499$

29. $\log N = 0.8663 - 4$ 30. $\log N = 0.8785 - 3$

31. $\log N = 0.7404 - 7$ 32. $\log N = 0.5888 - 8$

11.6 THE NEGATIVE CHARACTERISTIC

In Example 3 of the preceding section we found that

(29) $$\log 0.000952 = 0.9786 - 4.$$

The reader may well ask why we did not complete the computation in (29). If we perform the indicated subtraction, we obtain

$$-4.0000$$
$$+0.9786$$

(30) $$\log 0.000952 = -3.0214$$

Equation (30) is correct, but it is not convenient for computation because the fractional part -0.0214 is negative. The inconvenience will become clear to the reader after we have finished Section 11.8. In the meantime we make

GENTLEMAN'S
AGREEMENT

Each logarithm will be written as a positive fraction (the mantissa) plus an integer which may be positive, negative, or zero.

We always prefer the form (29) over (30). We often standardize the negative integer which appears in (29) by using some multiple of -10. For example, we may adjust the right side of (29) by adding 6 in front and subtracting 6 at the end. These operations give

(31) $$\log 0.000952 = 6.9786 - 10.$$

If we wish to be more generous with our additions and subtractions we may write

(32) $$\log 0.000952 = 16.9786 - 20$$

(33) $$\log 0.000952 = 36.9786 - 40$$

(34) $$\log 0.000952 = 76.9786 - 80.$$

Equations (31), (32), (33), and (34) are all equivalent. In Section 11.8 we will see why these various forms are useful.

11.7 INTERPOLATION

EXAMPLE 1   Find $\log 1.918$.

Solution: We can find $\log 1.910$ and $\log 1.920$ from Table C. We arrange these numbers as shown with the larger one first, to make subtraction easier.

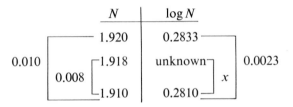

The difference between pairs of numbers are shown on the side. If we assume that the change in $\log N$ is proportional to the change in N, then the above scheme gives

(35) $$\frac{0.008}{0.010} = \frac{x}{0.0023}$$

From this equation we infer that

$$x = \frac{8}{10}(0.0023) = 0.00184 \approx 0.0018.$$

Hence, $\log 1.918 = 0.2810 + 0.0018 = 0.2828.$ ▲

The above work can be abbreviated if we drop the decimal points in the differences and work only in the "last column" on each side. This gives the equivalent scheme

N	log N
1.920	0.2833
1.918	unknown
1.910	0.2810

10 8 23 x

This time the proportion is

(36)
$$\frac{8}{10} = \frac{x}{23}$$

in place of (35). Equation (36) gives $x = 8(23/10) \approx 18$ in the last two columns on the right. Hence, $\log 1.918 = 0.2810 + 0.0018 = 0.2828$ as before.

EXAMPLE 2   Find $\log 813.3$.

Solution: From Table C we have

N	log N
814.0	2.9106
813.3	unknown
813.0	2.9101

This time the proportion is

$$\frac{3}{10} = \frac{x}{5}.$$

Hence, $x = 15/10 = 1.5$. We must make a choice between $x = 1$ and $x = 2$. There is no logical reason for selecting either number. For the sake of uniformity, we agree to round off so that the last digit in the final answer is even. Consequently, we set $x = 1$ and obtain $\log 813.3 = 2.9102$. ▲

The process illustrated in Examples 1 and 2 is called *interpolation*. It can be used in the reverse direction as indicated in

EXAMPLE 3   If $\log N = 5.4087$, find N.

Solution: For simplicity we ignore the characteristic. Table C gives

N	log N
2.570	0.4099
A	0.4087
2.560	0.4082

Consequently, we infer that

(37)
$$\frac{x}{10} = \frac{5}{17}.$$

Hence, $x = 50/17 \approx 3$. Therefore, $A = 2.563$. Since the characteristic of $\log N$ is 5, we find that $N = 256,300$. ▲

Now that we are familiar with interpolation, let us look at the theory which lies behind the process.

The graph of $y = \log x$ resembles the one in Figure 11.2. However, if we look at a very small portion of this curve and magnify that portion (use a larger scale), the curve will appear to be a straight line. For example, if we sketch that part of the graph for which $1.910 \leq x \leq 1.920$ we obtain the curve (?) shown in Figure 11.3. If we assume that the curve joining the points A and E in Figure 11.3 is indeed a straight line, then

(38) $\qquad \triangle ABC \quad$ is similar to $\quad \triangle ADE.$

From this similarity of triangles, we obtain

(39) $$\frac{|AB|}{|AD|} = \frac{|BC|}{|DE|},$$

for the lengths of the sides of the triangles. Equation (35) used for Example 1 is just equation (39) for that example.

Summary *Interpolation is based on the assumption that for a small range of the variable the graph of the function is a straight line.*

If the scheme is used in the manner indicated in the three examples, then there is no need to draw the similar triangles.

FIGURE 11.3

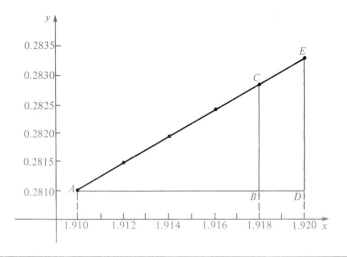

Exercise 5

In problems 1 through 4 express the logarithm of the given number in the form of equation (31).

1. 0.0543   2. 0.0000666   3. 7.89×10^{-6} 4. 9.41×10^{-9}

In problems 5 through 8 find N.

5. $\log N = 9.9222 - 10$ 6. $\log N = 17.6263 - 20$

7. $\log N = 24.2430 - 30$ 8. $\log N = 35.3636 - 37$

In problems 9 through 20 use interpolation in Table C to find the logarithm of the given number.

9. 54.32   10. 765.4   11. 0.8766   12. 0.009753

13. 2.345   14. 34,510,000   15. 0.0003007   16. 1999

17. 333.3   18. 77.77   19. 8997   20. 0.02467

In problems 21 through 30 use interpolation in Table C to find N to four significant figures.

21. $\log N = 5.4320$ 22. $\log N = 7.6543 - 10$

23. $\log N = 1.2345$ 24. $\log N = 3.4343$

25. $\log N = 6.5656 - 10$ 26. $\log N = 8.8888 - 10$

27. $\log N = 1.9192$ 28. $\log N = 1.1111$

29. $\log N = 8.1357 - 10$ 30. $\log N = 2.3232$

11.8 LOGARITHMS AS AN AID TO COMPUTATION

In this section we give six examples that show how logarithms can be used to facilitate computations.

EXAMPLE 1

Find the product $P = (234)(543)$.

Solution: By Theorem 1, $\log P = \log 234 + \log 543$.

$$\log 234 = 2.3692$$
$$\log 543 = 2.7348$$
$$\overline{\log P = 5.1040}$$
$$P = 127,100. \quad \blacktriangle$$

The exact value of P is 127,062, so that for this problem logarithmic computation gives the answer correct to four significant figures.

The work will go more smoothly if the student does as much of the problem as possible before looking in the tables. This means that he should write the logarithmic equation, arrange the work in a column, and fill in the characteristics. In short, he should write every part of the solution except the mantissas, and then turn to the tables.

EXAMPLE 2   Find the quotient $Q = 234/543$ to four significant figures.

Solution: By Theorem 2, $\log Q = \log 234 - \log 543$.

$$\log 234 = 2.3692$$
$$\log 543 = 2.7348.$$

But $\log 543$ is larger than $\log 234$, so subtraction will yield a negative quantity. To avoid this difficulty we adjust $\log 234$ by adding and subtracting 10. We have then

$$\log 234 = 12.3692 - 10$$
$$\log 543 = 2.7348$$
$$\overline{\phantom{\log 234 ={}}}$$
$$\log Q = 9.6344 - 10$$
$$Q = 0.4309. \quad \blacktriangle$$

Direct computation gives $Q = 0.4309392 \cdots$. Again, logarithmic computation gives the answer correct to four significant figures.

EXAMPLE 3   Find $\sqrt{0.1776}$ to four significant figures.

Solution: Let R be the desired root. Then by Theorem 3,

$$\log R = \log(0.1776)^{1/2} = \frac{1}{2}\log 0.1776$$

$$\log 0.1776 = 9.2494 - 10$$

$$\frac{1}{2}\log 0.1776 = 4.6247 - 5$$

$$R = 0.4214. \quad \blacktriangle$$

As a check we can square the answer. We find that

$$(0.4214)^2 = 0.17757796$$
$$(0.4215)^2 = 0.17766225.$$

Hence, logarithmic computation gives R correct to four significant figures.

EXAMPLE 4

Find $\sqrt[3]{0.1776}$ to four significant figures.

Solution: If we parallel the work of Example 3, we will run into trouble. Let us do so and see what happens.

$$\log R = \frac{1}{3}\log 0.1776 = \frac{1}{3}(9.2494 - 10) = 3.0831 - 3.3333.$$

This will give a negative fractional part after subtraction. To avoid this difficulty we may first alter the -10, which stands at the right in $\log 0.1776$, so that the new integer is divisible by 3. There are many ways of doing this, but the standard procedure is to change -10 into -30 by adding 20 in front and subtracting it at the back. We then have

$$\log R = \frac{1}{3}(9.2494 - 10) = \frac{1}{3}(29.2494 - 30) = 9.7498 - 10,$$

$$R = 0.5621. \quad \blacktriangle$$

As a check we cube the answer. We find that

$$(0.5621)^3 = 0.177599098061$$
$$(0.5622)^3 = 0.177693901848.$$

Clearly, $R = 0.5621$ is correct to four significant figures.

EXAMPLE 5

Compute $\dfrac{0.0044\sqrt[5]{666}}{5.65\sqrt[3]{10,000}}$ to four significant figures.

Solution: Before using the tables, we make an outline showing the procedure we intend to follow. Let Q denote the quantity sought. Then

$$\log Q = \log 0.0044 + \frac{1}{5}\log 666 - \left(\log 5.65 + \frac{1}{3}\log 10,000\right)$$

$\log 10,000 = 4.0000$	$\log 666 \quad = 2.$
$\frac{1}{3}\log 10,000 = 1.3333$	$\frac{1}{5}\log 666 \quad =$
$\log 5.65 \quad =$	$\log 0.0044 = 7. \qquad -10$
$\log D =$	$\log N =$
	$\log D =$

$$\log Q = \log N - \log D =$$

The student is invited to fill in the blanks and show that $Q = 0.0001327$. $\quad \blacktriangle$

Initially no name was given to the unknown quantity. We let N denote the numerator, D the denominator, and Q their quotient. The introduction of letters for unknown quantities makes it easier to think accurately about them. The student should form the habit of attaching names (letters) to any unnamed quantities that he meets in his scientific studies.

EXAMPLE 6   Compute $(3.456)^{2.753}$.

Solution: Let $P = (3.456)^{2.753}$. Then

$$\log P = 2.753 \log 3.456$$

$$\log(\log P) = \log 2.753 + \log(\log 3.456)$$

$$\log 3.456 = 0.5386$$

$$\log(\log 3.456) = 9.7313 - 10$$

$$\log 2.753 = 0.4398$$

$$\log(\log P) = 10.1711 - 10$$

$$\log P = 1.483$$

$$P = 30.41. \quad \blacktriangle$$

Because we used $\log P = 1.4830$, where the last digit was certainly in doubt, it is quite likely that the answer is accurate only to three significant figures. If we needed more accuracy, then we would look for a table of logarithms that gives the logarithms to 5 (or more) decimal places.

The reader is invited to try to work this problem on a calculator the next time one is available.

Exercise 6

In problems 1 through 20 use logarithms to compute the given quantity to three significant figures. (Do not interpolate.)

1. $(17.9)(0.196)$

2. $(0.00865)(1,990,000)$

3. $\dfrac{5,550}{0.0368}$

4. $\dfrac{0.0864}{6,840}$

5. $\dfrac{0.947}{9,740}$

6. $\dfrac{57,900}{0.759}$

7. $\dfrac{1}{(1.23)^6}$

8. $\dfrac{1}{(0.876)^8}$

9. $(234)(54.3)(0.687)$

10. $(4.32)(5,790)(8,520)$

11. $\dfrac{(24.6)(46.8)}{(7,530)(5,310)}$

12. $\dfrac{(8,460)(7,350)}{(4.86)(6.42)}$

13. $(1,110)^{3/2}$

14. $(2,320)^{5/4}$

15. $\sqrt[7]{200}$

16. $\sqrt[7]{2000}$

17. $\sqrt[7]{0.02}$

18. $\sqrt[7]{0.002}$

19. $\sqrt[3]{\dfrac{(243)(0.0888)}{95.5}}$

20. $\sqrt[4]{\dfrac{(832)(0.855)}{88,800}}$

In problems 21 through 47 use logarithms to compute the given quantity to four significant figures.

21. $(87.53)(0.0002347)$

22. $(3211)(0.06787)$

23. $\dfrac{0.008765}{45.67}$

24. $\dfrac{0.000004582}{0.0004953}$

25. $\dfrac{1}{(0.01351)^{2/3}}$

26. $\dfrac{10,000}{(2121)^3}$

27. $\dfrac{(130.5)^3}{0.0355}$

28. $\dfrac{(210)^3}{0.04242}$

29. $\sqrt[7]{0.009222}$

30. $\sqrt[6]{0.0009876}$

31. $\sqrt{20,420}$

32. $\sqrt{2042}$

33. $\left(\dfrac{3232}{0.5656}\right)^2$

34. $\dfrac{0.3333\sqrt{44.4}}{\sqrt[3]{(48)(685)}}$

35. $\sqrt{\dfrac{6321}{81.25\sqrt[3]{0.16}}}$

36. $\sqrt{\dfrac{(6123)(22.22)}{(81)(25)(0.16)^3}}$

37. $\dfrac{\sqrt{(1623)(33.56)}}{(2.815)(0.017)^2}$

38. $\dfrac{(52)(51)(50)(49)(48)}{(5)(4)(3)(2)(1)}$

★39. $\sqrt[3]{\dfrac{0.3010}{0.6990}}$

★40. $\sqrt{\dfrac{0.5441}{0.3979}}$

★41. $\sqrt[3]{\dfrac{\log 2}{\log 5}}$

★42. $\sqrt{\dfrac{\log 3.5}{\log 2.5}}$

★43. $(\log 1.776)(\log 1.984)$

★44. $(\log 25)(\log 50)$

★45. $\dfrac{\log 1,776}{\log 1.984}$

★46. $\dfrac{\log 50}{\log 25}$

★47. The notation $n!$ (read, "n factorial") means the product of all the integers from 1 to n inclusive. For example, in problem 38 the quantity could be written $52!/(5!\,47!)$ and gives the number of possible poker hands. The expression $B = 52!/(13!\,39!)$ gives the number of possible bridge hands. Simplify the right side by cancellation and then use logarithms to estimate B.

★48. Kepler's Law. This law states that the square of the time required for a planet to make one orbit around the sun varies as the cube of its average distance from the sun. If this average distance is 92,900,000 miles for the earth and 67,200,000 miles for Venus, find the number of days for Venus to make one orbit around the sun.

★49. Find the period of Mars to the nearest day if the average distance from the sun to Mars is 142,000,000 miles.

★11.9 **EXPONENTIAL EQUATIONS**

In certain equations the variable occurs in the exponent. Such an equation is called an *exponential equation*.

EXAMPLE 1   Solve the equation $5^x = 50$.

Solution: If we take logarithms of both sides we find

$$x \log 5 = \log 50$$

$$x = \frac{\log 50}{\log 5} \quad (\neq \log 10).$$

To compute x, we use logarithms (without interpolation).

$$\log 50 = 1.6990 \qquad \log(\log 50) = \log 1.699 = 10.2304 - 10$$

$$\log 5 \;= 0.6990 \qquad \log(\log 5) \;= \log 0.699 = \;9.8445 - 10$$

$$\log x = \;\;0.3859$$

Hence $x = 2.43$ (to 3 significant figures). ▲

EXAMPLE 2   If P dollars are invested at the rate of interest i per period, compounded, and A is the amount (accumulated value) after n periods, then

(40) $$A = P(1+i)^n.$$

Find the numbers of years (approximately) for \$1,000 to double itself at 4% per year.

Solution: Here $i = 0.04$, and equation (40) gives

$$2000 = 1000(1+0.04)^n$$

$$2 = (1.04)^n$$

$$\log 2 = n \log 1.04$$

$$n = \frac{\log 2}{\log 1.04} = \frac{0.3010}{0.0170} \approx 17.7.$$

To receive \$2,000, it would be necessary to invest the money for 18 years at 4%. ▲

Many natural phenomenon are governed by an exponential equation. The most common type of equation is

(41) $$Q = Q_0 e^{ct},$$

where t is time, Q_0 is the amount when $t = 0$, Q is the amount in general, c is a constant that depends on the particular phenomenon and $e \approx 2.718 \cdots$ is a certain transcendental number that plays a fundamental role in calculus. It can be shown that we can replace e by 10 if we make a suitable change in the value of the constant. For simplicity we replace (41) by the equivalent equation

(42) $$Q = Q_0 10^{kt}.$$

EXAMPLE 3   From past data, it appears that the population of Betatown is governed by equation (42), where Q is the population and $k = 0.02$. Estimate the number of years required for Betatown to grow from a population of 200,000 to half a million.

Solution: Equation (42) gives

$$500,000 = (200,000) 10^{0.02t} \quad \text{or} \quad 2.5 = 10^{0.02t}.$$

Hence, $\log 2.5 = 0.02t \log 10 = 0.02t$. Therefore,

$$t = \frac{\log 2.5}{0.02} = \frac{0.3979}{0.02} = 19.9 \text{ years.} \quad \blacktriangle$$

Exercise 7

In problems 1 through 6 solve for x without using tables.

1. $2^{3x} = 128$
        2. $9^{2x} = 3$
        3. $3^{5x+2} = 27$

4. $2^{x(x+2)} = 8$
        5. $3^{x(x-3)} = \dfrac{1}{3}$
        6. $6^{x(x+1)} = 36^{x(x-1)}$

In problems 7 through 12 find x to three significant figures. (Use tables if necessary.)

7. $2^x = 10$
        8. $3^x = 100$
        9. $4^{2x+7} = 1000$

10. $10^{3x+5} = 20$
        11. $10^{3x+5} = 40$
        12. $5^{3x+5} = 10$

13. Find the least number of years required for $100 to double itself if interest is compounded annually at (a) 3%, (b), 6%, and (c) 9%.

14. Assuming that the money was invested as described in problem 13, find the accumulated value to the nearest dollar at the end of (a) 24 years, (b) 12 years, (c) 9 years.

15. The decay of a radio active substance obeys the law given by equation (42), where the constant k is negative and depends on the particular material and the unit of time used. It is the custom to describe the behavior of the material by giving its half-life, the time required for half of the material to disappear. If the half-life of radium is 1590 years, find the value of k for radium.

16. A research foundation is carefully guarding a sample of pure radium weighing exactly 1 gram. In how many years will this sample contain only 0.9 gram of radium?

17. Find the half-life of Caesium if 11% disintegrates in a period of 5 years.

18. The absorption of x rays by a wall obeys the exponential law given by equation (42) when t is the thickness of the wall. If a one-inch thick wall of a certain material absorbs $1/2$ of the x rays, find k. Find the thickness of a wall made of the same material if it is to absorb 99% of the x rays.

19. In 1626 Peter Minuit paid the Indians 24 dollars for land in New York City. If in 1956 this same land was valued at 4.8×10^{10} dollars (48 billion dollars) find the rate of interest (compounded annually) at which the same investment would have given the same accumulated value.

20. If equations (41) and (42) govern the same phenomenon, find an equation that gives k as a function of c.

Review Questions

Try to answer the following questions as accurately as possible before consulting the text.

1. What is the domain and range of the function $y = 2^x$?

2. What is the domain and range of the function $y = \log_2 x$?

3. State the three laws of logarithms (see Theorems 1, 2, and 3).

4. What is scientific notation?

5. What is the characteristic and the mantissa of a logarithm?

6. Explain the theory on which interpolation is based.

Triangles

To solve a triangle means to find the length of each side and the measure of each angle. This chapter is devoted to methods for solving triangles and their various applications. The right triangle and related topics are covered in Sections 12.1 through 12.6. The general triangle is treated in Sections 12.7, 12.8, and 12.9.

The methods for solving triangles have been known for several hundred years, and the material has been developed over a period covering at least two thousand years. In fact, a Babylonian tablet (known‡ as Plimpton 322) is dated somewhere between 1900 B.C. and 1600 B.C. and appears to be a table of secants. A table of sines (although not called by this name) was computed by Hipparchus around 140 B.C.

The six trigonometric functions were first defined as ratios of the sides of a right triangle by Georg Joachim Rhaeticus (1514–1576). With the aid of hired computers he compiled a ten place table giving each of the trigonometric functions for every 10″ of arc.

Despite its ancient origin, this part of trigonometry may still be very attractive to anyone who is viewing it for the first time.

‡ See Howard Eaves, *An Introduction to the History of Mathematics* 2d ed. (New York: Holt, Rinehart, and Winston, 1953).

12.1 THE TRIGONOMETRIC FUNCTIONS IN A RIGHT TRIANGLE

As indicated in Figure 12.1 the triangle ABC is a right triangle with the right angle at C. The sides are labeled a, b, c, with the side a opposite the angle A, the side b opposite the angle B, and the hypotenuse c opposite the right angle C.

It should be observed that a symbol such as A does triple duty. Indeed in Figure 12.1, A represents (1) a point in the plane, (2) an angle, and (3) the measure of the angle. Similarly, b represents both the side opposite the angle B, and the measure (length) of the side. This multiple use of a symbol should never cause any confusion. In fact, it would be very confusing if we used three different symbols for the three different items represented by A.

To determine the trigonometric functions of the angle A in terms of the sides of the right triangle, we place the right triangle on a rectangular coordinate system, with the vertex A at the origin, and the side AC on the positive x-axis (see Figure 12.2). Let $P(x, y)$ be the point of intersection of the hypotenuse AB with the unit circle. (If $c < 1$, then the line segment AB must be extended until it meets the unit circle.) Finally, we let Q be the point of intersection of the line from P perpendicular to the x-axis.

Since ABC and APQ are both right triangles with a common angle at A, the two triangles are similar. Consequently, the ratios of corresponding sides are equal. Combining this with Definition 3 in Chapter 7 we have

$$\frac{a}{c} = \frac{y}{1} = y = \sin A, \qquad \frac{c}{a} = \frac{1}{y} = \csc A,$$

$$\frac{b}{c} = \frac{x}{1} = x = \cos A, \qquad \frac{c}{b} = \frac{1}{x} = \sec A,$$

$$\frac{a}{b} = \frac{y}{x} = \tan A, \qquad \frac{b}{a} = \frac{x}{y} = \cot A.$$

Thus, we have proved

FIGURE 12.1

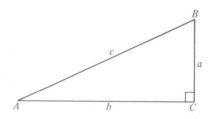

FIGURE 12.2

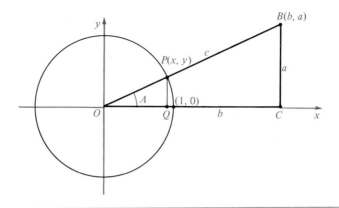

Theorem 1   *If ABC is a right triangle with the right angle at C, then*

$$\sin A = \frac{\text{side opposite the angle } A}{\text{hypotenuse}} = \frac{a}{c},$$

$$\cos A = \frac{\text{side adjacent to the angle } A}{\text{hypotenuse}} = \frac{b}{c},$$

(1)
$$\tan A = \frac{\text{side opposite the angle } A}{\text{side adjacent to the angle } A} = \frac{a}{b},$$

$$\cot A = \frac{\text{side adjacent to the angle } A}{\text{side opposite the angle } A} = \frac{b}{a},$$

$$\sec A = \frac{\text{hypotenuse}}{\text{side adjacent to the angle } A} = \frac{c}{b},$$

$$\csc A = \frac{\text{hypotenuse}}{\text{side opposite the angle } A} = \frac{c}{a}.$$

In the classical theory of trigonometry, equation set (1) was given as the definition of the trigonometric functions when A is an acute angle. Then the definition was extended to include arbitrary angles, as in Chapter 7.

For applications, it is helpful to learn the trigonometric functions as ratios of sides because one frequently must work with triangles that are not in standard position or do not have the standard lettering. For example, if the triangle MNQ shown in Figure 12.3 is a right triangle with the right angle at Q, then (as ratios of length of sides) we have the following.

FIGURE 12.3

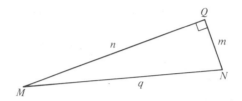

$$\sin N = \frac{\text{side opp. } N}{\text{hypotenuse}} = \frac{n}{q}, \qquad \csc N = \frac{\text{hypotenuse}}{\text{side opp. } N} = \frac{q}{n},$$

$$\cos N = \frac{\text{side adj. } N}{\text{hypotenuse}} = \frac{m}{q}, \qquad \sec N = \frac{\text{hypotenuse}}{\text{side adj. } N} = \frac{q}{m},$$

$$\tan N = \frac{\text{side opp. } N}{\text{side adj. } N} = \frac{n}{m}, \qquad \cot N = \frac{\text{side adj. } N}{\text{side opp. } N} = \frac{m}{n}.$$

Exercise 1

In problems 1 through 4 a right triangle is shown. For each triangle give each of the six trigonometric functions of the smallest angle as a ratio of the lengths of the sides.

1.

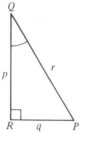

2.

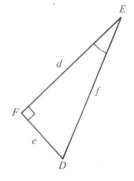

3.

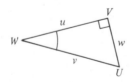

4.

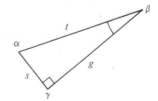

12.2 THE SOLUTION OF A RIGHT TRIANGLE

Usually a triangle is completely determined if we know three of the various items connected with the triangle. For example, if we know three sides, then the triangle is uniquely determined. If we know two sides and the included angle, then the triangle is completely determined. Since we already know one item (the right angle) in a right triangle, the triangle will be completely determined if we specify two more items, such as: (a) two sides, (b) one side and one angle, (c) one angle and the altitude from the vertex of the right angle, etc.

In this section we assume that all lengths are given to four significant figures, and all angles are given to the nearest minute. (See Appendix 6.) Consequently, we round off all answers, either to four significant figures or to the nearest minute.

EXAMPLE 1

Solve the right triangle ABC given $c = 3000$ and $A = 25°0'$, where the right angle is at C.

Solution: The angle B is easily found because in a right triangle $A + B = 90°$. Hence, $B = 90°0' - 25°0' = 65°0'$.

To find the other two sides it will be helpful to make a drawing, see Figure 12.4. It is good practice to make an accurate drawing to scale because the drawing will aid in checking the answers for large errors. Suppose now that we wish to find a. The side a occurs in four of the trigonometric functions of the angle A,

$$\tan A = \frac{a}{b}, \qquad \csc A = \frac{c}{a},$$

$$\cot A = \frac{b}{a}, \qquad \sin A = \frac{a}{c},$$

so which one shall we use? Clearly a and b are both unknown at this stage, so neither $\tan A$ nor $\cot A$ will help, since each one contains both unknowns. Also $\csc A$, can be rejected because Table B does not list it. We have left only

FIGURE 12.4

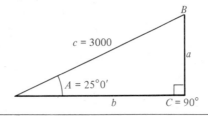

sin A. However, this will give us what we need because in the equation

$$\sin A = \frac{a}{c} \qquad \text{or} \qquad \sin 25°0' = \frac{a}{3000}$$

there is only one unknown. We multiply both sides of this equation by 3000 and use Table B for sin 25° 0'. We find that

$$a = 3000 \sin 25°0' = 3000(0.4226) = 1267.8 \approx 1268.$$

To find b we have,

$$\cos A = \frac{b}{c} \qquad \text{or} \qquad \cos 25°0' = \frac{b}{3000}.$$

Therefore,

$$b = 3000 \cos 25°0' = 3000(0.9063) \qquad \text{(Table B)}$$

$$= 2718.9 \approx 2719. \quad \blacktriangle$$

Notice that we could have found b from a using

$$\tan A = \frac{a}{b} \qquad \text{or} \qquad \tan 25°0' = \frac{1268}{b}.$$

Thus,

$$b = \frac{1268}{\tan 25°0'} = \frac{1268}{0.4663} = 2719.2 \cdots \approx 2719.$$

However, this is not good practice because an error in computing a in the first part would automatically give an error in b. It is much better to use $\tan A = a/b$ as a check. Thus,

$$\tan A = \frac{a}{b} = \frac{1268}{2719} = 0.46634 \cdots \approx 0.4663.$$

Since the table gives tan 25° 0' = 0.4663, the answers for a and b are checked simultaneously.

We could use logarithms for all of these computations. For simplicity, we refrain from using logarithms at this point but we will introduce them in the next section.

In Example 1 we were given the hypotenuse and one angle of the right triangle. Now suppose that we are given one side and an angle.

FIGURE 12.5

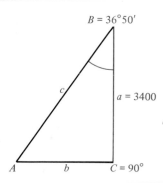

$B = 36° 50'$

$a = 3400$

c

A b $C = 90°$

EXAMPLE 2   Solve the right triangle ABC, given $a = 3400$ and $B = 36° 50'$. (See Figure 12.5.)

Solution: Since $A + B = 90°$, we find $A = 53° 10'$. To find c we have

$$\cos B = \frac{a}{c}.$$

Therefore,

$$c = \frac{a}{\cos B} = \frac{3400}{\cos 36° 50'} = \frac{3400}{0.8004} = 4247.8 \cdots \approx 4248.$$

To find b we could write $\cot B = a/b$, but it is easier to compute b if we use $\tan B = b/a$. Then,

$$b = a \tan B = 3400 \tan 36° 50' = 3400(0.7490) = 2546.6 \approx 2547. \quad \blacktriangle$$

As a check,

$$\sin B = \frac{b}{c} = \frac{2547}{4248} = 0.59957 \cdots \approx 0.5996.$$

From Table B, $\sin 36° 50' = 0.5995$. Since Table B gives only four-figure accuracy, this slight discrepancy is acceptable.

EXAMPLE 3   Solve the right triangle ABC, given $a = 1861$ and $c = 1918$. (See Figure 12.6.)

Solution:

$$\cos B = \frac{1861}{1918} = 0.97028 \cdots,$$

and since (from Table B) $\cos 14° 0' = 0.9703$, we have $B = 14° 0'$ to the nearest minute and $A = 90° - 14° 0' = 76° 0'$ with the same accuracy.

FIGURE 12.6

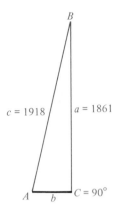

We could compute b using the Pythagorean Theorem, but this is not properly trigonometry. Instead we use $b/c = \sin B$. This gives

$$b = c \sin B = 1918 \sin 14°\,0' = 1918(0.2419) = 463.9642 \approx 464.0. \quad \blacktriangle$$

The Pythagorean Theorem could be used as a check, but it is quicker to use $b/a = \tan B$. Thus,

$$b = a \tan B = 1861 \tan 14°\,0' = 1861(0.2493) = 463.9473 \approx 463.9.$$

Exercise 2

In problems 1 through 10 solve the right triangle ABC, where the right angle is at C. In each problem give the sides to four significant figures and the angle to the nearest minute. Note that if the computations are done correctly, no interpolation is needed in Table B.

1. $c = 1776$, $A = 30°\,40'$ 2. $c = 1914$, $A = 36°\,50'$

3. $c = 1984$, $B = 12°\,50'$ 4. $c = 2000$, $B = 10°\,10'$

5. $a = 2001$, $c = 3000$ 6. $a = 1776$, $c = 4004$

7. $a = 1915$, $A = 39°\,40'$ 8. $a = 1861$, $A = 68°\,30'$

9. $a = 1898$, $b = 2000$ 10. $a = 1776$, $b = 1984$

12.3 LOGARITHMS OF THE
TRIGONOMETRIC FUNCTIONS

The computations involved in the solution of a triangle may require less time and energy if logarithms are used.

EXAMPLE 1   Use logarithms to solve the right triangle ABC, given $c = 98.70$ and $A = 23° 40'$.

Solution: Of course, $B = 90° - A = 66° 20'$ and logarithms are not needed for this computation. Since

$$(2) \qquad a = c \sin A, \qquad b = c \cos A,$$

we need $\log c$, $\log \sin A$, and $\log \cos A$. If we use Table B to find $\sin A$ and then use Table C to find the logarithm of this number, then we can obtain $\log \sin A$ in two steps. However, we can do still better because the result of the two steps is already listed in Table D. In other words, Table D is a table of logarithms of the trigonometric functions. For each entry in the table, 10 must be subtracted. Thus, we find in Table D (page 548) that

$$(3) \qquad \log \sin 23° 40' = 9.6036 - 10, \qquad \log \cos 23° 40' = 9.9618 - 10.$$

If we complete the computations dictated by equation (2), we have:

$$\log c = \log 98.70 = 1.9943 \qquad\qquad \log c = \log 98.70 = 1.9943$$
$$\underline{\log \sin 23° 40' = 9.6036 - 10} \qquad\qquad \underline{\log \cos 23° 40' = 9.9618 - 10}$$
$$\log a = 11.5979 - 10 \qquad\qquad\qquad \log b = 11.9561 - 10$$
$$a = 39.62 \qquad\qquad\qquad\qquad b = 90.38. \quad \blacktriangle$$

EXAMPLE 2   Find B in the right triangle ABC if $a = 34{,}900$ and $b = 27{,}100$.

Solution: Since $\tan B = b/a$, we have

$$\log b = \log 27{,}100 = 14.4330 - 10$$
$$\underline{\log a = \log 34{,}900 = 4.5428}$$
$$\log b/a = \log \tan B = 9.8902 - 10$$

Using Table D (page 550) and the column headed L. tan, we find that $B = 37° 50'$. ▲

12.4 INTERPOLATION IN TRIGONOMETRIC TABLES

The theory of interpolation in trigonometric tables agrees completely with that explained in Section 11.7 for logarithm tables. There is one minor mechanical problem. The functions sine and tangent are increasing functions.

Hence, in interpolation for these two functions we always place the larger angle at the top so that subtraction proceeds in the usual way. But cosine and cotangent are decreasing functions, so we must place the larger angle at the bottom when working with these functions if we wish to make subtraction of the function values proceed normally. This will be illustrated in the following examples. We first look at Table B.

EXAMPLE 1   If $\tan A = 0.5556$, find A to the nearest minute.

Solution: Using Table B we find entries just above and just below 0.5556 in the column headed Tan. Arranging these entries with the larger angle first (tangent is an increasing function) we have

$$
\begin{array}{c|c}
\theta & \tan\theta \\
\hline
29°\ 10' & 0.5581 \\
A & 0.5556 \\
29°\ 0' & 0.5543 \\
\end{array}
$$

10' ⌐ ⌐ A ⌐ 0.5556 ⌐ 13 ⌐ 38   x'

Consequently, we have the proportion

$$\frac{x}{10} = \frac{13}{38}.$$

Hence, $x = 10(13/38) = 3.42\cdots \approx 3$. Therefore, $A = 29°\,0' + 3' = 29°\,3'$ to the nearest minute. ▲

EXAMPLE 2   If $\cos B = 0.8998$, find B to the nearest minute.

Solution: Again we use Table B to find entries just above and below 0.8998. Since cosine is a decreasing function, we place the larger angle at the *bottom* in the following scheme.

$$
\begin{array}{c|c}
\theta & \cos\theta \\
\hline
25°\ 50' & 0.9001 \\
B & 0.8998 \\
26°\ 0' & 0.8988 \\
\end{array}
$$

10' ⌐ ⌐ B ⌐ 0.8998 ⌐ 10 ⌐ 13.   x'

Consequently, we have the proportion

$$\frac{x}{10} = \frac{10}{13}.$$

Hence, $x = 10(10/13) = 7.69 \cdots \approx 8$. Therefore, $B = 26°0 - 8' = 25°52'$ to the nearest minute. ▲

Note that here we subtract $8'$ from $26°0'$ to obtain B.

EXAMPLE 3   Find $\cot 19°43'$ to four significant figures.

Solution: Since cotangent is a decreasing function, we place the larger angle at the *bottom*. The entries in Table B give the scheme

θ	$\cot\theta$
19°40'	2.798
19°43'	
19°50'	2.773

$$\frac{7}{10} = \frac{x}{25}.$$

Hence, $x = 25(7/10) = 17.5$ units in the last column. We round off so that the last digit in the *final answer* is even. In other words, we take $x = 17$, so that $\cot 19°43' = 2.773 + 0.017 = 2.790$. ▲

Interpolation in Table D is performed in the same manner.

EXAMPLE 4   Find $\log \sin 10°34'$.

Solution: Using entries from Table D we have

θ	$\log\sin\theta$
10°40'	9.2674 − 10
10°34'	
10°30'	9.2606 − 10

$$\frac{4}{10} = \frac{x}{68}.$$

Hence, $x = 68(4/10) = 27.2 \approx 27$. Therefore

$$\log \sin 10°34' = 9.2606 - 10 + 0.0027 = 9.2633 - 10. \quad ▲$$

EXAMPLE 5   Given $\log \cos A = 9.3456 - 10$, find A.

Solution: We recall that if $A > 45°$ we use the right hand border for A and the headings at the bottom (footings).

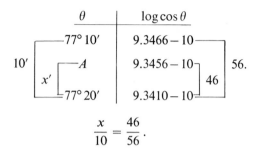

$$\frac{x}{10} = \frac{46}{56}.$$

Hence, $x = 8.21 \cdots \approx 8$. Therefore, $A = 77° 20' - 8' = 77° 12'$. ▲

Exercise 3

In problems 1 through 12 find the indicated trigonometric function to four significant figures by interpolation in Table B.

1. $\sin 19° 41'$	2. $\tan 41° 19'$	3. $\cos 19° 14'$	4. $\cot 14° 19'$
5. $\tan 25° 25'$	6. $\sin 31° 31'$	7. $\cot 84° 19'$	8. $\cos 42° 42'$
9. $\cot 53° 53'$	10. $\sin 25° 25'$	11. $\tan 76° 56'$	12. $\cos 8° 58'$

In problems 13 through 24 find the angle to the nearest minute by interpolation in Table B.

13. $\sin A = 0.1620$	14. $\cos B = 0.1776$	15. $\tan C = 1.812$
16. $\cot D = 1.861$	17. $\cos E = 0.1898$	18. $\sin F = 0.1914$
19. $\cot \alpha = 0.1933$	20. $\tan \beta = 0.1984$	21. $\cos \gamma = 0.4444$
22. $\sin \theta = 0.5555$	23. $\tan \psi = 6.666$	24. $\cos \rho = 0.7777$

In problems 25 through 36 find the indicated logarithm to four decimal places by interpolation in Table D.

25. $\log \sin 25° 34'$	26. $\log \sin 55° 44'$	27. $\log \cos 81° 19'$
28. $\log \cos 12° 34'$	29. $\log \tan 66° 44'$	30. $\log \tan 13° 57'$
31. $\log \cot 5° 6'$	32. $\log \cot 75° 31'$	33. $\log \sin 65° 43'$
34. $\log \sin 34° 56'$	35. $\log \cos 23° 45'$	36. $\log \cos 76° 54'$

In problems 37 through 48 find θ to the nearest minute by interpolation in Table D.

37. $\log \cos \theta = 9.9417 - 10$ 38. $\log \cot \theta = 11.1105 - 10$

39. $\log \tan \theta = 9.1988 - 10$
41. $\log \sin \theta = 9.9281 - 10$
43. $\log \cot \theta = 10.7654 - 10$
45. $\log \sin \theta = 9.5555 - 10$
47. $\log \cos \theta = 9.7532 - 10$

40. $\log \sin \theta = 9.5333 - 10$
42. $\log \cos \theta = 9.8042 - 10$
44. $\log \sin \theta = 9.3579 - 10$
46. $\log \cos \theta = 9.8765 - 10$
48. $\log \tan \theta = 11.1111 - 10$

12.5 THE SOLUTION OF A RIGHT TRIANGLE WITH LOGARITHMS

The theory has already been covered. One example will be sufficient to illustrate the proper arrangement of the computations.

EXAMPLE 1   Solve the right triangle given $a = 23{,}450$ and $b = 34{,}560$. (See Figure 12.7.)

Solution: $\tan A = \dfrac{a}{b}$.

$$\log a = \log 23{,}450 = 14.3702 - 10$$
$$\log b = \log 34{,}560 = 4.5386$$
$$\overline{ \log \tan A = 9.8316 - 10}$$
$$A = 34°\,10'$$
$$B = 90° - A = 55°\,50'$$

To find c we use $\dfrac{a}{c} = \sin A$. Hence $c = \dfrac{a}{\sin A}$.

$$\log a = \log 23{,}450 = 14.3702 - 10$$
$$\log \sin A = \log \sin 34°\,10' = 9.7494 - 10$$
$$\overline{ \log c = 4.6208}$$
$$c = 41{,}760. \quad \blacktriangle$$

FIGURE 12.7

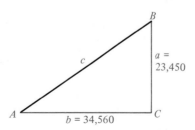

Check We can compute b from the values for A and c we have just obtained, using $b = c \cos A$.

$$\log c \quad = \quad \log 41{,}760 = \quad 4.6208$$
$$\log \cos A = \log \cos 34° 10' = \quad 9.9177 - 10$$
$$\overline{\qquad\qquad\qquad\qquad\qquad\qquad}$$
$$\log b = 14.5385 - 10$$
$$b = 34{,}550.$$

An alternate algebraic check can be made by using the Pythagorean Theorem in the form

$$(4) \qquad\qquad a^2 = c^2 - b^2 = (c+b)(c-b).$$

In this example the computations for a would appear as

$$c = 41{,}760 \qquad \log(c+b) = \log 76{,}320 = 4.8826$$
$$b = 34{,}560 \qquad \log(c-b) = \log \; 7{,}200 = 3.8573$$
$$\overline{\qquad\qquad\qquad}\qquad\qquad\overline{\qquad\qquad\qquad\qquad}$$
$$c + b = 76{,}320 \qquad\qquad\qquad \log a^2 = 8.7399$$
$$c - b = \; 7{,}200 \qquad\qquad\qquad \log a = 4.3700$$
$$a = 23{,}440.$$

Exercise 4

In problems 1 through 8 the triangle ABC is a right triangle with the right angle at C. Use Tables C and D to solve the triangle.

1. $a = 1620$, $b = 1984$
2. $a = 4321$, $b = 7654$
3. $c = 9.876$, $A = 43° 21'$
4. $c = 6.789$, $B = 12° 34'$
5. $b = 0.04343$, $A = 28° 28'$
6. $a = 0.02828$, $B = 43° 43'$
7. $a = 13.58$, $c = 24.68$
8. $b = 53.20$, $c = 86.02$

12.6 SOME TERMINOLOGY FOR APPLICATIONS

The *angle of elevation* (*depression*) of an object is the angle the line of sight to the object makes with the horizontal plane. For example, in Figure 12.8, A is the angle of elevation of the top of the lighthouse as viewed from the point O. B is the angle of depression of the base of the lighthouse as viewed from the same point.

FIGURE 12.8

The *bearing of a line* measures angles in a horizontal plane and is the acute angle this line makes with a north-south line. It is specified by writing first N or S, then the acute angle and E or W. For example, in Figure 12.9 the bearing of the line OA is N 10° E, the bearing of OB is N 30° W. For OC this bearing is S 50° W, and for the line OD it is S 70° E.

The bearing of a point B from a point A is the bearing of the line AB. If P and Q represents two extreme points of an object viewed from a point O, then the angle POQ is said to be the *angle subtended by the object at the point O*. For example, a 36-inch baseball bat held horizontally across home plate subtends an angle of 3° at the eye of the pitcher 60 feet away.

EXAMPLE 1   In order to determine the height of clouds covering the sky over an airport, a searchlight is focused on a cloud directly overhead. From a point on the ground 720 feet away from the searchlight, the angle of elevation of the spot of light on the cloud is 53°. Find the height of the cloud.

FIGURE 12.9

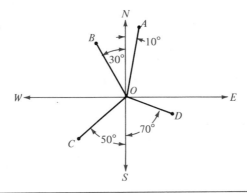

Solution: If *h* denotes this height, then

$$\frac{h}{720} = \tan 53°,$$

$$h = 720 \tan 53° = 720(1.327) = 955.44.$$

Since the angle is given to the nearest degree, we are not justified in reporting more than two significant figures (see Appendix 6). Hence, the cloud is 960 feet above the airport. ▲

EXAMPLE 2

An airplane takes off from an airfield at a point 750 feet from a power line directly ahead. If the top wire of the power line is 70 feet high and if the pilot wishes to clear these lines by 30 feet, what should be his angle of climb?

Solution: Let θ be the angle of climb. Then

$$\tan \theta = \frac{100}{750} = 0.1333.$$

From Table B, $\theta = 8°$ to the nearest degree. ▲

EXAMPLE 3

From a window 31 feet above the ground, the two curbstones on either side of the street have the angles of depression 35° and 22°. How wide is the street?

FIGURE 12.10

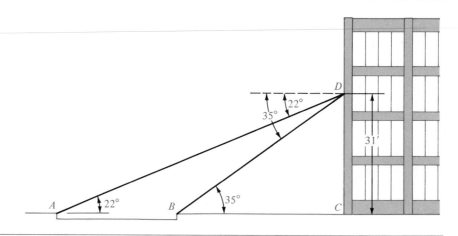

Solution: From Figure 12.10.

$$\frac{|AC|}{|DC|} = \cot 22°, \qquad \frac{|BC|}{|DC|} = \cot 35°,$$

$$|AC| = |DC| \cot 22° = 31(2.475) \qquad |BC| = |DC| \cot 35° = 31(1.428)$$

$$= 76.725, \qquad\qquad = 44.268.$$

$$|AB| = |AC| - |BC| = 76.725 - 44.268 = 32.457 \approx 32 \text{ feet}$$
(with two-figure accuracy).

Notice that if we round off $|AC|$ and $|BC|$ before subtraction, we obtain $|AB| = 77 - 44 = 33$ feet, a slightly different answer. Hence, it may be better not to round off until the last computation has been made. ▲

Exercise 5

In each of the following problems give the answer with the degree of accuracy (number of significant figures) justified by the data (see Appendix 6).

1. The largest tree in California is the General Sherman tree, estimated to be about 3,500 years old. At a point 195 feet from the center of its base and on the same elevation, the angle of elevation of the top of the tree is $54° 30'$. How tall is the tree?

2. The Eiffel Tower in Paris is 984.2 feet high. What will be the angle of elevation of the top from a point on the ground (assumed level) which is 300.0 feet from the center of its base?

3. A regular pentagon (five-sided figure) is inscribed in a circle of radius 15.00. What is the length of any one of its sides? What is the radius of the circle inscribed in this pentagon?

4. A regular decagon (ten-sided figure) is inscribed in a circle of radius 15.00. What is the length of any one of its sides? What is the radius of the circle inscribed in this decagon? Are the answers to this problem one-half the answers in problem 3?

5. A trapeze artist stands with his eyes 72 feet above one edge of a net which is 120 feet long. What is the angle of depression of the other end of the net as viewed by this artist?

6. A trapeze of length 33.5 feet swings through an angle of $18° 30'$ in reaching a vertical position. How far does a man sitting on the trapeze descend during this swing?

7. If the angle of swing in problem 6 is doubled, how far does the man descend? Is this answer twice the answer in problem 6?

8. Find, to the nearest minute, the angle between the diagonal of a cube and a diagonal of one of its faces, meeting the first diagonal at a corner.

9. A pilot flying due north estimates that the Washington Monument is 5.0 miles due east. After 3 minutes of flying, the Washington Monument

bears S 20° E. Estimate the speed of the airplane in miles per hour. How far is the airplane from the Washington Monument at the end of 3 minutes?

10. The University of Chicago is 222 miles north and 147 miles east of Washington University in St. Louis. What is the bearing of the University of Chicago from Washington University?

11. The University of Notre Dame in South Bend is 41 miles south and 126 miles west of the University of Michigan in Ann Arbor. The University of Michigan is 28 miles south and 36 miles east of Michigan State University in East Lansing. What is the bearing of Notre Dame from Michigan State University?

12. Find the angles of a rhombus if the two diagonals are 25.00 feet and 42.50 feet.

13. The angles of a rhombus are 70° 28′ and 109° 32′ and each side is 12.00 feet. How long are the two diagonals?

14. A fire truck has an extension ladder that is 70.0 feet long mounted on a swivel base 6.0 feet above the street level. Suppose that the closest the truck can approach the burning building puts the swivel base 15.0 feet from the wall. What is the highest point that the ladder can reach and still rest against the wall? How far does the top of the ladder descend if the fire forces the truck to move an additional 5.0 feet away from the building? What is the angle of elevation of the ladder in both positions?

★15. Two observers stationed 1,000 feet apart spot an approaching airplane. The angles of elevation of the airplane from these two points are 23° 40′ and 25° 40′. Assuming that the airplane and the two observers are in the same vertical plane, compute the altitude of the airplane to three significant figures.

12.7 THE LAW OF COSINES

We now turn our attention to triangles that are not right triangles. Such triangles are often called *oblique triangles*. In this section and the following section we will prove two useful and very pretty theorems known as the law of cosines and the law of sines, respectively. Together they permit us to solve any triangle, when enough sides and angles are given to determine the triangle uniquely. The various possibilities are listed in Table 1.

In Case V one should not expect to solve for the sides, because two triangles can have the same angles (say 20°, 50°, and 110°) and still not be congruent. Of course, they must be similar triangles. Our inability to solve the triangle in this case can also be explained by the fact that *three* items were not really given but only *two*, because given any two angles of a triangle, the third angle is automatically determined by the equation $A + B + C = 180°$.

TABLE 1

Case	Symbol	Items given	Law to be used in solving the triangle	Discussed in Section
I	SSS	Three sides	Cosine law	12.7
II	SAS	Two sides and the included angle	Cosine law	12.7
III	SSA	Two sides and the angle opposite one of the sides	Sine law	12.9
IV	ASA	Two angles and the included side	Sine law	12.8
V	AAA	Three angles	Cannot be solved	

Theorem 2   **The Law of Cosines**   *In any triangle*

$$(5) \qquad\qquad c^2 = a^2 + b^2 - 2ab\cos C.$$

This states that the square of one side is equal to the sum of the squares of the other two sides minus twice the product of these two sides and the cosine of their included angle. In (5) the angle C is the angle opposite the side c.

Since the unknown side could be either a, b, or c, there are two other forms for (5), namely

$$(6) \qquad\qquad a^2 = b^2 + c^2 - 2bc\cos A$$

and

$$(7) \qquad\qquad b^2 = c^2 + a^2 - 2ca\cos B.$$

These two formulas can be obtained from formula (5) by cyclic permutations of the letters a, b, and c. In a *cyclic permutation* of these letters, a is changed into b, b is changed into c, and c is changed into a. The same sort of changes

FIGURE 12.11

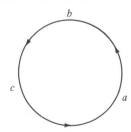

FIGURE 12.12

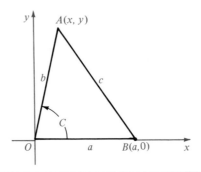

are also made on the letters A, B, and C. The diagram of Figure 12.11 should prove helpful in making these changes. Observe that by a cyclic permutation of the letters, (5) will give (6), and (6) will give (7). Thus, these last two formulas will be proved once we have proved (5).

Proof of Theorem 2:   Place the triangle on a rectangular coordinate system with the vertex C at the origin and the side a on the positive x-axis as shown in Figure 12.12. Obviously the coordinates of B are $(a,0)$. To find the co-ordinates (x, y) of A, recall that

(8) $$\sin C = \frac{y}{b} \quad \text{and} \quad \cos C = \frac{x}{b}.$$

From (8) we have $x = b\cos C$ and $y = b\sin C$. Thus the coordinates of A are $(b\cos C, b\sin C)$. Now apply the formula of Theorem 1 in Chapter 9 to find the square of the distance from B to A. We find

$$c^2 = (b\cos C - a)^2 + (b\sin C - 0)^2$$
$$= b^2\cos^2 C - 2ab\cos C + a^2 + b^2\sin^2 C$$
$$= a^2 + b^2(\cos^2 C + \sin^2 C) - 2ab\cos C$$
$$= a^2 + b^2 - 2ab\cos C. \quad \blacksquare$$

If three sides of a triangle are given, then the angle C can be found using

Theorem 3   *In any triangle*

(9) $$\cos C = \frac{a^2 + b^2 - c^2}{2ab}.$$

Cyclic permutations of this formula give

(10)
$$\cos A = \frac{b^2 + c^2 - a^2}{2bc}$$

and

(11)
$$\cos B = \frac{c^2 + a^2 - b^2}{2ca}.$$

Proof: By transposing terms in (5) we find $2ab\cos C = a^2 + b^2 - c^2$. Dividing both sides of this equation by $2ab$ we obtain (9). ∎

Formula (5) is easy to remember because it is a generalization of the Pythagorean Theorem. Indeed this formula differs from $c^2 = a^2 + b^2$, only in the extra term at the end. When $C = 90°$, $\cos C = 0$ and equation (5) *is* the Pythagorean Theorem for a right triangle. When $0 < C < 90°$, $\cos C$ is positive and something must be *subtracted* from $a^2 + b^2$ to get c^2. When $90° < C < 180°$, $\cos C$ is negative and now something is *added* to $a^2 + b^2$. These possibilities are illustrated in Figure 12.13.

FIGURE 12.13

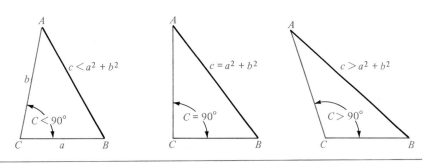

EXAMPLE 1   Given $a = 5$, $b = 8$, $C = 60°$, find the third side c of the triangle.

Solution:
$$c^2 = a^2 + b^2 - 2ab\cos C$$
$$= 5^2 + 8^2 - 2(5)(8)\cos 60$$
$$= 25 + 64 - 40 = 49,$$
$$c = 7. \quad ▲$$

EXAMPLE 2   Find to the nearest minute the angles A and B in the triangle of the preceding example.

Solution: Using formulas (10) and (11) we have

$$\cos A = \frac{b^2 + c^2 - a^2}{2bc} = \frac{8^2 + 7^2 - 5^2}{2(8)(7)} = \frac{64 + 49 - 25}{112} = \frac{88}{112} = 0.7857,$$

$$A = 38° 13'.$$

$$\cos B = \frac{c^2 + a^2 - b^2}{2ca} = \frac{7^2 + 5^2 - 8^2}{2(7)(5)} = \frac{49 + 25 - 64}{70} = \frac{10}{70} = 0.1429,$$

$$B = 81° 47'. \quad \blacktriangle$$

Check $A + B + C = 38° 13' + 81° 47' + 60° = 180° 0'.$

Exercise 6

1. How does equation (5) simplify when $C = 180°$? Draw the triangle in this case. Is equation (5) still true?

2. Do problem 1 with $C = 0°$.

In problems 3 through 8 find the third side to two significant figures.

3. $a = 7$, $b = 8$, $C = 120°$ 4. $a = 10$, $b = 6$, $C = 120°$

5. $b = 5\sqrt{2}$, $c = 7$, $A = 135°$ 6. $b = 17$, $c = 7\sqrt{2}$, $A = 135°$

7. $a = 0.012$, $c = 0.023$, $B = 24° 30'$ 8. $a = 0.080$, $c = 0.070$, $B = 72°$

In problems 9 and 10 find all of the angles of the triangle to the nearest ten minutes.

9. $a = 5$, $b = 7$, $c = 11$ 10. $a = 600$, $b = 700$, $c = 1200$

★★11. There are exactly six different nondegenerate triangles with integer sides, for which $a = 16$ and $B = 60°$. Find them.

★★12. There are exactly three different nondegenerate triangles with integer sides, for which $c = 18$ and $A = 60°$. Find them.

13. In Washington, D.C., Constitution Avenue and Pennsylvania Avenue intersect at an angle of 19°. The White House is on Pennsylvania Avenue 5,600 feet from this point of intersection. The National Academy of Sciences is on Constitution Avenue 8,600 feet from this intersection. Both buildings lie on the same side of 4th Street, which runs perpendicular to Constitution Avenue at the intersection. How far is the White House from the National Academy of Sciences?

14. The Lincoln Memorial in Washington, D.C. is 4,300 feet due west of the Washington Monument. The Pentagon is 7,100 feet S15°W from the Lincoln Memorial. How far is the Pentagon from the Washington Monument?

15. Two cars start at the same time from an intersection of two highways. The car on one highway averages 32 miles per hour while the car on the other highway is driven steadily at 44 miles per hour. If the highways are straight and the angle of intersection is 28°, how far apart are the cars at the end of 1 hour and 15 minutes?

16. If in problem 15 the slower car left the intersection at noon and the faster car left the intersection at 12:30 P.M., how far apart would they be at 3 P.M. of the same day?

★ 17. Starting from the law of cosines prove that in any triangle

$$a^2 + b^2 + c^2 = 2(ab\cos C + bc\cos A + ca\cos B).$$

★ 18. Prove that in any triangle

$$\frac{\cos A}{a} + \frac{\cos B}{b} + \frac{\cos C}{c} = \frac{a^2 + b^2 + c^2}{2abc}.$$

12.8 THE LAW OF SINES

The law of sines is given in

Theorem 4   *In any triangle*

(12)
$$\frac{a}{\sin A} = \frac{b}{\sin B} = \frac{c}{\sin C}.$$

This states that in any triangle the sides are proportional to the sines of the opposite angles.

Proof: Since the lettering of the sides of an oblique triangle is arbitrary, it is sufficient to prove only the first half of the theorem, namely

(13)
$$\frac{a}{\sin A} = \frac{b}{\sin B}.$$

FIGURE 12.14

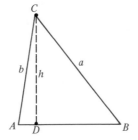 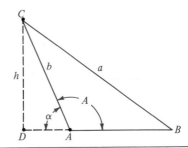

There are two cases to consider, either (1) both of the angles A and B are acute, or (2) one of the angles is obtuse, and in this case we label the obtuse angle A. The two possibilities are shown in Figure 12.14.

We first drop a perpendicular from the vertex of the angle C to the opposite side, which must be extended in Case 2. Let h be the length of this perpendicular line segment, and let D be the point of intersection with the opposite side.

CASE 1   Triangles ADC and BDC are right triangles. Therefore,

(14)
$$\frac{h}{a} = \sin B, \qquad \frac{h}{b} = \sin A.$$

Hence,

(15)
$$h = a \sin B, \qquad h = b \sin A.$$

Since h is the same in both equations,

(16)
$$a \sin B = b \sin A.$$

If we divide both sides of (16) by the product $\sin A \sin B$, we obtain (13).

CASE 2   In this case triangles ADC and BDC are still right triangles, but instead of (14) we must write

(17)
$$\frac{h}{a} = \sin B, \qquad \frac{h}{b} = \sin \alpha,$$

since α is the angle in triangle ADC. However, $\alpha + A = 180°$, so α is the related angle of A. Since A is in Q.II, $\sin A$ is positive. Therefore, $\sin \alpha = \sin A$. When this substitution is made in the second equation of the set (17), this set is identical with the set (14). The rest of the proof in Case 2 proceeds exactly as in Case 1. ▮

EXAMPLE 1   Solve the triangle if $a = 28$, $A = 135°$, and $B = 30°$.

Solution: By the law of sines $b/\sin B = a/\sin A$. Therefore,

$$b = \frac{a \sin B}{\sin A} = \frac{28 \sin 30°}{\sin 135°} = \frac{28(1/2)}{\sqrt{2}/2} = \frac{28}{\sqrt{2}} = 14\sqrt{2} \approx 20$$

to two significant figures. Since $C = 180° - (A + B) = 180° - 165° = 15°$, then

$$c = \frac{a \sin C}{\sin A} = \frac{28 \sin 15°}{\sin 135°} = \frac{28(0.2588)}{\sqrt{2}/2} = 28\sqrt{2}(0.2588) \approx 10. \quad \blacktriangle$$

As a check we can compute c using b, thus

$$c = \frac{b \sin C}{\sin B} = \frac{20 \sin 15°}{\sin 30°} = \frac{20(0.2588)}{1/2} \approx 10.$$

EXAMPLE 2   Solve the triangle, given $b = 1984$, $A = 19° 14'$, and $C = 25° 52'$.

Solution: Here the information is given with four significant figures so it is desirable to use logarithms. First we find the third angle.

$$B = 180° - (A+C) = 180° - 45° 6' = 134° 54'.$$

$$a = \frac{b \sin A}{\sin B} = \frac{1984 \sin 19° 14'}{\sin 134° 54'}. \qquad c = \frac{b \sin C}{\sin B} = \frac{1984 \sin 25° 52'}{\sin 134° 54'}.$$

$\log 1984$	$= 3.2976$		$\log 1984$	$= 3.2976$
$\log \sin 19° 14'$	$= 9.5177 - 10$		$\log \sin 25° 52'$	$= 9.6397 - 10$
$\log$ numerator	$= 12.8153 - 10$		$\log$ numerator	$= 12.9373 - 10$
$\log \sin 45° 6'$	$= 9.8502 - 10$		$\log \sin 45° 6'$	$= 9.8502 - 10$
$\log a =$	2.9651		$\log c =$	3.0871
$a = 922.8$			$c = 1222.$ ▲	

Exercise 7

In problems 1 through 4 find the sides of the triangle to two significant figures without using logarithms. Check your results by making an accurate drawing.

1. $a = 100$, $B = 60°$, $C = 45°$ 2. $a = 80$, $B = 45°$, $C = 30°$

3. $b = 50$, $C = 120°$, $A = 45°$ 4. $c = 25$, $A = 130°$, $B = 20°$

In problems 5 through 8 use logarithms to find the sides of the triangle to four significant figures.

5. $b = 0.04646$, $A = 28° 39'$, $C = 39° 28'$

6. $b = 888.8$, $A = 12° 34'$, $B = 23° 45'$

7. $c = 25.25$, $A = 43° 21'$, $B = 54° 32'$

8. $c = 7676$, $B = 19° 59$, $C = 18° 12'$

9. A vertical building stands on a street that slopes downward at an angle of 8° 20′. At a point 125 feet down this street from the base of the building, the angle of elevation of the top of the building is 59° 30′. How tall is the building?

10. A tree on a sloping hill casts a shadow 135 feet straight down the hill. If the hill slopes downward at an angle of 12° 30′ and the angle of elevation of the sun is 25° 40′, how tall is the tree?

11. The angles of elevation of an approaching airplane from two antiaircraft guns are 78° 15′ and 53° 25′. If these guns are 878 feet apart, and if the airplane is in a vertical plane with these guns, how far is the airplane from the nearest gun?

12. Two lookout towers *A* and *B* are 5.00 miles apart with *A* due west of *B*. A column of smoke is sighted from *A* with a bearing N 75° 10′ E, and from *B* with a bearing N 53° 20′ E. How far is the fire from the nearest tower?

13. Two points *A* and *B* on one bank of a river are 95 feet apart. A point *C* across the river is located so that angle *CAB* is 75° and angle *CBA* is 80°. How far is *C* from *A*?

12.9 THE AMBIGUOUS CASE

Suppose that we are asked to solve a triangle, given two sides and the angle opposite one of them. To be specific, suppose that we are given $a = 35.00$, $b = 50.00$, and $A = 30° 0′$, as illustrated in Figure 12.15. Our attempt to draw the corresponding triangle is shown in Figure 12.16. We first draw angle *A* with the base line extended. We then lay off the known side $b = AC = 50.00$. To locate the side opposite the angle *A*, we take point *C* as a center and with a radius $a = 35.00$ we describe an arc of a circle. This circular arc represents all possible positions for the end-point *B* of the line segment opposite the angle *A*. But *B* must also be on the base line, so *B* is at the point of intersection of this circular arc and the base line. In this case there are two such points of intersection, which we denote by *B* and *B′* (read, "*B* prime"). Therefore (to our surprise) there are two triangles, *ABC* and *AB′C*, not congruent to each other, and each one has $a = 35.00$, $b = 50.00$, and $A = 30° 0′$. To complete the solution of this problem, we must solve each of these triangles.

FIGURE 12.15

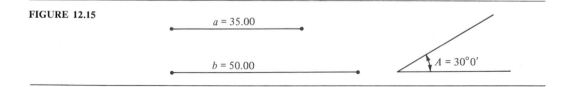

FIGURE 12.16

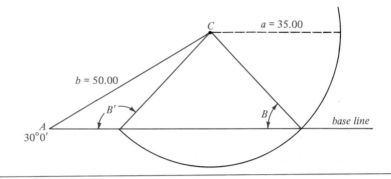

CASE 1

Triangle ABC. From (13) we have

(18) $$\sin B = \frac{b \sin A}{a}$$

$$= \frac{50.00 \sin 30°}{35.00} = 0.7143.$$

Since in this case the angle B is acute, we find that $B = 45°\,35'$. Then $C = 180° - (30°\,0' + 45°\,35') = 104°\,25'$. Finally,

$$c = \frac{a \sin C}{\sin A} = \frac{35.00 \sin 104°\,25'}{\sin 30°} = 70.00(0.9685) = 67.795 \approx 67.80.$$

CASE 2

Triangle $AB'C$. In this case the angle B' is obtuse. The law of sines is still applicable, so with B' replacing B, equation (18) is still true. In other words, without any new computation $\sin B' = \sin B$. Therefore, B is the related angle of B'. This can also be seen from the fact that the triangle $B'CB$ is isosceles. Consequently $B' = 180' - B = 180° - 45°\,35' = 134°\,25'$. Finally, $C = 180° - (A + B') = 180° - 164°\,25' = 15°\,35'$, so that

$$c = \frac{a \sin C}{\sin A} = 70.00(0.2686) = 18.802 \approx 18.80.$$

Now that we have solved this specific example, let us return to the general situation. Suppose, again, that we are given the sides a and b and the angle A, and we are to determine the number of different (noncongruent) triangles with the given sides and angle. Figure 12.17 shows the various possibilities when the angle A is acute. The important thing to observe is that the number of triangles depends on the magnitude of the side a.

FIGURE 12.17

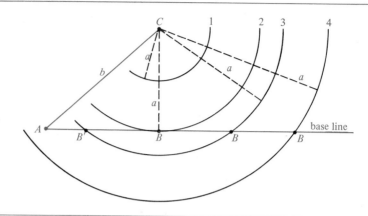

1. If a is too short, there will be no triangles.
2. If a is just right, there will be one triangle.
3. If a is somewhat longer, there will be two triangles.
4. If a is too long, there will be one triangle.

Of course, these statements about the magnitude of a are vague, but it is easy to make them precise.

Consider Case 2 where a is "just right." This means that the circular arc is tangent to the base line so that the triangle ABC is a right triangle. Then $a/b = \sin A$, and "just right" means $a = b \sin A$. Consequently, in Case 1 "too short" means $a < b \sin A$, and in Cases 3 and 4, $a > b \sin A$. Obviously, in Case 4, $a \geqq b$; so in Case 3 we have $b \sin A < a < b$. These results are summarized in Table 2. Because the number of triangles may be none, one, or two, depending on a, the case SSA is called the *ambiguous case.*

TABLE 2

If angle A is acute,

Case	Condition on a	Number of triangles
1	$a < b \sin A$	None
2	$a = b \sin A$	One right triangle
3	$b \sin A < a < b$	Two
4	$b \leqq a$	One

The situation is considerably simpler if the angle A is obtuse. We leave it to the student to examine Figure 12.18, which shows the two possibilities in this case, and to convince himself that Table 3 is correct.

FIGURE 12.18

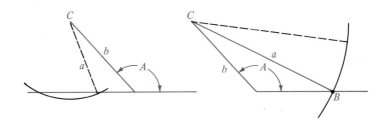

TABLE 3

If angle A is obtuse,

Case	Condition on a	Number of triangles
1	$a \leqq b$	None
2	$b < a$	One

EXAMPLE 1

How many different (noncongruent) triangles can be drawn with $A = 76° 10'$, $a = 107.5$, and $b = 111.1$?

Solution:

$$b \sin A = 111.1(0.9710) = 107.87 \cdots$$

$$a = 107.5 < 107.8 < b \sin A.$$

Therefore, there are no triangles with the given angle and sides. ▲

It is not necessary to memorize the condition $a < b \sin A$, because if we try to solve the triangle in this case we will soon run into a dead end, which automatically signals to us that there are no triangles. If we try to solve the triangle of this example, we have

$$\sin B = \frac{b \sin A}{a}$$

$$= \frac{111.1 \sin 76° 10'}{107.5}$$

$\log 111.1$	$= 2.0457$
$\log \sin 76° 10'$	$= 9.9872 - 10$
$\log \text{numerator}$	$= 12.0329 - 10$
$\log 107.5$	$= 2.0314$
$\log \sin B$	$= 10.0015 - 10$

Since $\log \sin B = 0.0015$, a positive number, $\sin B > 1$. Therefore the angle B does not exist.

EXAMPLE 2

How many different triangles can be drawn with $A = 45°$, $b = \sqrt{5}$, $a = 3\sqrt{2/7}$?

Solution: Since $18/7 < 5$, taking square roots on both sides gives $3\sqrt{2/7} < \sqrt{5}$ or $a < b$, so Case 4 is excluded.

We next examine Case 1, putting the equation in the form of a question:

(19)
$$a \overset{?}{<} b \sin A$$

(read, "is a less than $b \sin A$?"). Making the appropriate substitutions we have

(20)
$$3\sqrt{\frac{2}{7}} \overset{?}{<} \sqrt{5}\frac{\sqrt{2}}{2}$$
or on squaring
(21)
$$\frac{18}{7} \overset{?}{<} \frac{10}{4}.$$

But $18 \times 4 > 7 \times 10$. Since all of our steps are reversible, the inequality signs in (19), (20), and (21) should be reversed. Thus, we have Case 3, and there are two triangles. ▲

Exercise 8

In problems 1 through 10 determine the number of different (noncongruent) triangles that can be drawn with the given data. Whenever the angle B exists, compute it to the nearest minute.

1. $a = 75$, $b = 85$, $A = 135°$
2. $a = 50$, $b = 70$, $A = 120°$
3. $a = 90$, $b = 45$, $A = 150°$
4. $a = 75$, $b = 50\sqrt{2}$, $A = 135°$
5. $a = 40$, $b = 70$, $A = 30°$
6. $a = 30$, $b = 25\sqrt{2}$, $A = 45°$
7. $a = 7\sqrt{13}$, $b = 16\sqrt{5}$, $A = 45°$
8. $a = 9\sqrt{17}$, $b = 12\sqrt{13}$, $A = 60°$
9. $a = 11\sqrt{3}$, $b = 19$, $A = 60°$
10. $a = 15$, $b = 14$, $A = 30°$

In problems 11 through 14 determine the number of different triangles that can be drawn with the given data. Solve each triangle completely, using logarithms.

11. $b = 4646$, $c = 8080$, $B = 33° 33'$
12. $b = 2020$, $c = 5556$, $B = 18° 18'$
13. $a = 5.556$, $c = 1.930$, $C = 20° 20'$
14. $a = 0.005013$, $c = 0.001994$, $C = 23° 28'$

15. A vertical tree 68.5 feet tall grows on a sloping hill. From a point 157 feet from the base of this tree, measured straight down the hill, the tree subtends an angle of 21° 20′. Find the angle that the sloping hill makes with the horizontal plane.

16. In order to measure the angle a sloping hill makes with the horizontal plane, a man sights his transit on a small stone S lying on the hill and finds the angle of elevation to be $7° 40'$. The transit is 5.00 feet above a point T on the ground. The distance TS straight up the hill is 123 feet. What is the angle that the hill makes with the horizontal plane?

17. If the angle subtended at the earth by a line joining Venus and the sun is $31°$, and if Venus is 6.8×10^7 miles from the sun, and the earth is 9.3×10^7 miles from the sun, what is the distance from the earth to Venus? (Two answers are possible.)

18. A lighthouse is 23 miles $N 55° E$ of a dock. A ship leaves the dock at $1:00$ P.M. and sails due east at a speed of 15 miles per hour. Find the time to the nearest minute when the ship will be 18 miles from the lighthouse. (Two answers are possible.)

19. The sides of a parallelogram are 15.0 inches and 11.0 inches and the longer diagonal makes an angle of $18° 50'$ with the longer side. Find the length of the longer diagonal.

Review Questions

Try to answer the following questions as accurately as possible before consulting the text.

1. Give the definition of the six trigonometric functions as ratios of lengths of sides in a right triangle.

2. State the Law of Sines.

3. State the Law of Cosines.

4. Give a formula for finding $\cos A$ if the three sides a, b, and c are known.

5. If we put $C = 90°$ in the law of cosines, we obtain $c^2 = a^2 + b^2$ as a special case. Can this be regarded as a proof of the Pythagorean Theorem?

6. What is the ambiguous case that may arise in solving triangles?

Systems of Equations

13

A set of equations $E_1 = 0,\ E_2 = 0, \ldots, E_n = 0,$ where each expression involves one or more variables, is regarded as a system of equations, if we are interested in finding a set of values for the variables such that each of the equations is true for this set of values. In this chapter we will concentrate on the simplest systems: (a) systems of two linear equations in two variables and (b) systems of three linear equations in three variables. These two simple cases are sufficient to indicate the central features in the solution of systems of n linear equations in n variables.

If some equations of the system are not linear, then the theory is more complicated. However, the theory of systems of linear equations is often helpful in the study of nonlinear systems.

13.1 SYSTEMS OF LINEAR EQUATIONS IN TWO VARIABLES

Before giving the theory, let us look at

EXAMPLE 1   Solve the system of linear equations in two variables

(1)
$$2x + y = 4$$

(2)
$$5x - 2y = 37.$$

Solution: We suppose that x and y have been assigned values x_0 and y_0 (if there are any) such that both equations are true. Thus, we have the true statements

(3) $$2x_0 + y_0 = 4$$

and

(4) $$5x_0 - 2y_0 = 37.$$

If we multiply equation (3) by 5 and multiply equation (4) by 2, we obtain the true statements

(5) $$10x_0 + 5y_0 = 20$$

and

(6) $$10x_0 - 4y_0 = 74.$$

If we subtract equation (6) from equation (5), we obtain

(7) $$9y_0 = -54.$$

Notice that x_0 is no longer present in equation (7). Indeed, we selected the multipliers 5 and 2 so that x_0 would disappear when we subtracted equation (6) from equation (5). Equation (7) gives $y_0 = -54/9 = -6$. If we put $y_0 = -6$ in equation (3), we find that $2x_0 - 6 = 4$. Hence, $2x_0 = 4 + 6$ or $x_0 = 5$. Thus, the only solution to the system of equations (1) and (2) is $x = 5$ and $y = -6$, which we write for brevity as the ordered pair $(5, -6)$. ▲

Check We substitute these values for the variables in each of the equations (1) and (2). Then

$$2(5) + (-6) = 10 - 6 = 4$$

and

$$5(5) - 2(-6) = 25 + 12 = 37.$$

We now discuss the process used in solving this example. In the first place it was logically necessary to add subscripts to x and y in order to distinguish the value assigned to a variable from the variable. Once this logical distinction is understood, we may drop the subscripts because the mechanical steps used in finding the solution will be the same. The reader can recopy equations

(3), (4), (5), (6), and (7) without the subscripts and arrive at the same conclusion: $y = -6$ is necessary for a solution. In the future we will drop the subscripts unless there is a compelling reason not to.

To solve Example 1 we replaced the given system (equations (1) and (2)) by a new system (equations (3) and (7)), which has the same solution set as the original system. We summarize the process used with two definitions and one theorem.

DEFINITION 1

A system of equations of the form

(8)
$$a_1 x + b_1 y = c_1, \qquad |a_1| + |b_1| > 0,$$

(9)
$$a_2 x + b_2 y = c_2, \qquad |a_2| + |b_2| > 0,$$

is called a system of two linear equations in two variables. An ordered pair (x, y) for which both (8) and (9) are true is called a solution of the system. The solution set of the system is the set of all such pairs.

DEFINITION 2

Two systems of equations are said to be equivalent systems if both systems have the same solution set.

In manipulating with the equations (8) and (9) it is convenient to leave the constants c_1 and c_2 on the right side, but in discussing the theory it is advantageous to put c_1 and c_2 on the left side. Hence, we define the expressions

(10) $E_1 \equiv a_1 x + b_1 y - c_1$ and $E_2 \equiv a_2 x + b_2 y - c_2.$

However, in the theorem below the expressions E_1 and E_2 are not necessarily linear but can be completely arbitrary.

Theorem 1
PLE

Let E_1 and E_2 be arbitrary expressions, and suppose that $k \neq 0$. Then the system of equations

(11)
$$E_1 = 0, \qquad E_2 = 0$$

is equivalent to any one of the systems:

(12)
$$E_2 = 0, \qquad E_1 = 0,$$

(13)
$$k E_1 = 0, \qquad E_2 = 0,$$

(14)
$$E_1 = 0, \qquad E_2 + k E_1 = 0,$$

(15)
$$E_1^n = 0, \qquad E_2 = 0.$$

The system (12) is obtained from the system (11) by interchanging the equations. The system (13) is obtained from (11) by multiplying one of the equations (the first one) by a nonzero constant. The system (14) is obtained from the system (11) by adding to one of the equations (the second one) a constant times another equation (the first one). Each of these transformations is called an *elementary transformation* of the system. To solve a system, such as that given in Example 1, we look for a sequence of elementary transformations that leads to a simpler system, equivalent to the given one. The ultimate goal is one equation in which all the coefficients are zero except one. We then say that *the other variables have been eliminated.* It is an easy matter to solve for the remaining variable. In Example 1 we eliminated x and solved for the remaining variable y.

The transformation from the system (11) to the system (15) is not needed for systems of linear equations, but may be very useful for solving more complicated systems.

By an interchange of equations we can select either of the two given equations to be the first one ($E_1 = 0$). Consequently, the transformations can be applied to any equation of a given system. In Theroem 1 we concentrated on $E_1 = 0$ for simplicity.

EXAMPLE 2   Solve the system of linear equations

(16) $$3x + 2y = 9$$

(17) $$7x + 11y = 2.$$

Solution: Multiply the first equation by 7, multiply the second equation by 3, and then subtract as indicated. We get

(18) $$21x + 14y = 63$$

(19) $$\begin{aligned} 21x + 33y &= 6 \\ \hline 0x - 19y &= 57 \end{aligned} \quad \text{(subtracting).}$$

Hence, $y = 57/(-19) = -3$. Then from equation (16)

$$3x = 9 - 2y = 9 - 2(-3) = 9 + 6 = 15,$$

so $x = 5$. The solution is $x = 5$, $y = -3$. We may also say that the solution set is the ordered pair $(5, -3)$. ▲

If we multiply equation (16) by 11 and equation (17) by 2, and then take the difference, we will eliminate y, and find that $x = 5$.

Whenever a system of linear equations has a unique solution (as in Examples 1 and 2) we say that the system is an *independent system.*

EXAMPLE 3

Solve the system of linear equations

(20) $$4x - 6y = -7$$

(21) $$10x - 15y = 27.$$

Solution: The multipliers seem to be 10 and 4, but 5 and 2 will suffice. If we multiply equation (20) by 5, multiply equation (21) by 2, and then subtract as indicated, we get

$$\begin{array}{r} 20x - 30y = -35 \\ 20x - 30y = 54 \\ \hline 0 = 19 \end{array} \quad \text{(subtracting)}.$$

Since $0 \neq 19$, we infer that there is no pair (x, y) for which (20) and (21) are both true. In the next section we will see a geometric reason for this. ▲

Whenever a system of equations has no solution (as in Example 3) we say that the system is *inconsistent*.

EXAMPLE 4

Solve the system of linear equations

(22) $$4x - 6y = -10$$

(23) $$10x - 15y = -25.$$

Solution: As in Example 3, we multiply the first equation by 5, we multiply the second equation by 2, and subtract. We obtain

$$\begin{array}{r} 20x - 30y = -50 \\ 20x - 30y = -50 \\ \hline 0 = 0. \end{array}$$

Everything is gone! We arrived at $0 = 0$, a true statement, but we have no information about the solution of the given system. Inspection shows that equation (23) is merely 5/2 times equation (22) and therefore, any solution of (22) is automatically a solution of (23). It is easy to see that equation (22) has infinitely many solutions because we can select any number for x and then solve for y. Consequently, the given system has infinitely many solutions. Some of these solutions are $(-4, -1)$, $(-1, 1)$, $(2, 3)$, $(5, 5)$, $\cdots$. ▲

Geometrically, the graph of equation (22) is a straight line. The graph of equation (23) is the same straight line. Since there are infinitely many points on this line, it is geometrically obvious that the system has infinitely many solutions.

Whenever a system of linear equations has more than one solution (as in Example 4) we say that the system is a *dependent system*.

Exercise 1

In problems 1 through 10 solve the given system of linear equations in two variables. Check each solution.

1. $2x - 3y = 12$
   $4x + 5y = 2$

2. $3x - 2y = 29$
   $5x + 6y = 11$

3. $3x - 2y = 7$
   $5x - y = -7$

4. $6x + 3y = 6$
   $5x + 4y = -1$

5. $3x + 11y = -3$
   $2x + 5y = 5$

6. $5x + 7y = 9$
   $-11x - 5y = 1$

7. $4A + 3B = 15$
   $-A + 4B = 20$

8. $3u + 2v = 5$
   $6u - 10v = 3$

9. $4x - y = 10$
   $-8x + 2y = 13$

10. $5z - 3w = 4$
    $-15z + 9w = -1$

11. Now I am 30 years older than my son. In another 5 years I will be 4 times as old as my son. Find our present ages.

12. Mary is now 3 times as old as John was when John was one half as old as Mary was 6 years ago. How old are they both now?

13. In a certain chess game there was a lively exchange in which black lost 3 men, and white lost 4 men. A kibitzer observed that before the exchange white had twice as many men as black had after the exchange, and that 3 times the number of white men after the exchange was 5 more than the number of black men before the exchange. Find the number of men on each side just prior to the exchange.

14. Explain the reason for the condition $|a_1| + |b_1| > 0$ in Definition 1.

15. Prove that if t is any real number, then $x = 2 + 3t$, $y = 3 + 2t$ is a solution of the system given in Example 4.

13.2 GEOMETRIC INTERPRETATION OF SYSTEMS OF EQUATIONS

Let G_1 be the graph of the equation $E_1 = 0$, and let G_2 be the graph of $E_2 = 0$. If (x, y) is a solution of the system of equations $E_1 = 0$, $E_2 = 0$, then the point $P(x, y)$ is a point of both graphs. Thus, P is a point of intersection of G_1 and G_2. Conversely, any point in $G_1 \cap G_2$ gives a solution for the system of equations. This information will guide us in solving systems of equations.

EXAMPLE 1   Solve the system of equations

(24) $$y = -(x+4)$$

(25) $$y = 6 + 2x - x^2.$$

Indicate the geometric interpretation of the solution.

Solution: We substitute the value of y from equation (24) in (25) (or subtract equation (24) from (25)). This gives

$$-(x+4) = 6 + 2x - x^2$$
$$x^2 - 3x - 10 = 0$$
$$(x-5)(x+2) = 0.$$

Thus, $x = -2$ or $x = 5$. Computing y from equation (24) we obtain the two solutions $(-2, -2)$ and $(5, -9)$. These must be coordinates of points P and Q that are intersection points of the graphs of these equations. These points are shown in Figure 13.1. ▲

If we first sketch the graphs of equations (24) and (25), we could anticipate two solutions, since the straight line and the parabola intersect in two points.

FIGURE 13.1

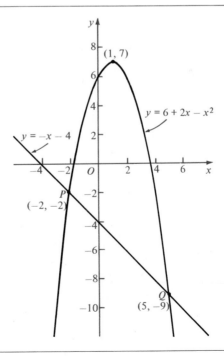

Further, if the sketch is made with some care, we could obtain good approximations to the solutions from the picture. This method is extremely useful when the equations are complicated, so that the algebraic solution is difficult.

EXAMPLE 2   Solve the system of equations

(25) $$y = 6 + 2x - x^2$$

(26) $$y = -x + 9.$$

Indicate the geometric interpretation of the solution.

Solution: For this example we look first at the geometric interpretation. The equation $y = 6 + 2x - x^2$ is the same in both Example 1 and Example 2, and the graph is shown in Figure 13.1. The reader is invited to add to Figure 13.1, the straight line $y = -x + 9$ which is thirteen units above the line $y = -x - 4$ and to show that the old parabola and the new straight line do not meet. Hence, we suspect that the system consisting of equations (25) and (26) has no real roots.

If we attempt to solve the system algebraically we are led to the equation $x^2 - 3x + 3 = 0$, which has the complex roots $x = (3 \pm \sqrt{3i})/2$. Hence, the system has no real solution. ▲

EXAMPLE 3   Give a geometric explanation for our failure to find a solution for the system consisting of equations (20) and (21). (See page 317.)

Solution: Each of these equations is the equation of a straight line. A quick sketch, shown in Figure 13.2, seems to indicate that the lines are parallel. If so, one could hardly expect to find a solution for the system because there is no point of intersection for pairs of parallel lines. Since we proved in Section 13.1 that there is no solution for the system, we have also proved that the graphs are indeed parallel lines. ▲

FIGURE 13.2

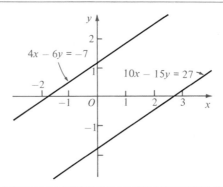

Exercise 2

In problems 1 through 16 solve the given system of equations. Make a sketch of the graph of each equation. Show each solution as a point of intersection of the graphs.

1. $y = 6x - 3 - x^2$
   $y = 2$

2. $y = x^2 - 5x - 2$
   $y = 4$

3. $y = 9 + 2x - x^2$
   $y = -3x + 13$

4. $y = x^2 - 2x - 3$
   $y = 3 - x$

5. $y = 1 + x^2$
   $y = 1 - 4x - x^2$

6. $y = 2x^2 - 3x - 10$
   $y = x^2 - 2x + 10$

7. $y = 2x - 4$
   $y = \sqrt{x^2 - 4x + 7}$

8. $y = 3x$
   $2y = \sqrt{9x^2 + 12}$

★ 9. $y = -x + 2$
   $y = x^3 - x^2 - 5x + 6$

★ 10. $y = x + 4$
   $y = x^3 - 3x^2 + x + 4$

11. $2y = x + 3$
   $y = |x|$

12. $2y = x + 5$
   $y = |x + 2|$

13. $\dfrac{x^2}{18} + \dfrac{y^2}{8} = 1$

14. $x^2 + y^2 = 16$
   $2x^2 - 3y^2 = 12$

$\dfrac{x^2}{3} - \dfrac{y^2}{2} = 1$

15. $x + y = 1$
   $x^2 + y^2 = 1$

16. $x + y = 10$
   $x^2 + y^2 = 10$

17. Solve graphically each of the systems given in Exercise 1. In other words, draw the two straight lines carefully on coordinate paper, and from the drawings determine the solution as the coordinates of the point of intersection of the two lines.

13.3 THE THREE-DIMENSIONAL RECTANGULAR COORDINATE SYSTEM

The rectangular coordinate system for the plane supplied us with a useful geometric interpretation for the solution of a system of two equations in two variables. We may expect that a rectangular coordinate system for three-dimensional space would be equally useful for the study of a system of three equations in three variables. The extension of the rectangular coordinate system from the plane to three-dimensional space presents no difficulty.

FIGURE 13.3

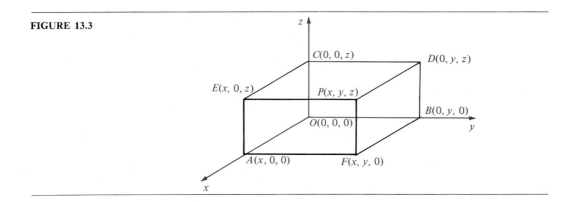

As indicated in Figure 13.3, we select a point O for the origin and three mutually perpendicular directed lines that meet at O. These three lines are called the x-axis, y-axis, and z-axis, respectively. It is customary and convenient to place the x-axis and y-axis in a horizontal plane and to select the positive direction upward on the z-axis. Further it is customary to label the axes so that the system is a right-handed system. This means that if the thumb of the right hand points in the positive direction of the x-axis and the index finger points in the positive direction of the y-axis, then the middle finger points in the positive direction of the z-axis.

The plane determined by the x-axis and the y-axis is called the *xy-plane*. Similarly, the two vertical planes are called the *xz-plane* and the *yz-plane*, respectively. If P is any point in space, it has three coordinates (x, y, z) defined thus:

x is the directed distance $\overline{DP}$ of P from the yz-plane,

y is the directed distance $\overline{EP}$ of P from the xz-plane,

z is the directed distance $\overline{FP}$ of P from the xy-plane.

As indicated in Figure 13.3, the segments DP, EP, and FP are edges of a rectangular parallelepiped (hereafter called a box). Consequently, we have an alternate definition for the coordinates of P:

x is the directed distance $\overline{OA}$,

y is the directed distance $\overline{OB}$,

z is the directed distance $\overline{OC}$.

A number of points, their coordinates, and associated boxes are shown in Figure 13.4.

FIGURE 13.4

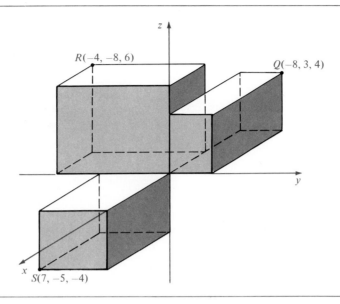

A detailed exploration of this three-dimensional coordinate system belongs properly to the domain of solid analytic geometry. Here we considered only the few items that we need for systems of linear equations in three variables.

DEFINITION 3   *Let E be an expression in three variables, x, y, and z. The graph of the equation E = 0 is the set of all points P whose coordinates (x, y, z) satisfy the equation.*

Again there is the usual distinction between a graph, which is a collection of points, and a sketch of the graph, which is a physical representation or picture of the set of points. Usually (but not always) the graph of $E = 0$ will be a surface.

EXAMPLE 1   Sketch the graph of

(27) $$2x + y + 4z - 8 = 0.$$

Solution: If we can solve the equation for z in terms of x and y, then we can find points of the graph by selecting x and y and computing z. We find that

(28) $$z = \frac{8 - 2x - y}{4}.$$

If we compute coordinates and plot points, the true nature of the graph may not appear; therefore, an alternate approach is desirable.

FIGURE 13.5

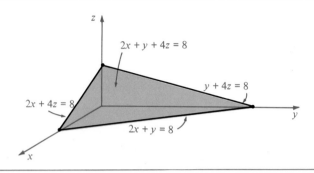

What is the intersection of the graph of (27) with the xy-plane? We find this intersection by setting $z = 0$ in (27). When we do this we find $2x + y - 8 = 0$, the equation of a straight line. Similarly, the intersection of the graph of (27) with the yz-plane is the straight line $y + 4z - 8 = 0$; and with the xz-plane, it is the straight line $2x + 4z - 8 = 0$. Portions of these three straight lines are shown in Figure 13.5. This drawing suggests that the graph of (27) is a plane, a portion of which is shaded in color in Figure 13.5. The reader can now use (28) to compute the coordinates for more points of the graph and check (optically) to see if they seem to lie on the indicated plane. ▲

Since the graph of $2x + y + 4z = 8$ seems to be a plane and since (Theorem 3 in Chapter 6), the graph of $ax + by + c = 0$ is a line. This suggests

Theorem 2
PWO

If $|a| + |b| + |c| > 0$, then the graph of

(29) $$ax + by + cz + d = 0$$

is a plane. Conversely, if G is a plane, then there is an equation of the form (29) for which G is the graph.

We omit this proof because we have not really defined a plane. At the moment, we have nothing on which to base a proof. One way out of the dilemma is to convert Theorem 2 into a definition. In other words, the set of all points that satisfy an equation of the form (29) is called a *plane*. Such a definition is logically legitimate and certainly saves us time and energy. However, it is not completely satisfactory.

Suppose now that we are concerned with the solution of the system of three linear equations in three variables

(30) $$a_1 x + b_1 y + c_1 z = d_1$$

FIGURE 13.6

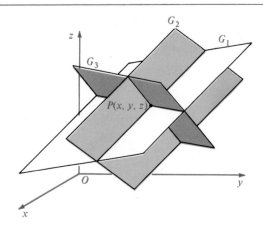

(31) $$a_2 x + b_2 y + c_2 z = d_2$$

(32) $$a_3 x + b_3 y + c_3 z = d_3,$$

where $|a_k| + |b_k| + |c_k| > 0$ for $k = 1, 2, 3$.

By Theorem 2, the graph of each equation is a plane. If the equations are chosen at random or arise in connection with some reasonable physical problem, then it is very likely that the planes G_1, G_2, and G_3 will intersect in a single point. (See Figure 13.6.) In this case the system will have a unique solution, the coordinates (x, y, z) of the unique point P common to all three planes. This is the usual situation. However, if two of the three planes are parallel, then clearly there will be no common point, and the system will not have a solution. This is pictured in Figure 13.7.

FIGURE 13.7

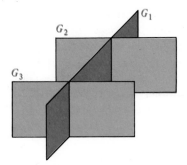

FIGURE 13.8

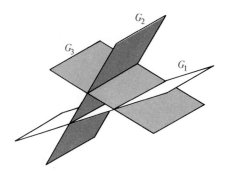

There are other special cases for the positions of the three planes. As indicated in Figure 13.8, it may happen that no two of the planes are parallel, and yet the system of linear equations has no solution. This occurs if the pairs of planes intersect in three lines that are parallel.

In the other direction the system of equations can have infinitely many solutions. This occurs if two of the planes coincide, or if the three planes intersect in a common line L. This latter possibility is shown in Figure 13.9. In this case every point on L gives a solution of the system.

EXAMPLE 2   Solve the system of three linear equations in three variables

(33) $$x - 2y + 3z = 10$$

(34) $$x + 2y - z = 6$$

(35) $$-2x + y + 2z = -2.$$

FIGURE 13.9

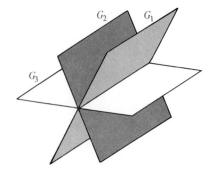

Solution: We recall that in Section 13.1 we introduced the concept of elementary transformations of systems of two equations. We saw that these transformations were quite helpful in solving such systems (see Theorem 1). These concepts and Theorem 1 can be extended to cover any system of n equations in m variables. We proceed to use elementary transformations on the given system.

If we subtract equation (33) from equation (34) and add twice equation (33) to equation (35), we obtain the system

$$(33) \qquad x - 2y + 3z = 10 \qquad\qquad\qquad x + 2y - z = 6$$

$$x - 2y + 3z = 10$$

$$(36) \qquad\qquad 4y - 4z = -4 \quad\longleftarrow\quad\qquad \overline{ 4y - 4z = -4}$$

$$-2x + y + 2z = -2$$

$$2x - 4y + 6z = 20$$

$$(37) \qquad\qquad -3y + 8z = 18, \quad\longleftarrow\quad \overline{ -3y + 8z = 18}$$

which is equivalent to the original system.

Next we divide equation (36) by 4 and add 3 times the resulting equation to equation (37). This gives a new system

$$(33) \qquad x - 2y + 3z = 10$$

$$3y - 3z = -3$$

$$(38) \qquad\qquad y - z = -1$$

$$-3y + 8z = 18$$

$$(39) \qquad\qquad\qquad 5z = 15, \quad\longleftarrow\quad \overline{ 5z = 15}$$

which is equivalent to the original system. This system is easy to solve. From equation (39), $z = 3$. Equation (38) gives $y = -1 + z = -1 + 3 = 2$. Finally, equation (33) yields

$$x = 10 + 2y - 3z = 10 + 2(2) - 3(3) = 10 + 4 - 9 = 5. \quad \blacktriangle$$

Check   We substitute $x = 5$, $y = 2$, $z = 3$ in the given equations. We find

$$x - 2y + 3z = 5 - 2(2) + 3(3) = 5 - 4 + 9 = 10$$

$$x + 2y - z = 5 + 2(2) - 3 = 5 + 4 - 3 = 6$$

$$-2x + y + 2z = -2(5) + 2 + 2(3) = -10 + 2 + 6 = -2.$$

Consequently, the given system has a unique solution which we can write in the form of an ordered triple $(5, 2, 3)$. The graphs of the three equations form three planes which intersect in the point $(5, 2, 3)$.

The method used in solving Example 2 is called the *echelon method* or the method of *elimination of variables*. The key idea is to select one variable (possibly x), and eliminate it from all equations except one, etc.

EXAMPLE 3   Solve the system of linear equations

(40) $$x - 2y + 3z = 5$$

(41) $$3x - 6y + 9z = 21$$

(42) $$3x - 2y + z = 7.$$

Solution: If we multiply equation (40) by three and subtract the product from equation (41), we obtain the new system

$$3x - 6y + 9z = 21$$

(40) $\quad x - 2y + 3z = 5$ $\qquad\qquad 3x - 6y + 9z = 15$

(43) $\qquad\qquad 0 = 6 \longleftarrow \quad\qquad\qquad 0 = 6$

(42) $\quad 3x - 2y + z = 7.$

Clearly, equation (43) is never true, so we conclude that the given system has no solution. Hence, this system is inconsistent. ▲

We arrived at this conclusion without using equation (42). Consequently, no matter how the third equation is changed, the system still has no solution. From this we infer that equations (40) and (41) are the equations of parallel planes. In this system we have the type of situation illustrated in Figure 13.7.

EXAMPLE 4   Solve the system of linear equations

(44) $$x - y + z = 5$$

(45) $$x + 2y - 2z = -4$$

(46) $$2x + y - z = 10.$$

Solution: If we subtract equation (44) from equation (45), and then subtract two times equation (44) from equation (46), we arrive at the equivalent system

(44) $$x - y + z = 5$$

(47) $$3y - 3z = -9$$

(48) $$3y - 3z = 0.$$

Equations (47) and (48) tell us that if the original system has a solution, then $-9 = 0$. Since this is false, the original system is inconsistent. ▲

We leave it for the reader to show that each pair of equations from the system does have a solution. Consequently, no two of the planes are parallel. Hence, we infer that this system of equations is of the type illustrated in Figure 13.8.

A system of the type illustrated in Figure 13.9 will be given as problem 13 in the next exercise.

Exercise 3

In problems 1 through 8 solve the given system of linear equations in three variables.

1. $x + y - z = 6$
    $2x - y - 3z = 9$
    $3x - 2y + z = -4$

2. $x + 3y + 3z = -8$
    $3x + 2y - 5z = 11$
    $-5x - 6y + z = -7$

3. $x - y + z = 4$
    $3x + y + 2z = 20$
    $4x - y + 5z = 29$

4. $2x + y - z = 5$
    $3x - y + 4z = 7$
    $-4x + y + z = -15$

5. $A + 2B = 12$
    $3B + 4C = 3$
    $5A - 6C = 28$

6. $5u + 3v + 5w = 3$
    $3u + 5v + w = -5$
    $u + 2v + w = 2$

7. $x - 5y + 7z = 13$
    $2x - 6y + 8z = 16$
    $-3x + 8y - 6z = -17$

★8. $x - 2y + 3z = 15$
    $5x + 7y - 11z = -29$
    $-13x + 17y \quad 19z = 37$

In problems 9 and 10 solve the given system of '   ⸱ in four variables.

9. $x + y + z - u = 9$
    $-x + y + z + u = 1$
    $x - y + z + u = -1$
    $x + y - z + u = -3$

10. 3p⸱
    2

11. According to Euler's formula, in
    of vertices), E (the number of ℯ
    satisfy the equation $V - E + $⸍
    number of edges is three ⸍
    number of faces is one lℯ⸍
    of vertices, edges, and f
    that satisfies the cond⸍

12. Let G_1, G_2, and G_3 be the graphs of equations (44), (45), and (46), respectively. Prove that no pair of these planes are parallel by finding at least one point on each of the straight lines: (a) $G_1 \cap G_2$, (b) $G_2 \cap G_3$, (c) $G_1 \cap G_3$.

13. Consider the system of equations

$$\begin{aligned}
x + y - 2z &= 6 \\
2x + 3y - 3z &= 14 \\
3x + 4y - 5z &= 20.
\end{aligned}$$

Prove that any set (x_1, y_1, z_1) that satisfies the first two equations automatically satisfies the third equation. Prove that if t is any real number, then $x = 4 + 3t$, $y = 2 - t$, $z = t$, satisfies all three equations. This system is of the type illustrated in Figure 13.9. The points $(4 + 3t, 2 - t, t)$ all lie on a line, the common line of all three planes.

★ 14. Find and prove a formula that gives the distance of the point $P(x, y, z)$ from the origin. *Hint:* See equation (14) in Chapter 6.

★ 15. Find and prove a formula that gives the distance from the point $P(x_1, y_1, z_1)$ to the point $Q(x_2, y_2, z_2)$. *Hint:* See Theorem 1 in Chapter 9.

★ 16. Find an equation for a sphere with center at (a, b, c) and radius $r > 0$.

★ 17. Let $A\star(a_1, a_2, a_3)$ and $B\star(b_1, b_2, b_3)$ be two fixed points. Let G be the set of all points P such that $|PA\star| = |PB\star|$. Prove that the set G is the graph of an equation of the form $Ax + By + Cz + D = 0$.

★★ 18. Can the result in problem 17 be used as the definition of a plane? If so, how would you phrase the definition?

19. Find a, b, and c so that the graph of $y = ax^2 + bx + c$ goes through the points $(1, 2)$, $(2, 3)$, and $(3, 6)$.

20. Find a, b, and c so that the graph of $y = ax^2 + bx + c$ goes through the points $(-1, 9)$, $(3, -3)$, $(5, 15)$.

21. Find an equation for the plane that goes through the points $(3, 2, 1)$, $(1, -3, -3)$, and $(4, 3, 2)$. *Hint:* Assume that one of the coefficients is 1.

22. Find an equation for the plane that passes through the points $(0, 0, 0)$, $(-1, 3, 4)$, and $(3, 3, -2)$.

13.4 MATRICES

If we review the material on solving systems of linear equations we might observe that the solution of a system depends on the coefficients, and that the letters x, y, z can be regarded as place holders, that may be suppressed if we ___ to shorten the labor of writing. For example, in the system (see Section ___ mple 2)

(33)
$$x - 2y + 3z = 10$$

(34)
$$x + 2y - z = 6$$

(35)
$$-2x + y + 2z = -2$$

we can represent the left-hand side by the array of coefficients

(49)
$$M = \begin{bmatrix} 1 & -2 & 3 \\ 1 & 2 & -1 \\ -2 & 1 & 2 \end{bmatrix}.$$

If we wish to include the right-hand side of the system, we would expand the array to form

(50)
$$M^\star = \left[\begin{array}{ccc|c} 1 & -2 & 3 & 10 \\ 1 & 2 & -1 & 6 \\ -2 & 1 & 2 & -2 \end{array} \right],$$

where the vertical line in (50) serves to remind us of the equal signs and the different roles played by the numbers on the two sides of the line. Either of these arrays is called a *matrix*, and $M^\star$ is called an *augmented matrix*.

Although we can motivate the study of matrices by their relation to systems of linear equations, it must be realized that matrices have many other applications. Further matrices have many interesting properties that stand apart from their source. Consequently, it is worthwhile to study the algebra (and calculus) of matrices purely as a mathematical subject.

DEFINITION 4   *A* 3×3 *matrix M is an array of real numbers of the form*

(51)
$$M = \begin{bmatrix} a_1 & b_1 & c_1 \\ a_2 & b_2 & c_2 \\ a_3 & b_3 & c_3 \end{bmatrix}.$$

A matrix of the form

(52)
$$M^\star = \left[\begin{array}{ccc|c} a_1 & b_1 & c_1 & d_1 \\ a_2 & b_2 & c_2 & d_2 \\ a_3 & b_3 & c_3 & d_3 \end{array} \right]$$

is called an augmented matrix of M.

These are the two matrices that we naturally associate with the system of linear equations: (30), (31), and (32). In this chapter we will concentrate on matrices of this kind, but we want to be free to consider arbitrary matrices with m rows and n columns. Clearly, if n is large ($n > 26$) we must forsake the alphabet, $a, b, c, \ldots$, and use some other notation. The simplest procedure for an arbitrary matrix is to use a double subscript a_{ij}. Here i denotes the row, and j denotes the column in which the element a_{ij} stands. Thus, the subscripts give the location of the element a_{ij}.

DEFINITION 5   *A matrix A is a rectangular array of real numbers of the form*

(53)
$$A = \begin{bmatrix} a_{11} & a_{12} & a_{13} & \cdots & a_{1n} \\ a_{21} & a_{22} & a_{23} & \cdots & a_{2n} \\ \vdots & \vdots & \vdots & \vdots & \vdots \\ a_{m1} & a_{m2} & a_{m3} & \cdots & a_{mn} \end{bmatrix}.$$

The matrix A is called an $m \times n$ (read, "m by n") matrix if it has m rows and n columns. The number a_{ij} is called an element of the matrix and is the element in the ith row and jth column. If $m = n$, the matrix is said to be a square matrix, and the order of the matrix is n. If A has only one row, then it is called a row matrix (or a row vector). If A has only one column, it is called a column matrix (or a column vector). If all the elements of A are zero, then A is called the null matrix or the zero matrix and is denoted by 0. The dimension of A is $m \times n$.

As examples of matrices we have

$$A = [3, 2, -1], \qquad B = \begin{bmatrix} 1 \\ 7 \\ 11 \end{bmatrix}, \qquad C = \begin{bmatrix} 4, & 3 \\ -1, & -5 \end{bmatrix},$$

$$D = \begin{bmatrix} 1 & 3 \\ 5 & 7 \\ -2 & 4 \\ 3 & -6 \end{bmatrix}, \qquad E = \begin{bmatrix} -4, & 11, & \sqrt[3]{2} & -9 \\ 1/2, & \sqrt{98}, & 72 & 0 \end{bmatrix}.$$

The matrices A, B, C, D, and E are 1×3, 3×1, 2×2, 4×2, and 2×4 matrices, respectively. The matrix $A = [3, 2, -1]$ is also a row vector, and the matrix B is also a column vector.

EXAMPLE 1   For the example matrices find the elements $a_{13}, b_{21}, c_{22}, d_{32}, d_{41}, e_{23}$, and e_{24}.

Solution: By inspection of the matrices A, B, C, D, and E we find that $a_{13} = -1$, $b_{21} = 7$, $c_{22} = -5$, $d_{32} = 4$, $d_{41} = 3$, $e_{23} = 72$, and $e_{24} = 0$.   ▲

The matrix A, given by equation (53), represents the left side of the system of m linear equations in the n variables $x_1, x_2, \ldots, x_n$:

$$
\begin{aligned}
a_{11}x_1 + a_{12}x_2 + \cdots + a_{1n}x_n &= k_1 \\
a_{21}x_1 + a_{22}x_2 + \cdots + a_{2n}x_n &= k_2 \\
\vdots \qquad \vdots \qquad \vdots \qquad \vdots \qquad &\quad \vdots \\
a_{m1}x_1 + a_{m2}x_2 + \cdots + a_{mn}x_n &= k_m.
\end{aligned}
$$

(54)

The entire system (54) is represented by the augmented matrix

(55)
$$
A^\star = \left[
\begin{array}{ccccc|c}
a_{11} & a_{12} & a_{13} & \cdots & a_{1n} & k_1 \\
a_{21} & a_{22} & a_{23} & \cdots & a_{2n} & k_2 \\
\vdots & \vdots & \vdots & \vdots & \vdots & \vdots \\
a_{m1} & a_{m2} & a_{m3} & \cdots & a_{mn} & k_m
\end{array}
\right].
$$

Any elementary transformation that carries a system of linear equations into an equivalent system has its counterpart in a row operation on a matrix.

DEFINITION 6   *Any one of the following operations on the rows of a matrix is called an elementary row operation:*

   (I) *the interchange of any two rows,*
   (II) *multiplication of any row by a nonzero constant,*
   (III) *the termwise addition of any multiple of one row to any other row.*

DEFINITION 7   *If A can be transformed into B by a finite number of elementary row operations, then we say that A is row equivalent to B.*

Obviously we can introduce elementary column operations of a matrix and the column equivalence of two matrices. At the moment, column operations do not appear to be helpful in solving equations, but their usefulness will be established later.

We now return to systems of three linear equations in three variables. Let $M^\star$, given by (52), be the matrix for such a system. If we can transform $M^\star$ by a sequence of elementary row operations into a matrix

(56)
$$
I^\star = \left[
\begin{array}{ccc|c}
1 & 0 & 0 & e_1 \\
0 & 1 & 0 & e_2 \\
0 & 0 & 1 & e_3
\end{array}
\right],
$$

then the unique solution of the system is $x = e_1$, $y = e_2$, and $z = e_3$. This is exactly the technique used in Sections 13.1 and 13.3, except that now we have dropped the variables x, y, and z, and in this way we save quite a bit of mechanical labor.

EXAMPLE 2   Use the new technique to solve the system given in Example 2 of Section 13.3.

Solution: The augmented matrix for the system is

$$M^\star = \left[\begin{array}{ccc|c} 1 & -2 & 3 & 10 \\ 1 & 2 & -1 & 6 \\ -2 & 1 & 2 & -2 \end{array} \right].$$

We subtract the first row from the second row (add to the second row -1 times the first row). Then we add to the third row twice the first row. This gives

$$N^\star = \left[\begin{array}{ccc|c} 1 & -2 & 3 & 10 \\ 0 & 4 & -4 & -4 \\ 0 & -3 & 8 & 18 \end{array} \right],$$

a matrix that is row equivalent to $M^\star$. We multiply the second row by $1/4$. Then we add to the third row 3 times the second row. This gives

$$P^\star = \left[\begin{array}{ccc|c} 1 & -2 & 3 & 10 \\ 0 & 1 & -1 & -1 \\ 0 & 0 & 5 & 15 \end{array} \right].$$

In $P^\star$, we multiply the third row by $1/5$. Then we add to the first row 2 times the second row and -1 times the third row. Lastly, we add the third row to the second row. The student should write the intermediate row equivalent matrices and show that the last step gives

$$I^\star = \left[\begin{array}{ccc|c} 1 & 0 & 0 & 5 \\ 0 & 1 & 0 & 2 \\ 0 & 0 & 1 & 3 \end{array} \right].$$

Consequently, the solution of the system of equations (33), (34), and (35) is $x = 5$, $y = 2$, $z = 3$, the system represented by $I^\star$. ▲

To condense the writing still further, we use the symbol $\sim$ (read, "is equivalent to") to denote row equivalence. Then, in the example just completed, $M^\star \sim N^\star \sim P^\star \sim I^\star$. We also place an indication of the operations under the equivalence sign. Thus, the first two steps in the above example could be condensed by writing

$$\begin{bmatrix} 1 & -2 & 3 & | & 10 \\ 1 & 2 & -1 & | & 6 \\ -2 & 1 & 2 & | & -2 \end{bmatrix} \underset{\substack{\text{(R2)} - \text{(R1)}, \\ \text{(R3)} + 2\,\text{(R1)}}}{\sim} \begin{bmatrix} 1 & -2 & 3 & | & 10 \\ 0 & 4 & -4 & | & -4 \\ 0 & -3 & 8 & | & 18 \end{bmatrix}$$

$$\underset{\substack{\frac{1}{4}\,\text{(R2)}, \\ \text{(R3)} + 3\,\text{(R2)}}}{\sim} \begin{bmatrix} 1 & -2 & 3 & | & 10 \\ 0 & 1 & -1 & | & -1 \\ 0 & 0 & 5 & | & 15 \end{bmatrix},$$

where the meaning of each symbol is obvious from the way it is used. For example, $\text{(R2)} - \text{(R1)}$ means, subtract the first row termwise from the second row.

EXAMPLE 3   Solve the system of four linear equations in four variables

$$2x + y - z - u = 1$$
$$4x + y + 3z - 3u = -10$$
$$-2x + 2y + z + 2u = 12$$
$$6x - y - 2z + 3u = 14.$$

Solution: We apply the techniques and the notation just explained

$$\begin{bmatrix} 2 & 1 & -1 & -1 & | & 1 \\ 4 & 1 & 3 & -3 & | & -10 \\ -2 & 2 & 1 & 2 & | & 12 \\ 6 & -1 & -2 & 3 & | & 14 \end{bmatrix}$$

$$\underset{\substack{\text{(R2)} - 2\,\text{(R1)}, \\ \text{(R3)} + \text{(R1)}, \\ \text{(R4)} - 3\,\text{(R1)}}}{\sim} \begin{bmatrix} 2 & 1 & -1 & -1 & | & 1 \\ 0 & -1 & 5 & -1 & | & -12 \\ 0 & 3 & 0 & 1 & | & 13 \\ 0 & -4 & 1 & 6 & | & 11 \end{bmatrix}$$

$$\begin{array}{c} \sim \\ \textcircled{R1} + \textcircled{R2}, \\ \textcircled{R3} + 3\,\textcircled{R2}, \\ \textcircled{R4} - 4\,\textcircled{R2} \end{array} \left[\begin{array}{cccc|c} 2 & 0 & 4 & -2 & -11 \\ 0 & -1 & 5 & -1 & -12 \\ 0 & 0 & 15 & -2 & -23 \\ 0 & 0 & -19 & 10 & 59 \end{array}\right]$$

$$\begin{array}{c} \sim \\ \textcircled{R3} + \textcircled{R4}, \\ -\dfrac{1}{4}\,\textcircled{R3}, \\ -\,\textcircled{R2} \end{array} \left[\begin{array}{cccc|c} 2 & 0 & 4 & -2 & -11 \\ 0 & 1 & -5 & 1 & 12 \\ 0 & 0 & 1 & -2 & -9 \\ 0 & 0 & -19 & 10 & 59 \end{array}\right]$$

$$\begin{array}{c} \sim \\ \textcircled{R4} + 19\,\textcircled{R3}, \\ -\dfrac{1}{28}\,\textcircled{R4} \end{array} \left[\begin{array}{cccc|c} 2 & 0 & 4 & -2 & -11 \\ 0 & 1 & -5 & 1 & 12 \\ 0 & 0 & 1 & -2 & -9 \\ 0 & 0 & 0 & 1 & 4 \end{array}\right]$$

$$\begin{array}{c} \sim \\ \textcircled{R1} - 4\,\textcircled{R3}, \\ \textcircled{R2} + 5\,\textcircled{R3}, \\ \textcircled{R3} + 2\,\textcircled{R4} \end{array} \left[\begin{array}{cccc|c} 2 & 0 & 0 & 6 & 25 \\ 0 & 1 & 0 & -9 & -33 \\ 0 & 0 & 1 & 0 & -1 \\ 0 & 0 & 0 & 1 & 4 \end{array}\right]$$

$$\begin{array}{c} \sim \\ \textcircled{R1} - 6\,\textcircled{R4}, \\ \textcircled{R2} + 9\,\textcircled{R4} \end{array} \left[\begin{array}{cccc|c} 2 & 0 & 0 & 0 & 1 \\ 0 & 1 & 0 & 0 & 3 \\ 0 & 0 & 1 & 0 & -1 \\ 0 & 0 & 0 & 1 & 4 \end{array}\right].$$

Restoring the variables we see that $2x = 1$. Hence, $x = 1/2$, $y = 3$, $z = -1$, $u = 4$ is the unique solution of the given system. ▲

Exercise 4

1. Solve the systems given in Exercise 1 by the methods of this section.

In problems 2 through 18 solve the given system.

2. $2x - y + z = 3$
  $-x + y + 2z = -5$
  $2x - 2y + 3z = 3$

3. $x + 2y + 5z = -4$
  $x + 3y + 5z = -3$
  $x + 2y + 6z = -5$

4. $x+2y-3z = -1$
    $5x+3y-4z = 2$
    $-3x+ y-2z = -8$

5. $2x-y+ z = 1$
    $3x +2z = -1$
    $4x+y+2z = 2$

6. $2x+4y- 9z = 3$
    $x-8y+15z = 3$
    $5x-2y-12z = 5$

7. $x-2y+3z = 2$
    $2x+3y-2z = 4$
    $4x- y+4z = 1$

8. $x-2y+3z = 22$
    $-3x+5y+4z = 2$
    $5x+3y+3z = 11$

9. $2x+2y+ 6z = 5$
    $4x-5y+13z = 6$
    $x+5y- 8z = -1$

10. $5x+6y+8z = -7$
    $6x+8y+5z = 7$
    $4x+9y+7z = 1$

11. $x+y+z+w = 1$
    $x-y+z+w = 1$
    $y+z-w = 1$
    $y +w = 1$

12. $5x+4y+5z = 6$
    $2x+3y +2w = -3$
    $y- z+ w = -3$
    $5z-2w = 7$

13. $3x- 6y+ z+ w = -2$
    $y- z = -1$
    $6x-12y+2z+3w = 0$
    $4x- 8y+ z+2w = -1$

14. $2x+3y+ z+5w = 11$
    $3x+2y+4z-3w = 6$
    $x+3y+2z- w = 5$
    $10x+8y-8z-6w = 4$

15. $x^2+y^2+z^2 = 14$
    $x^2-y^2+z^2 = -4$
    $2x^2-y^2+z^2 = 0$

16. $5x^2+3y^2- z^2 = 6$
    $x^2-4y^2+5z^2 = 7$
    $3x^2+9y^2-2z^2 = 8$

17. $\dfrac{1}{x}+\dfrac{2}{y+4}-\dfrac{1}{z-2} = 4$

    $\dfrac{1}{y+4}+\dfrac{2}{z-2} = 5$

    $\dfrac{1}{z-2} = 1$

18. $\dfrac{1}{x+1}+\dfrac{2}{y^2}+\dfrac{5}{z-1} = -4$

    $\dfrac{2}{x+1}+\dfrac{3}{y^2}+\dfrac{6}{z-1} = -5$

    $\dfrac{3}{x+1}+\dfrac{5}{y^2}+\dfrac{6}{z-1} = -4$

19. Prove that if $a_1 b_2 - a_2 b_1 \neq 0$, then the system

$$a_1 x + b_1 y = k_1$$
$$a_2 x + b_2 y = k_2$$

has the unique solution

(57) $$x = \frac{k_1 b_2 - k_2 b_1}{a_1 b_2 - a_2 b_1}, \qquad y = \frac{a_1 k_2 - a_2 k_1}{a_1 b_2 - a_2 b_1}.$$

13.5 DETERMINANTS

We now give a second method for solving systems of linear equations. In some cases, this new method may be superior to the echelon method. In other cases, it may be longer or more cumbersome. However, for any problem, we want to have under our command as many methods as possible.

In Exercise 4, problem 19 the student found formulas for the solution of the general system of two linear equations in two variables. A study of these formulas, equation set (57), reveals a pattern, which we now investigate.

Given a 2×2 matrix

$$(58) \qquad M = \begin{bmatrix} a_1 & b_1 \\ a_2 & b_2 \end{bmatrix}$$

we associate with M a real number called the *determinant* of M and denoted by $D(M)$. For the 2×2 matrix, (58), this number is defined by

$$D(M) = a_1 b_2 - a_2 b_1.$$

There is an alternate notation in which we replace the brackets by vertical lines. Thus, if M is given by (58), then by definition

$$(59) \qquad D(M) \equiv \begin{vmatrix} a_1 & b_1 \\ a_2 & b_2 \end{vmatrix} \equiv a_1 b_2 - a_2 b_1.$$

In this form it is easy to remember the definition because we form products along diagonal elements (indicated below by arrows) attaching a plus sign if we descend from left to right, and a negative sign if we descend from right to left. Thus,

$$(60) \qquad \begin{vmatrix} a_1 & b_1 \\ a_2 & b_2 \end{vmatrix} = a_1 b_2 - a_2 b_1.$$

EXAMPLE 1   Find the determinant of each of the matrices:

$$A = \begin{bmatrix} 1 & 3 \\ 2 & 9 \end{bmatrix}, \qquad B = \begin{bmatrix} 2 & -5 \\ 3 & 7 \end{bmatrix}, \qquad C = \begin{bmatrix} -4 & -3 \\ -6 & 1/2 \end{bmatrix}.$$

Solution: Using either (59) or (60) we find

$$D(A) = \begin{vmatrix} 1 & 3 \\ 2 & 9 \end{vmatrix} = 1(9) - 3(2) = 9 - 6 = 3$$

$$D(B) = \begin{vmatrix} 2 & -5 \\ 3 & 7 \end{vmatrix} = 2(7) - (-5)3 = 14 + 15 = 29$$

$$D(C) = \begin{vmatrix} -4 & -3 \\ -6 & 1/2 \end{vmatrix} = (-4)\left(\frac{1}{2}\right) - (-3)(-6) = -2 - 18 = -20. \quad \blacktriangle$$

Using determinants we can give the solution of a system of linear equations in a neat and compact form.

Theorem 3
PLE

The system of linear equations

$$a_1 x + b_1 y = k_1$$

$$a_2 x + b_2 y = k_2$$

has the unique solution

(61)
$$x = \frac{\begin{vmatrix} k_1 & b_1 \\ k_2 & b_2 \end{vmatrix}}{\begin{vmatrix} a_1 & b_1 \\ a_2 & b_2 \end{vmatrix}}, \quad y = \frac{\begin{vmatrix} a_1 & k_1 \\ a_2 & k_2 \end{vmatrix}}{\begin{vmatrix} a_1 & b_1 \\ a_2 & b_2 \end{vmatrix}},$$

provided that the determinant in the denominator is not zero.

The student has already proved this theorem in problem 19 of Exercise 4. All that remains is to show that the quantities in (57) are indeed the determinants in (61).

The rule for forming the determinants in (61) is quite simple. The determinant in the denominator is formed from the matrix of the coefficients of the variables. To find x we replace the column a_1, a_2 by k_1, k_2. To find y we replace b_1, b_2 by k_1, k_2. This rule (together with Theorem 3) is called Cramer's rule.

EXAMPLE 2

Solve the system of linear equations

(62)
$$2x - 3y = 9$$

$$-x + 9y = -2.$$

Solution: By Cramer's rule (Theorem 3),

$$x = \frac{\begin{vmatrix} 9 & -3 \\ -2 & 9 \end{vmatrix}}{\begin{vmatrix} 2 & -3 \\ -1 & 9 \end{vmatrix}} = \frac{9(9) - (-3)(-2)}{2(9) - (-3)(-1)} = \frac{81 - 6}{18 - 3} = \frac{75}{15} = 5$$

and

$$y = \frac{\begin{vmatrix} 2 & 9 \\ -1 & -2 \end{vmatrix}}{\begin{vmatrix} 2 & -3 \\ -1 & 9 \end{vmatrix}} = \frac{2(-2)-9(-1)}{15} = \frac{-4+9}{15} = \frac{5}{15} = \frac{1}{3}. \quad \blacktriangle$$

Check   Using $x = 5$ and $y = 1/3$ we find that

$$2x - 3y = 2(5) - 3\left(\frac{1}{3}\right) = 10 - 1 = 9$$

$$-x + 9y = -5 + 9\left(\frac{1}{3}\right) = -5 + 3 = -2.$$

If Cramer's rule applied only to systems in two variables, it would be almost worthless, but the beauty of the rule is that it applies to any system of n linear equations in n variables. In this text we will be content with proving the rule when $n = 3$, but we will indicate the material needed in the proof for arbitrary n.

DEFINITION 8   *A determinant is a certain real-valued function defined over the set of all square matrices. If the matrix M is of nth order, the determinant $D(M)$ is said to be of nth order. If M is the 2×2 matrix (58), then $D(M)$ is defined by (59). If M is the 3×3 matrix*

(63)
$$M = \begin{bmatrix} a_1 & b_1 & c_1 \\ a_2 & b_2 & c_2 \\ a_3 & b_3 & c_3 \end{bmatrix},$$

then

(64) $D(M) \equiv a_1 b_2 c_3 + a_3 b_1 c_2 + a_2 b_3 c_1 - a_1 b_3 c_2 - a_2 b_1 c_3 - a_3 b_2 c_1.$

The right side of (59) or (64) is called the expansion of the determinant.

We also use the notation

$$\begin{vmatrix} a_1 & b_1 & c_1 \\ a_2 & b_2 & c_2 \\ a_3 & b_3 & c_3 \end{vmatrix}$$

for the determinant of M.

The right side of (64) appears to be rather complicated. There are several different ways of arriving at this expression. We give one here, and indicate two more ways in problems 17 and 18 of the next exercise. We need

DEFINITION 9   ***Minor***   *The minor M_{ij} of a square matrix M is the determinant of the matrix obtained by deleting the ith row and the jth column of M.*

EXAMPLE 3   Find the minors M_{11}, M_{12}, M_{13} and M_{32} of

(65)
$$M \equiv \begin{bmatrix} 3 & 2 & 7 \\ -1 & 5 & 3 \\ 2 & -3 & 6 \end{bmatrix}.$$

Solution: For M_{11} we delete the first row and the first column of M. This gives

$$M_{11} = \begin{vmatrix} 5 & 3 \\ -3 & 6 \end{vmatrix} = 5(6) - (3)(-3) = 39.$$

For the other three minors we have:

$$M_{12} = \begin{vmatrix} -1 & 3 \\ 2 & 6 \end{vmatrix} = (-1)(6) - 3(2) = -12$$

$$M_{13} = \begin{vmatrix} -1 & 5 \\ 2 & -3 \end{vmatrix} = (-1)(-3) - 5(2) = -7$$

$$M_{32} = \begin{vmatrix} 3 & 7 \\ -1 & 3 \end{vmatrix} = 3(3) - 7(-1) = 16. \quad \blacktriangle$$

We can express a third-order determinant in terms of minors.

Theorem 4   *If M is the 3 × 3 matrix given by (63), then*

(66)
$$\begin{vmatrix} a_1 & b_1 & c_1 \\ a_2 & b_2 & c_2 \\ a_3 & b_3 & c_3 \end{vmatrix} = a_1 M_{11} - b_1 M_{12} + c_1 M_{13}.$$

Proof: By the definition of the symbols, the right side of (66) is

$$a_1 \begin{vmatrix} b_2 & c_2 \\ b_3 & c_3 \end{vmatrix} - b_1 \begin{vmatrix} a_2 & c_2 \\ a_3 & c_3 \end{vmatrix} + c_1 \begin{vmatrix} a_2 & b_2 \\ a_3 & b_3 \end{vmatrix}$$

$$= a_1(b_2 c_3 - b_3 c_2) - b_1(a_2 c_3 - a_3 c_2) + c_1(a_2 b_3 - a_3 b_2)$$

$$= a_1 b_2 c_3 + a_3 b_1 c_2 + a_2 b_3 c_1 - a_1 b_3 c_2 - a_2 b_1 c_3 - a_3 b_2 c_1.$$

This is identical with the right side of (64). ∎

The expression (66) is called a *Laplace expansion of the determinant by minors*. In this case, it is an expansion by minors of the first row, since the coefficients are a_1, b_1, c_1, the elements of the first row. One can also expand by minors of any other row or by minors of any column. In such an expansion the sign $(-1)^{i+j}$ is attached to each minor M_{ij}. For example, expanding by minors of the second row we obtain

$$(67) \quad \begin{vmatrix} a_1 & b_1 & c_1 \\ a_2 & b_2 & c_2 \\ a_3 & b_3 & c_3 \end{vmatrix} = -a_2 \begin{vmatrix} b_1 & c_1 \\ b_3 & c_3 \end{vmatrix} + b_2 \begin{vmatrix} a_1 & c_1 \\ a_3 & c_3 \end{vmatrix} - c_2 \begin{vmatrix} a_1 & b_1 \\ a_3 & b_3 \end{vmatrix}.$$

If we expand by minors of the first column, the same matrix gives

$$(68) \quad D(M) = a_1 \begin{vmatrix} b_2 & c_2 \\ b_3 & c_3 \end{vmatrix} - a_2 \begin{vmatrix} b_1 & c_1 \\ b_3 & c_3 \end{vmatrix} + a_3 \begin{vmatrix} b_1 & c_1 \\ b_2 & c_2 \end{vmatrix}.$$

EXAMPLE 4

Compute $D(M)$ for the matrix given in Example 3.

Solution: We have already computed the minors for the elements in the first row. Using an expansion by minors of the first row, we find that

$$D(M) = 3(39) - 2(-12) + 7(-7) = 117 + 24 - 49 = 92.$$

The definition of a fourth-order determinant is given in Section 13.7, but we postpone consideration of nth-order determinants until the next chapter. For the present we concentrate on proving theorems about third-order determinants. However, all of these theorems (5 through 12) are true for determinants of any order $n \geq 2$.

Theorem 5

Let M be given by (63) and let N be the transpose of M; i.e., let

$$(69) \quad N \equiv \begin{vmatrix} a_1 & a_2 & a_3 \\ b_1 & b_2 & b_3 \\ c_1 & c_2 & c_3 \end{vmatrix},$$

the matrix obtained by reflecting the elements of M across the main diagonal (a_1, b_2, c_3). Then $D(M) = D(N)$.

Proof: Expanding by minors of the first row (Theorem 4), we find

$$D(N) = a_1 \begin{vmatrix} b_2 & b_3 \\ c_2 & c_3 \end{vmatrix} - a_2 \begin{vmatrix} b_1 & b_3 \\ c_1 & c_3 \end{vmatrix} + a_3 \begin{vmatrix} b_1 & b_2 \\ c_1 & c_2 \end{vmatrix}$$

$$= a_1 b_2 c_3 - a_1 b_3 c_2 - a_2 b_1 c_3 + a_2 b_3 c_1 + a_3 b_1 c_2 - a_3 b_2 c_1.$$

This is the same as the right side of equation (64). ∎

We will use this theorem as follows: We prove a property of the columns of M. Then this property will also be true of the rows of M, because the rows of M are the columns of N. As a first example of this technique we have

Theorem 6   *If in one column of M every element is zero, then $D(M) = 0$. If in any one row of M every element is zero, then $D(M) = 0$.*

Proof: Suppose that every element in the first column is zero. Since each term in the expansion (64) contains an a_i, each term is zero. Hence, the sum is zero. Each term also contains a b_i and a c_i. Next suppose that in M there is a row with every element equal to zero. If N is the transpose of M, then in N there is a column with every element equal to zero. Hence, $D(N) = 0$. By Theorem 5 we have $D(M) = D(N) = 0$. ∎

For example, by Theorem 6 we see immediately that

$$\begin{vmatrix} 3 & -2 & 0 \\ 5 & 7\pi & 0 \\ 9 & \sqrt{53} & 0 \end{vmatrix} = 0 \quad \text{and} \quad \begin{vmatrix} 1 & 3 & \sqrt{42} \\ 0 & 0 & 0 \\ -3 & 7 & \pi\sqrt[3]{19} \end{vmatrix} = 0.$$

Theorem 7   *Suppose that the matrix P is obtained from M by the interchange of any two rows or by the interchange of any two columns. Then, $D(P) = -D(M)$.*

Proof: By Theorem 5, it is sufficient to consider only the interchange of columns. Suppose that we interchange the first two columns. If M is given by (63), then

$$D(P) = \begin{vmatrix} b_1 & a_1 & c_1 \\ b_2 & a_2 & c_2 \\ b_3 & a_3 & c_3 \end{vmatrix} = b_1 \begin{vmatrix} a_2 & c_2 \\ a_3 & c_3 \end{vmatrix} - a_1 \begin{vmatrix} b_2 & c_2 \\ b_3 & c_3 \end{vmatrix} + c_1 \begin{vmatrix} b_2 & a_2 \\ b_3 & a_3 \end{vmatrix}$$

$$= b_1 a_2 c_3 - b_1 a_3 c_2 - a_1 b_2 c_3 + a_1 b_3 c_2 + c_1 b_2 a_3 - c_1 a_2 b_3$$

$$= -(a_1 b_2 c_3 + a_3 b_1 c_2 + a_2 b_3 c_1 - a_1 b_3 c_2 - a_2 b_1 c_3 - a_3 b_2 c_1)$$

$$= -D(M).$$

The proof is similar, if we interchange the second and third columns, or interchange the first and third columns. ∎

Theorem 8

If two columns of M are identical, then $D(M) = 0$. If two rows of M are identical, then $D(M) = 0$.

For example,

$$\begin{vmatrix} \sqrt[3]{5} & 12\pi & \sqrt[3]{5} \\ -4 & \sqrt{17} & -4 \\ 181 & -3 & 181 \end{vmatrix} = 0.$$

Proof: Let P be obtained from M by the interchange of the two identical columns. Then $P = M$ and, hence, $D(P) = D(M)$. But by Theorem 7 we have $D(P) = -D(M)$. Hence, $D(M) = -D(M)$ or $2D(M) = 0$. ∎

An examination of the right side of (64) gives immediately

Theorem 9
PLE

If R is obtained from M by multiplying each element in a fixed column by k (or each element in a fixed row by k), then $D(R) = kD(M)$.

In practice, we use this theorem to simplify determinants by "factoring out" common factors in a column or row. For example, if we factor 3 from the third column and 4 from the third row, we have

$$\begin{vmatrix} 5 & 4 & 9 \\ 2 & -1 & 3 \\ 20 & 8 & -12 \end{vmatrix} = 3 \begin{vmatrix} 5 & 4 & 3 \\ 2 & -1 & 1 \\ 20 & 8 & -4 \end{vmatrix} = 12 \begin{vmatrix} 5 & 4 & 3 \\ 2 & -1 & 1 \\ 5 & 2 & -1 \end{vmatrix}.$$

By combining Theorems 8 and 9, we derive

Theorem 10
PLE

If two columns of M are proportional or two rows of M are proportional, then $D(M) = 0$.

For example,

$$\begin{vmatrix} -5 & \sqrt{\pi} & 20 \\ -1 & 10{,}000 & 4 \\ 7 & \sqrt[3]{63} & -28 \end{vmatrix} = 0,$$

because the third column is proportional to the first column with -4 as the constant of proportionality ($c_i = (-4)a_i$ for $i = 1, 2, 3$).

Theorem 11

If S is obtained from M by adding to any one column (termwise) k times some other column, then $D(S) = D(M)$. If S is obtained by adding to any one row (termwise) k times some other row, then $D(S) = D(M)$.

Proof: We give the proof for the first two rows. Any other case can be reduced to this case by an interchange of rows (Theorem 7) or by considering the transpose of M (Theorem 5).

Let M_{11}, M_{12}, and M_{13} be the minors of M, given by (63). Then

$$D(S) = \begin{vmatrix} a_1 + ka_2 & b_1 + kb_2 & c_1 + kc_2 \\ a_2 & b_2 & c_2 \\ a_3 & b_3 & c_3 \end{vmatrix}$$

$$= (a_1 + ka_2)M_{11} - (b_1 + kb_2)M_{12} + (c_1 + kc_2)M_{13}$$

$$= a_1 M_{11} - b_1 M_{12} + c_1 M_{13} + k[a_2 M_{11} - b_2 M_{12} + c_2 M_{13}]$$

$$= \begin{vmatrix} a_1 & b_1 & c_1 \\ a_2 & b_2 & c_2 \\ a_3 & b_3 & c_3 \end{vmatrix} + k \begin{vmatrix} a_2 & b_2 & c_2 \\ a_2 & b_2 & c_2 \\ a_3 & b_3 & c_3 \end{vmatrix} = D(M) + 0. \quad \blacksquare$$

To illustrate the application of Theorem 11, we note that

$$(70) \qquad \begin{vmatrix} 7 & 2 & 147 \\ -15 & -5 & 16 \\ 9 & 3 & -10 \end{vmatrix} = \begin{vmatrix} 1 & 2 & 147 \\ 0 & -5 & 16 \\ 0 & 3 & -10 \end{vmatrix},$$

because the second determinant can be obtained from the first one by multiplying each element in the second column by -3 and adding it to the first column. Now the right side of (70) is easy to compute. We expand by minors of the first column and find that

$$D = 1 \begin{vmatrix} -5 & 16 \\ 3 & -10 \end{vmatrix} + 0 \begin{vmatrix} 2 & 147 \\ 3 & -10 \end{vmatrix} + 0 \begin{vmatrix} 2 & 147 \\ -5 & 16 \end{vmatrix}$$

$$= 50 - 48 + 0 + 0 = 2.$$

EXAMPLE 5   Find $D(M)$ for the matrix

$$M = \begin{bmatrix} 36 & 18 & 17 \\ 183 & 101 & 78 \\ 45 & 24 & 20 \end{bmatrix}.$$

Solution: We use our theorems to replace the large numbers in M by smaller ones and then eventually by zeros. We also use the shorthand introduced in Section 13.4 for the elementary row operations, but now we use either an R or C to indicate whether the operation is on a row or column.

$$D(M) = \begin{vmatrix} 36 & 18 & 17 \\ 183 & 101 & 78 \\ 45 & 24 & 20 \end{vmatrix} \underset{\text{(R2)} - 4\text{(R3)}}{=} \begin{vmatrix} 36 & 18 & 17 \\ 3 & 5 & -2 \\ 45 & 24 & 20 \end{vmatrix}$$

$$\underset{\frac{1}{3}\text{(C1)}}{=} 3\begin{vmatrix} 12 & 18 & 17 \\ 1 & 5 & -2 \\ 15 & 24 & 20 \end{vmatrix} \underset{\text{(C2)} - \text{(C3)}}{=} 3\begin{vmatrix} 12 & 1 & 17 \\ 1 & 7 & -2 \\ 15 & 4 & 20 \end{vmatrix}$$

$$\underset{\text{(R3)} - \text{(R1)}}{=} 3\begin{vmatrix} 12 & 1 & 17 \\ 1 & 7 & -2 \\ 3 & 3 & 3 \end{vmatrix} \underset{\frac{1}{3}\text{(R3)}}{=} 9\begin{vmatrix} 12 & 1 & 17 \\ 1 & 7 & -2 \\ 1 & 1 & 1 \end{vmatrix}$$

$$\underset{\substack{\text{(C2)} - \text{(C1)}, \\ \text{(C3)} - \text{(C1)}}}{=} 9\begin{vmatrix} 12 & -11 & 5 \\ 1 & 6 & -3 \\ 1 & 0 & 0 \end{vmatrix} \underset{\substack{\text{(Theorem 7,} \\ \text{twice)}}}{=} 9\begin{vmatrix} 1 & 0 & 0 \\ 12 & -11 & 5 \\ 1 & 6 & -3 \end{vmatrix}$$

$$\underset{\text{(Theorem 4)}}{=} 9\begin{vmatrix} -11 & 5 \\ 6 & -3 \end{vmatrix} = 9[(-11)(-3) - 5(6)] = 9(33 - 30) = 27. \quad \blacktriangle$$

For those who prefer brute force, the evaluation of $D(M)$ directly from the definition, equation (64), yields

$$a_1 b_2 c_3 = 72{,}720 \qquad -a_1 b_3 c_2 = -67{,}392$$

$$a_3 b_1 c_2 = 63{,}180 \qquad -a_2 b_1 c_3 = -65{,}880$$

$$a_2 b_3 c_1 = \underline{74{,}664} \qquad -a_3 b_2 c_1 = \underline{-77{,}265}$$

$$210{,}564 \qquad -210{,}537$$

$$D(M) = 210{,}564 - 210{,}537 = 27.$$

Exercise 5

In problems 1 through 10 evaluate the given determinant.

1. $\begin{vmatrix} 5 & 1 & 3 \\ 2 & -5 & 6 \\ 6 & -9 & 5 \end{vmatrix}$ 2. $\begin{vmatrix} -3 & 4 & 2 \\ 5 & 0 & -1 \\ 6 & -5 & -4 \end{vmatrix}$

3. $\begin{vmatrix} 1/2 & -3/2 & 1/3 \\ -6 & 9 & 1 \\ 5 & -3 & -2 \end{vmatrix}$ 4. $\begin{vmatrix} 1 & 2 & 3 \\ 4 & 5 & 6 \\ 7 & 8 & 9 \end{vmatrix}$

5. $\begin{vmatrix} 1 & 4 & 9 \\ 4 & 9 & 16 \\ 9 & 16 & 25 \end{vmatrix}$ 6. $\begin{vmatrix} 12 & 15 & 11 \\ 6 & 2 & 7 \\ 13 & 16 & 12 \end{vmatrix}$

7. $\begin{vmatrix} 17 & 4 & 5 \\ 16 & 10 & 9 \\ 8 & 1 & 3 \end{vmatrix}$ 8. $\begin{vmatrix} 2 & 3 & 5 \\ 7 & 11 & 13 \\ 17 & 19 & 23 \end{vmatrix}$

9. $\begin{vmatrix} 171 & 110 & 60 \\ 141 & 71 & 70 \\ 344 & 221 & 121 \end{vmatrix}$ 10. $\begin{vmatrix} 28 & 30 & -35 \\ -15 & -14 & 19 \\ 6 & 11 & -5 \end{vmatrix}$

In problems 11 through 14 prove the given assertion.

11. $\begin{vmatrix} 1 & 1 & 1 \\ x & y & z \\ x^2 & y^2 & z^2 \end{vmatrix} = (z-x)(z-y)(y-x)$

12. $\begin{vmatrix} a & b & c \\ c & a & b \\ b & c & a \end{vmatrix} = a^3 + b^3 + c^3 - 3abc$

13. $\begin{vmatrix} b+c & a-c & a-b \\ b-c & c+a & b-a \\ c-b & c-a & a+b \end{vmatrix} = 8abc$

14. $\begin{vmatrix} b+c & c & b \\ c & c+a & a \\ b & a & a+b \end{vmatrix} = 4abc$

15. The numbers 228, 589, and 779 are all divisible by 19. Use this fact to prove that the determinant

$$D = \begin{vmatrix} 2 & 2 & 8 \\ 5 & 8 & 9 \\ 7 & 7 & 9 \end{vmatrix}$$

is divisible by 38, without finding the value of D.

16. Prove each of the Theorems 5 through 11 when $n = 2$ (second order determinants).

★17. As indicated in the following diagram, we (a) copy the first two columns again on the right side, (b) take products of terms on the slanted lines, and (c) assign a factor ± 1.

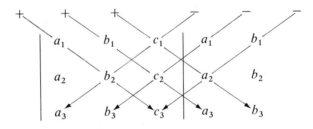

Prove that this scheme also gives $D(M)$. *Caution:* If this scheme is used to compute determinants of order $n \geq 4$, it will not give the right value for $D(M)$. The Laplace expansion by minors (Theorem 4) will give the correct result for determinants of any order.

★ 18. Another scheme for finding $D(M)$ is indicated below.

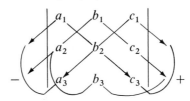

With a suitable interpretation of these symbols, prove that this scheme also gives $D(M)$. *Caution:* This scheme has the same defect as the one given in problem 17.

13.6 CRAMER'S RULE

For $n = 3$, Cramer's rule is given in

Theorem 12   *If the determinant*

(71)
$$D \equiv \begin{vmatrix} a_1 & b_1 & c_1 \\ a_2 & b_2 & c_2 \\ a_3 & b_3 & c_3 \end{vmatrix} \neq 0,$$

then the system of linear equations

(72)
$$a_1 x + b_1 y + c_1 z = k_1$$
$$a_2 x + b_2 y + c_2 z = k_2$$
$$a_3 x + b_3 y + c_3 z = k_3$$

has a unique solution given by

(73) $$x = \frac{\begin{vmatrix} k_1 & b_1 & c_1 \\ k_2 & b_2 & c_2 \\ k_3 & b_3 & c_3 \end{vmatrix}}{D}, \quad y = \frac{\begin{vmatrix} a_1 & k_1 & c_1 \\ a_2 & k_2 & c_2 \\ a_3 & k_3 & c_3 \end{vmatrix}}{D}, \quad z = \frac{\begin{vmatrix} a_1 & b_1 & k_1 \\ a_2 & b_2 & k_2 \\ a_3 & b_3 & k_3 \end{vmatrix}}{D}.$$

The rule for forming the determinants in (73) is simple. The denominator D is formed from the coefficients of the variables (see equation set (72)). To solve for x, we replace the first column of D by a column of k's to obtain the numerator (see equation set (73)). To solve for y, we replace the second column by k's, and to solve for z, we replace the third column by k's.

Proof: We first prove that if $D \neq 0$, then the system has a solution. We recall from Section 13.1 that we can interchange any two equations of a system without changing the solution of the system. By Theorem 7 this elementary transformation either reverses the sign of the determinant (multiplies it by -1) or leaves it unchanged. We can also add any constant multiple of one equation to any other equation of the system without changing the solution. By Theorem 11 this elementary transformation does not change the determinant of the system.

By a suitable sequence of elementary transformations of the two types just described, we can transform the given system into a system of the form

(74) $$A_1 x + B_1 y + C_1 z = K_1$$

(75) $$B_2 y + C_2 z = K_2$$

(76) $$C_3 z = K_3,$$

which has the same solution as the system (72). The determinant of this system is $A_1 B_2 C_3$, and by our previous remarks $A_1 B_2 C_3 \neq 0$. Then $A_1 \neq 0$, $B_2 \neq 0$, and $C_3 \neq 0$. Hence, we can solve equation (76) for z and find that $z = K_3/C_3$. We can use this result in equation (75) and solve for y. Finally, we can use these results for z and y in equation (74) and solve for x. Since these steps are also reversible, we have proved that if $D \neq 0$, then the system (72) has a unique solution.

To show that this solution is given by equation set (73), we let (x, y, z) represent the solution. Starting with the product xD we have

$$xD = x \begin{vmatrix} a_1 & b_1 & c_1 \\ a_2 & b_2 & c_2 \\ a_3 & b_3 & c_3 \end{vmatrix}$$

$$= \begin{vmatrix} a_1 x & b_1 & c_1 \\ a_2 x & b_2 & c_2 \\ a_3 x & b_3 & c_3 \end{vmatrix} \begin{matrix} \text{C1} + y\,\text{C2} \\ \text{C1} + z\,\text{C3} \end{matrix} = \begin{vmatrix} a_1 x + b_1 y + c_1 z & b_1 & c_1 \\ a_2 x + b_2 y + c_2 z & b_2 & c_2 \\ a_3 x + b_3 y + c_3 z & b_3 & c_3 \end{vmatrix}.$$

Since (x, y, z) is a solution of the system (72), the elements in the first column of the last determinant are k_1, k_2, k_3. Hence,

$$xD = \begin{vmatrix} k_1 & b_1 & c_1 \\ k_2 & b_2 & c_2 \\ k_3 & b_3 & c_3 \end{vmatrix}.$$

Dividing both sides of this equation by D (which, by hypothesis, is not zero), we obtain the first of the equations in (73). The other two equations in that set are obtained similarly starting with yD and zD, respectively. ∎

EXAMPLE 1   Solve the system of linear equations

$$14x + 2y - 6z = 9$$
(77) $$-4x + y + 9z = 3$$
$$6x - 4y + 3z = -4.$$

Solution: We first evaluate the determinants in (73).

$$D = \begin{vmatrix} 14 & 2 & -6 \\ -4 & 1 & 9 \\ 6 & -4 & 3 \end{vmatrix} = \begin{vmatrix} 26 & -6 & 0 \\ -22 & 13 & 0 \\ 6 & -4 & 3 \end{vmatrix}$$

$$= 3 \begin{vmatrix} 26 & -6 \\ -22 & 13 \end{vmatrix} = 3 \begin{vmatrix} 4 & 7 \\ -22 & 13 \end{vmatrix}$$

$$= 3(52 + 154) = 618.$$

For the numerator of x in (73) we have

$$\begin{vmatrix} 9 & 2 & -6 \\ 3 & 1 & 9 \\ -4 & -4 & 3 \end{vmatrix} = \begin{vmatrix} 1 & -6 & 0 \\ 15 & 13 & 0 \\ -4 & -4 & 3 \end{vmatrix} = 3(13 + 90) = 309.$$

Whence,

$$x = \frac{309}{D} = \frac{309}{618} = \frac{1}{2}.$$

Similarly, for the numerator of y we find

$$\begin{vmatrix} -14 & 9 & -6 \\ -4 & 3 & 9 \\ 6 & -4 & 3 \end{vmatrix} = \begin{vmatrix} 26 & 1 & 0 \\ -22 & 15 & 0 \\ 6 & -4 & 3 \end{vmatrix} = 3 \begin{vmatrix} 26 & 1 \\ -22 & 15 \end{vmatrix} = 1236.$$

Whence,

$$y = \frac{1236}{D} = \frac{1236}{618} = 2.$$

Finally, for the numerator of z we have

$$\begin{vmatrix} 14 & 2 & 9 \\ -4 & 1 & 3 \\ 6 & -4 & -4 \end{vmatrix} = 2\begin{vmatrix} 14 & 2 & 9 \\ -4 & 1 & 3 \\ 3 & -2 & -2 \end{vmatrix} = 2\begin{vmatrix} 22 & 0 & 3 \\ -4 & 1 & 3 \\ -5 & 0 & 4 \end{vmatrix} = 2\begin{vmatrix} 22 & 3 \\ -5 & 4 \end{vmatrix}$$

$$= 2(88+15) = 206.$$

$$z = \frac{206}{618} = \frac{1}{3}. \quad \blacktriangle$$

Check   If we set $x = 1/2$, $y = 2$, and $z = 1/3$ in the system (77), we find that

$$14x + 2y - 6z = 14\left(\frac{1}{2}\right) + 2(2) - 6\left(\frac{1}{3}\right) = 7 + 4 - 2 = 9$$

$$-4x + y + 9z = -4\left(\frac{1}{2}\right) + 2 + 9\left(\frac{1}{3}\right) = -2 + 2 + 3 = 3$$

$$6x - 4y + 3z = 6\left(\frac{1}{2}\right) - 4(2) + 3\left(\frac{1}{3}\right) = 3 - 8 + 1 = -4.$$

Exercise 6

In problems 1 through 6 use Cramer's rule to solve the given system.

1. $2x + y - z = 3$
   $x + y + z = 1$
   $x - 2y - 3z = 4$

2. $4x \quad + 5z = 6$
   $\quad y - 6z = -2$
   $3x \quad + 4z = 3$

3. $x + y + z = 1$
   $2x + 3y + 4z = 1$
   $x - y - z = 0$

4. $3x - y + z = 1$
   $x - 2y + z = 2$
   $2x + y + 3z = 0$

5. $x + 2y = 4z - 4$
   $5x - 3y = 7z + 6$
   $3x - 2y = -3z + 11$

6. $3y - z = -2x + 1$
   $3x + 2z = 8 - 5y$
   $3z - 1 = x - 2y$

★ 13.7 FOURTH-ORDER DETERMINANTS

The definition of an nth-order determinant will be given in the next chapter after we have studied mathematical induction. However, it is sufficient to study fourth-order determinants because the methods used when $n = 4$ and when $n > 4$ are the same.

We recall that M_{ij} is the determinant of the matrix obtained from M when the elements in the ith row and the jth column are deleted. If M is a 4×4 matrix, the definition of $D(M)$ is given in terms of third-order determinants. Indeed by definition,

$$
\begin{vmatrix}
a_1 & b_1 & c_1 & d_1 \\
a_2 & b_2 & c_2 & d_2 \\
a_3 & b_3 & c_3 & d_3 \\
a_4 & b_4 & c_4 & d_4
\end{vmatrix}
= a_1 M_{11} - b_1 M_{12} + c_1 M_{13} - d_1 M_{14}.
$$

It can be proved that with this definition, Theorems 5 through 12 still hold. Further, a Laplace expansion by minors of any row (not just the first row) and a Laplace expansion by minors of any column give the same value for $D(M)$. Some of these theorems are easy to prove; others are more complicated. For brevity, we omit the proofs if $n > 3$.

EXAMPLE 1   Find $D(M)$ for the matrix

$$
M = \begin{bmatrix}
1 & 3 & 5 & 7 \\
1 & 5 & 3 & 7 \\
1 & 7 & 5 & 3 \\
1 & 3 & 7 & 5
\end{bmatrix}.
$$

Solution: We subtract the first row from each of the following rows (Theorem 11). Then,

$$
D(M) = \begin{vmatrix}
1 & 3 & 5 & 7 \\
0 & 2 & -2 & 0 \\
0 & 4 & 0 & -4 \\
0 & 0 & 2 & -2
\end{vmatrix}.
$$

If we expand by minors of the first column,

$$
D(M) = 1 \begin{vmatrix}
2 & -2 & 0 \\
4 & 0 & -4 \\
0 & 2 & -2
\end{vmatrix} - 0 M_{21} + 0 M_{31} - 0 M_{41}
$$

$$
\underset{C2 \ + \ C1}{=} \begin{vmatrix}
2 & 0 & 0 \\
4 & 4 & -4 \\
0 & 2 & -2
\end{vmatrix} = 2 \begin{vmatrix}
4 & -4 \\
2 & -2
\end{vmatrix} = 0. \quad \blacktriangle
$$

Exercise 7

In problems 1 through 8 evaluate the given determinant. Assume that Theorems 5 through 12 are true for $n = 4$.

1.
$$\begin{vmatrix} 4 & 2 & 2 & 2 \\ 2 & 4 & 2 & 2 \\ 2 & 2 & 4 & 2 \\ 2 & 2 & 2 & 4 \end{vmatrix}$$

2.
$$\begin{vmatrix} 1 & 1 & 1 & 1 \\ 1 & 2 & 3 & 4 \\ 1 & 3 & 6 & 10 \\ 1 & 4 & 10 & 20 \end{vmatrix}$$

3.
$$\begin{vmatrix} -6 & 1 & 2 & 3 \\ 1 & -6 & 2 & 3 \\ 2 & 1 & -6 & 3 \\ 2 & 1 & 3 & -6 \end{vmatrix}$$

4.
$$\begin{vmatrix} 2 & -3 & 0 & 4 \\ -4 & 2 & 3 & -5 \\ 2 & 0 & -2 & 4 \\ 3 & -4 & 5 & 2 \end{vmatrix}$$

5.
$$\begin{vmatrix} 1 & 1 & 1 & 1 \\ 1 & -1 & 1 & -1 \\ 1 & 1 & -1 & -1 \\ 1 & -1 & -1 & 1 \end{vmatrix}$$

6.
$$\begin{vmatrix} 1 & 2 & 3 & 4 \\ 4 & 1 & 2 & 3 \\ 3 & 4 & 1 & 2 \\ 2 & 3 & 4 & 1 \end{vmatrix}$$

7.
$$\begin{vmatrix} 1 & 3 & 5 & 7 \\ 1 & 4 & 8 & 12 \\ 1 & 8 & 17 & 27 \\ 1 & 12 & 25 & 40 \end{vmatrix}$$

8.
$$\begin{vmatrix} 3 & 7 & -1 & 8 \\ 2 & -5 & 16 & 3 \\ -8 & 3 & 10 & 13 \\ 13 & -1 & 5 & -2 \end{vmatrix}$$

In problems 9 through 12 solve the given system. Assume that Cramer's rule is true for systems of four linear equations in four variables.

9.
$$\begin{aligned}
y \quad\ \ - w &= 0 \\
2z - w &= 0 \\
x + y + 2z \qquad\ &= 1 \\
x + 2y - z - 2w &= 1
\end{aligned}$$

10.
$$\begin{aligned}
y \quad - 2w &= -1 \\
x - y \ + w &= 1 \\
2y + z - 3w &= 1 \\
x + 2y - z + 2w &= 3
\end{aligned}$$

11.
$$\begin{aligned}
2x + y + z - 4w &= 2 \\
x - y + z + 3w &= 1 \\
x + y - z + w &= -2 \\
3x + 2y + z + w &= 6
\end{aligned}$$

12.
$$\begin{aligned}
x \qquad - 2z \qquad &= 0 \\
- z + w &= 0 \\
x + y - z \qquad &= 1 \\
2x + 5y + z - 3w &= 1
\end{aligned}$$

In problems 13 through 16 prove the given assertion.

13.
$$\begin{vmatrix} 1 & 1 & 1 & 1 \\ x & y & z & u \\ x^2 & y^2 & z^2 & u^2 \\ x^3 & y^3 & z^3 & u^3 \end{vmatrix} = (u-x)(u-y)(u-z)(z-x)(z-y)(y-x)$$

14.
$$\begin{vmatrix} 1+a & 1 & 1 & 1 \\ 1 & 1+b & 1 & 1 \\ 1 & 1 & 1+c & 1 \\ 1 & 1 & 1 & 1+d \end{vmatrix} = abcd\left(1 + \frac{1}{a} + \frac{1}{b} + \frac{1}{c} + \frac{1}{d}\right)$$

15.
$$\begin{vmatrix} x & 1 & 1 & 1 \\ 1 & x & 1 & 1 \\ 1 & 1 & x & 1 \\ 1 & 1 & 1 & x \end{vmatrix} = -(x+3)(x-1)^3$$

16.
$$\begin{vmatrix} 0 & 1 & 1 & 1 \\ 1 & y+z & x & x \\ 1 & y & z+x & y \\ 1 & z & z & x+y \end{vmatrix} = x^2 + y^2 + z^2 - 2(xy+yz+zx)$$

★ 13.8 THE ALGEBRA OF MATRICES

The matrix

(78)
$$A = \begin{bmatrix} a_{11} & a_{12} & a_{13} & \cdots & a_{1n} \\ a_{21} & a_{22} & a_{23} & \cdots & a_{2n} \\ \vdots & \vdots & \vdots & \vdots & \vdots \\ a_{m1} & a_{m2} & a_{m3} & \cdots & a_{mn} \end{bmatrix}$$

of dimension $m \times n$ can be indicated by writing $[a_{ij}]_{m,n}$ using the dimension as a subscript. Similarly $[b_{ij}]_{p,q}$ is a condensed notation for the $p \times q$ matrix

(79)
$$B = \begin{bmatrix} b_{11} & b_{12} & b_{13} & \cdots & b_{1q} \\ b_{21} & b_{22} & b_{23} & \cdots & b_{2q} \\ \vdots & \vdots & \vdots & \vdots & \vdots \\ b_{p1} & b_{p2} & b_{p3} & \cdots & b_{pq} \end{bmatrix}.$$

DEFINITION 10   **Equality**   *The matrices A and B are said to be equal, and we write $A = B$ if they are of the same dimension ($p = m$ and $q = n$) and if*

(80) $$a_{ij} = b_{ij}$$

for every pair i, j with $i = 1, 2, ..., m$ and $j = 1, 2, ..., n$. (They are termwise equal.)

For example,

$$\begin{bmatrix} 3 & 5 \\ 7 & 11 \end{bmatrix} \neq \begin{bmatrix} 3 & 7 \\ 5 & 11 \end{bmatrix} \neq \begin{bmatrix} 3 & 7 & 0 \\ 5 & 11 & 0 \end{bmatrix}.$$

However, if $x = 2$, then

$$\begin{bmatrix} x & x^2 \\ x-1 & x^2-2 \end{bmatrix} = \begin{bmatrix} 2 & 4 \\ 1 & 2 \end{bmatrix}.$$

Addition of two matrices is defined only if they have the same dimension.

DEFINITION 11   **Addition**   *If $A = [a_{ij}]$ and $B = [b_{ij}]$ are both $m \times n$ matrices, then the sum $A + B$ is the $m \times n$ matrix $C = [c_{ij}]$ where*

(81) $$c_{ij} = a_{ij} + b_{ij}$$

for every pair i, j with $i = 1, 2, ..., m$ and $j = 1, 2, ..., n$.

For example,

(82) $$\begin{bmatrix} 1 & 3 & 5 \\ 6 & -4 & -2 \end{bmatrix} + \begin{bmatrix} 3 & -7 & 9 \\ 0 & 4 & 8 \end{bmatrix} = \begin{bmatrix} 4 & -4 & 14 \\ 6 & 0 & 6 \end{bmatrix}.$$

Since addition is commutative and associative for real numbers, we have the same property for matrices.

Theorem 13
PLE

If A and B are matrices of the same dimension, then

(83) $$A + B = B + A.$$

Theorem 14
PLE

If A, B, and C are matrices of the same dimension, then

(84) $$(A+B) + C = A + (B+C).$$

DEFINITION 12　　***Multiplication of a Matrix by a Real Number***　*If $A = [a_{ij}]$ and c is a real number, then cA is the matrix $[d_{ij}]$, where $d_{ij} = ca_{ij}$ for every pair i,j. Further $Ac \equiv cA$.*

In other words, to multiply a matrix A by a real number c we multiply each element of A by c.

EXAMPLE 1　　Given

$$A = \begin{bmatrix} 1 & 2 \\ 4 & 7 \end{bmatrix} \quad \text{and} \quad B = \begin{bmatrix} 3 & -5 \\ -1 & 2 \end{bmatrix},$$

find $A + B$, $2A + 3B$, and $3A - 2B$.

Solution:

$$A + B = \begin{bmatrix} 1 & 2 \\ 4 & 7 \end{bmatrix} + \begin{bmatrix} 3 & -5 \\ -1 & 2 \end{bmatrix} = \begin{bmatrix} 4 & -3 \\ 3 & 9 \end{bmatrix}$$

$$2A + 3B = \begin{bmatrix} 2 & 4 \\ 8 & 14 \end{bmatrix} + \begin{bmatrix} 9 & -15 \\ -3 & 6 \end{bmatrix} = \begin{bmatrix} 11 & -11 \\ 5 & 20 \end{bmatrix}$$

$$3A - 2B = \begin{bmatrix} 3 & 6 \\ 12 & 21 \end{bmatrix} + \begin{bmatrix} -6 & +10 \\ +2 & -4 \end{bmatrix} = \begin{bmatrix} -3 & 16 \\ 14 & 17 \end{bmatrix}. \quad \blacktriangle$$

We recall from Definition 5 (page 332) that a zero matrix, denoted by 0, is a matrix in which all elements are zero.

Theorem 15　　*Let A be a matrix and let 0 denote the zero matrix of the same dimension. Then,*
PLE

(85)　　　　　　　　　　　$A + 0 = 0 + A = A$

(86)　　　　　　　　　　　$0A = 0$

(87)　　　　　　　　　　　$c0 = 0.$

Further, there is a unique matrix which we denote by $-A$ such that

(88)　　　　　　　　　　　$A + (-A) = 0.$

The definition of multiplication for two matrices is more complicated and perhaps less natural than the definition of addition. We begin with two rather simple matrices: a row vector, which we denote by R, and a column vector, which we denote‡ by K. Using the corresponding lower case letters for the elements, we have

(89)
$$R = [r_1, r_2, \ldots, r_n], \qquad K = \begin{bmatrix} k_1 \\ k_2 \\ \vdots \\ k_p \end{bmatrix}.$$

The product RK of these two matrices is not defined unless $n = p$. If $n = p$, then by definition

(90)
$$RK = r_1 k_1 + r_2 k_2 + \cdots + r_n k_n.$$

For example,

$$[1, 3, -4, -2] \begin{bmatrix} 3 \\ 9 \\ -3 \\ 10 \end{bmatrix} = 1(3) + 3(9) + (-4)(-3) + (-2)10$$

$$= 3 + 27 + 12 - 20 = 22.$$

Hence, the product RK is a number. This product is called the *scalar product* of the vectors R and K.

To extend the definition to arbitrary matrices, we observe that a matrix can be regarded as being composed of row vectors, or can be regarded as being composed of column vectors. Thus A, given by (78), can be written as

$$A = \begin{bmatrix} R_1 \\ R_2 \\ \vdots \\ R_m \end{bmatrix} \quad \text{where} \quad \begin{aligned} R_1 &= [a_{11}, a_{12}, \ldots, a_{1n}] \\ R_2 &= [a_{21}, a_{22}, \ldots, a_{2n}] \\ \vdots & \quad \vdots \quad \vdots \quad \vdots \quad \vdots \\ R_m &= [a_{m1}, a_{m2}, \ldots, a_{mn}]. \end{aligned}$$

Similarly the matrix B, given by (79), can be written as $B = [K_1, K_2, \ldots, K_q]$ where

$$K_1 = \begin{bmatrix} b_{11} \\ b_{21} \\ \vdots \\ b_{p1} \end{bmatrix}, \qquad K_2 = \begin{bmatrix} b_{12} \\ b_{22} \\ \vdots \\ b_{p2} \end{bmatrix}, \ldots, \qquad K_q = \begin{bmatrix} b_{1q} \\ b_{2q} \\ \vdots \\ b_{pq} \end{bmatrix}.$$

‡ We use K for a column vector in place of the more natural C because we want to reserve C for use in the product $AB = C$.

To define the product of A and B we require that the number of columns in A be the same as the number of rows in B. Then we have

DEFINITION 13   **_Multiplication of Two Matrices_**   _Let $A = [a_{ij}]_{m,n}$ and $B = [b_{ij}]_{n,p}$. The product AB is the matrix $C = [c_{ij}]_{m,p}$, where the element c_{ij} in the ith row and the jth column of C is the product of the ith row vector of A and the jth column vector of B._

If we write $AB = C$ with all the elements displayed, we have

$$\begin{bmatrix} a_{11} & a_{12} & \cdots & a_{1n} \\ \vdots & \vdots & \vdots & \vdots \\ a_{i1} & a_{i2} & \cdots & a_{in} \\ \vdots & \vdots & \vdots & \vdots \\ a_{m1} & a_{m2} & \cdots & a_{mn} \end{bmatrix} \begin{bmatrix} b_{11} & \cdots & b_{1j} & \cdots & b_{1p} \\ b_{21} & \cdots & b_{2j} & \cdots & b_{2p} \\ \vdots & & \vdots & & \vdots \\ b_{n1} & \cdots & b_{nj} & \cdots & b_{np} \end{bmatrix}$$

$$= \begin{bmatrix} c_{11} & \cdots & c_{1j} & \cdots & c_{1p} \\ \vdots & & \vdots & & \vdots \\ c_{i1} & \cdots & c_{ij} & \cdots & c_{ip} \\ \vdots & & \vdots & & \vdots \\ c_{m1} & \cdots & c_{mj} & \cdots & c_{mp} \end{bmatrix}.$$

Using the elements shown in color we have, by definition,

(91) $$c_{ij} = R_i K_j = a_{i1}b_{1j} + a_{i2}b_{2j} + a_{i3}b_{3j} + \cdots + a_{in}b_{nj}.$$

EXAMPLE 2   Compute AB and BA if

$$A = \begin{bmatrix} 3 & -1 \\ -2 & 5 \end{bmatrix} \quad \text{and} \quad B = \begin{bmatrix} -2 & 1 & 3 \\ 3 & -1 & 2 \end{bmatrix}.$$

Solution: A is composed of two row vectors and B is composed of three column vectors. The six products of these pairs of vectors give

$$AB = \begin{bmatrix} 3 & -1 \\ -2 & 5 \end{bmatrix} \begin{bmatrix} -2 & 1 & 3 \\ 3 & -1 & 2 \end{bmatrix}$$

$$= \begin{bmatrix} 3(-2)+(-1)3 & 3(1)+(-1)(-1) & 3(3)+(-1)2 \\ (-2)(-2)+5(3) & (-2)1+5(-1) & (-2)3+5(2) \end{bmatrix}$$

$$= \begin{bmatrix} -9 & 4 & 7 \\ 19 & -7 & 4 \end{bmatrix}. \quad \blacktriangle$$

The number of columns in B is three, and the number of rows in A is two. Since $3 \neq 2$, the product BA is not defined.

To combine both addition and multiplication, the matrices involved must be square matrices.

DEFINITION 14   *The square matrix*

$$(92) \qquad I = \begin{bmatrix} 1 & 0 & 0 & \cdots & 0 \\ 0 & 1 & 0 & \cdots & 0 \\ 0 & 0 & 1 & \cdots & 0 \\ \vdots & \vdots & \vdots & & \vdots \\ 0 & 0 & 0 & \cdots & 1 \end{bmatrix}$$

with one in the main diagonal and zero everywhere else, is called the identity matrix.

Theorem 16
PWO

Let A, B, and C be square matrices of order n, and let I be the identity matrix of order n. Then

$$(93) \qquad IA = AI = A,$$

and

$$(94) \qquad A(B+C) = AB + AC.$$

One might expect two more laws for matrix multiplication. The associative law

$$(95) \qquad A(BC) = (AB)C$$

does hold but is somewhat complicated to prove. The commutative law is false (see Exercise 8, problem 21). In other words, there are pairs of square matrices A and B such that

$$(96) \qquad AB \neq BA.$$

We can also look for a multiplicative inverse for the matrix A. If one exists, then we denote it by A^{-1}. The facts are contained in

Theorem 17
PWO

If $D(A) \neq 0$, then there is a unique matrix A^{-1} such that

$$(97) \qquad A^{-1}A = I = AA^{-1}.$$

Conversely, if (97) holds, then $D(A) \neq 0$.

If a matrix does not have an inverse, it is said to be *singular*. If it has an inverse, then it is *nonsingular*. From Theorem 17 we see that a matrix is nonsingular if and only if $D(A) \neq 0$.

Let A be the matrix associated with the system of n linear equations in n variables,

$$
\begin{aligned}
a_{11} x_1 + a_{12} x_2 + \cdots + a_{1n} x_n &= k_1 \\
a_{21} x_1 + a_{22} x_2 + \cdots + a_{2n} x_n &= k_2 \\
\vdots \qquad \vdots \qquad \vdots \qquad \vdots \qquad \vdots \\
a_{n1} x_1 + a_{n2} x_2 + \cdots + a_{nn} x_n &= k_n.
\end{aligned}
$$

(98)

Let K be the column vector whose components are the k_i on the right side of (98). By the definition of matrix multiplication, this system can be written as the matrix equation

(99) $$AX = K,$$

where X is the column vector whose components are $x_1, x_2, \ldots, x_n$.

If A has an inverse A^{-1}, then we multiply equation (99) on both sides by A^{-1} and obtain

$$A^{-1}(AX) = A^{-1} K.$$

But $A^{-1}(AX) = (A^{-1} A) X = IX = X$. Hence, (99) yields

(100) $$X = A^{-1} K.$$

This equation gives each x_i ($i = 1, 2, \ldots, n$). Hence, if we know the inverse of A, it is easy to find the solution of the system (98).

We recall that the linear equation in one variable $ax = k$ has a unique solution if $a \neq 0$. The statement, that the system (98) has a unique solution if $D(A) \neq 0$, is a natural generalization.

Exercise 8

In problems 1 through 18 let

$$
A = \begin{bmatrix} 1 & 2 \\ 3 & 4 \end{bmatrix}, \qquad
B = \begin{bmatrix} 5 & -3 \\ -1 & 2 \end{bmatrix}, \qquad
C = \begin{bmatrix} -2 & 4 \\ 1 & -5 \end{bmatrix}.
$$

Find the required matrix.

1. $A + B$
2. $B + C$
3. $(A + B) + C$
4. $A + (B + C)$

5. $2A - 5B$

6. $3B - 4C$

7. AB

8. BC

9. BA

10. CB

11. $(AB)C$

12. $A(BC)$

13. $A(B+C)$

14. $C(A+B)$

15. $A^2 - 5A - 2I$

16. $B^2 - 7B + 7I$

17. Find X so that $A + X = B$.

18. Find Y so that $Y + C = A$.

19. Prove that if c is a real number and A and B are matrices of the same dimension, then $c(A+B) = cA + cB$.

20. Prove that for square matrices $IA = AI = A$. (See Theorem 16.)

21. Find square matrices A and B such that $AB \neq BA$.

22. Prove the last part of Theorem 15.

23. Suppose that $\triangle \equiv a_1 b_2 - a_2 b_1 \neq 0$. Prove that

$$\frac{1}{\triangle} \begin{bmatrix} b_2 & -b_1 \\ -a_2 & a_1 \end{bmatrix} \begin{bmatrix} a_1 & b_1 \\ a_2 & b_2 \end{bmatrix} = \begin{bmatrix} 1 & 0 \\ 0 & 1 \end{bmatrix}.$$

24. Find square matrices A and B such that $AB = 0$, but $A \neq 0$ and $B \neq 0$.

25. Prove the second part of Theorem 16 when $n = 2$.

26. Prove that the matrix equation (99) is equivalent to the system (98). For simplicity, set $n = 3$.

Review Questions

Try to answer the following questions as accurately as possible before consulting the text.

1. What is a matrix?

2. Name three elementary operations on the rows of a matrix. Explain how these operations are used to solve systems of equations.

3. Define the determinant of a 2×2 matrix.

4. What is the minor M_{ij} of a matrix?

5. Give the Laplace expansion of a third order determinant by minors of (a) the first row, (b) the second row, (c) the third row.

6. State Cramer's rule for a system of three equations in three variables.

7. Give the rules for the addition of two matrices and the multiplication of a matrix by a number. State as many theorems as you can about these two operations with matrices.

8. Give the definition of the product RK where R is a row vector and K is a column vector.

9. Give the definition of the product of two matrices. State as many theorems as you can about multiplication of matrices.

10. What is the 0 matrix, the identity matrix, and the inverse of a matrix?

Mathematical Induction

One of the most powerful methods of proving theorems carries the title "mathematical induction." Although this venerable method has been used for centuries, it is still used today in some of the most recent and deepest mathematical research.

Suppose that we have a proposition $P(n)$ that depends on a positive integer n. We want to prove that $P(n)$ is true for every $n \in \mathbf{N}$. How shall we proceed? Of course the assertion may not be true, so first we might test it for selected values of n, for example, $n = 1, 2, 3,$ and 4. With the check of each succeeding case, our belief that $P(n)$ is true for all $n \in \mathbf{N}$ increases. But checking particular cases, no matter how numerous, does not prove that $P(n)$ is always true. Mathematical induction is a simple and elegant method for proving that $P(n)$ is true for every positive integer‡.

14.1 AN EXAMPLE

Many of the simpler formulas are discovered purely by experiment. In this spirit, we form the sum of the first n odd integers. We will find (to our surprise!) that the sum is *always* a perfect square. Indeed, direct computation

---

‡ We cannot expect the method to work in every case because: (a) $P(n)$ might be false for some n or (b) the assertion $P(n)$ may be extremely difficult to prove and may even represent an open problem.

gives

$$\text{if } n = 1, \quad 1 = 1 \qquad\qquad\qquad = 1^2;$$
$$\text{if } n = 2, \quad 1 + 3 = 4 \qquad\qquad\qquad = 2^2;$$
$$\text{if } n = 3, \quad 1 + 3 + 5 = 9 \qquad\qquad = 3^2;$$
$$\text{if } n = 4, \quad 1 + 3 + 5 + 7 = 16 \qquad = 4^2;$$
$$\text{if } n = 5, \quad 1 + 3 + 5 + 7 + 9 = 25 \qquad = 5^2;$$
$$\text{if } n = 6, \quad 1 + 3 + 5 + 7 + 9 + 11 = 36 = 6^2.$$

Clearly each sum is the square of an integer (see the extreme right-hand column), and further it seems to be the square of the number of terms in the sum. If n denotes the number of terms, then the last term of the sum seems to be $2n - 1$. To check this we observe that

$$\text{if } n = 1, \quad 2n - 1 = 2 - 1 = 1;$$
$$\text{if } n = 2, \quad 2n - 1 = 4 - 1 = 3;$$
$$\text{if } n = 3, \quad 2n - 1 = 6 - 1 = 5;$$
$$\text{if } n = 4, \quad 2n - 1 = 8 - 1 = 7;$$

and so on. Thus, purely by experiment we arrive at the formula

(1) $$1 + 3 + 5 + \cdots + (2n - 1) = n^2.$$

This is a proposition $P(n)$ which we now conjecture to be true for all $n \in \mathbf{N}$.

Can we prove $P(n)$ is always true by checking a few more cases? Suppose we try $n = 11$. In this case the last term of the sum is $2 \cdot 11 - 1 = 21$ and we find that

$$1 + 3 + 5 + 7 + 9 + 11 + 13 + 15 + 17 + 19 + 21 = 121 = 11^2.$$

The formula (1) is true in this case. The reader is invited to test the formula with several other values for n selected at random. He will soon be convinced that $P(n)$ is true for every positive integer. However, with this approach we can not prove that (1) is always true, because there are infinitely many cases, and no amount of pure computation can settle them all. What we need is some logical argument that will prove that (1) is always true.

14.2 A WARNING

We must always be on guard that the formula we have discovered experimentally may be false for some value of n. We give two examples.

Let us examine integers of the form

$$(2) \qquad\qquad A = 2^{2^n} + 1.$$

We find by direct computation that

$$
\begin{aligned}
&\text{if } n = 0, &&A = 2^1 \ + 1 = 3; \\
&\text{if } n = 1, &&A = 2^2 \ + 1 = 5; \\
&\text{if } n = 2, &&A = 2^4 \ + 1 = 17; \\
&\text{if } n = 3, &&A = 2^8 \ + 1 = 257; \\
&\text{if } n = 4, &&A = 2^{16} + 1 = 65{,}537.
\end{aligned}
$$

The first four integers 3, 5, 17, and 257 are prime numbers. A little labor will show that 65,537 is also a prime. This might lead us to conjecture that A, given by equation (2), is a prime for every $n \in \mathbf{N}$. In fact, Fermat (1601–1665), the great French mathematician, did conjecture exactly that, but about a hundred years later Euler (1707–1783), the leading Swiss mathematician, showed that for $n = 5$,

$$A = 2^{2^5} + 1 = 4{,}294{,}967{,}297 = 641 \times 6{,}700{,}417,$$

and, hence, A is not a prime for $n = 5$. Consequently, the proposition $P(n)$, that $A = 2^{2^n} + 1$ is always a prime, is true for $n = 0, 1, 2, 3$, and 4 but is false when $n = 5$.

We next consider the assertion

$$(3) \qquad\qquad 2^n < n^{10} + 2.$$

We test this assertion for a few values of n. Direct computation gives

$$
\begin{aligned}
&\text{if } n = 1, &&\text{we have} &&2 < 1 + 2 = 3; \\
&\text{if } n = 2, &&\text{we have} &&4 < 1024 + 2 = 1026; \\
&\text{if } n = 3, &&\text{we have} &&8 < 59{,}051; \\
&\text{if } n = 4, &&\text{we have} &&16 < 1{,}048{,}578.
\end{aligned}
$$

It seems that the inequality (3) is true for every positive integer n. If we try a large value of n, say $n = 20$, then the inequality (3) asserts that

$$126{,}976 < 10{,}240{,}000{,}000{,}002,$$

which is certainly true. Even this computation does not prove that (3) is always true. In fact, such an assertion is *false*, for if we set $n = 59$, we find that (approximately)

$$2^{59} = 5.764 \times 10^{17} \quad \text{and} \quad 59^{10} + 2 = 5.111 \times 10^{17}.$$

It is not hard to prove that for every $n \geqq 59$, *the inequality* (3) *is false.*

These two examples illustrate the following principle: *We cannot conclude that an assertion involving an integer n is true for all positive values of the integer n, merely by checking specific values of n, no matter how many we check.*

14.3 INTUITIVE PROOF OF THE GENERAL PRINCIPLE

Let $P(n)$ be any meaningful assertion that involves a positive integer n. The assertion may be true for some values of n and false for others. With this assertion P fixed, we let **T** be the set of positive integers for which $P(n)$ is true, and we let **F** be the set of positive integers for which $P(n)$ is false. To prove that $P(n)$ is always true, it is sufficient to show that the set **F** is empty. We concentrate on **F**.

It is obvious that if **F** is a nonempty set, then it has a smallest positive integer. Two cases can occur.

(A) The smallest integer in **F** is 1.

(B) If the smallest integer in **F** is not 1, we denote the smallest integer by $k + 1$. Then by the definition of $k + 1$, $P(k)$ is true, and $P(k + 1)$ is false.

To show that the set **F** is empty, it is sufficient to show that neither of the two cases listed above can occur. First, we must prove that $P(1)$ is true, so that 1 is in **T** and not in **F**. Second, we must prove that for each positive integer k, if $P(k)$ is true, then $P(k + 1)$ is true. If we have proved this, then the second case cannot occur. If we have proved these two statements, then **F** must be empty, and **T** must contain all of the positive integers.

Thus the proof that $P(n)$ is true for all positive integers can be given in two steps and these two steps form the

Principle of Mathematical Induction *If for a given assertion $P(n)$ we can prove that:*

(A) The assertion is true for $n = 1$.

(B) If it is true for index $n = k$, then it is also true for index $n = k + 1$.

Then the assertion is true for every positive integer n.

EXAMPLE 1

Prove that for every positive integer n

(1)
$$1 + 3 + 5 + \cdots + (2n - 1) = n^2.$$

Solution: We use the principle of mathematical induction. Let $P(n)$ be the assertion given by equation (1).

(A) We have already seen in Section 14.1 that $P(1)$ is true. (In fact, we proved much more.)

(B) We now assume that $P(n)$ is true for the index $n = k$; i.e., we assume that

(4) $$1 + 3 + 5 + \cdots + (2k-1) = k^2.$$

To obtain the sum of the first $k+1$ odd integers, we merely add the next odd one, $2k+1$, to both sides of (4). This gives

$$1 + 3 + 5 + \cdots + (2k-1) + (2k+1) = k^2 + (2k + 1)$$
$$= (k+1)^2.$$

But this equation is precisely equation (1), when the index n is $k+1$. Hence, we have shown that if the assertion is true for index k, it is also true for index $k+1$. By the principle of mathematical induction this completes the proof that equation (1) is true for every positive integer n. ▲

EXAMPLE 2

Prove that for every positive integer n

(5) $$\frac{1}{1\cdot 2} + \frac{1}{2\cdot 3} + \frac{1}{3\cdot 4} + \cdots + \frac{1}{n(n+1)} = \frac{n}{n+1}.$$

Solution: (A) For $n = 1$, equation (5) asserts that

$$\frac{1}{1\cdot 2} = \frac{1}{1+1},$$

and this is true.

(B) We assume that (5) is true for index k. Thus we assume that

(6) $$\frac{1}{1\cdot 2} + \frac{1}{2\cdot 3} + \frac{1}{3\cdot 4} + \cdots + \frac{1}{k(k+1)} = \frac{k}{k+1}.$$

To obtain the left side of equation (5) for the index $n = k+1$, we must add $1/(k+1)(k+2)$ to the left side of equation (6). This gives

(7) $$\frac{1}{1\cdot 2} + \frac{1}{2\cdot 3} + \frac{1}{3\cdot 4} + \cdots + \frac{1}{k(k+1)} + \frac{1}{(k+1)(k+2)}$$

$$= \frac{k}{k+1} + \frac{1}{(k+1)(k+2)} = \frac{k(k+2)+1}{(k+1)(k+2)}$$

$$= \frac{k^2 + 2k + 1}{(k+1)(k+2)} = \frac{(k+1)^2}{(k+1)(k+2)}$$

$$= \frac{k+1}{k+2} = \frac{k+1}{(k+1)+1}.$$

But this is just equation (5) when the index n is $k+1$. Hence, by the principle of mathematical induction, (5) is true for every positive integer n. ▲

An assertion $P(n)$ may be false or meaningless for certain small values of n. In this case, the assertion would necessarily be modified to state only what is actually true. Then the principle of mathematical induction would also be altered to meet the situation. Thus in step 1 of the process we would not set $n = 1$, but we would use instead some integer n_0 such that $P(n)$ is true for every $n \geq n_0$.

EXAMPLE 3

Find those positive integers for which

(8) $$2^n > 2n + 1.$$

Solution: We try a few values of n in (8).
For $n = 1$, equation (8) states that $2 > 3$, and this is false.
For $n = 2$, equation (8) states that $4 > 5$, and this is false.
For $n = 3$, equation (8) states that $8 > 7$, and this is true.
(B) We assume that equation (8) is true for $n = k$, where $k > 2$. Thus, we assume that

(9) $$2^k > 2k + 1.$$

For $k > 2$ we also have

(10) $$2^k > 2.$$

If we add the inequalities (9) and (10) we obtain

$$2^k + 2^k > (2k+1) + 2$$

$$2(2^k) > (2k+2) + 1$$

(11) $$2^{k+1} > 2(k+1) + 1.$$

But this is just the inequality (8), when the index n is $k+1$. Hence, using the principle of mathematical induction, we have proved that *the inequality* (8) *is true for every integer $n \geq 3$.* ▲

The principle of mathematical induction can also be applied to geometric problems, but they are a little more complicated, so we do not pursue the matter here.

Exercise 1

In problems 1 through 27 prove that the given assertion is true for every integer $n \geq 1$.

1. The nth positive even integer is $2n$.

2. The nth positive odd integer is $2n-1$.

3. $1 + 2 + 3 + 4 + \cdots + n = \dfrac{n(n+1)}{2}$

4. $1 + 5 + 9 + \cdots + (4n-3) = n(2n-1)$

5. $2 + 5 + 8 + \cdots + (3n-1) = \dfrac{n(3n+1)}{2}$

6. $5 + 11 + 17 + \cdots + (6n-1) = n(3n+2)$

7. $1^2 + 2^2 + 3^2 + 4^2 + \cdots + n^2 = \dfrac{n(n+1)(2n+1)}{6}$

8. $1 \cdot 2 + 2 \cdot 3 + 3 \cdot 4 + \cdots + n(n+1) = \dfrac{n(n+1)(n+2)}{3}$

9. $1^2 + 3^2 + 5^2 + \cdots + (2n-1)^2 = \dfrac{n(2n-1)(2n+1)}{3}$

10. $3 \cdot 4 + 4 \cdot 7 + 5 \cdot 10 + \cdots + (n+2)(3n+1) = n(n+2)(n+3)$

11. $\dfrac{1}{1 \cdot 3} + \dfrac{1}{3 \cdot 5} + \dfrac{1}{5 \cdot 7} + \cdots + \dfrac{1}{(2n-1)(2n+1)} = \dfrac{n}{2n+1}$

12. $\dfrac{1}{1 \cdot 4} + \dfrac{1}{4 \cdot 7} + \dfrac{1}{7 \cdot 10} + \cdots + \dfrac{1}{(3n-2)(3n+1)} = \dfrac{n}{3n+1}$

13. $1 + x + x^2 + \cdots + x^n = \dfrac{1 - x^{n+1}}{1 - x}$, if $x \neq 1$

14. $\dfrac{1}{x(x+1)} + \dfrac{1}{(x+1)(x+2)} + \cdots + \dfrac{1}{(x+n-1)(x+n)} = \dfrac{n}{x(x+n)}$, if $x > 0$

15. $1^3 + 2^3 + 3^3 + \cdots + n^3 = \dfrac{n^2(n+1)^2}{4}$

16. $1^3 + 3^3 + 5^3 + \cdots + (2n-1)^3 = n^2(2n^2-1)$

★17. $\dfrac{1}{n} + \dfrac{1}{n+1} + \dfrac{1}{n+2} + \cdots + \dfrac{1}{2n-1} = 1 - \dfrac{1}{2} + \dfrac{1}{3} - \dfrac{1}{4} + \cdots + \dfrac{1}{2n-1}$

★18. $1^3 + 2^3 + 3^3 + \cdots + n^3 = (1 + 2 + 3 + \cdots + n)^2$

19. $1 \cdot 3 + 3 \cdot 5 + 5 \cdot 7 + \cdots + (2n-1)(2n+1) = \dfrac{n(4n^2 + 6n - 1)}{3}$

20. $2 \cdot 2 + 3 \cdot 2^2 + 4 \cdot 2^3 + \cdots + (n+1)2^n = n2^{n+1}$

★21. $\dfrac{1}{1 \cdot 2 \cdot 3} + \dfrac{1}{2 \cdot 3 \cdot 4} + \dfrac{1}{3 \cdot 4 \cdot 5} + \cdots + \dfrac{1}{n(n+1)(n+2)} = \dfrac{n(n+3)}{4(n+1)(n+2)}$

22. $1 - x + x^2 - x^3 + \cdots + x^{2n} = \dfrac{1 + x^{2n+1}}{1 + x}$, if $x \neq -1$

★23. $n^3 + 5n$ is divisible by 3.

★24. $3^{2n} + 7$ is divisible by 8.

★★25. $\sin x + \sin 3x + \sin 5x + \cdots + \sin(2n-1)x = \dfrac{\sin^2 nx}{\sin x}$

★★26. $\cos y + \cos 3y + \cos 5y + \cdots + \cos(2n-1)y = \dfrac{\sin 2ny}{2 \sin y}$

★★27. $\sin t + \sin 2t + \sin 3t + \cdots + \sin nt = \dfrac{\sin \dfrac{n+1}{2} t \sin \dfrac{nt}{2}}{\sin \dfrac{t}{2}}$

28. Find all positive integers for which $2^n > n^2$.

29. Find all positive integers for which $2^n > n^3$.

30. Prove that if $x > -1$ and n is an integer greater than 1, then we have the inequality $(1+x)^n > 1 + nx$.

31. Prove that for $n \geq 2$

$$\frac{1}{n+1} + \frac{1}{n+2} + \frac{1}{n+3} + \cdots + \frac{1}{2n} > \frac{13}{24}.$$

★14.4 A RE-EXAMINATION OF
THE GENERAL PRINCIPLE

If we review the proof of the principle of mathematical induction given in Section 14.3, we find that it is based on the statement:

S. *Every nonempty set of positive integers has a smallest integer.*

Certainly this statement is reasonable, and most students are willing to accept S as a starting point for the proof of the principle of mathematical induction. However, a careful and critical scholar may ask for a proof of S. If he insists, then we will need some other statement T (carefully selected) that we can use to prove S. If now the scholar asks for a proof of T, then this must be based on some other statement U. Clearly, this process can not go on indefinitely. Eventually we must arrive at a suitable set of statements which we agree to accept without proof. Such statements are called *axioms*. We have already touched on axiom systems in Section 1.5. In Chapters 2 and 3 we gave a particular system of axioms for the real numbers.

Since the principle of mathematical induction is a statement about the set **N** of natural numbers (positive integers), we must agree on a system of axioms for **N** before attempting a proof of this principle.

There are various systems of axioms for **N** but the most popular and the most natural system is the one set forth by G. Peano and known as the Peano axioms. These axioms are explained in Appendix 3 and the interested reader can pursue the subject there.

Axiom 5 of the Peano axioms is the principle of mathematical induction. By listing it as an axiom, we do not need to prove it. This is not a defect because this principle is an essential item in the description of **N**. Further, we cannot prove Axiom 5 as a consequence of the other four axioms. This is covered in problems 8 and 9 in the exercise for Appendix 3.

14.5 THE $\sum$ NOTATION

The frequent occurrence of sums in Exercise 1 suggests that it might be worthwhile to have a compact notation to indicate a sum. The three dots notation is adequate, but it is time to introduce a new notation which is more compact, more precise, and provides for greater flexibility.

The symbol $\sum$ is a capital sigma in the Greek alphabet and corresponds to our English S. Thus, it reminds us of the word sum. The symbol

$$(12) \qquad\qquad \sum_{i=1}^{5} i^2$$

(read, "the sum of i^2 as i runs from 1 to 5") means

$$1^2 + 2^2 + 3^2 + 4^2 + 5^2 = 1 + 4 + 9 + 16 + 25 = 55.$$

The numbers 1 and 5 are called the *lower* and *upper limits of summation*, respectively. The collection of symbols in (12) tells us to substitute in i^2 each of the integers from 1 to 5, inclusive, and to add the results. The index i in equation (12) is called the *index of summation*. The sum need not start at 1 or end at 5. Further, any letter can be used instead of i. Finally, any function may be used in place of i^2. We indicate the sum $f(1) + f(2) + \cdots + f(k)$ by writing

$$\sum_{i=1}^{k} f(i).$$

The following examples indicate the various possibilities. The new shorthand notation is on the left-hand side, and its meaning is on the right-hand side in each of these equations.

$$\sum_{i=0}^{4} i^3 = 0 + 1 + 8 + 27 + 64 = 100.$$

$$\sum_{k=2}^{7} k^2 = 4 + 9 + 16 + 25 + 36 + 49 = 139.$$

$$\sum_{j=1}^{5} j(j+1) = 1(2) + 2(3) + 3(4) + 4(5) + 5(6)$$
$$= 2 + 6 + 12 + 20 + 30 = 70.$$

$$\sum_{i=1}^{8} 1 = 1 + 1 + 1 + 1 + 1 + 1 + 1 + 1 = 8.$$

$$\sum_{k=1}^{n} f(k) = f(1) + f(2) + f(3) + \cdots + f(n).$$

$$\sum_{t=1}^{n} f(t) = f(1) + f(2) + f(3) + \cdots + f(n).$$

$$\sum_{\theta=1}^{n} g(\theta) = g(1) + g(2) + g(3) + \cdots + g(n).$$

Sometimes the terms to be added involve subscripts or combinations of functions with subscripts. These possibilities are illustrated below.

$$\sum_{n=3}^{8} a_n = a_3 + a_4 + a_5 + a_6 + a_7 + a_8.$$

$$\sum_{n=1}^{5} nb_n = b_1 + 2b_2 + 3b_3 + 4b_4 + 5b_5.$$

$$\sum_{k=1}^{n} \frac{a_k}{k} = a_1 + \frac{a_2}{2} + \frac{a_3}{3} + \cdots + \frac{a_n}{n}.$$

EXAMPLE 1   Find $\sum_{k=2}^{6} k^3$.

Solution: By the meaning of the symbols, we have

$$\sum_{k=2}^{6} k^3 = 8 + 27 + 64 + 125 + 216 = 440. \quad \blacktriangle$$

EXAMPLE 2   Use the new notation to write the formulas that we proved in problems 3, 4, 7, and 8 of Exercise 1.

Solution: From page 371 we have

(13) $$\sum_{i=1}^{n} i = \frac{n(n+1)}{2}, \qquad \text{(problem 3)}$$

(14) $$\sum_{i=1}^{n} (4i-3) = n(2n-1), \qquad \text{(problem 4)}$$

(15) $$\sum_{i=1}^{n} i^2 = \frac{n(n+1)(2n+1)}{6}, \qquad \text{(problem 7)}$$

and

(16) $$\sum_{i=1}^{n} i(i+1) = \frac{n(n+1)(n+2)}{3}, \qquad \text{(problem 8).} \quad \blacktriangle$$

EXAMPLE 3   Prove that equation (15) is true for each positive integer n, using the new notation for a sum.

Solution: We are still using mathematical induction. If $n = 1$, then (15) gives

$$1^2 = \frac{1(2)(3)}{6}$$

and this is true.

We next assume that (15) is true when n is replaced by k. This gives

(17) $$\sum_{i=1}^{k} i^2 = \frac{k(k+1)(2k+1)}{6}.$$

If we add $(k+1)^2$ to both sides, we find

$$(k+1)^2 + \sum_{i=1}^{k} i^2 = \frac{k(k+1)(2k+1)}{6} + (k+1)^2$$

$$\sum_{i=1}^{k+1} i^2 = \frac{1}{6}(k+1)[k(2k+1) + 6(k+1)]$$

$$= \frac{1}{6}(k+1)(2k^2 + 7k + 6)$$

(18) $$\sum_{i=1}^{k+1} i^2 = \frac{1}{6}(k+1)(k+2)(2k+3).$$

But the right side of (18) is the right side of (15) when n is replaced by $k+1$. $\blacktriangle$

Exercise 2

In problems 1 through 6 show that the computation is correct.

1. $\displaystyle\sum_{k=4}^{8} k = 30$

2. $\displaystyle\sum_{k=2}^{6} k^2 = 90$

3. $\displaystyle\sum_{i=1}^{10} 3i = 165$

4. $\displaystyle\sum_{n=0}^{6} \frac{n(n-1)}{2} = 35$

5. $\displaystyle\sum_{j=1}^{5} \frac{1}{j} = \frac{137}{60}$

6. $\displaystyle\sum_{i=1}^{6} \frac{1}{i(i+1)} = \frac{6}{7}$

In problems 7 through 12 compute the indicated sum.

7. $\displaystyle\sum_{i=1}^{7} (3i+2)$

8. $\displaystyle\sum_{j=0}^{6} (3j+5)$

9. $\displaystyle\sum_{i=0}^{10} i(i-1)$

10. $\displaystyle\sum_{k=1}^{5} (-1)^k k^2$

11. $\displaystyle\sum_{k=1}^{6} (-1)^{k+1} k(k+1)$

12. $\displaystyle\sum_{i=2}^{8} 2^i$

In problems 13 through 27 state whether the given assertion is always true or sometimes may be false.

13. $\displaystyle\sum_{i=1}^{n} a_i = \sum_{j=1}^{n} a_j$

14. $\displaystyle d\sum_{k=1}^{n} a_k = \sum_{k=1}^{n} da_k$

15. $\displaystyle\sum_{i=1}^{N} i(i+2) = \sum_{j=0}^{N-1} (j+1)(j+3)$

16. $\displaystyle\sum_{i=0}^{n} b_i = \sum_{i=0}^{n} b_{n-i}$

17. $\displaystyle\sum_{t=1}^{n} ct^4 = c\sum_{t=1}^{n} t^4$

18. $\displaystyle\left(\sum_{i=1}^{n} a_i\right)\left(\sum_{i=1}^{n} b_i\right) = \sum_{i=1}^{n} a_i b_i$

19. $\displaystyle\left(\sum_{k=1}^{5} a_k\right)^2 = \sum_{k=1}^{5} a_k^2 + \sum_{k=1}^{5} 2a_k + \sum_{k=1}^{5} 1$

20. $\displaystyle\sum_{k=1}^{n} a = na$

21. $\displaystyle\sum_{k=1}^{n} (a_k+b_k) = \sum_{k=1}^{n} a_k + \sum_{k=1}^{n} b_k$

22. $\displaystyle\sum_{k=1}^{n-1} \frac{n-k}{k} = \sum_{k=1}^{n-1} \frac{k}{n-k}$

23. $\displaystyle\sum_{k=1}^{2n+1} (-1)^k = 0$

24. $\displaystyle\sum_{k=1}^{2n} (-1)^k = 1$

25. $\displaystyle\sum_{k=0}^{2n} (-1)^k = 1$

26. $\displaystyle\sum_{k=0}^{2n} (-1)^{k^2} = 1$

27. $\displaystyle\sum_{k=1}^{n+1} (-1)^{k(k+1)} = n+1$

28. Use the summation notation with k as the index of summation to write the formulas proved in Exercise 1, problems 11, 13, 15, and 22.

★ **14.6 SEQUENCES**

An array of numbers of the form

(19) $$a_1, a_2, a_3, \ldots, a_n,$$

is called a *finite sequence* or a *sequence*. As indicated by the subscript, a_1 is the first term, a_2 is the second term, a_k is the kth term, and a_n is the last term, or the nth term. If the sequence has no last term, it is called an *infinite sequence*. An infinite sequence is indicated by writing

(20) $$a_1, a_2, a_3, \ldots, a_k, \ldots,$$

where the three dots at the end show that there is a corresponding number a_n for every positive integer n.

In a sequence each integer k gives a corresponding term a_k, so that a_k is really a function of k. Consequently, the kth term could be written $f(k), g(k), h(k), \ldots$, using a suitable letter for the function. However, the universal custom is to use a subscript on a suitable letter: $a_k, b_k, c_k, \ldots$.

DEFINITION 1

Sequence *An infinite sequence is a function whose domain is the set of all positive integers. The image of k under this function is called the kth term of the sequence. If the domain of the function is the set of integers $\{k \mid 1 \leqq k \leqq n\}$, then the function is called a finite sequence.*

EXAMPLE 1

Suppose that the first term of a sequence is 5, and each term thereafter is 4 more than the preceding term. Find a relation between terms of the sequence, and find the first six terms of the sequence.

Solution: We are given that $a_1 = 5$ and that for each $k \geqq 1$

(21) $$a_{k+1} = a_k + 4.$$

Using equation (21) with $k = 1, 2, 3, 4$, and 5, we find that $a_2 = 5+4 = 9$, $a_3 = 9+4 = 13, \ldots$. Hence, the first six terms of the sequence are 5, 9, 13, 17, 21, 25. ▲

An equation, such as (21), that gives each term of a sequence as a function of one or more of the preceding terms is called a *recursion formula*. When a sequence is defined by such a formula, we say that the sequence is defined *recursively*.

EXAMPLE 2   If $a_1 = 3$ and the sequence is defined recursively by the formula $a_{k+1} = 2a_k$ for $k \geq 1$, find the first six terms of the sequence.

Solution: Computation with the formula yields 3, 6, 12, 24, 48, 96. ▲

DEFINITION 2   **Arithmetic Sequence**   *A sequence defined by the recursion formula*

(22) $a_{k+1} = a_k + d$

is called an arithmetic sequence. It is also called an arithmetic progression. The constant d is called the common difference.

Thus, the sequence given in Example 1 is an arithmetic sequence for which the common difference $a_{k+1} - a_k$ is 4.

Theorem 1   *If d is the common difference, then the nth term in an arithmetic sequence is given by*

(23) $a_n = a_1 + (n-1)d.$

If S_n denotes the sum of the first n terms, then

(24) $$S_n = \frac{n}{2}[2a_1 + (n-1)d] = \frac{n}{2}(a_1 + a_n).$$

Proof: Equation (23) is simple and is left for the reader. To prove (24) we use (23) and problems 21 and 14 of Exercise 2. Then,

$$S_n = \sum_{k=1}^{n} a_k = \sum_{k=1}^{n} (a_1 + (k-1)d) = \sum_{k=1}^{n} a_1 + \sum_{k=1}^{n} d(k-1)$$

$$= na_1 + d \sum_{k=1}^{n} (k-1)$$

$$= na_1 + d\frac{(n-1)n}{2} \qquad \text{(problem 3, Exercise 1)}$$

$$= \frac{n}{2}[2a_1 + (n-1)d].$$

This is the first part of (24). The second part follows from the relation $2a_1 + (n-1)d = a_1 + a_1 + (n-1)d = a_1 + a_n$. ∎

DEFINITION 3   **Geometric Sequence**   *A sequence defined by the recursion formula*

(25) $a_{k+1} = ra_k$

is called a geometric sequence. It is also called a geometric progression. The constant r is called the common ratio.

Thus, the sequence given in Example 2 is a geometric sequence for which the common ratio a_{k+1}/a_k is 2.

From Definitions 2 and 3, we see that:

(A) In an arithmetic sequence (or progression) each term (except the first) is obtained by adding d to the preceding term.

(B) In a geometric sequence (or progression) each term (except the first) is obtained by multiplying the preceding term by r.

Theorem 2

If $r \neq 0$ is the common ratio, then the nth term in a geometric sequence is given by

$$(26) \qquad a_n = r^{n-1} a_1.$$

If S_n denotes the sum of the first n terms and $r \neq 1$, then

$$(27) \qquad S_n = a_1 \frac{1-r^n}{1-r}.$$

Proof: Equation (26) is simple and is left for the reader. To prove (27) we use (26) and have

$$(28) \qquad S_n = a_1 + a_1 r + a_1 r^2 + a_1 r^3 + \cdots + a_1 r^{n-1}.$$

Multiplying both sides of (28) by r, we have

$$(29) \qquad S_n r = a_1 r + a_1 r^2 + a_1 r^3 + \cdots + a_1 r^{n-1} + a_1 r^n.$$

We subtract equation (29) from equation (28) and obtain

$$(30) \qquad S_n - S_n r = S_n(1-r) = a_1 - a_1 r^n = a_1(1-r^n).$$

If we divide both sides of (30) by $1-r \; (\neq 0)$, we obtain (27). ∎

EXAMPLE 3

Find the sum of the first eight terms of an arithmetic sequence if the first term is -9 and the common difference is 5.

Solution: By equation (24)

$$S_8 = \frac{8}{2}[2(-9) + (8-1)5] = 4(-18+35) = 68. \quad \blacktriangle$$

EXAMPLE 4

Find the sum of the first five terms of a geometric sequence if the first term is 6 and the common ratio is 1/3.

Solution: By equation (27)

$$S_5 = 6\frac{1-(1/3)^5}{1-1/3} = 6\frac{1-(1/3)^5}{1-1/3}\cdot\frac{3}{3}$$

$$= 6\frac{3-(1/3)^4}{3-1} = 3\left[3 - \left(\frac{1}{3}\right)^4\right] = 9 - \frac{1}{27} = 8\frac{26}{27}. \quad \blacktriangle$$

Exercise 3

In problems 1 through 10 a sequence is given by specifying the first term and a recursion formula for $k \geq 1$. In each case, find the first 5 terms of the sequence. In problems 1 through 6 find S_5 both directly and from the formulas of Theorems 1 or 2.

1. $a_1 = 7,\quad a_{k+1} = a_k + 2$ 2. $a_1 = -5,\quad a_{k+1} = a_k + 3$

3. $a_1 = 6,\quad a_{k+1} = a_k - 2$ 4. $a_1 = 2,\quad a_{k+1} = 3a_k$

5. $a_1 = -4,\quad a_{k+1} = a_k/2$ 6. $a_1 = 8,\quad a_{k+1} = -2a_k$

7. $a_1 = 2,\quad a_{k+1} = a_k + k$ 8. $a_1 = 1,\quad a_{k+1} = \sqrt{k}\,a_k$

9. $a_1 = 1,\quad a_{k+1} = a_k + (-1)^k k^2$ 10. $a_1 = 12,\quad a_{k+1} = a_k/(k+1)$

11. Given the first term and the sum of the first n terms in an arithmetic sequence, find a formula for computing the common difference.

12. For a geometric sequence find a formula for computing the common ratio when the first term and the nth term are known. Assume that $r > 0$.

13. What is S_n in Theorem 2, if $r = 1$?

14. In a certain arithmetic sequence the first term is -3. Find d if (a) $S_6 = 57$, (b) $S_9 = 99$, (c) $S_{11} = -143$.

15. Find the intermediate terms of an arithmetic sequence if (a) $a_1 = 5$ and $a_6 = 20$, (b) $a_4 = 19$ and $a_9 = -16$.

16. Find the intermediate terms of a geometric sequence if (a) $a_1 = 7$ and $a_5 = 112$, (b) $a_3 = -2$ and $a_6 = 54$, (c) $a_5 = 48$ and $a_{12} = 3/8$.

17. A large auditorium for a mathematics lecture has 17 seats in the first row and 18 rows. Each row has one more seat than the preceding row. Find the total number of seats in the auditorium.

18. Find the sum of all the integers between 2 and 322 that are divisible by 3.

19. Find the sum of all of the odd integers between 100 and 500.

20. At the end of each year the value of a car is 20% less than at the beginning of that year. If a new car costs $5,000, find its value at the end of 4 years.

21. Each time that a pair of oily workclothes goes through the washing machine, one half of the oil present is removed. Find the number of washings required to remove at least 95% of the oil.

★ 22. Use mathematical induction to prove Theorems 1 and 2.

23. Show that problems 3, 4, 5, and 6 of Exercise 1 are special cases of Theorem 1.

24. Show that problems 13 and 22 of Exercise 1 are special cases of Theorem 2.

25. An industrious student found that his 6 exam grades formed an increasing arithmetic sequence. If the first grade was 55 and the last grade was 95, find his average for the course.

26. Prove that no matter how many exams the student in problem 25 takes, if his first grade is 55, his last grade is 95 and the exam grades form an increasing arithmetic sequence, then his average is always the same.

27. Three numbers form an arithmetic sequence. If the sum of the numbers is 27 and the sum of their squares is 293, find the numbers. *Hint:* Let x be the middle term.

28. The sum of three numbers in an arithmetic sequence is 33, and their product is 1,276. Find the three numbers.

29. The sum of three numbers in a geometric sequence is 21, and their product is 216. Find the numbers.

★ **14.7 DETERMINANTS OF ARBITRARY ORDER**

We recall that a matrix of nth order is an array of numbers of the form

(31)
$$M = \begin{bmatrix} a_{11} & a_{12} & a_{13} & \cdots & a_{1n} \\ a_{21} & a_{22} & a_{23} & \cdots & a_{2n} \\ \vdots & \vdots & \vdots & \cdots & \vdots \\ a_{n1} & a_{n2} & a_{n3} & \cdots & a_{nn} \end{bmatrix}.$$

The determinant of M is denoted by $D(M)$ or by the same array with the brackets replaced by vertical lines. We have defined $D(M)$ for $n = 2$, $n = 3$, and $n = 4$. We now use mathematical induction to obtain the definition of an nth-order determinant. Thus we assume that for all matrices of order $2, 3, \ldots, n-1$, the determinant has been defined.

If we delete one row and one column from an nth order matrix, we obtain a matrix of order $n-1$, and such a matrix has a determinant. Consequently, for the matrix M given by (31), all of the minors M_{ij} are defined. Thus, we can use a Laplace expansion by minors of the first row on this matrix and obtain

$$(32) \qquad D(M) = a_{11}M_{11} - a_{12}M_{12} + a_{13}M_{13} - \cdots + (-1)^{1+n}M_{1n}$$

$$= \sum_{j=1}^{n} a_{1j}(-1)^{1+j}M_{1j}.$$

Equation (32) is the definition of the determinant of the matrix M given by (31).

With this definition of $D(M)$, Theorems 5 through 12 in Chapter 13 are true for nth-order determinants. Further, the Laplace expansion by minors of any other row or by minors of any column also gives $D(M)$. The formula for expansion by minors of the ith row is

$$(33) \qquad D(M) = (-1)^i \sum_{j=1}^{n} a_{ij}(-1)^j M_{ij}.$$

If we expand by minors of the jth column, we have

$$(34) \qquad D(M) = (-1)^j \sum_{i=1}^{n} a_{ij}(-1)^i M_{ij}.$$

EXAMPLE 1   Find $D(M)$ if

$$M = \begin{bmatrix} 1 & 1 & 1 & 1 & 1 \\ 1 & 2 & 3 & 4 & 5 \\ 0 & 1 & 3 & 5 & 7 \\ 0 & 0 & 1 & 4 & 10 \\ 0 & 0 & 0 & 1 & 5 \end{bmatrix}.$$

Solution: As indicated below, we subtract the first row from the second row, then expand by minors of the first column. A sequence of such steps gives

$$\begin{vmatrix} 1 & 1 & 1 & 1 & 1 \\ 1 & 2 & 3 & 4 & 5 \\ 0 & 1 & 3 & 5 & 7 \\ 0 & 0 & 1 & 4 & 10 \\ 0 & 0 & 0 & 1 & 5 \end{vmatrix} = \begin{vmatrix} 1 & 1 & 1 & 1 & 1 \\ 0 & 1 & 2 & 3 & 4 \\ 0 & 1 & 3 & 5 & 7 \\ 0 & 0 & 1 & 4 & 10 \\ 0 & 0 & 0 & 1 & 5 \end{vmatrix} = \begin{vmatrix} 1 & 2 & 3 & 4 \\ 1 & 3 & 5 & 7 \\ 0 & 1 & 4 & 10 \\ 0 & 0 & 1 & 5 \end{vmatrix}$$

$$= \begin{vmatrix} 1 & 2 & 3 & 4 \\ 0 & 1 & 2 & 3 \\ 0 & 1 & 4 & 10 \\ 0 & 0 & 1 & 5 \end{vmatrix} = \begin{vmatrix} 1 & 2 & 3 \\ 1 & 4 & 10 \\ 0 & 1 & 5 \end{vmatrix} = \begin{vmatrix} 1 & 2 & 3 \\ 0 & 2 & 7 \\ 0 & 1 & 5 \end{vmatrix} = \begin{vmatrix} 2 & 7 \\ 1 & 5 \end{vmatrix}$$

$$= 10 - 7 = 3. \quad \blacktriangle$$

Review Questions

Answer each question as accurately as possible before consulting the text.

1. State the principle of mathematical induction.

2. Give the definition of an arithmetic sequence.

3. For an arithmetic sequence give the formula for the nth term and the formula for S_n.

4. For a geometric sequence give the formula for the nth term and the formula for S_n.

5. A few of the formulas for sums are used quite often. Try to recall (or reconstruct) the formulas for the following sums. (See Exercise 1.)

$$\sum_{k=1}^{n} k \text{ (problem 3)}, \qquad \sum_{k=1}^{n} k^2 \text{ (problem 7)},$$

$$\sum_{k=1}^{n} k^3 \text{ (problem 15)}, \qquad \sum_{k=1}^{n} \frac{1}{k(k+1)}, \text{ (Example 2)}.$$

The Binomial Theorem

15

The expression $(a+b)^n$ occurs so frequently in mathematics that it is desirable to have a formula for the expansion. Since $a+b$ is a binomial, the formula for the expansion of $(a+b)^n$ is called the *Binomial Theorem*. We discuss this theorem and give a proof when n is a positive integer. We also piscuss the extension of the formula when n is not a positive integer.

15.1 THE FACTORIAL NOTATION

The symbol $n!$ (read, "n factorial") is used to represent the product of all the positive integers from 1 to n inclusive. For example,

$$2! = 1 \cdot 2 = 2,$$

$$3! = 1 \cdot 2 \cdot 3 = 6,$$

$$4! = 1 \cdot 2 \cdot 3 \cdot 4 = 24,$$

$$5! = 1 \cdot 2 \cdot 3 \cdot 4 \cdot 5 = 120,$$

etc. This description of $n!$ is not meaningful if $n = 0$ or $n = 1$. We include these cases by making the two definitions

(1) $$1! = 1 \quad \text{and} \quad 0! = 1.$$

The fact that $1! = 1$ seems natural enough, but the agreement to set $0! = 1$ may come as a slight shock. In due time the skeptical reader will see that this definition is really the correct one.

We often write that

(2)
$$n! = 1 \cdot 2 \cdot 3 \cdots (n-1)n,$$

or

(3)
$$n! = n(n-1)(n-2) \cdots 3 \cdot 2 \cdot 1,$$

with suitable agreements about the meaning of the dots if n is small.

In general $(n+m)! \neq n! + m!$. For example, if $n = 3$ and $m = 5$, the left side is $8! = 40{,}320$, and the right side is $3! + 5! = 6 + 120 = 126$. Clearly, these are not equal.

Similarly we observe that $(2n)!$ and $2n!$ have different meanings. In the first expression the factorial applies to $2n$, while in the second expression we are to double $n!$. For example, if $n = 3$, then $(2n)! = 6! = 720$. On the other hand, $2n! = 2 \cdot 3! = 2 \cdot 6 = 12$.

Exercise 1

1. Compute (a) $6!$, (b) $7!$, and (c) $8!$.

2. Compute (a) $9!/6!$, (b) $10!/7!$, and (c) $11!/8!$.

3. Prove that if $n > 2$, then $(n^2 - n)(n-2)! = n!$.

4. Prove that if $n \geq k$, then $(n-k+1)n! + k \cdot n! = (n+1)!$.

★ 5. Find all positive integers such that $(2n)! = 2n!$.

★ 6. Find all pairs of positive integers m and n such that $(m+n)! = m! + n!$.

★ 7. Find all pairs of integers m, $n \geq 0$ such that $(m!)(n!) = (mn)!$.

8. Prove that $(n+1)! = (n+1)n!$ if $n \geq 1$. Show that if we are permitted to put $n = 0$ in this equation, then we can deduce that $0! = 1$.

9. Prove that if $n \geq 2$, then

$$\frac{1}{(n-2)!} + \frac{1}{(n-1)!} = \frac{n}{(n-1)!}.$$

★ 10. Prove that if $2 \leq r \leq k-1$, then

$$\frac{k!}{(k-r)!\,r!} + \frac{k!}{(k-r+1)!\,(r-1)!} = \frac{(k+1)!}{(k-r+1)!\,r!}.$$

11. Prove that the equation in problem 10 is also true when $r = 1$ and $r = k$.

12. Use mathematical induction to prove that for $n \in \mathbf{N}$.

$$1 \cdot 1! + 2 \cdot 2! + 3 \cdot 3! + \cdots + n \cdot n! = (n+1)! - 1.$$

★13. Prove that for $n \geq 3$,

$$\frac{3!}{1!} + \frac{4!}{2!} + \frac{5!}{3!} + \cdots + \frac{n!}{(n-2)!} = \frac{n^3 - n - 6}{3}.$$

15.2 THE PASCAL TRIANGLE

By direct computation we expand $(a+b)^n$. For $n = 0, 1, 2, 3, 4,$ and 5 we find

$$(a+b)^0 = \qquad\qquad 1$$

$$(a+b)^1 = \qquad\qquad a + b$$

$$(a+b)^2 = \qquad\qquad a^2 + 2ab + b^2$$

$$(a+b)^3 = \qquad\qquad a^3 + 3a^2 b + 3ab^2 + b^3$$

$$(a+b)^4 = \qquad a^4 + 4a^3 b + 6a^2 b^2 + 4ab^3 + b^4$$

$$(a+b)^5 = a^5 + 5a^4 b + 10a^3 b^2 + 10a^2 b^3 + 5ab^4 + b^5.$$

Can we predict the expansion (the right side) for arbitrary positive integer n, by studying these first few cases? Certain items appear at once. The terms of the expansion all have the form $a^q b^r$, where the sum of the exponents $q+r$ is always n. Hence, we can set $q = n - r$ and write $a^{n-r}b^r$. Further, we can arrange the terms in order of increasing powers of b. Consequently, except for the unknown coefficients, which we denote temporarily by ■ to indicate the gap in our knowledge, we have

(4) $\quad (a+b)^n = a^n + \blacksquare a^{n-1}b + \blacksquare a^{n-2}b^2 + \cdots$

$$+ \blacksquare a^{n-r}b^r + \cdots + \blacksquare ab^{n-1} + b^n.$$

Therefore, in analyzing the results of the first six expansions of $(a+b)^n$, we can ignore the variables a and b and concentrate on the coefficients. These coefficients form the array shown in Figure 15.1.

This array of numbers can be extended downward indefinitely, and when extended, the complete array is called Pascal's triangle. The triangle has many fascinating properties, but we consider only the most elementary.

A careful inspection of Pascal's triangle will suggest a law of formation. Clearly, the element at the vertex is 1. Each element in each of the lateral sides

FIGURE 15.1

$$(a+b)^0 \qquad\qquad 1$$
$$(a+b)^1 \qquad\qquad 1 \quad 1$$
$$(a+b)^2 \qquad\qquad 1 \quad 2 \quad 1$$
$$(a+b)^3 \qquad\qquad 1 \quad 3 \quad 3 \quad 1$$
$$(a+b)^4 \qquad\qquad 1 \quad 4 \quad 6 \quad 4 \quad 1$$
$$(a+b)^5 \qquad\qquad 1 \quad 5 \quad 10 \quad 10 \quad 5 \quad 1$$

is 1. Each element inside the triangle is the sum of the two numbers which appear on the line above, just to the left and just to the right of the element. For example, in the sixth row $5 = 1+4$ and $10 = 4+6$. If this law of formation is correct, then the next row is obtained by the calculation

$$1 + 5 = 6, \qquad 5 + 10 = 15, \qquad 10 + 10 = 20, \qquad \text{etc.}$$

Consequently we expect that

$$(5) \quad (a+b)^6 = a^6 + 6a^5 b + 15a^4 b^2 + 20a^3 b^3 + 15a^2 b^4 + 6ab^5 + b^6,$$

and a brief computation of $(a+b)(a+b)^5$ shows that equation (5) is indeed ,correct.

In the next section we will state and prove the binomial theorem. In preparation we introduce a symbol for the coefficient of $a^{n-r}b^r$ in the expansion of $(a+b)^n$. This symbol merely replaces the ■ in equation (4).

DEFINITION 1   *Let n be a positive integer and let $0 \leqq r \leqq n$. The symbol $\binom{n}{r}$ (read, "n above r") is the coefficient of $a^{n-r}b^r$ in the expansion of $(a+b)^n$.*

In other words (by the definition of the symbol)

$$(6) \quad (a+b)^n = \binom{n}{0}a^n + \binom{n}{1}a^{n-1}b + \binom{n}{2}a^{n-2}b^2 + \cdots$$

$$+ \binom{n}{r}a^{n-r}b^r + \cdots + \binom{n}{n}b^n.$$

The reader must keep in mind that for given n and r we still do not know the value of $\binom{n}{r}$ for fixed n and r, but we will soon prove

Theorem 1   If $0 \leq r \leq n$, then

(7)
$$\binom{n}{r} = \frac{n!}{(n-r)!\,r!} = \frac{n(n-1)\cdots(n-r+1)}{r!}.$$

EXAMPLE 1   Find the value of

$$\binom{5}{3}, \quad \binom{13}{9}, \quad \binom{57}{56}, \quad \binom{123}{123}, \quad \binom{278}{0}.$$

Solution: We use equation (7), although the proof is given in the next section.

$$\binom{5}{3} = \frac{5!}{2!\,3!} = \frac{5 \cdot 4 \cdot 3 \cdot 2 \cdot 1}{2 \cdot 1 \cdot 3 \cdot 2 \cdot 1} = 10,$$

$$\binom{13}{9} = \frac{13!}{4!\,9!} = \frac{13 \cdot 12 \cdot 11 \cdot 10}{4 \cdot 3 \cdot 2 \cdot 1} = 715,$$

$$\binom{57}{56} = \frac{57!}{1!\,56!} = \frac{57 \cdot 56 \cdots 2 \cdot 1}{1 \cdot 56 \cdots 2 \cdot 1} = \frac{57}{1} = 57,$$

$$\binom{123}{123} = \frac{123!}{0!\,123!} = \frac{123!}{1 \cdot 123!} = 1,$$

$$\binom{278}{0} = \frac{278!}{278!\,0!} = \frac{278!}{278! \cdot 1} = 1. \quad \blacktriangle$$

Exercise 2

1. Using the rule of formation for Pascal's triangle (adding the two elements directly above on each side) find the 8th, 9th, and 10th rows of Pascal's triangle. *Hint:* These are the rows for $(a+b)^7$, $(a+b)^8$, and $(a+b)^9$, respectively.

2. Select at random six of the numbers found above and show that equation (7) gives the same value for the coefficient of $a^{n-r}b^r$.

3. Compute the value of

   (a) $\binom{10}{4}$,   (b) $\binom{11}{3}$,   (c) $\binom{10}{6}$,   (d) $\binom{11}{8}$.

4. Compute the value of

   (a) $\binom{12}{5}$,   (b) $\binom{16}{3}$,   (c) $\binom{12}{7}$,   (d) $\binom{16}{13}$.

5. Prove that if $0 \leq r \leq n$, then $\binom{n}{r} = \binom{n}{n-r}$.

6. Find $\binom{n}{0}$, $\binom{n}{1}$, $\binom{n}{n-1}$, and $\binom{n}{n}$.

7. Prove that if $1 \leq r \leq k$, then $\binom{k}{r} + \binom{k}{r-1} = \binom{k+1}{r}$.

8. Compute the value of

(a) $\binom{3}{2}\binom{4}{3}\binom{5}{4}$,　　(b) $\binom{6}{4}\binom{7}{5}$,　　(c) $\binom{9}{5} / \binom{8}{4}$.

★ 9. Find the value of

(a) $\binom{6}{\binom{3}{2}}$,　　(b) $\binom{\binom{6}{3}}{2}$,　　(c) $\binom{7}{\binom{4}{2}}$,　　(d) $\binom{\binom{7}{4}}{2}$.

★ 10. Prove that if $1 \leq r \leq n$, then $\binom{n}{\binom{r}{1}} = \binom{\binom{n}{r}}{1}$.

★ 11. Prove that $\binom{2}{1}\binom{3}{2}\binom{4}{3}\cdots\binom{n-1}{n-2}\binom{n}{n-1} = n!$.

15.3   THE BINOMIAL THEOREM

In this section we prove Theorem 1. In other words, we prove that for each positive integer n

$$(6) \qquad (a+b)^n = \binom{n}{0}a^n + \binom{n}{1}a^{n-1}b + \binom{n}{2}a^{n-2}b^2 + \cdots$$

$$+ \binom{n}{r}a^{n-r}b^r + \cdots + \binom{n}{n}b^n,$$

where the symbols for the coefficients are defined by

$$(7) \qquad \binom{n}{r} = \frac{n!}{(n-r)!\,r!} = \frac{n(n-1)\cdots(n-r+1)}{r!}.$$

We use mathematical induction. We have already settled the case $n = 1$. Although not logically required, we have also obtained the expansion of $(a+b)^n$ for $n = 2, 3, 4$, and 5 and it is an easy matter to check that equation (7) gives the correct coefficients for these values of n as well.

We assume that (6) is true for index k. Thus, we rewrite equation (6) replacing n by k and obtain

$$(8) \quad (a+b)^k = \binom{k}{0}a^k + \binom{k}{1}a^{k-1}b + \binom{k}{2}a^{k-2}b^2 + \cdots$$
$$+ \binom{k}{r}a^{k-r}b^r + \cdots + \binom{k}{k}b^k.$$

To go from the index k to $k+1$, we write $(a+b)^{k+1} = (a+b)(a+b)^k$. In other words, we multiply both sides of equation (8) by $a+b$. The left side gives $(a+b)^{k+1}$ and thus the right side gives two sums which we denote by S_1 and S_2. Thus, by definition

$$(9) \quad (a+b)^{k+1} = S_1 + S_2 \equiv a(a+b)^k + b(a+b)^k.$$

Using equation (8), we find that

$$(10) \quad S_1 = \binom{k}{0}a^{k+1} + \binom{k}{1}a^k b + \binom{k}{2}a^{k-1}b^2 + \cdots$$
$$+ \binom{k}{r}a^{k-r+1}b^r + \cdots + \binom{k}{k}ab^k$$

and

$$(11) \quad S_2 = \binom{k}{0}a^k b + \binom{k}{1}a^{k-1}b^2 + \binom{k}{2}a^{k-2}b^3 + \cdots$$
$$+ \binom{k}{r}a^{k-r}b^{r+1} + \cdots + \binom{k}{k}b^{k+1}.$$

Let us look at the simplest items first. For the first term in S_1 we have

$$(12) \quad \binom{k}{0}a^{k+1} = 1 \cdot a^{k+1} = \binom{k+1}{0}a^{k+1}.$$

For the last term in S_2 we have

$$(13) \quad \binom{k}{k}b^{k+1} = 1 \cdot b^{k+1} = \binom{k+1}{k+1}b^{k+1}.$$

The terms given by (12) and (13) correspond to the left and right edges of the Pascal triangle.

Every other term in S_1 can be paired with a corresponding term in S_2. We look for those terms that involve $a^{k+1-r}b^r$ in S_1 and S_2. These are the $(r+1)$th term in S_1 and the rth term in S_2. Combining these we find the term

(14) $$T_r \equiv \binom{k}{r} a^{k+1-r} b^r + \binom{k}{r-1} a^{k-(r-1)} b^r$$

(15) $$T_r = \left[\binom{k}{r} + \binom{k}{r-1} \right] a^{k+1-r} b^r, \qquad 1 \leq r \leq k.$$

We apply the identity proved in Exercise 2, problem 7. Then (15) yields

(16) $$T_r = \binom{k+1}{r} a^{k+1-r} b^r, \qquad 1 \leq r \leq k.$$

If we combine the results obtained in equations (9), (12), (13), and (16), we obtain

(17) $$(a+b)^{k+1} = \binom{k+1}{0} a^{k+1} + \binom{k+1}{1} a^k b + \binom{k+1}{2} a^{k-1} b^2 + \cdots$$
$$+ \binom{k+1}{r} a^{k+1-r} b^r + \cdots + \binom{k+1}{k+1} b^{k+1}.$$

But equation (17) is identical with equation (6) when $n = k+1$. ∎

Since the identity used in going from equation (15) to (16) is exactly the rule of formation of the Pascal triangle (adding two adjacent terms in one line to get the term in the next line), we have also proved that Pascal's triangle gives the coefficients of $(a+b)^n$ for arbitrary $n \in \mathbf{N}$.

EXAMPLE 1   Expand $(2x-y)^4$.

Solution: We use the fifth line in the Pascal triangle or equation (7) with $n = 4$. Consequently, the coefficients are 1, 4, 6, 4, 1. Further, we set $a = 2x$ and $b = -y$. Hence, $(2x-y)^4 = [2x+(-y)]^4$. Thus,

$$(2x-y)^4 = 1(2x)^4 + 4(2x)^3(-y) + 6(2x)^2(-y)^2 + 4(2x)(-y)^3 + 1(-y)^4$$
$$= 16x^4 - 32x^3 y + 24x^2 y^2 - 8xy^3 + y^4. \ \blacktriangle$$

EXAMPLE 2   In the expansion of $(x^2 + 1/3x)^9$ find the coefficient of (a) x^6, (b) x^4.

Solution: The general term of the expansion is

$$\binom{9}{r} (x^2)^{9-r} \left(\frac{1}{3x} \right)^r = \binom{9}{r} \frac{1}{3^r} x^{18-3r}.$$

To find the coefficient of x^6, set $18 - 3r = 6$. Then $r = 4$, and the coefficient is

$$\binom{9}{4}\frac{1}{3^4} = \frac{9 \cdot 8 \cdot 7 \cdot 6}{1 \cdot 2 \cdot 3 \cdot 4} \cdot \frac{1}{3^4} = \frac{14}{9}.$$

Since $18 - 3r \neq 4$ for any integer r, the term x^4 does not appear in the expansion. We may say that the coefficient is zero. ▲

Exercise 3

1. In the expansion of $(u+v)^{11}$ find the coefficient of (a) $u^8 v^3$, (b) $u^7 v^4$, and (c) $u^5 v^5$.

2. In the expansion of $(x-y)^{12}$ find the coefficient of (a) $x^9 y^3$, (b) $x^8 y^4$, and (c) $x^6 y^7$.

3. In the expansion of $(2A-3B)^5$ find the coefficient of (a) $A^2 B^3$, and (b) $A^3 B^2$.

4. In the expansion of $(2r-s/3)^7$ find the coefficient of (a) $r^4 s^3$, (b) $r^3 s^4$, and (c) $r^2 s^5$.

5. Prove that for each positive integer n

$$\binom{n}{0} + \binom{n}{1} + \binom{n}{2} + \cdots + \binom{n}{n} = 2^n.$$

Check this result by adding the terms in the nth row of the Pascal triangle for $n = 2, 3, 4, 5,$ and 6.

6. Prove that for each positive integer n,

$$\binom{n}{0} - \binom{n}{1} + \binom{n}{2} - \binom{n}{3} + \cdots + (-1)^n \binom{n}{n} = 0.$$

★ 7. What can you say about the sum

$$\binom{n}{0} + 2\binom{n}{1} + 4\binom{n}{2} + 8\binom{n}{3} + \cdots + 2^n \binom{n}{n} ?$$

Check your conjecture with the Pascal triangle.

In problems 8 through 11 find the coefficient of the specified terms in the expansion of the given binomial.

8. $(x^2 - 2x^3)^5$, coefficient of x^9, x^{11}, x^{13}

9. $(x^2 + 3/x^3)^6$, coefficient of x^2, x^{-2}, x^{-3}

★10. $(x+2y+3z)^3$, coefficient of xyz, xy^2, z^3

★11. $(A-2B-C)^4$, coefficient of A^3B, A^2BC, AB^2C, AC^3

12. Find the first four terms in the expansion of $(a+b/2)^{40}$, and $(a-2b)^{20}$.

13. Find the first four terms in the expansion of $(x-y/3)^{27}$, and $(1-5x)^{20}$.

★14. Prove that for each $n \in \mathbf{N}$

$$\left(1+\frac{1}{n}\right)^n \leqq \frac{1}{0!} + \frac{1}{1!} + \frac{1}{2!} + \frac{1}{3!} + \cdots + \frac{1}{n!}.$$

15. Use the binomial theorem to compute (a) $99^4 = (100-1)^4$, and (b) $(1.1)^6$.

16. Use the binomial theorem to compute (a) $(0.98)^3$, and (b) $(1.05)^4$.

17. Expand $(\sqrt{x}+1/\sqrt{x})^5$.

18. Expand $(\sqrt[3]{x}-1/\sqrt[3]{x^2})^6$.

19. Prove that $(2+\sqrt{3})^4 + (2-\sqrt{3})^4$ is an integer.

20. Find $[(3+\sqrt{5})^3 - (3-\sqrt{5})^3]/\sqrt{5}$.

★15.4 EXTENSIONS OF THE
BINOMIAL THEOREM

If the Binomial Theorem applied only when n is a positive integer, it would still be a beautiful and important theorem. The remarkable fact is that under certain conditions, it is true for arbitrary n. The proof lies beyond the scope of this book, but the patient reader will encounter it in one of his advanced courses. Here we present the facts.

We return to the binomial coefficients

(7)
$$\binom{n}{r} = \frac{n!}{(n-r)!\,r!} = \frac{n(n-1)\cdots(n-r+1)}{r!}$$

and suppose now that n is not an integer, while r is still a positive integer. Since $n!$ is undefined the middle item in (7) is meaningless. But, the extreme right term in (7) has meaning for arbitrary n, as long as r is a positive integer. Under these conditions we drop the middle term and make the right side the definition of $\binom{n}{r}$. It should be observed that the numerator of the right side of (7) always has r factors because it can be put in the form

$$(n-0)(n-1)(n-2)\cdots(n-[r-1]).$$

EXAMPLE 1

Compute explicitly $\binom{-1}{3}$, $\binom{-1}{4}$, $\binom{3/2}{3}$, and $\binom{1/2}{5}$.

Solution: Using the definition just given we have

$$\binom{-1}{3} = \frac{(-1)(-2)(-3)}{3 \cdot 2 \cdot 1} = -1$$

$$\binom{-1}{4} = \frac{(-1)(-2)(-3)(-4)}{4 \cdot 3 \cdot 2 \cdot 1} = 1$$

$$\binom{3/2}{3} = \frac{(3/2)(1/2)(-1/2)}{3 \cdot 2 \cdot 1} = -\frac{3}{8 \cdot 3!} = -\frac{1}{16}$$

$$\binom{1/2}{5} = \frac{(1/2)(-1/2)(-3/2)(-5/2)(-7/2)}{5 \cdot 4 \cdot 3 \cdot 2 \cdot 1} = \frac{7}{256}. \quad \blacktriangle$$

We continue with the theory of $(a+b)^n$. If $a = 0$ or $b = 0$, the expansion is of no great interest. We assume that $|a| \neq |b|$ and we select the notation so that $|a| > |b| > 0$ (put the larger number first). Then,

(18) $$(a+b)^n = \left(a\left[1 + \frac{b}{a}\right]\right)^n = a^n\left(1 + \frac{b}{a}\right)^n = a^n(1+x)^n,$$

where $x \equiv b/a$. Consequently, $|x| < 1$. Hence, we can expand $(a+b)^n$ by expanding $(1+x)^n$ and then multiplying by a^n. We now concentrate on $(1+x)^n$. If we use the Binomial Theorem on $(1+x)^n$ and note that $1^{n-r} = 1$ for all n, r, then the expansion (6) gives

(19) $$(1+x)^n = 1 + \binom{n}{1}x + \binom{n}{2}x^2 + \binom{n}{3}x^3 + \cdots + \binom{n}{r}x^r + \cdots.$$

Be careful! If n is not an integer, then the sum on the right side of (19) has infinitely many terms. A meaning can be given for the sum of infinitely many terms, but this long digression also lies outside the scope of this book. The essential fact is this: If $|x| < 1$, then the terms in (19) get smaller and smaller as r gets larger and larger, and the infinite sum is a number. The correct expression is "the infinite series converges." Further, it converges to $(1+x)^n$. From a practical point of view all we need to know is that if we take enough of the first terms, we get a good approximation for $(1+x)^n$. We summarize in

Theorem 2
PWO

If $|x| < 1$, then equation (19) is true for every real number $n \neq 0$.

This is the general binomial theorem. The right side of (19) is known as the *binomial series.*

EXAMPLE 2   Find an approximate value for $\sqrt{1.2}$.

Solution: We first find a few terms of the expansion of $(1+x)^{1/2}$. The reader should use equations (7) and (19), and show that

$$(20) \qquad (1+x)^{1/2} = 1 + \frac{1}{2}x - \frac{1}{8}x^2 + \frac{1}{16}x^3 - \frac{5}{128}x^4 + \frac{7}{256}x^5 + \cdots.$$

We put $x = 0.2$ and use the first six terms of the infinite series (20). Separating the positive and negative terms we have

$$
\begin{array}{ll}
+\,1.00000000 & -\,0.00500000 \\
+\,0.10000000 & -\,0.00006250 \\
+\,0.00050000 & \overline{-\,0.00506250} \\
+\,0.00000875 & \\
\overline{+\,1.10050875} & \\
-\,0.00506250 & \\
\overline{1.09544625} &
\end{array}
$$

Examination of the last term used from (20) shows that we can not trust more than the first five digits. The work indicates that $\sqrt{1.2} \approx 1.0954$. A laborious computation shows that

$$(1.09544)^2 < 1.19999$$

and

$$(1.09545)^2 > 1.20001.$$

Therefore, $1.09544 < \sqrt{1.2} < 1.09545$. Hence, (20) gives $\sqrt{1.2}$ to five significant figures. ▲

EXAMPLE 3   Find an approximate value for $\sqrt[4]{14}$.

Solution: Since $\sqrt[4]{16} = 2$, we write

$$(21) \qquad \sqrt[4]{14} = \sqrt[4]{16 - 2} = \sqrt[4]{16\left(1 - \frac{2}{16}\right)} = 2\sqrt[4]{1 + x},$$

where $x = -1/8$. We then find the binomial series for $\sqrt[4]{1+x}$. Using equation (7) we find that

$$(22) \qquad (1+x)^{1/4} = 1 + \frac{1}{4}x - \frac{3}{32}x^2 + \frac{7}{128}x^3 - \frac{77}{2048}x^4 + \cdots.$$

If we put $x = -1/8$ and keep only terms that affect the first five decimal places, we have

$$\sqrt[4]{1 - 1/8} \approx 1 - 0.03125 - 0.00146 - 0.00011 - 0.00001 = 0.96717.$$

Hence, $\sqrt[4]{14} \approx 2 \times (0.96717) = 1.93434$. From the method used we might expect five significant figures, but in this particular case $\sqrt[4]{14} = 1.93434$ to six significant figures. ▲

Exercise 4

In problems 1 through 4 find the first five terms of the binomial series for the indicated expression.

1. $\sqrt{1-x}$ 2. $\sqrt[3]{1+x}$

3. $\sqrt[5]{1+x}$ 4. $(1+x)^{2/3}$

5. $(1+x)^{-1}$ 6. $(1-x)^{-2}$

7. Find the coefficient of x^n in the binomial series for $(1+x)^{-1}$.

8. Find the coefficient of x^n in the binomial series for $(1+x)^{-2}$.

9. Obtain the solution to problem 1 by replacing x by $-x$ in equation (20). Does this replacement seem legitimate to you?

10. Assuming that $|x| > |y|$, use the result of problem 2 to find the first 5 terms of the expansion for $(x+y)^{1/3}$.

11. Find the first five terms of the square of the polynomial found in problem 1.

12. Find the first five terms of the product of $(1+x)$ and the polynomial found in problem 5.

In problems 13 to 22 estimate the indicated quantity to four significant figures

13. $(1.05)^{1/3}$ 14. $(1.06)^{2/3}$ 15. $(1.1)^{2/3}$

16. $(1.12)^{1/3}$ 17. $\sqrt{15}$ 18. $\sqrt{18}$

19. $\sqrt[3]{60}$ 20. $\sqrt[3]{24}$ 21. $\sqrt[5]{40} = \sqrt[5]{32+8}$

22. $\sqrt[5]{27}$

Review Questions

Answer each question as accurately as possible before consulting the text.

1. State the law of formation for the Pascal triangle.

2. Give the definition of the symbol $\binom{n}{k}$. Why is this symbol important?

3. Does the symbol $\binom{n}{k}$ have a meaning when n is not a positive integer? If so, what is the meaning?

Permutations, Combinations, and Probability

16

Nearly everyone is familiar with the concept of probability. Statements such as, "He is almost certain to succeed," "Very likely it will not rain today," and "You will hardly ever have four aces in a poker game" are all statements about the probability of certain events. In the theory of probability we attempt to assign a number which measures the likelihood of a given event. For some events, it is difficult to assign the proper number or even a close approximation. In other cases, notably in games such as dice, poker, and roulette, it is relatively easy to determine the numerical value for the probability of a given event. When applied to such games, the theory of probability is remarkably accurate and extremely useful. Further, the theory is useful in a wide variety of scientific investigations that have nothing to do with games of chance.

The elementary portion of probability theory leans heavily on counting the elements in a set. Consequently, we devote the first three sections of this chapter to counting. Although counting may sound trivial and boring, it is quite the contrary. When the set is complicated, counting the elements may be rather difficult, and discovering the proper method may be an exciting achievement.

401

**16.1 A FUNDAMENTAL
COUNTING PRINCIPLE**

We recall from Section 6.9 that the Cartesian product of two sets **A** and **B** is denoted by **A × B** and is the set of all pairs (x, y) where $x \in$ **A** and $y \in$ **B**.

EXAMPLE 1

Let **A** = $\{1, 2, 3\}$, and let **B** = $\{a, b, c, d, e\}$. Exhibit all of the elements in the Cartesian product **A × B**, and determine the number of elements in **A × B**.

Solution: We use the elements of **A × B** to form a rectangular array, putting all elements of the form $(1, y)$ in the first row, all elements of the form $(2, y)$ in the second row, etc. In this way we obtain the array shown in Table 1.

TABLE 1

$(1, a)$	$(1, b)$	$(1, c)$	$(1, d)$	$(1, e)$
$(2, a)$	$(2, b)$	$(2, c)$	$(2, d)$	$(2, e)$
$(3, a)$	$(3, b)$	$(3, c)$	$(3, d)$	$(3, e)$

Consequently, the set **A × B** has $3 \times 5 = 15$ elements. ▲

This example suggests

Theorem 1

Let $n(\mathbf{A})$, $n(\mathbf{B})$, and $n(\mathbf{A \times B})$ denote the number of elements in the sets **A**, **B**, *and* **A × B**, *respectively. If* **A** *and* **B** *are finite sets, then*

$$n(\mathbf{A \times B}) = n(\mathbf{A})n(\mathbf{B})$$

Proof: Just as in the example, the elements of **A × B** can be placed in a rectangular array with $n(\mathbf{A})$ rows and $n(\mathbf{B})$ columns. Hence, the number of elements in **A × B** is the product of the "height" $n(\mathbf{A})$ and the "width" $n(\mathbf{B})$. ∎

The idea of a Cartesian product can be generalized to any number of sets. Here it is convenient to use subscripts to denote the sets and the elements. The elements (x, y) of **A × B** are called *ordered pairs*. This concept is generalized in

DEFINITION 1

An ordered k-tuple is a collection of k elements of the form $(x_1, x_2, ..., x_k)$, in which the order is important. The first element is x_1, the second element is x_2, etc. Two ordered k-tuples $(x_1, x_2, ..., x_k)$ and $(y_1, y_2, ..., y_k)$ are said to be equal if and only if

(1) $$x_1 = y_1, \quad x_2 = y_2, ..., x_k = y_k.$$

Thus, $(x_1, x_2, ..., x_k) = (y_1, y_2, ..., y_k)$ if and only if the corresponding elements are equal.

DEFINITION 2 *The Cartesian product of k sets $\mathbf{A}_1, \mathbf{A}_2, ..., \mathbf{A}_k$ is denoted by $\mathbf{A}_1 \times \mathbf{A}_2 \times \cdots \times \mathbf{A}_k$ and is the set of all ordered k-tuples $(x_1, x_2, ..., x_k)$ in which $x_1 \in \mathbf{A}_1$, $x_2 \in \mathbf{A}_2, ..., x_k \in \mathbf{A}_k$.*

In this definition, the sets $\mathbf{A}_1, \mathbf{A}_2, ..., \mathbf{A}_k$ may all be the same, or may be different. If we follow the notation and methods of Theorem 1, we obtain

Theorem 2
PWO

The Fundamental Counting Principle   *If $\mathbf{A}_1, \mathbf{A}_2, ..., \mathbf{A}_k$ are finite sets, then*

(2) $$n(\mathbf{A}_1 \times \mathbf{A}_2 \times \cdots \times \mathbf{A}_k) = n(\mathbf{A}_1)n(\mathbf{A}_2)\cdots n(\mathbf{A}_k).$$

EXAMPLE 2   Let $\mathbf{A}_1 = \{1, 2, 3, 4, 5\}$, $\mathbf{A}_2 = \{1, 2, 3, 4\}$ and $\mathbf{A}_3 = \{\text{dog, cat, rat, cow, horse}\}$. Exhibit a few elements in the set $\mathbf{A}_1 \times \mathbf{A}_2 \times \mathbf{A}_3$, and find $n(\mathbf{A}_1 \times \mathbf{A}_2 \times \mathbf{A}_3)$.

Solution: Selecting elements more or less at random we have $(1, 2, \text{cat})$, $(5, 4, \text{dog})$, $(3, 3, \text{rat})$, and $(4, 4, \text{horse})$. Observe that the ordered k-tuple $(4, 5, \text{dog})$ is not in $\mathbf{A}_1 \times \mathbf{A}_2 \times \mathbf{A}_3$, but it is in $\mathbf{A}_2 \times \mathbf{A}_1 \times \mathbf{A}_3$.

Clearly $n(\mathbf{A}_1) = 5$, $n(\mathbf{A}_2) = 4$, and $n(\mathbf{A}_3) = 5$. Consequently, the number of elements in $\mathbf{A}_1 \times \mathbf{A}_2 \times \mathbf{A}_3$ is the product $5 \cdot 4 \cdot 5 = 100$. ▲

16.2   PERMUTATIONS

Two types of sets occur frequently in counting problems: (1) the set of all *permutations* of k things selected from a set of n things, and (2) the set of all *combinations* of k things selected from a set of n things. Briefly, in a permutation the order is important, and in a combination the order is not important.

DEFINITION 3   ***Permutation***   *An arrangement of k things in a definite order is called a permutation of the things. If the things are selected from a collection of n things, then the arrangement is called a permutation of n things taken k at a time.*

If the things are all different, we can regard them as elements of a set. Then a permutation is an arrangement of the elements of some subset of a universal set $\mathbf{U}$ of n elements. In this case, it is more accurate to call it a permutation of n *elements* taken k at a time. We shall frequently use the word "element" in place of "thing." Occasionally some of the objects to be arranged may contain duplicates and, hence, cannot be regarded as elements of a set. In such a case, the term "thing" must be used.

A permutation of k things can be regarded as an ordered k-tuple $(x_1, x_2, ..., x_k)$ where each x_i is taken from the collection of n things.

EXAMPLE 1   Find the number of permutations of 5 different things taken 2 at a time.

Solution: We will soon have a formula that will give us this number immediately. For the present, we merely list the various permutations and count them. The number of permutations does not depend on the nature of the things. In this problem we set $\mathbf{U} = \{a, b, c, d, e\}$ and list all of the permutations of these 5 things taken 2 at a time in Table 2. Notice that repetitions, such as (a, a), (b, b), etc., are not listed because in this case the things a, b, c, d, and e are all different.

TABLE 2	First element	Permutations of a, b, c, d, and e taken two at a time			
	a	(a, b)	(a, c)	(a, d)	(a, e)
	b	(b, a)	(b, c)	(b, d)	(b, e)
	c	(c, a)	(c, b)	(c, d)	(c, e)
	d	(d, a)	(d, b)	(d, c)	(d, e)
	e	(e, a)	(e, b)	(e, c)	(e, d)

Since there are 5 rows and 4 columns, the number of such permutations is $5 \cdot 4 = 20$. ▲

Theorem 3   *Let $P(n, k)$ denote the number of permutations of n different things taken k at a time. Then*

$$(3) \qquad P(n, k) = n(n-1)(n-2) \cdots (n-k+1).$$

Proof: We first observe that the product always has k factors. Thus, by the definition of the symbol $P(n, k)$ we have

$$P(n, 1) = n, \qquad P(n, 2) = n(n-1), \qquad P(n, 3) = n(n-1)(n-2),$$

etc. When $k = 1$, there are n different ways of selecting one thing from a set of n things, and, hence, $P(n, 1) = n$. If $k = 2$, the first element in (x_1, x_2) can be selected in n different ways. Once the first element has been chosen, there are only $n-1$ different ways of selecting the second element. These pairs can be arranged in a rectangular array with n rows and $n-1$ columns (just as in Table 2 for Example 1). Hence, the number of such pairs is $n(n-1)$, or $P(n, 2) = n(n-1)$.

To find $P(n, 3)$ we observe that for each fixed permutation of 2 things (x_1, x_2), there are $n-2$ things left to select for x_3 in (x_1, x_2, x_3). Consequently, each such permutation of 2 things gives rise to $n-2$ permutations of 3 things. Hence,

$$P(n, 3) = (n-2)P(n, 2) = (n-2)n(n-1) = n(n-1)(n-2).$$

The general case is similar. After selecting $k-1$ things from a set of n different things to form $(x_1, x_2, ..., x_{k-1})$, there are $n-(k-1)$ things left. Any one of these may be used as the kth element in an ordered k-tuple. Consequently,

(4) $\qquad P(n,k) = [n-(k-1)]\,P(n, k-1) = P(n, k-1)(n-k+1).$

Then equation (3) will follow from equation (4). ∎

Equation (3) can be expressed quite simply using factorial notation. Indeed, cancellation of common factors in $n!/(n-k)!$ will give the right side of (3). Hence,

(5) $\qquad\qquad\qquad\qquad P(n,k) = \dfrac{n!}{(n-k)!}.$

As a special case of equation (3) or (5) we note that the number of permutations of n different things taken n at a time is given by

(6) $\qquad\qquad\qquad\qquad P(n,n) = n!.$

EXAMPLE 2

In how many different ways can 5 persons stand in a line for a group picture?

Solution: Here we assume that the order is important. Hence, this is the number of permutations of 5 things taken 5 at a time. From equation (6) $P(5,5) = 5! = 5 \cdot 4 \cdot 3 \cdot 2 \cdot 1 = 120.$ ▲

EXAMPLE 3

First, second, third, and fourth prizes are to be awarded at a science fair in which 13 exhibits have been entered. In how many different ways can the prizes be awarded?

Solution: Certainly the order of selection is important (at least to the contestants) and repetitions are not allowed (no one can receive more than one prize). Hence, the answer is the number of permutations of 13 things taken 4 at a time. Using either equation (5) or equation (3) we find

$$P(13,4) = \frac{13!}{(13-4)!} = \frac{13!}{9!} = 13 \cdot 12 \cdot 11 \cdot 10 = 17{,}160. \quad ▲$$

EXAMPLE 4

How many three letter words can be formed using the letters of (a) square, (b) circle?

Solution: Any arrangement of letters is called a *word*, even though it may be difficult or impossible to pronounce and may have no meaning in the English language.

(a) This is the number of permutations of 6 letters taken 3 at a time: $P(6, 3) = 6 \cdot 5 \cdot 4 = 120$.

(b) In "circle" we have only 5 different letters. Using only different letters we can make $P(5, 3) = 5 \cdot 4 \cdot 3 = 60$ different three-letter words. If we use two c's and one other letter, which we denote by x, then the number of different permutations of ccx is $3!/2!$, because in any one arrangement, an interchange of the two c's leads to the same arrangement. Since x can be selected in 4 different ways, the number of different three-letter words is

$$P(5, 3) + 4 \cdot \frac{3!}{2!} = 5 \cdot 4 \cdot 3 + 4 \cdot 3 = 60 + 12 = 72. \quad \blacktriangle$$

The method used in handling the two c's is a special case of

Theorem 4
PWO

If a collection of n things can be partitioned into r classes such that the first class has n_1 things all alike, the second class has n_2 things all alike, etc., then the number of different permutations of the n things taken n at a time is

(7)
$$\frac{n!}{n_1! n_2! \cdots n_r!}.$$

EXAMPLE 5   Find the number of different permutations of the letters in Tennessee.

Solution: Here $n = 9$, $n_1 = 4$ (for e), $n_2 = 2$ (for n), $n_3 = 2$ (for s), and $n_3 = 1$ (for t). By Theorem 4, the number of permutations is

$$\frac{9!}{4! 2! 2! 1!} = \frac{9 \cdot 8 \cdot 7 \cdot 6 \cdot 5}{2 \cdot 2} = 3780. \quad \blacktriangle$$

Exercise 1

In problems 1 through 6 compute the indicated quantity.

1. $P(7, 3)$ 2. $P(8, 4)$ 3. $P(9, 3)$

4. $P(9, 4)$ 5. $P(8, 5)$ 6. $P(11, 5)$

7. Prove that $P(n, n-1) = P(n, n)$.

8. Solve the equation $P(n, 3) = P(n-1, 4)$.

9. Solve the equation $6P(n, 3) = P(n-1, 4)$.

10. Find the number of four-digit numbers in which each digit is even and there are no zeros.

11. Find the number of five-digit numbers in which each digit is odd.

12. How many fraternity names can be made using the letters of the Greek alphabet (there are 24 letters) if the name consists of 3 letters and repetitions are allowed? How many if repetitions are not allowed?

13. The chief designer for a large automobile company is considering 3 different radiator grilles, 3 different styles of headlights, and 4 different rear fender designs. With respect to these items alone, how many different style cars can be made?

14. A manufacturer makes shirts in 6 different patterns, each in 5 different neck sizes, and each neck size with 3 different sleeve lengths. How many different shirts must a store stock in order to carry a complete line of these shirts?

15. A college freshman is to have a program of four courses. He is to take any one of 4 social science courses, a beginning course in any one of 5 different foreign languages, either one of 2 different mathematics courses, and any one of 5 different physical science courses. How many different programs of study are possible?

16. Having selected his program, the student in problem 15 finds that there are 6 different sections of the social science, 3 different sections of the foreign language, 7 different sections of mathematics, and 4 different sections of the physical science course. Assuming there are no conflicts in time, how many different schedules are possible?

17. License plates in Zambalia consist of 2 letters followed by a dash and 3 digits. If the letter O is not to be used, find the number of different license plates that can be made.

18. Four flags are to be displayed on a vertical flagpole in order to transmit a message from one boat to another by a prearranged code. If the four flags can be selected from 11 different flags, how many different messages can be transmitted?

19. In how many different ways can 6 people be arranged for a group photograph if 3 are in the back row and 3 are in the front row?

20. A Scrabble player with seven different letters on his rack decides to test all possible five-letter arrangements before making his next play. If he tests one arrangement each second, how long will it take before he is ready to play?

21. In how many different ways can a club with 21 members select a president, secretary, and treasurer if no member can hold two offices and each member is eligible for any office?

22. A true-false test has 12 questions. In how many different ways can a student guess at the answers?

★ 23. Find the number of different ways that 5 persons can be seated around a poker table. Two arrangements are considered the same if and only if each person has the same neighbor on his left and the same neighbor on his right.

★ 24. Find the number of ways that 6 persons can line up for a photograph if 2 of them are not on speaking terms and refuse to stand next to each other.

★ 25. In how many different ways can 4 boys and 5 girls enter a restaurant if the boys are gentlemen and let all of the girls go first? In how many different ways if they alternate?

26. Find the number of permutations of the letters in the word arsenic. How many of these end in a vowel? How many end in two vowels?

27. How many three-letter words can be formed using the letters of (a) line, (b) round, (c) hyperbola.

★ 28. How many three-letter words can be formed using the letters of (a) begin, (b) start, (c) Arizona, (d) sleeper.

★ 29. Using factorial notation give the number of permutations of the letters in (a) ellipse, (b) parabola, (c) papaya, (d) Mississippi.

★ 30. Using factorial notation give the number of permutations of the letters in (a) thirteen, (b) parallel, (c) Massachusetts, (d) abracadabra.

31. Find the number of permutations that start with a consonant and end with a vowel using the letters of (a) end, (b) begin, (c) triangle, (d) rectangle.

32. Find the number of different ways in which 5 girls and 8 boys can form couples at a dance if all of the girls have partners.

33. A die (singular of dice) is a cube with one to 6 dots on the 6 faces. On a throw of 2 dice, the number rolled is the sum of the number of dots showing on the 2 top faces. Find the number of different ways of throwing 2 dice and obtaining (a) a 7, (b) an even number, (c) an even number or a 7, (d) a number $S \leqq 5$.

16.3 COMBINATIONS

An arrangement of k things selected from n different things is called a combination, if the order is unimportant. For example, a poker hand is a combination of 5 things (cards) selected from 52 different things because the order in which the cards are held is not important. Thus, a combination is merely a subset.

DEFINITION 4   ***Combination***   *Let* **U** *be a set of n elements and let* **A** *be a subset of* **U***, where* **A** *has k elements. Then* **A** *is a combination of n things (elements) taken k at a time.*

Theorem 5　　Let $C(n, k)$ denote the number of combinations of n things taken k at a time. Then

(8) $$C(n, k) = \frac{n(n-1)(n-2) \cdots (n-k+1)}{k!} = \frac{n!}{k!(n-k)!}.$$

Proof: We can obtain the set of all permutations of n different things taken k at a time in two steps: (1) we select a set **A** of k elements from **U** and (2) we form a permutation of the elements of **A**. The first step can be done in $C(n, k)$ different ways because, by definition, $C(n, k)$ is the number of k-element subsets of **U**. The second step can be done in $k!$ different ways because from equation (6), $P(k, k) = k!$ is the number of different permutations of **A**. Consequently, the number of permutations of n different things taken k at a time is the product of $C(n, k)$ and $k!$ (see Theorem 1)

(9) $$P(n, k) = C(n, k)k!.$$

But $P(n, k) = n!/(n-k)!$ Using this in (9) and dividing both sides by $k!$, we obtain (8). ∎

EXAMPLE 1　　Let **U** be a set of 7 points in the plane with no three on a straight line. Find the number of triangles that have points of **U** as vertices.

Solution: Every selection of 3 points from **U** determines a unique triangle. The order of the points selected is unimportant. Hence, the desired number is the number of combinations of 7 things taken 3 at a time. From equation (8)

$$C(7, 3) = \frac{7!}{3!(7-3)!} = \frac{7!}{3!4!} = \frac{7 \cdot 6 \cdot 5}{3 \cdot 2 \cdot 1} = 35. ▲$$

EXAMPLE 2　　Find the number of different poker hands.

Solution: This is the number of combinations of 52 cards taken 5 at a time. From equation (8)

$$C(52, 5) = \frac{52!}{5!47!} = \frac{52 \cdot 51 \cdot 50 \cdot 49 \cdot 48}{5 \cdot 4 \cdot 3 \cdot 2 \cdot 1} = 2,598,960.$$

EXAMPLE 3　　How many different 3-card hands can be formed (a) that contain a pair, (b) that contain either a pair of 3 of a kind.

Solution: By a pair we mean two cards of the same rank, such as 2 aces, 2 kings, etc. Suppose first that the hand contains 2 aces and a third card that is not an ace. Since there are 4 aces in a standard deck, the 2 aces may be selected in $C(4, 2)$ different ways. The third card may be any one of $52 - 4 = 48$ different cards. Hence, this type of hand can be obtained in $48C(4, 2)$ different ways. Since there are 13 different ranks $2, 3, 4, ...,$ queen, king, ace, the number of 3-card hands with a pair is

$$13(48)C(4, 2) = 13(48)\frac{4 \cdot 3}{2 \cdot 1} = 3744.$$

By three of a kind we mean three cards of the same rank, such as three aces, three kings, etc. The number of 3-card hands with 3 of a kind is $13C(4, 3) = 13(4) = 52$. Hence, the answer to part (b) is $3744 + 52 = 3796$. ▲

Equation (8) gives immediately

Theorem 6
PLE

If $1 \leq k \leq n$, then

(10)
$$C(n, k) = \binom{n}{k}.$$

Exercise 2

In problems 1 through 9 compute the indicated quantity.

1. $C(8, 3)$ 2. $C(9, 6)$ 3. $C(8, 5)$

4. $C(9, 3)$ 5. $C(13, 3)$ 6. $C(11, 7)$

7. $C(13, 10)$ 8. $C(11, 4)$ 9. $C(n, 4)$

10. Problems 1 through 8 illustrate the fact that $C(n, k) = C(n, n-k)$. Prove that this is always true in two different ways: (a) by using equation (8) and (b) by considering the fact that every selection of a set of k elements from a set of n elements leaves a set of $n-k$ elements.

11. How many triangles are formed by 9 lines in the plane if no two lines are parallel and no three lines are concurrent? Note that some triangles may be inside others. Generalize this problem to n lines.

12. Given 11 points in space with no 4 lying in the same plane, how many different planes are there containing 3 of the given points? Generalize this problem to n points.

13. In how many different ways can the Supreme Court give a 6 to 3 decision upholding a lower court? In how many ways can it give a majority decision reversing a lower court?

14. How many different sums of money can be made using a penny, a nickel, a dime, a quarter, a half dollar, and a dollar?

In problems 15 through 20 a hand consists of 3 different cards drawn from a standard deck of 52 cards. Find the number of different hands of the type described.

15. No restrictions or conditions

16. Does not contain a pair

17. Contains a flush (all cards from the same suit)

18. Each card is 5 or lower (ace is high)

19. Each card is 9 or higher.

20. Contains 2 spades and one heart

21. Given a set **U** with 7 elements. Find the number of different subsets of **U**, counting the empty set and **U** itself.

22. If **U** has 8 elements, find the number of subsets of **U** (a) with an odd number of elements, (b) with 2 or fewer elements, (c) with 6 or more elements.

In problems 23 through 27 a poker hand consists of 5 cards from a standard deck. A flush means all cards are from the same suit. A straight means that 5 cards are in a sequence in rank (the ace can be high or low). Find the number of different poker hands of the type described.

23. Three of a kind, and the remaining cards do not form a pair

24. A straight flush (including an ace high flush)

25. A flush that is not a straight flush

26. A full house (3 of a kind together with a pair)

27. Two pairs

28. A coin is tossed 7 times in a row and in each case lands either heads H or tails T. How many different sequences of H's and T's are possible? Find the number of different sequences that contain exactly 3 H's and 4 T's.

29. In how many different ways can a club of 12 boys and 10 girls select a committee of 2 boys and 2 girls?

30. Ten persons enter a chess tournament. In how many different ways can they be paired to play the first round?

31. Given n points on a circle, we draw all the chords with these points as endpoints. If no three chords are concurrent, find the number of intersection points of these chords that lie inside the circle.

32. Suppose that in a multiple choice examination a student must select any one of three possible answers as the correct one. If there are 5 questions, find the number of different ways in which the student can guess at the answers.

33. Suppose that in the examination described in problem 32 each correct answer is worth 20 points. In how many different ways can the student score (a) 60, (b) 60 or better?

16.4 THE BASIC CONCEPTS OF PROBABILITY

We use an example to illustrate the basic concepts and terminology of probability.

EXAMPLE 1   One die is thrown. Assuming that the die is a good one (not loaded) compute the probability of throwing a 4.

Since "probability" has not yet been defined, we can not work the problem, but this example may itself suggest a suitable definition. Throwing a die is an example of an *experiment*. Other experiments are "Draw a card from a standard deck of cards," "Toss a coin," or "Ask a person for his preference in the next election." An experiment may be very complicated, but in this chapter we will consider rather simple ones. Further, an experiment may be compounded from more elementary ones. For example, "Throw a pair of dice 3 times," or "Draw 4 cards from a deck, and then toss a coin" are both examples of a compound experiment.

In any experiment there are a variety of outcomes. When we throw a die the number of dots showing on the top face is an outcome, and this may be any one of the numbers from 1 to 6.

DEFINITION 5   ***Sample Space***   *In any experiment the set* **U** *of all possible outcomes is called the sample space for the experiment. Each individual outcome is an element or point of* **U**. *If* **E** *is any subset of* **U**, *then* **E** *is called an event.*

When we throw a die, the space **U** is the set $\{1, 2, 3, 4, 5, 6\}$. Throwing a 4 is an event. Throwing an odd number is also an event. In this latter case **E** is the subset $\{1, 3, 5\}$.

If the specified event actually occurs, this is called a *success* (or *favorable outcome*). If the event does not occur, this is called a *failure* (or *unfavorable outcome*).

Our task is to assign a suitable number to the event **E** that measures the probability that the event **E** will occur. We must first select a scale, and experience has shown that it is most convenient to insist that the probability of an event is a number between 0 and 1. Further, if two events are equally likely, then the probabilities (numbers) should be the same. If an event $\mathbf{E}_1$ is less likely than another event $\mathbf{E}_2$, then the probability of $\mathbf{E}_1$ should be less than the probability of $\mathbf{E}_2$.

We will use the symbol $P[\mathbf{E}]$ to denote the probability of $\mathbf{E}$ (the number assigned to measure the probability of the event $\mathbf{E}$). Then the above natural requirements are:

(a) For every event $\mathbf{E}$, we have $0 \leqq P[\mathbf{E}] \leqq 1$.
(b) If $\mathbf{E}_1$ and $\mathbf{E}_2$ are equally likely, then $P[\mathbf{E}_1] = P[\mathbf{E}_2]$.
(c) If $\mathbf{E}_1$ is less likely than $\mathbf{E}_2$, then $P[\mathbf{E}_1] < P[\mathbf{E}_2]$.

The following additional requirements also seem natural.

(d) If the event $\mathbf{E}$ cannot occur, then $P[\mathbf{E}] = 0$.
(e) If the event $\mathbf{E}$ must certainly occur, then $P[\mathbf{E}] = 1$.

These five conditions suggest

DEFINITION 6   *If the experiment has n possible outcomes, each equally likely, then the probability of any one outcome is $1/n$. If $\mathbf{E}$ is a set of k different outcomes, then*

(11) $$P[\mathbf{E}] = \frac{number\ of\ outcomes\ in\ \mathbf{E}}{number\ of\ possible\ outcomes} = \frac{n(\mathbf{E})}{n(\mathbf{U})} = \frac{k}{n}.$$

It is difficult to make the phrase "equally likely" precise. When we throw a die it would appear that all of the numbers from 1 to 6 are equally likely. If the die is a "good" die, all tests seem to justify the assumption that all of the numbers are equally likely. However if the die is loaded, then such an assumption is certainly false.

Solution to Example 1:   The phrase "good die" indicates that all numbers from 1 to 6 are equally likely, then Definition 6 is applicable and, hence,‡

$$P[4] = \frac{1}{6}. \quad \blacktriangle$$

EXAMPLE 2   Two cards are drawn from a well-shuffled deck of cards. Compute the probability that both cards are spades.

Solution: The term "well-shuffled" implies that all of the cards are equally likely. Since a standard deck has 52 cards, the number of different ways of drawing 2 cards is $C(52, 2)$. Hence, the sample space $\mathbf{U}$ has $C(52, 2)$ points.

‡ For simplicity we condense $P[$the die shows a 4$]$ to $P[4]$. We will use such simplifications whenever the meaning is clear. Thus, if we are drawing a card, then $P[J]$ means the probability that the card drawn is a jack.

Further, there are 13 spades, so the number of ways of drawing 2 spades is $C(13, 2)$. For this event

(12) $$P[2 \text{ spades}] = \frac{C(13, 2)}{C(52, 2)} = \frac{13 \cdot 12/2 \cdot 1}{52 \cdot 51/2 \cdot 1} = \frac{1}{17}. \quad \blacktriangle$$

Notice that in this example we assumed that the order of drawing the cards is unimportant, and we used $C(n, k)$. We could also solve the problem on the assumption that the order is important. In this case we would have $P[2 \text{ spades}] = P(13, 2)/P(52, 2)$. This differs from equation (12) by a common factor 2 in the numerator and denominator. Hence, we still find $P[\mathbf{E}] = 1/17$.

A variety of terms are used to indicate that the outcomes of an experiment are equally likely. For example, in a single toss of a coin the sample space is $\{H, T\}$. If we say that the coin is a "good coin," a "true coin," or an "unbiased coin," we mean that $P[H] = P[T] = 1/2$. Similar terms are used for throwing several coins or for throwing dice. The phrase "draw a card at random" means that all cards are equally likely.

EXAMPLE 3

Two good dice are thrown. Find the probability of throwing (a) a 13, (b) a number S such that $1 \le S \le 15$, (c) $6 \le S \le 10$.

Solution: (a) Since the sum $S = 13$ is impossible, $P[S = 13] = 0$.
(b) In this case $\mathbf{E}$ is the entire sample space $\mathbf{U}$, so that $k = n$. Then equation (11) gives $P[1 \le S \le 15] = 1$.
(c) Here we must pay close attention to the sample space. At first glance, we might select for $\mathbf{U}$, the set of possible sums $\{2, 3, 4, 5, 6, 7, 8, 9, 10, 11, 12\}$, but a little thought will show that these sums are not all equally likely. When throwing a pair of dice, the dice are usually indistinguishable. However, if we paint one die red and the other green (this will not alter the way they land), we will be able to tell them apart. Thus, the sample space consists of the set of order pairs (x, y), where x is the number of dots showing on the red die (the first one) and y is the number of dots showing on the green die (the second one). These pairs (the points of the sample space) are shown in Table 3. The set $\mathbf{E}$ of points for which $6 \le S \le 10$, is enclosed in a box.

TABLE 3

$(1, 1)$	$(1, 2)$	$(1, 3)$	$(1, 4)$	$(1, 5)$	$(1, 6)$
$(2, 1)$	$(2, 2)$	$(2, 3)$	$(2, 4)$	$(2, 5)$	$(2, 6)$
$(3, 1)$	$(3, 2)$	$(3, 3)$	$(3, 4)$	$(3, 5)$	$(3, 6)$
$(4, 1)$	$(4, 2)$	$(4, 3)$	$(4, 4)$	$(4, 5)$	$(4, 6)$
$(5, 1)$	$(5, 2)$	$(5, 3)$	$(5, 4)$	$(5, 5)$	$(5, 6)$
$(6, 1)$	$(6, 2)$	$(6, 3)$	$(6, 4)$	$(6, 5)$	$(6, 6)$

A direct count of the number of points in **E** gives

$$P[6 \leq S \leq 10] = \frac{k}{n} = \frac{23}{36}. \quad \blacktriangle$$

When the probability of an event is determined by some logical argument, then the number obtained is called an *a priori* probability. This was the case in Examples 1, 2, and 3.

On the other hand, suppose that the dice in Example 3 were loaded (or perhaps magnetized on one side and thrown on a steel table). Clearly, the probability $P[6 \leq S \leq 10] = 23/36$ computed in the solution would have no meaning for an experiment with these modified dice. In such a case, the usual procedure is to rely on observation. The experiment is performed n times (where n is large), and if k is the number of successes actually observed, then the number k/n is often accepted as a reasonable approximation for the probability. When the probability of an event is estimated in this way, by direct observation, the number obtained is called an *a posteriori* probability, a *statistical* probability, or an *empirical* probability. Life expectancy tables used by insurance companies are tables of statistical probabilities.

Exercise 3

In the following problems assume that all outcomes are equally likely. In each problem find $n(\mathbf{U})$, the number of points in the sample space **U**.

1. Two good coins are tossed. Compute the probability of obtaining (a) 2 tails, (b) 1 head and 1 tail, (c) at least 1 head, (d) no heads.

2. Three good coins are tossed. Compute the probability of obtaining (a) 3 heads, (b) 2 heads and 1 tail, (c) 2 heads or more.

3. Two cards are drawn from a well-shuffled deck. Find the probability that (a) both cards are jack or higher (ace is high), (b) both cards are 9 or lower, (c) both cards are tens, (d) the 2 cards form a pair (same rank).

4. Three cards are drawn from a well-shuffled deck. Compute the probability that (a) they form a flush (3 cards from the same suit), (b) they are three of a kind (same rank), (c) they form a pair, but not three of a kind, (d) there is one card from each of three different suits, (e) each card is 10 or higher.

5. Two good dice are thrown. Let S be the sum of the dots showing. Compute (a) $P[S = 7]$, (b) $P[S \leq 4)$, (c) $P[S \geq 10]$, (d) $P[3$ divides $S]$, (e) $P[5$ divides $S]$.

6. Two good dice are thrown. Compute the probability of throwing (a) a double (both dice show the same number), (b) a 4 on one of the dice but not on both, (c) a difference of 3 between the numbers on the 2 dice.

★ 7. Three good dice are thrown. Let S be the sum of the dots showing on the 3 dice. Find (a) $P[S = 4]$, (b) $P[S = 5]$, (c) $P[S = 6]$, (d) $P[S = 16]$.

★ 8. Three good dice are thrown. Find the probability that (a) two of the faces are odd and one is even, (b) two of the faces show the same number and one is different.

★ 9. Nine balls numbered 1 to 9 are placed in urn A. Eight balls numbered 1 to 8 are placed in urn B. Two balls are drawn at random, 1 from each urn. If S is the sum of the numbers on the 2 balls drawn, find (a) $P[S = 7]$, (b) $P[S \geq 15]$, (c) $P[S$ is even$]$, (d) $P[S = 11, 13, $ or $17]$.

★10. Suppose that the two sets of balls from problem 9 are put in the same urn, thoroughly mixed, and two balls are drawn at random without replacement. Find (a) $P[S = 7]$, (b) $P[S \geq 15]$, (c) $P[S$ is even$]$.

11. An urn contains 3 red balls, 5 green balls, and 7 orange balls. Two balls are drawn at random without replacement. Find the probability of drawing (a) 2 red balls, (b) 2 balls of the same color, (c) not a green ball, (d) one red ball and one orange ball.

12. Suppose that in problem 11, one ball is drawn at random, then replaced, and the balls are thoroughly mixed before the second ball is drawn. Compute the probabilities of the events described in problem 11.

★13. From a set of 100 cards numbered 1 to 100, a card is drawn at random, and the number x is noted. Compute the probability that (a) 2 divides x, (b) 5 divides x, (c) both 2 and 5 divide x, (d) either 2 or 5 divides x.

★14. In the experiment of problem 13 compute the probability that (a) x divides 100, (b) x divides 128, (c) x divides 1728, (d) either x divides 64 or x divides 81.

★15. Four cards are drawn at random from a standard deck. Compute the probability of drawing (a) 4 spades, (b) 2 spades and 2 hearts, (c) 4 of a kind, (d) 2 pairs.

★16. A standard deck of cards is modified by removing all jacks, queens, and kings. Four cards are drawn at random from this modified deck. Compute the probabilities of the events described in problem 15.

★★17. An urn contains 5 red balls numbered 1 to 5, 4 green balls numbered 1 to 4, and 3 orange balls numbered 1 to 3. Two balls are drawn at random without replacement. Let S be the sum of the numbers on the 2 balls drawn. Find the probability that (a) both balls are red and S is odd, (b) both balls are red or S is odd, (c) both balls are not red and $S \geq 5$, (d) the two balls drawn are of different color.

★★18. Suppose that in problem 17 one ball is drawn at random, replaced, and the balls are thoroughly mixed before the second ball is drawn. Compute the probabilities of the events described in problem 17.

★ 16.5 PROBABILITY FUNCTIONS

Suppose that for some experiment the various outcomes are not all equally likely. This is certainly the case when we throw a loaded die, or when we are given the top card from a deck that has been "stacked." In any such case it may be difficult to determine $P[\mathbf{E}]$ for the various events. However, the function $P[\mathbf{E}]$ must satisfy certain natural conditions, before it can be used as a measure of probability. Before we give these conditions, we recall that outcomes are regarded as points of a space $\mathbf{U}$, and events are subsets of $\mathbf{U}$. Hence, each operation on sets corresponds to some statement about the possible outcome of an experiment. Some of the simpler set operations together with their corresponding statements are given in Table 4.

TABLE 4		
$\mathbf{A} \cup \mathbf{B}$,	Either $\mathbf{A}$ or $\mathbf{B}$ occurs.	
$\mathbf{A} \cap \mathbf{B}$,	Both $\mathbf{A}$ and $\mathbf{B}$ occur.	
$\mathbf{A} \subset \mathbf{B}$,	If $\mathbf{A}$ occurs, then $\mathbf{B}$ occurs.	
$\mathbf{A}'$,	If $\mathbf{A}'$ occurs, then $\mathbf{A}$ does not occur.	
$\mathbf{A} \cup \mathbf{B} = \mathbf{U}$,	Either $\mathbf{A}$ or $\mathbf{B}$ must occur.	
$\mathbf{A} \cap \mathbf{B} = \varnothing$	$\mathbf{A}$ and $\mathbf{B}$ cannot both occur. If one occurs, the other cannot occur.	

DEFINITION 7   ***Probability Function*** *Let* $\mathbf{U}$ *be a sample space with a finite number of points. A function P defined for all subsets of* $\mathbf{U}$ *is said to be a probability function if it has the following three properties:*

(I) *For every subset* $\mathbf{A}$ *of* $\mathbf{U}$

$$(13) \qquad 0 \leqq P[\mathbf{A}] \leqq 1.$$

(II) *For the universal set*

$$(14) \qquad P[\mathbf{U}] = 1.$$

(III) *If* $A \cap \mathbf{B} = \varnothing$, *then*

$$(15) \qquad P[\mathbf{A} \cup \mathbf{B}] = P[\mathbf{A}] + P[\mathbf{B}].$$

If we define the function P by setting $P[\mathbf{A}] = n(\mathbf{A})/n(\mathbf{U})$, as we did in Definition 6, then we get a probability function. However, there are many other ways of defining $P[\mathbf{A}]$, and in certain cases a different selection for $P[\mathbf{A}]$ may come closer to fitting the facts.

EXAMPLE 1   Prove that if P is a probability function, then $P[\emptyset] = 0$.

Solution: Let $P[\emptyset] = x$. Since $\emptyset \cap \emptyset = \emptyset$, we can apply equation (15) to the identity $\emptyset \cup \emptyset = \emptyset$. We find that

$$P[\emptyset \cup \emptyset] = P[\emptyset] + P[\emptyset],$$
$$x = x + x.$$

Hence, $x = 2x$. Consequently $x = 0$.   ▲

Using the third condition repeatedly it is easy to prove

Theorem 7   *If P is a probability function and $\mathbf{E}_1, \mathbf{E}_2, ..., \mathbf{E}_n$ are pairwise disjoint sets whose*
PLE   *union is $\mathbf{U}$, then*

(16) $$P[\mathbf{E}_1] + P[\mathbf{E}_2] + \cdots + P[\mathbf{E}_n] = 1.$$

If $\mathbf{E}_1, \mathbf{E}_2, ..., \mathbf{E}_k$ are pairwise disjoint sets and $\mathbf{A} = \mathbf{E}_1 \cup \mathbf{E}_2 \cup \cdots \cup \mathbf{E}_k$, then

(17) $$P[\mathbf{A}] = P[\mathbf{E}_1] + P[\mathbf{E}_2] + \cdots + P[\mathbf{E}_k].$$

In particular, if $\mathbf{E}_1, \mathbf{E}_2, .., \mathbf{E}_k$ are distinct points of $\mathbf{U}$ and $\mathbf{A}$ is the set consisting of these points, then equation (17) holds.

Exercise 4

1. A coin is loaded so that $P[H] = 2/3$. Find $P[T]$.

2. A die is loaded so that $P[6] = 1/2$, and all other outcomes are equally likely. Find the probability function for this die. If x is the number thrown compute (a) $P[x$ is even$]$, (b) $P[x$ is odd$]$, (c) $P[x \leq 4]$.

3. Suppose that a die is loaded so that $P[1] = P[3] = P[5] = 1/12$, $P[2] = P[6] = 1/8$, and $P[4] = 1/2$. Find (a) $P[x$ is odd$]$, (b) $P[x$ is even$]$, (c) $P[x$ is prime$]$, and (d) $P[x \neq 5]$.

4. A die is loaded so that the probability of throwing an even number is twice the probability of throwing an odd number. All of the odd numbers are equally likely, and all of the even numbers are equally likely. If this die is thrown once, find (a) $P[x = 2]$, (b) $P[x = 2,$ or $4]$, (c) $P[x = 5$ or $6]$, (d) $P[x \geq 3]$.

5. A die is loaded so that $P[1] = 1/12$, and the probabilities of throwing 1, 2, 3, 4, 5, and 6 form an arithmetic progression. Find (a) $P[x = 3]$, (b) $P[x$ is even$]$, (c) $P[x = 2,$ or $3]$.

6. *A*, *B*, and *C* are running for mayor. The probability that either *A* or *B* wins is 1/2. The probability that either *A* or *C* wins is 5/6. Can you find the probability that *A* wins?

7. *A*, *B*, *C*, and *D* are candidates for governor in an election. A newspaper poll gives *C* and *D* equal chances. Further, *A* is only 1/2 as likely to win as *C*, but *B* is three times more likely to win than *D*. Find the probability of winning for each contestant. What is the probability that either *B* or *C* will win?

In problems 8 through 13 assume that *P* is a probability function and prove the given assertion. *Hint:* A Venn diagram may be helpful

8. $P[\mathbf{A}] = P[\mathbf{A} - \mathbf{B}] + P[\mathbf{A} \cap \mathbf{B}]$

9. $P[\mathbf{A} \cup \mathbf{B}] = P[\mathbf{A}] + P[\mathbf{B}] - P[\mathbf{A} \cap \mathbf{B}]$

10. $P[\mathbf{A} \cup \mathbf{B}] \leqq P[\mathbf{A}] + P[\mathbf{B}]$

11. If $\mathbf{A} \subset \mathbf{B}$, then $P[\mathbf{A}] \leqq P[\mathbf{B}]$

12. $P[\mathbf{A}'] = 1 - P[\mathbf{A}]$

13. $P[\mathbf{A} \cap \mathbf{B}] \leqq P[\mathbf{B}]$

14. State the converse of the assertion in problem 11. Is it always true?

15. If $P[\mathbf{A}] > 1/2$ and $P[\mathbf{B}] > 1/2$, what can you say about $\mathbf{A} \cap \mathbf{B}$?

16.6 INDEPENDENT EVENTS

According to the theory, the probability of throwing a seven with a particular pair of dice on any one throw is always the same, and therefore it is independent of the behavior of the pair of dice in the past. Briefly put, dice have no memory. Now this particular doctrine has often been challenged by skeptics, but long and careful tests have always sustained this principle of independence. To exlpore this more carefully we turn to a much simpler situation in

EXAMPLE 1

Five balls numbered 1 to 5 are placed in an urn. One ball is withdrawn, and the number *x* is recorded. The ball is replaced in the urn. After thoroughly mixing a second ball is drawn, and the number *y* is recorded. Find the probability that $x \leqq 3$ and $y \leqq 2$.

Solution: The sample space is the set of all ordered pairs (x, y), where *x* and *y* are arbitrary integers from 1 to 5. It is convenient to represent these outcomes as points in a rectangular coordinate system. As indicated in Figure 16.1, the event $\mathbf{E}_1$ in which $x = 1, 2$, or 3 is represented by the dots in the shaded rectangle whose base is 3 units. The event $\mathbf{E}_2$ in which $y = 1$ or 2 is represented

FIGURE 16.1

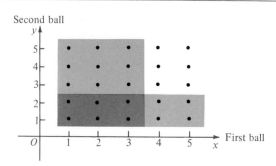

by the dots in the shaded rectangle whose base is 5 units. The event $E_1 \cap E_2$ is represented by the dots in the heavily shaded rectangle; that is, the intersection of these 2 rectangles. The event $E_1 \cap E_2$ is also the set of pairs (x, y) for which $x = 1, 2$, or 3, and $y = 1$ or 2. We are asked to find $P[E_1 \cap E_2]$. Since we selected the balls at random, each outcome is equally likely. A direct count of the dots (or the ordered pairs) gives

(18) $$P[E_1 \cap E_2] = \frac{6}{25}. \quad \blacktriangle$$

We now observe that this probability can be computed in a second way. The event E_1 is the set of all pairs (x, y) with $x = 1, 2$, or 3. A direct count gives

(19) $$P[E_1] = \frac{15}{25} = \frac{3}{5}.$$

Similarly, E_2 is the set of pairs (x, y) with $y = 1$ or 2. Thus,

(20) $$P[E_2] = \frac{10}{25} = \frac{2}{5}.$$

It is important to notice that the right side of (18) is the product of the probabilities computed in (19) and (20). Indeed $6/25 = (3/5)(2/5)$. Hence, in this example

(21) $$P[E_1 \cap E_2] = P[E_1]P[E_2].$$

Although equation (21) was obtained through an example, this equation is true in any case where the experiment is conducted in two steps, and the outcome of the second step is independent of the outcome of the first step.

DEFINITION 8   **Independent Events** *If equation* (21) *is true, then* $\mathbf{E}_1$ *and* $\mathbf{E}_2$ *are said to be independent events.*

This definition can be extended to any number of events. Thus $\mathbf{E}_1, \mathbf{E}_2, ..., \mathbf{E}_n$ are said to be *independent* if

(22) $P[\mathbf{E}_1 \cap \mathbf{E}_2 \cap \cdots \cap \mathbf{E}_n] = P[\mathbf{E}_1] P[\mathbf{E}_2] \cdots P[\mathbf{E}_n].$

EXAMPLE 2   We roll a good pair of dice four times. Compute the probability that we do not obtain a 7 on any roll.

Solution: For a single roll $P[S \neq 7] = 30/36 = 5/6$. Hence, from equation (22) with $n = 4$

(23) $P[S \neq 7 \text{ in four rolls}] = \dfrac{5}{6} \cdot \dfrac{5}{6} \cdot \dfrac{5}{6} \cdot \dfrac{5}{6} = \dfrac{625}{1296}.$ ▲

EXAMPLE 3   We roll a good pair of dice four times. Compute the probability of getting at least one 7.

Solution: Let $\mathbf{E}$ be the event described in Example 2, and let $\mathbf{E}'$ be the complementary event ($\mathbf{E}'$ is the complement of the set $\mathbf{E}$ in the sample space $\mathbf{U}$). Then
$$P[\mathbf{E}] + P[\mathbf{E}'] = P[\mathbf{U}] = 1.$$
Hence,

(24)

$P[\text{At least one 7}] = P[\mathbf{E}'] = 1 - P[\mathbf{E}] = 1 - \dfrac{625}{1296} = \dfrac{671}{1296} > \dfrac{1}{2}.$ ▲

EXAMPLE 4   Two identical urns A and B contain colored balls as follows: A contains 3 red balls and 2 green balls, B contains 1 red ball and 2 green balls. An urn is selected at random, and then a ball is selected at random from that urn. Find $P[R]$ the probability that the ball chosen is red.

Solution: Clearly the probability of drawing a red ball depends on the particular urn selected. Hence, equation (21) cannot be applied directly. However, once the urn is selected then the probability of drawing a red ball from that urn is independent of the choice. We use the notation (A, R) to indicate that urn A was selected, and a red ball was drawn. Then

$$P[(A, R)] = P[A]P[R, \text{from urn } A] = \frac{1}{2} \cdot \frac{3}{5} = \frac{3}{10},$$

$$P[(B, R)] = P[B]P[R, \text{from urn } B] = \frac{1}{2} \cdot \frac{1}{3} = \frac{1}{6}.$$

Since the two events (A, R) and (B, R) are disjoint we add the probabilities. Hence,

$$P[R] = P[(A, R)] + P[(B, R)] = \frac{3}{10} + \frac{1}{6} = \frac{9+5}{30} = \frac{7}{15}. \quad \blacktriangle$$

Observe that if we put all the balls in one urn and selected a ball at random, then

$$P[R] = \frac{3+1}{5+3} = \frac{4}{8} = \frac{1}{2} \neq \frac{7}{15}.$$

16.7 BERNOULLI TRIALS

Any sequence of experiments in which each experiment is the same and is run under the same conditions is called a sequence of *repeated trials*. In such a sequence, the probability function is the same for each experiment. Hence, the outcome of any one experiment is independent of the outcomes of the preceding experiments. Under these circumstances we can apply equation (22), with the understanding that $\mathbf{E}_k$ refers to a certain event on the kth trial, and $\mathbf{E}_1 \cap \mathbf{E}_2 \cap \cdots \cap \mathbf{E}_n$ means that $\mathbf{E}_1$ occurs on the first trial, $\mathbf{E}_2$ occurs on the second trial, ..., and $\mathbf{E}_n$ occurs on the nth trial.

Whenever we have a sequence of repeated trials, each trial may have a number of possible results. For example, in the throw of a pair of dice there are 11 possible sums. But in most applications these results can be divided into 2 disjoint subsets labeled **S** (success) and **F** (failure). In such a case, the trials are called Bernoulli trials.

DEFINITION 9   ***Bernoulli Trials*** *A sequence of Bernoulli trials is a sequence of repeated trials under identical conditions in which on each trial there are only two possible outcomes* **S** *and* **F**.

It is customary to set $P[\mathbf{S}] = p \geq 0$ and $P[\mathbf{F}] = q \geq 0$, where, of course, $p + q = 1$. The central result for Bernoulli trials is

Theorem 8   *Let $b(n, k, p)$ denote the probability of exactly k successes in a sequence of n Bernoulli trials, where $P[\mathbf{S}] = p$. Then*

(25) $$b(n, k, p) = C(n, k) p^k q^{n-k}.$$

Proof: If X is any one sequence such as $(\mathbf{F}, \mathbf{S}, \mathbf{S}, ..., \mathbf{S}, \mathbf{F}, \mathbf{S})$ with exactly k successes and, hence, $n - k$ failures, equation (22) gives

(26) $$P[X] = qpp \cdots pqp = p^k q^{n-k}.$$

In forming X, we can select locations for the k **S**'s in $C(n,k)$ different ways. We have exactly k successes if and only if we have one of the $C(n,k)$ sequences of type X. For each such sequence, $P[X]$ is given by (26). Thus $b(n,k,p)$ is the sum of $C(n,k)$ terms all equal to $p^k q^{n-k}$. ∎

EXAMPLE 1

A student is faced with a true-false examination consisting of 10 questions. He plans on guessing but he estimates that with his knowledge of the material he has a probability of 2/3 of guessing correctly on each question. Find the probability of his guessing at least 7 correct answers. Find the probability for the same result if the guessing is truely random, $P[\mathbf{S}] = 1/2$.

Solution: We regard the 10 consecutive guesses as Bernoulli trials. The probability of exactly k correct answers is

(27) $$b(10, k, 2/3) = C(10, k)\left(\frac{2}{3}\right)^k \left(\frac{1}{3}\right)^{10-k} = C(10, k)\frac{2^k}{3^{10}}.$$

Using equation (27) with $k = 7, 8, 9,$ and 10 we find that

$$P[k \geq 7] = C(10, 7)\frac{2^7}{3^{10}} + C(10, 8)\frac{2^8}{3^{10}} + C(10, 9)\frac{2^9}{3^{10}} + C(10, 10)\frac{2^{10}}{3^{10}}$$

$$= \frac{1}{3^{10}}(120 \cdot 2^7 + 45 \cdot 2^8 + 10 \cdot 2^9 + 2^{10}) = \frac{33024}{59049} \approx 0.559.$$

If $p = 1/2$, the same type of computation gives

$$P[k \geq 7] = \frac{1}{2^{10}}(120 + 45 + 10 + 1) = \frac{176}{2^{10}} = \frac{11}{64} \approx 0.172. \quad \blacktriangle$$

Exercise 5

1. A good coin is tossed 4 times. Compute the probability of obtaining (a) no heads, (b) exactly 3 heads, (c) more heads than tails, (d) the same number of heads and tails.

2. Compute the probabilities of the events described in problem 1 if a good coin is tossed 5 times.

3. Do problem 1 if the coin is loaded so that $P[H] = 2/3$.

4. The coin described in problem 3 is tossed 5 times. Compute the probability of obtaining (a) all heads, (b) more heads than tails, and (c) heads on the last two throws.

5. Two dice are rolled three times. Compute the probability of throwing (a) a 7 exactly once, (b) a 7 at least once, (c) a number $S > 7$ every time, (d) a number $S < 8$ at least once.

6. For the experiment described in problem 5 compute the probability of throwing (a) a number $S \leq 4$ exactly once, (b) a number $S \geq 10$ at least once, (c) $5 \leq S \leq 9$ every time, (d) $S < 5$ or $S > 9$ every time.

7. Five balls numbered 1 to 5 are placed in an urn. Two balls are selected in sequence without replacement (see Example 1 of Section 16.6). Let x be the number on the first ball and y the number on the second ball. Make a diagram for the sample space similar to Figure 16.1. Find (a) $P[x \leq 3]$, (b) $P[y \leq 2]$, (c) $P[x \leq 3$ and $y \leq 2]$, (d) $P[x \leq 3$ or $y \leq 2]$.

8. Are the events (a) and (b) of problem 7 independent?

9. Two evenly matched baseball teams play in the World Series, so that for any one game either team has probability $1/2$ of winning. The series is over as soon as one team wins 4 games. Find the probability that the series is over in just 4 games.

★ 10. For the teams of problem 9 find the probability that all seven games are necessary to decide the world championship.

11. Assuming that in birth either sex is equally probable, find the probability that a family of 6 children will consist of 3 boys and 3 girls.

12. Suppose that the probability that a couple will get a divorce within the next 8 years is $1/5$. Given six couples, selected at random, what is the probability that they will all have the same mates after 8 years?

13. A batting average for a baseball player is a statistical approximation to his probability of getting a hit. It is customary to multiply $P[\text{Hit}]$ by 1,000, so that a 300 hitter is one for which $P[\text{Hit}] = 0.3$. For such a hitter find the probability of getting 3 hits or more in 4 times at bat. Find the probability, if his batting average is 250.

14. Opinion polls can often give erroneous results when a small sample is taken. Suppose that 60% of the population of a town are opposed to a certain action and that the others are for it. If five people are asked for their opinion, what is the probability that this small poll will show that the majority favor the action (contrary to fact).

15. Suppose that an airplane dropping a bomb has probability $1/3$ of hitting the target. Find the smallest number of (identical) airplanes making the same run that are necessary to be at least 95% certain of hitting the target at least once.

16. A student taking mathematics, science, English, and philosophy estimates that his probability of earning an A is 1/5, 2/5, 3/5, and 4/5, respectively. Further, since the four professors involved never talk to each other, the student assumes that the grades can be regarded as independent events. Based on these assumptions, compute the probability that he receives (a) no A's, and (b) exactly one A.

16.8 EXPECTED VALUE

Let $s(n)$ denote the number of times that an event **A** occurs in n repeated trials of a certain experiment. If n is small, then the ratio $s(n)/n$ may be far from $P[\mathbf{A}]$, but if n is large, then we may expect that $s(n)/n$ is close to $P[\mathbf{A}]$. The exact statement is

$$(28) \qquad \lim_{n \to \infty} \frac{s(n)}{n} = P[\mathbf{A}],$$

where the symbols $\lim_{n \to \infty}$ are read, "limit as n approaches infinity of." These symbols merely indicate that as n gets larger and larger, the ratio $s(n)/n$ gets closer and closer to $P[\mathbf{A}]$.

Now, equation (28) is a statement and, hence, has the appearance of a mathematical theorem. However, it is really a statement about the outcome of an infinite set of physical trials. As such, no mathematical proof is possible. Further, we can not check the truth of (28) by running the experiment an infinite number of times (there is not enough time). If we perform the experiment a large number of times, we might hope to discern a pattern and use equation (28) to guess at $P[\mathbf{A}]$. But in any long sequence there might be unusual runs that would lead us to the "wrong" value. Worse still, the left-hand side of (28) may not have a limit. There is a theorem closely related to equation (28) that can be proved, but both the proof and the statement are rather complicated and we will omit them. However, equation (28) can be used as a guide to further topics of interest. One of these is the expected value of an experiment.

To introduce E, the expected value, we adopt the point of view of John Doe playing against the house. The experiment may be regarded as any gambling game (dice, roulette, poker, etc.) in which there are a finite number of mutually exclusive and exhaustive events, $\mathbf{A}_1, \mathbf{A}_2, ..., \mathbf{A}_m$. We suppose that for each event $\mathbf{A}_k$ there is an exchange of $|V_k|$ dollars (or any other convenient monetary unit), where V_k is positive if the house pays J.D., and V_k is negative if J.D. pays the house.

Suppose that in n plays of the game the event $\mathbf{A}_k$ occurs $s_k(n)$ times $k = 1, 2, ..., m$. Then M, the total amount of money exchanged, is given by the sum

$$(29) \qquad M = \sum_{k=1}^{m} V_k s_k(n) = V_1 s_1(n) + V_2 s_2(n) + \cdots + V_m s_m(n).$$

If $M > 0$, then J.D. has won a total of M dollars after playing the game n times. If $M < 0$, then J.D. has lost $|M|$ dollars. Dividing by n gives the average gain (or loss) per game

$$(30) \qquad \frac{M}{n} = V_1 \frac{s_1(n)}{n} + V_2 \frac{s_2(n)}{n} + \cdots + V_m \frac{s_m(n)}{n}.$$

We now let n approach infinity in equation (30). From (28) we can replace each ratio $s_k(n)/n$ by $P[A_k]$. At the same time we replace the ratio M/n by E and obtain

$$(31) \qquad E = V_1 P[A_1] + V_2 P[A_2] + \cdots + V_m P[A_m].$$

The *expected value* E is defined by the right side of equation (31). If $E > 0$, it represents the average gain per game that J.D. may expect if he plays the game for a long time. If $E < 0$, then $|E|$ is the average loss per game that J.D. may expect.

EXAMPLE 1

The game of roulette is played with a wheel containing 37 slots. The slots are numbered with the integers from 0 to 36, inclusive. A player may place a $1 bet (or any other amount) on any number. If the ball falls in the slot bearing that number, the player recovers his bet and receives in addition 35 times the amount wagered. Otherwise, he loses his bet. Using a $1 bet, find the expected value of this game.

Solution: The probability of the ball rolling into any given slot is $1/37$, because (presumably) all of the slots are the same size. Hence, from equation (31) with $m = 2$,

$$(32) \qquad E = 35 \cdot \frac{1}{37} + (-1) \cdot \frac{36}{37} = -\frac{1}{37} \approx -0.027. \quad \blacktriangle$$

This means that if one plays roulette a long time using a $1 bet, his loss will be about 2.7 cents per game. For example, if he plays 370 games he may expect that his loss will be near to $370 \times 1/37 = 10$ dollars.

We have introduced E as a quantity associated with a gambling game, but the expected value can be applied to any function that associates with each event a real number.

EXAMPLE 2

A student estimates that in his mathematics course the probabilities of earning A, B, C, D, and F are 0.20, 0.50, 0.20, 0.10, and 0.00, respectively. These grades contribute to his grade point average, 4, 3, 2, 1, and 0, respectively. Find the expected value of his grade point average for this one course.

Solution: We use equation (31) with $m = 5$. Then

$$E = 4 \cdot \frac{2}{10} + 3 \cdot \frac{5}{10} + 2 \cdot \frac{2}{10} + 1 \cdot \frac{1}{10} + 0 \cdot \frac{0}{10} = \frac{8+15+4+1}{10} = 2.8. \quad \blacktriangle$$

We could take into account his other courses. The principle would be the same, but the computations would be much more involved.

We return to the expected value of a game. We say that the game is *fair* if $E = 0$. The game is *biased* in favor of one player if for that player $E > 0$. In quoting the odds that a certain event will happen, the gambler offering the odds usually quotes two integers. These represent ratios of the bets involved, but the bets may be very large or very small. For simplicity we use one dollar for our unit. In accordance with standard usage, if a gambler states that he will give odds of x to y that an event $\mathbf{A}$ will happen, he means that if $\mathbf{A}$ does occur he will receive y dollars, and that if $\mathbf{A}$ does not occur, he will pay x dollars. These odds favor the gambler if the expected value for the gambler is positive. For the gambler

$$(33) \qquad E = yP[\mathbf{A}] - xP[\mathbf{A}'] = yP[\mathbf{A}] - x(1 - P[\mathbf{A}]).$$

From this equation we see that $E = 0$ if and only if

$$(34) \qquad \frac{x}{y} = \frac{P[\mathbf{A}]}{1 - P[\mathbf{A}]}.$$

DEFINITION 10   ***Fair Bet*** *If a gambler is giving odds of x to y that an event* $\mathbf{A}$ *will occur, he receives y dollars if* $\mathbf{A}$ *does occur, and he pays x dollars if* $\mathbf{A}$ *does not occur. The bet is a fair bet if and only if equation (34) is satisfied. The bet favors the gambler if and only if*

$$(35) \qquad \frac{x}{y} < \frac{P[\mathbf{A}]}{1 - P[\mathbf{A}]}.$$

EXAMPLE 3   A gambler bets that he will draw a heart from a standard deck of cards. What odds should he give for a fair bet?

Solution: $P[\mathbf{A}] = 13/52 = 1/4$ and $P[\mathbf{A}'] = 1 - P[\mathbf{A}] = 1 - 1/4 = 3/4$. Hence,

$$\frac{x}{y} = \frac{1/4}{3/4} = \frac{1}{3}.$$

The gambler should give odds of 1 to 3. $\blacktriangle$

In some games the player receives money for each one of several possible outcomes, but he must pay for the privilege of playing the game. For a *fair game*, the player should pay the expected value of the game each time he plays.

Exercise 6

1. A gambler claims that he can draw a 10 or higher in one draw from a standard deck of cards. What odds should he give for a fair game?

2. Do problem 1 if he claims that he can draw 10 or higher at least once in two successive tries, replacing the first card drawn and shuffling the deck before the second try.

3. A player rolls a die and receives a number of dollars equal to the number of dots showing on the face of the die. What should the player pay each time to play this game in order for the game to be a fair one?

4. In roulette a player may bet on red or black. The numbers from 1 to 36 are evenly divided between these two colors. If a player bets on red and the ball comes to rest in a red slot, he wins the amount he has bet. If the ball comes to rest in a black slot or in zero (green), he loses. Using a $1 bet, find the expected value of this game.

5. A dealer offers to pay $45 to any player who draws the ace of spades from a well-shuffled standard deck of cards. The dealer is willing to pay $5 for any other spade drawn provided the player will pay the dealer $3 if he fails to draw a spade. Find the expected value of this game. Who has the advantage in this game?

6. One popular game played with 3 dice allows the player to select as his point any number x from 1 to 6. If on the roll of 3 dice he gets three x's, he wins $10. For a pair of x's he receives $7. For one x he receives $5. Otherwise he receives nothing. If he pays $3 each time to play the game, find an approximate value for his average gain (or loss) per game if he plays a large number of times.

7. A good coin is tossed 6 times in a row. Find the expected value of the number of tails.

8. Suppose that N tickets are sold for a lottery and that k prizes (1st, 2nd, ..., kth) are to be awarded. If the winning numbers are drawn at random without replacement, prove that for fixed j the probability of winning the jth prize with one ticket is $1/N$.

9. Find the expected value of a lottery ticket, if 1 million tickets were sold and the prizes are 1 prize of $50,000, 3 prizes of $25,000, 10 prizes of $5000, 100 prizes of $100, and 1,000 prizes of $10. Would it be to your advantage to buy tickets at $1 apiece?

10. The Mutual Help Insurance Company carries insurance on 10,000 cars. They estimate that in any one year $P[X]$ the probability of paying a liability claim for X dollars is as follows: $P[50,000] = 0.001$, $P[25,000] = 0.003$, $P[5000] = 0.006$, and $P[100] = 0.05$. Find the expected value of the amount of money to be payed per car. What premium should the drivers pay Mutual Help for a liability policy?

11. A drunk has 6 keys. He tries them at random on his front door, but if a key does not work, he is sober enough not to try it again. Find the expected value of the numbers of keys he must try before the door will open (assuming that only 1 key works).

12. Suppose that on the draw of a card, the dealer will pay $15 for an ace, $10 for a king or queen, and $2 for a jack or a 10. What is the expected value of this game?

In problems 13 through 16, determine the odds for a fair bet on the event described.

13. The roll of a number $S \geq 7$ with a pair of dice

14. The roll of an odd number with a pair of dice

15. The draw of a 7 or better from a standard deck of cards

16. The draw of at least one card that is higher than an 8 when two cards are selected from a standard deck

17. Prove that if the odds are x to y that $\mathbf{E}$ occurs, then $P[\mathbf{E}] = x/(x+y)$.

18. Arsenic, Billy Boy, Careful Colt, and Droopy Dan are entered in a horse race. Just before the race, the odds posted on the board for a win are A, 3 to 2; B, 1 to 4; C, 1 to 9; and D, 1 to 24. Find the probability for each horse to win. Do these probabilities sum to 1?

19. The odds that $\mathbf{A}$ will occur are 1 to 3, and the odds that $\mathbf{B}$ will occur are 1 to 2. If $\mathbf{A} \cap \mathbf{B} = \varnothing$, find $P[\mathbf{A} \cup \mathbf{B}]$.

20. A box contains 2 red balls, 3 green balls, and 5 orange balls. A gambler selects 2 balls at random in sequence (1 after the other without replacement). He wins $50 if he gets 2 red balls, $15 if he gets 2 green balls, and $4 if he gets 2 orange balls. Find the expected value of this game. For a fair game, how much should G pay each time to play this game?

★**16.9 GAMES**

EXAMPLE 1   Compute the probability of drawing some hand of interest on the deal in a poker game.

Solution: The deal consists of 5 cards selected at random from a deck of 52 cards. The probability of drawing each hand of interest is given in Table 5. Many of the items tabulated have already been computed. (See Examples 2 and 3 of Section 16.3 and problems 23 through 27 of Exercise 2.)

TABLE 5	*Rank*	E, *Type of hand*	$n(E)$	$P[E] = n(E)/n(U)$
	1	Royal flush: Ace, king, queen, jack, ten in the same suit	4	0.0000015
	2	Straight flush: Five cards in a sequence, in the same suit, but not a royal flush	36	0.000014
	3	Four of a kind (for example, four jacks or four sevens)	624	0.00024
	4	Full house: Three of a kind together with a pair	3744	0.0014
	5	Flush: Five cards in a single suit, but not a straight	5108	0.0020
	6	Straight: Five cards in a sequence not all in the same suit	10,200	0.0039
	7	Three of a kind	54,912	0.0211
	8	Two pairs	123,552	0.0475
	9	One pair	1,098,240	0.4226
	10	Nothing of interest	1,302,540	0.5012

EXAMPLE 2   The standard game of dice is played according to the following rules. One player throws a pair of dice. If he rolls a 7 or 11 (known as *naturals*) on the first throw, he wins. If on the first throw, he rolls a 2, 3, or 12, he loses. If he rolls any other number, 4, 5, 6, 8, 9, or 10, this number becomes his "*point*", and he must continue to roll until he rolls his "point" again or rolls a 7, whichever occurs first. If his point occurs first, he wins. If the 7 occurs first, he loses. Compute the probability that the one throwing the dice will win. Find the expected value.

Solution: We assume that the dice are not loaded. Then the probability of each event is easy to determine. The results of the computations are given in Table 6. The first four columns refer to the outcome on the first throw, and the last four columns refer to the outcomes on subsequent throws.

TABLE 6

	1	2	3	4	5	6	7	8
	Event on first throw	$P[E]$	$P[L]$	$P[W]$	$P[7]$	$P[point]$	$P[L]$	$P[W]$
	2	$\frac{1}{36}$	$\frac{1}{36}$					
	3	$\frac{1}{18}$	$\frac{1}{18}$					
	4	$\frac{1}{12}$			$\frac{2}{3}$	$\frac{1}{3}$	$\frac{2}{36}$	$\frac{1}{36}$
	5	$\frac{1}{9}$			$\frac{3}{5}$	$\frac{2}{5}$	$\frac{3}{45}$	$\frac{2}{45}$
	6	$\frac{5}{36}$			$\frac{6}{11}$	$\frac{5}{11}$	$\frac{30}{396}$	$\frac{25}{396}$
	7	$\frac{1}{6}$		$\frac{1}{6}$				
	8	$\frac{5}{36}$			$\frac{6}{11}$	$\frac{5}{11}$	$\frac{30}{396}$	$\frac{25}{396}$
	9	$\frac{1}{9}$			$\frac{3}{5}$	$\frac{2}{5}$	$\frac{3}{45}$	$\frac{2}{45}$
	10	$\frac{1}{12}$			$\frac{2}{3}$	$\frac{1}{3}$	$\frac{2}{36}$	$\frac{1}{36}$
	11	$\frac{1}{18}$		$\frac{1}{18}$				
	12	$\frac{1}{36}$	$\frac{1}{36}$					

On any one throw there are 36 possible outcomes. The various sums are listed in column 1, and the probability of each sum is given in column 2. Those that are losses on the first throw are listed in column 3, and those that are wins on the first throw are listed in column 4.

We now consider what happens if the player does not throw a 2, 3, 7, 11, or 12 on his first throw. Suppose, for example, that he rolls a 4 on his first throw. This is now his point, and he continues to throw the dice until either a 4 or a

7 shows. The sequence of rolls before he obtains either a 4 or a 7 may be very long. For example, after the 4 he may throw the sequence 5, 9, 11, 2, 8, 3, 5, 6, 2, 11, 8, 2, 6, 6 .... The analysis of all possible sequences that end in either a 4 or a 7 can be performed, but it is very complicated and requires considerable technique. We can avoid this labor if we are willing to grant that once the point 4 has been established, then only the 4 and the 7 are important and the other sums "do not count" and can be ignored. Since the 4 can be thrown in 3 different ways and the 7 can be thrown in 6 different ways, $P[7] = 6/(3+6) = 2/3$, and $P[4] = 3/(3+6) = 1/3$. These are the entries in columns 5 and 6, respectively, when 4 is the point. The probability of losing via the sequence $(4, 7)$ is the product of the entries in columns 2 and 5. Thus,

$$(36) \qquad P[(4,7)] = \frac{1}{12} \cdot \frac{2}{3} = \frac{2}{36}, \qquad P[(4,4)] = \frac{1}{12} \cdot \frac{1}{3} = \frac{1}{36},$$

and these are the entries in columns 7 and 8, respectively, in the third row of the body of Table 6. All of the other entries in the last 4 columns of Table 6 are obtained in the same way.

Since the probability of winning is the sum of the probabilities of winning in the various ways, we have

$$(37) \qquad P[W] = \frac{1}{6} + \frac{1}{18} + 2\left(\frac{1}{36} + \frac{2}{45} + \frac{25}{396}\right) = \frac{244}{495} < \frac{1}{2}.$$

To compute the expected value we assume that the person rolling the dice receives \$1 if he wins and pays \$1 if he loses. Then

$$(38) \qquad E = 1 \cdot \frac{244}{495} + (-1)\frac{251}{495} = -\frac{7}{495} \approx -0.0141.$$

Consequently, the player who continues to throw the dice for a long time will certainly lose unless he can "control" the dice.

Review Questions

Answer each question as accurately as possible before consulting the text.

1. Give the definition of (a) Cartesian product, (b) ordered k-tuple, (c) permutation, (d) combination.

2. Give a formula for (a) $P(n, k)$, (b) $C(n, k)$, (c) the number of elements in $A \times B \times C$.

3. What is "a priori probability," and what is "a posteriori probability"?

4. If a function P is a probability function, what three conditions must it satisfy?

5. When are two events said to be independent?

6. What is a sequence of Bernoulli trials?

7. Give the formula for computing the expected value. Explain the meaning of the symbols in the formula.

8. Give the formula for the odds for an event. Explain the meaning of the symbols in the formula.

Vectors

Many quantities that occur in nature have both direction and magnitude. A force is such a quantity. It is customary to represent a force by an arrow which points in the direction of the force. The length gives the magnitude of the force. Thus in Figure 17.1, the two arrows could represent forces of 5 and 10 pounds, respectively, acting in the direction indicated by the arrows. The same pair of arrows could also represent the velocities of moving particles, where the particle moving in the direction from C to D has twice the speed of the one moving from A to B.

Initially the theory of vectors was constructed to handle problems involving forces and velocities, and in the first organization of the material, a vector was merely a directed line segment. Later the theory was generalized and refined by the purists until today a vector is defined as any element of a vector space,

FIGURE 17.1

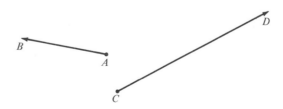

and a vector space is any collection of elements that satisfy a certain set of axioms. In this advanced theory a vector may have no recognizable relation to a directed line segment. For example, the set of all polynomials of degree $n \leq 10$ forms a vector space (of dimension 11), and in this set the polynomial $3 + 7x - \sqrt{2}x^3 + 43x^9$ is a vector.

Since we will consider only the most elementary applications, we prefer to stay with the initial concept of a vector as a directed line segment. The student who has been warned will have no trouble readjusting his definition as he goes further in mathematics.

17.1 THE ADDITION AND DECOMPOSITION OF VECTORS

Following the classical approach we have

DEFINITION 1   *A vector is a directed line segment.*

We indicate a vector by the symbol $\overrightarrow{AB}$, where the vector is the directed line segment from the point A to the point B. The point A is called the *initial point* or the beginning point of the vector. The point B is called the *terminal point* or ending point of the vector. We may also use a single letter with an arrow such as $\vec{F}, \vec{U}, \vec{V}, \ldots$ to indicate a vector, whenever it is convenient.

DEFINITION 2   *Two vectors are equal if and only if they have the same length and the same direction.*

For example, among the vectors shown in Figure 17.2 we see that $\overrightarrow{CD} = \overrightarrow{EF}$. On the other hand, $\overrightarrow{CD} \neq \overrightarrow{GH}$ because the directions are not the same. Further, $\overrightarrow{CD} \neq \overrightarrow{JK}$ because they do not have the same length, although they have the same direction.

FIGURE 17.2

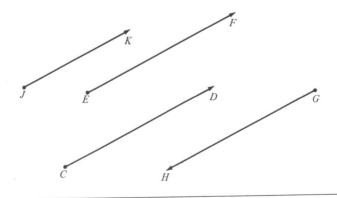

FIGURE 17.3

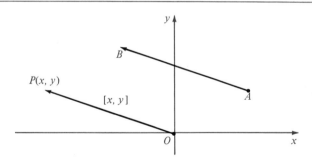

If the initial point and terminal point coincide, then the "vector" has zero length, but the direction is not well defined. Such a "directed line segment" is not covered in Definition 1, but we enlarge our definition to include this exceptional case.

DEFINITION 3   *There is a unique zero vector denoted by* $\vec{0}$. *Further,* $\vec{0} = \overrightarrow{AB}$ *if and only if the points A and B coincide.*

Given any vector $\overrightarrow{AB}$ in the plane there is always a vector $\overrightarrow{OP}$ that has the origin for its initial point and is equal to $\overrightarrow{AB}$ (See Figure 17.3.)

The vector $\overrightarrow{OP}$ is completely specified by the coordinates of its terminal point P. Consequently, we can use the ordered pair $[x, y]$ to denote the vector $\overrightarrow{OP}$. We use brackets for the vector $[x, y]$ rather than parentheses, because logically the point P with coordinates (x, y) is not the same as the vector from O to P.

The length of the line segment $\overrightarrow{AB}$ is called the *magnitude* of the vector $\overrightarrow{AB}$ and is denoted by $|AB|$. Clearly if $\overrightarrow{AB} = \overrightarrow{OP} = [x, y]$, then

(1) $$|AB| = |OP| = r = \sqrt{x^2 + y^2}.$$

If $\overrightarrow{OP}$ is not the zero vector, then the angle θ from the positive x-axis to $\overrightarrow{OP}$ is well defined, and

(2) $$\frac{x}{r} = \cos\theta, \qquad \frac{y}{r} = \sin\theta.$$

Hence,

(3) $$x = r\cos\theta, \qquad y = r\sin\theta.$$

FIGURE 17.4

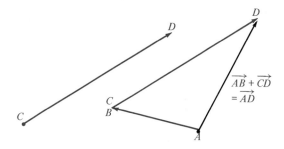

Clearly, these latter equations also hold when $r = 0$. Consequently, the vector $\overrightarrow{OP} = [x, y]$ can also be put in the form $[r\cos\theta, r\sin\theta]$.

DEFINITION 4   *To add the vectors $\overrightarrow{AB}$ and $\overrightarrow{CD}$ we place $\overrightarrow{CD}$ so that the initial point C coincides with B the terminal point of $\overrightarrow{AB}$. Then the sum $\overrightarrow{AB} + \overrightarrow{CD}$ is the vector from A to D.*

This definition is illustrated in Figure 17.4.

A number of theorems flow from this definition, but since our interest lies in the elementary applications to forces and velocities, we will merely state the facts. The addition of vectors is commutative and associative. The zero vector is the identity element for addition: thus,

(4)
$$\vec{0} + \overrightarrow{AB} = \overrightarrow{AB} + \vec{0} = \overrightarrow{AB},$$

for every vector $\overrightarrow{AB}$. Further, each vector $\overrightarrow{AB}$ has a unique additive inverse $\overrightarrow{BA}$ for which

(5)
$$\overrightarrow{AB} + \overrightarrow{BA} = \vec{0}.$$

The definition of the sum of two vectors can be put in a different way that may be more familiar to the reader. Given $\overrightarrow{AB}$ and $\overrightarrow{CD}$, we place the two vectors so that the two initial points coincide. We next construct the parallelogram in which the given vectors are adjacent sides. Then (as indicated in Figure 17.5) the sum $\overrightarrow{AB} + \overrightarrow{CD}$ is the diagonal of the parallelogram directed from the common initial point to the opposite vertex.

The sum of two (or more) vectors is often called the *resultant* of the vectors. Conversely, if we are given a vector, we can represent it as a sum of other vectors, which are then referred to as *components* of the given vector. The precise statement is

FIGURE 17.5

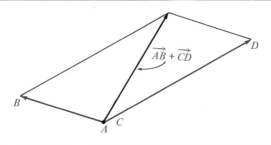

Theorem 1

Let $\overrightarrow{AC}$ be an arbitrary vector in the plane, and let L_1 and L_2 be two nonparallel lines in the same plane. Then there are two uniquely determined vectors $\overrightarrow{AB}$ and $\overrightarrow{BC}$ such that $\overrightarrow{AB}$ is parallel to L_1, $\overrightarrow{BC}$ is parallel to L_2, and

(6) $$\overrightarrow{AC} = \overrightarrow{AB} + \overrightarrow{BC}.$$

The vectors $\overrightarrow{AB}$ and $\overrightarrow{BC}$ are called the *components* in the directions of L_1 and L_2, respectively. The most usual case occurs when L_1 is the x-axis and L_2 is the y-axis, but the theorem is true for any pair of nonparallel lines L_1 and L_2. The process of finding $\overrightarrow{AB}$ and $\overrightarrow{BC}$ is called the *resolution* of $\overrightarrow{AC}$ into its components or the decomposition of $\overrightarrow{AC}$ into its components.

Proof: As illustrated in Figure 17.6, construct lines through A and C parallel to L_1 and lines through A and C parallel to L_2. These four lines form a parallelogram (which may collapse to a single line segment). If the construction gives a proper parallelogram (as shown in Figure 17.6), then any two adjacent sides may be taken as $\overrightarrow{AB}$ and $\overrightarrow{BC}$. The cases in which the parallelogram is not proper are left for the student. ∎

FIGURE 17.6

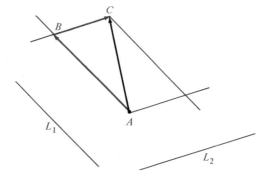

Exercise 1

In problems 1 through 8 the points A and B are given. Plot the points, and draw the vector $\overrightarrow{AB}$. Find $P(x, y)$ such that $\overrightarrow{OP} = \overrightarrow{AB}$. *Hint:* Find the components of $\overrightarrow{AB}$ in the directions of the x-axis and y-axis.

1. $A(1, 2)$, $B(11, 12)$ 2. $A(-2, 5)$, $B(3, 9)$

3. $A(1, 8)$, $B(-4, 17)$ 4. $A(3, 6)$, $B(-2, -8)$

5. $A(-3, 9)$, $B(6, 4)$ 6. $A(2, -8)$, $B(-3, -2)$

7. $A\left(\dfrac{5}{2}, \dfrac{-11}{3}\right)$, $B\left(\dfrac{-11}{2}, \dfrac{4}{3}\right)$ 8. $A\left(\dfrac{7}{5}, \dfrac{19}{4}\right)$, $B\left(\dfrac{22}{5}, \dfrac{-5}{4}\right)$

In problems 9 through 14 find r and θ (where $0° \leqq \theta < 360°$) for the given vector $\overrightarrow{OP}$.

9. $[7, 7]$ 10. $[\sqrt{3}, -1]$

11. $[-\sqrt{2}, \sqrt{6}]$ 12. $[-30, 40]$

13. $[50, -120]$ 14. $[-3, -3]$

In problems 15 through 22 find the vector $\overrightarrow{OP} = [x, y]$ given r and θ.

15. $r = 10$, $\theta = 150°$ 16. $r = 18$, $\theta = 300°$

17. $r = 50$, $\theta = 225°$ 18. $r = 100$, $\theta = 100°$

19. $r = 40$, $\theta = 340°$ 20. $r = 7$, $\theta = 3\pi/2$

21. $r = \sqrt{11}$, $\theta = \pi$ 22. $r = \pi$, $\theta = \pi$

★ 23. Suppose that $\overrightarrow{AB} = [x_1, y_1]$ and $\overrightarrow{CD} = [x_2, y_2]$. Prove geometrically that $\overrightarrow{AB} + \overrightarrow{CD} = [x_1 + x_2, y_1 + y_2]$. In other words, to add two vectors we merely add the corresponding components of the vectors to obtain the components of the sum.

24. Prove that if $\overrightarrow{AB} = [x, y]$, then $\overrightarrow{BA} = [-x, -y]$. The vector $\overrightarrow{BA}$ is often written as $-\overrightarrow{AB}$ or $(-1)\overrightarrow{AB}$. We use this relation to define subtraction. Indeed, the difference $\overrightarrow{AB} - \overrightarrow{CD}$ is defined to be $\overrightarrow{AB} + (-1)\overrightarrow{CD} = \overrightarrow{AB} + \overrightarrow{DC}$.

★ 25. The vector $\overrightarrow{OP}$ from the origin to the point P is called the *position vector* of the point P. Prove that if A and B are any two points then

(7) $$\overrightarrow{AB} = \overrightarrow{OB} + (-1)\overrightarrow{OA} \equiv \overrightarrow{OB} - \overrightarrow{OA}.$$

In words, $\overrightarrow{AB}$ is the position vector of B minus the position vector of A.

17.2 FORCES

It is an *experimental fact* that forces behave like vectors. This means that if a system of forces acts on a particle, that particle will move as though it were acted on by a single force equal to the vector sum of the forces. This single force is called the *resultant* of the system of forces. If the particle does not move, then the resultant is zero. Under these circumstances we say that the particle is in *equilibrium*. We also say that the forces are in equilibrium.

EXAMPLE 1   Find the resultant $\vec{R}$ of two forces, one a 30-pound force acting northward, the other a 40-pound force acting eastward.

FIGURE 17.7

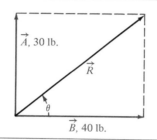

Solution: The resultant is the diagonal of the rectangle shown in Figure 17.7. The magnitude of the resultant is

$$|R| = \sqrt{30^2 + 40^2} = 50 \text{ pounds},$$

and $\tan \theta = 3/4$ so that $\theta = 36° 50'$. Hence $\vec{R}$ is a force of 50 pounds acting in the direction N 53° 10′ E. ▲

EXAMPLE 2   What single force must be added to the system of forces shown in Figure 17.8 to obtain a system that is in equilibrium?

FIGURE 17.8

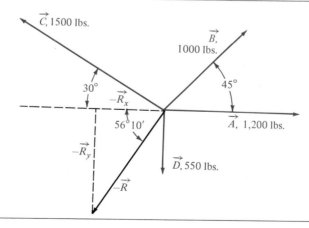

Solution: We first find $\vec{R}$ the resultant of the given system. The simplest procedure is to find the horizontal component (or *x*-component) of each force and the vertical component (or *y*-component) of each force.

Horizontal component (to the nearest pound)

$$A_x = |A|\cos\alpha = 1200(1) \qquad\qquad = \qquad 1200$$
$$B_x = |B|\cos\beta = 1000(0.7071) \qquad = \qquad 707$$
$$C_x = |C|\cos\gamma = 1500(-0.8660) \qquad = -1299$$
$$D_x = |D|\cos\delta = 550(0) \qquad\qquad = \qquad 0$$

$R_x \equiv$ horizontal component of resultant $= \qquad 608.$

Vertical component (to the nearest pound)

$$A_y = |A|\sin\alpha = 1200(0) \qquad\qquad = \qquad 0$$
$$B_y = |B|\sin\beta = 1000(0.7071) \qquad = \qquad 707$$
$$C_y = |C|\sin\gamma = 1500(0.5000) \qquad = \qquad 750$$
$$D_y = |D|\sin\delta = 550(-1) \qquad\qquad = -550$$

$R_y \equiv$ vertical component of resultant $= \qquad 907$

For the resultant, $\theta = \text{Arc tan } 907/608 = \text{Arc tan } 1.492 = 56°\,10'$.

We can compute $|R|$ by using $|R| = \sqrt{R_x^2 + R_y^2}$. If we are using logarithms or a slide rule, then either the relation

$$(8) \qquad\qquad |R| = \frac{R_x}{\cos\theta} = \frac{608}{0.5568} = 1092 \text{ pounds}$$

or the relation

$$(9) \qquad\qquad |R| = \frac{R_y}{\sin\theta} = \frac{907}{0.8307} = 1092 \text{ pounds}$$

is simpler.

The force that must be added to obtain equilibrium is the negative of $\vec{R}$. This is a force of 1,092 pounds and $-\vec{R}$ makes an angle of $180° + 56°\,10' = 236°\,10'$ with the positive *x*-axis. See Figure 17.8 ▲

EXAMPLE 3 A car weighing 4,550 pounds is held on an inclined plane by a tow rope which runs parallel to the plane. If the plane makes an angle of $15°\,40'$ with the horizontal, find the tension in the tow rope. Assume that the plane is smooth.

FIGURE 17.9

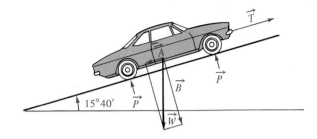

Solution: If the inclined plane is smooth, then the only force that the plane can exert on the car is perpendicular to the plane. This is indicated in Figure 17.9 by the vectors drawn to the wheels of the car. Further, we regard the weight of the car as concentrated at the center of gravity of the car. The force $\vec{W}$ exerted by gravity is then resolved into two components, one $\vec{A}$ parallel to the inclined plane and the other $\vec{B}$ perpendicular to it. Then $\theta = 15° 40'$. (Why?) Since the car does not move, the force $\vec{A}$ is balanced by the pull $\vec{T}$ of the tow rope. Therefore

$$|T| = |W| \sin \theta = 4550 \sin 15° 40' = 4550(0.2700) \approx 1{,}230 \text{ pounds.}$$

The force $\vec{B}$ is balanced by the force exerted by the ramp on the wheels of the car and the magnitude of this force is

$$|W| \cos \theta = 4550 \cos 15° 40' = 4550(0.9628) \approx 4{,}380 \text{ pounds}$$

distributed over the four wheels of the car. ▲

17.3   VELOCITY

Suppose that an airplane headed due east travels with an *airspeed* (speed in still air) of 225 miles per hour. If a wind is blowing from the south at 45 miles per hour, the airplane will be blown off its course. In order to determine its *ground velocity* $\vec{R}$ (its velocity with respect to the ground) we might argue thus: In one hour the airplane would travel 225 miles due east if there is no wind. But during this hour the entire mass of air in which the airplane is moving, has moved 45 miles to the north carrying the plane with it. We would expect that after one hour the plane would be at a point 225 miles east and 45 miles north of its starting point. Then the ground velocity $\vec{R}$ of the airplane would be the vector sum as indicated in Figure 17.10. It is an experimental fact that in such a case the resultant velocity is just the vector sum as indicated in the figure. Therefore,

(10) $$\vec{R} = \vec{V} + \vec{W}.$$

FIGURE 17.10

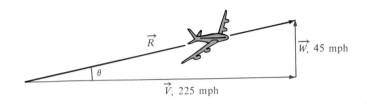

Hence for the ground speed $|R|$ we have

$$|R| = \sqrt{225^2 + 45^2} = \sqrt{52,650} \approx 229 \text{ miles per hour,}$$

and $\tan \theta = 45/225 = 0.2000$. Consequently $\theta = 11° 20'$. Therefore $\vec{R}$ has the bearing N 78° 40′ E.

The same type of argument applies to boats traveling in water, and, in general, to any object traveling in a medium (air or water) that is also moving. Experimental evidence substantiates the following

PRINCIPLE   *If an object travels with the velocity $\vec{V}$ in a medium which is moving with the velocity $\vec{W}$ with respect to the earth, then $\vec{R}$, the velocity of the object with respect to the earth, is the vector sum of $\vec{V}$ and $\vec{W}$.*

EXAMPLE 1   Suppose that the aviator in the preceding discussion wishes to travel due east. In what direction must he head his plane? What will be his *ground speed* (the magnitude of his ground velocity)?

Solution: We again use equation (10) but this time $\vec{R}$ is a vector due east, and the direction of $\vec{V}$ is unknown. The vector diagram is shown in Figure 17.11. We find

$$\sin \alpha = \frac{45}{225} = 0.2000, \qquad \alpha = 11° 30'.$$

$$|R| = \sqrt{225^2 - 45^2} = \sqrt{48,600} \approx 220 \text{ miles per hour.} \quad \blacktriangle$$

FIGURE 17.11

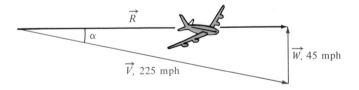

Observe that the angle α is not equal to the angle θ of Figure 17.10. Also notice that in this example the ground speed is less than the air speed, although in the preceding example the ground speed is greater than the air speed.

Exercise 2

In problems 1 through 14 find the forces to the nearest pound and the angles to the nearest 10 minutes.

1. Find the resultant of two forces, one of 25 pounds acting due south and one of 32 pounds acting due west.

2. Find $\vec{R}$ if the second force in problem 1 is doubled.

3. A force of 66 pounds acts in the direction N 66° W. Find its east and north components.

4. Find the east and north components of a 77-pound force acting in the direction S 22° W.

5. Find the resultant of the following system of forces acting simultaneously at a point: $\vec{A}$, 200 pounds in the direction N 36° E; $\vec{B}$, 300 pounds in the direction N 22° W; $\vec{C}$, 400 pounds in the direction S 80° W; $\vec{D}$, 600 pounds in the direction S 25° E.

6. Find the resultant of the following system of forces acting at a point: $\vec{A}$, 200 pounds in the direction N 45° E; $\vec{B}$, 600 pounds in the direction N 70° W; $\vec{C}$, 500 pounds in the direction S 60° W; $\vec{D}$, 300 pounds in the direction S 10° E.

7. Two forces of 200 pounds and 300 pounds act at the origin and make angles of 15° and 35°, respectively, with the positive x-axis. Find the resultant (a) by using components and (b) using the law of cosines.

8. Do problem 7 if the forces have the same magnitude, but the angle each force makes with the x-axis is doubled.

9. A barrel full of water is held on an inclined plane by a man who pushes on it parallel to the plane. If the angle of inclination of the plane is 10° 10′ and the barrel together with its contents weight 550 pounds, what force does the man exert? With what force does the barrel press on the supporting plane?

10. Solve problem 9 if the barrel is empty so that the weight is only 135 pounds, and the angle of inclination of the plane is increased to 22° 20′.

11. If a young boy can push on the barrel with a force of at most 76 pounds, find the weight of the heaviest barrel he can just hold on the inclined plane of problem 10:

12. Solve problem 11 if the angle of inclination of the plane is 7° 30′.

13. A tightrope walker weighing 165 pounds is balanced on a wire ABC at point B midway between the support points A and C. If these supports are at the same height and if each of the lines AB and BC makes an angle of 24° 20′ with the horizontal, find the tension in the wire.

14. The wire in problem 13 is tightened so that under the artist's weight the tension is now 1,500 pounds. Find the angle that AB and BC make with the horizontal.

In problems 15 through 20 find the speed to the nearest mile per hour and the angle to the nearest 10 minutes.

15. A 30 miles per hour wind is blowing from the south. At what speed should a plane head due west to be on course N 80° 50′ W? Find the ground speed of the plane.

16. Suppose that the pilot in problem 15 flies with an airspeed of 240 miles per hour. In what direction should the pilot head his plane if he wishes to travel due west? Find his ground speed.

17. A pilot flies with an airspeed of 450 miles per hour heading N 44° W. The wind is blowing in the direction N 15° E. If the resulting course is N 40° W find the speed of the wind and the ground speed of the airplane.

18. A pilot wishes to average 300 miles per hour in the direction S 12° 10′ E, and he decides to fly on a heading S 10° E. Find the airspeed the pilot must maintain and the speed of the wind.

19. A pilot flying with an airspeed of 200 miles per hour in the direction N 7° E finds that after one hour he has actually traveled 250 miles in the direction N 11° W. Find the velocity of the wind.

20. A pilot flying with an airspeed of 300 miles per hour in the direction N 28° E, is in a 60 miles per hour wind that is headed S 34° E. Find his ground velocity.

★21. A weight of 1,000 pounds is supported by two cables that make angles of 30° and 40° with the horizontal. Find the tension in each cable.

★22. Do problem 21 if the angles are 20° and 60° for the two cables.

★23. Do problem 21 if the angles are 10° and 80° for the two cables.

★★24. Suppose that in problem 21, the cables make angles θ_1 and θ_2 with the horizontal, where $0° < \theta_1 < 90°$ and $0° < \theta_2 < 90°$. Find formulas for the tension in the two cables. Note that these formulas solve problems 21, 22, and 23 (and many others) in one stroke.

Review Questions

Try to answer the following questions as accurately as possible before consulting the text.

1. What do we mean by the magnitude of a vector?

2. Give one definition for the sum of two vectors.

3. State the theorem about the resolution of a plane vector into components.

4. Give a geometric picture for the sum of three vectors.

5. What is the definition of $\overrightarrow{AB} - \overrightarrow{CD}$?

6. Give a geometric picture for subtraction.

7. What is the resultant of a system of forces?

8. Describe a method for finding the resultant of a system of forces (a) geometrically and (b) analytically.

Complex Numbers

18

If x is any real number, then $x^2 \geq 0$. Consequently, the equation $x^2 + 1 = 0$ has no root in the set of real numbers (has no real root). This is not an isolated example. We recall from Theorem 3 in Chapter 5 that if the quadratic equation

$$(1) \qquad\qquad ax^2 + bx + c = 0, \qquad a \neq 0,$$

has a real root, then it may have at most two real roots and these are given by the formula

$$(2) \qquad\qquad r = \frac{-b \pm \sqrt{b^2 - 4ac}}{2a}\,.$$

Hence, if $b^2 - 4ac$ is negative, then equation (1) does not have any real roots. However, our desire for order and beauty in mathematics leads us to demand that each second-degree equation has two roots.

In this chapter we introduce a new symbol i for $\sqrt{-1}$ and make certain definitions regarding the algebraic properties of this new number. With this new symbol the real number system $\mathbf{R}$ can be enlarged to form a new system $\mathbf{C}$ of complex numbers. In this enlarged system every second-degree equation (1) has two distinct roots if $b^2 - 4ac \neq 0$ and has a twice repeated root if $b^2 - 4ac = 0$. In short, a second degree equation always has two roots, and these roots are given by the quadratic formula (2).

449

Our objective is to study the elementary algebra of the enlarged number system **C**, obtained by adjoining $i = \sqrt{-1}$ to the system **R** of real numbers. The letter i is used because it is the first letter of the word "imaginary." To dispel the feeling that numbers involving i are imaginary, many careful writers prefer to suppress i and to introduce complex numbers as an ordered pair of real numbers. In this treatment, the number i is eventually readmitted to gracious society after the logical foundations have been adequately treated.

In Section 18.1 we use the symbol i without any apology for the suspect nature of its origin. In Section 18.2 we introduce complex numbers as an ordered pair (a, b) of real numbers. In due time i appears as the ordered pair $(0, 1)$.

Although i is called an imaginary number, it is no more a creation of the imagination than any other number used in mathematics. Moreover, the use of i in mathematics has very important, very real, and very practical applications.

18.1 THE INTUITIVE CONCEPT OF A COMPLEX NUMBER

We let $i \equiv \sqrt{-1}$. Thus, i is a number such that

$$(3) \qquad\qquad i^2 = -1.$$

DEFINITION 1 *A number of the form $a + bi$, where a and b are real numbers, is called a complex number. The set **C** of all such numbers is called the set of complex numbers or the complex number system.*

It is convenient to use a single letter such as z or w to denote a complex number.

DEFINITION 2 *If $z = a + bi$ is a complex number, than a is called the real part of z, and b is called the imaginary part of z. If $a = 0$, z is called an imaginary number. If $b = 0$, z is called a real number.*

Thus, if we identify the real number a and the complex number $z = a + 0i$, we see that the set of real numbers is contained in the set of complex numbers.

DEFINITION 3 ***Equality*** *Two complex numbers $z_1 = a_1 + b_1 i$ and $z_2 = a_2 + b_2 i$ are equal if and only if*

$$(4) \qquad\qquad a_1 = a_2 \quad \text{and} \quad b_1 = b_2.$$

Thus $z_1 = z_2$ if and only if their real parts are equal ($a_1 = a_2$) and their imaginary parts are equal ($b_1 = b_2$).

DEFINITION 4

Addition *If $z_1 = a_1 + b_1 i$ and $z_2 = a_2 + b_2 i$, the sum $z_1 + z_2$ is defined by*

$$z_1 + z_2 = (a_1 + a_2) + (b_1 + b_2)i.$$

Thus to add two complex numbers, we add their real parts to obtain the real part of $z_1 + z_2$ and we add their imaginary parts to obtain the imaginary part of $z_1 + z_2$.

It should be observed that $a - bi$ is merely a brief way of writing $a + (-b)i$. For example, if $z_1 = 3 + 5i$ and $z_2 = 7 - 2i = 7 + (-2)i$, then by Definition 4 we have

$$\cdot \quad z_1 + z_2 = (3 + 5i) + (7 - 2i) = (3 + 7) + \big(5 + (-2)\big)i = 10 + 3i.$$

Theorem 1
PLE

Let z_1, z_2, and z_3 be arbitrary complex numbers. Then

(5) $z_1 + z_2 = z_2 + z_1$,   *(Commutative Law of Addition),*

and

(6) $(z_1 + z_2) + z_3 = z_1 + (z_2 + z_3)$,   *(Associative Law of Addition).*

The proof of this theorem follows immediately from the corresponding laws for the addition of real numbers.

In search of an identity element for addition we observe that

$$(0 + 0i) + (a + bi) = (0 + a) + (0 + b)i = a + bi.$$

This leads to

DEFINITION 5

The complex number $0 + 0i$ is called zero. It may be denoted briefly by 0.

Theorem 2
PLE

For every complex number z

$\cdot$ (7) $0 + z = z + 0 = z.$

Conversely, if $w + z = z$, then $w = 0$.

Every z has a unique additive inverse denoted by $-z$ or $(-z)$ such that

(8) $z + (-z) = 0$

Indeed, if $z = a + bi$, we take for $-z$ the complex number $-a + (-b)i$. Then equation (8) becomes

$$a + bi + \big(-a + (-b)i\big) = (a - a) + (b - b)i = 0 + 0i = 0.$$

Conversely, if $z + w = 0$, then $w = -z = -a + (-b)i$.

To multiply two complex numbers we assume that the ordinary laws of multiplication hold and that in addition $i \times i = i^2 = -1$. Thus,

$$(3+5i)(7-2i) = 3 \times 7 + 5i \times 7 + 3 \times (-2i) + 5i \times (-2i)$$
$$= 21 + 35i - 6i - 10i^2 = 21 + 35i - 6i + 10 = 31 + 29i.$$

This example suggests

DEFINITION 6   **Multiplication**   *If $z_1 = a_1 + b_1 i$ and $z_2 = a_2 + b_2 i$, then the product $z_1 z_2$ is defined by the equation*

(9) $$z_1 z_2 = a_1 a_2 - b_1 b_2 + (a_1 b_2 + b_1 a_2)i.$$

Theorem 3
PLE

Let z_1, z_2, and z_3 be arbitrary complex numbers. Then

(10) $z_1 z_2 = z_2 z_1$,   (*Commutative Law of Multiplication*),

(11) $(z_1 z_2)z_3 = z_1(z_2 z_3)$,   (*Associative Law of Multiplication*),

and

(12) $z_1(z_2 + z_3) = z_1 z_2 + z_1 z_3$,   (*Distributive Law*).

As an indication of the method used, we prove that (12) always holds. Let $z_1 = a_1 + b_1 i$, $z_2 = a_2 + b_2 i$, and $z_3 = a_3 + b_3 i$. Then for L, the left side of (12), we have

(13) $L = (a_1 + b_1 i)(a_2 + b_2 i + a_3 + b_3 i)$
$$= (a_1 + b_1 i)[a_2 + a_3 + (b_2 + b_3)i]$$
$$= a_1(a_2 + a_3) - b_1(b_2 + b_3) + [a_1(b_2 + b_3) + b_1(a_2 + a_3)i].$$

Since a_1, a_2, a_3, b_1, b_2, and b_3 are real numbers, we have

$$L = a_1 a_2 + a_1 a_3 - b_1 b_2 - b_1 b_3 + [a_1 b_2 + a_1 b_3 + b_1 a_2 + b_1 a_3]i$$
$$= [a_1 a_2 - b_1 b_2 + (a_1 b_2 + b_1 a_2)i] + [a_1 a_3 - b_1 b_3 + (a_1 b_3 + b_1 a_3)i]$$
$$= (a_1 + b_1 i)(a_2 + b_2 i) + (a_1 + b_1 i)(a_3 + b_3 i) = z_1 z_2 + z_1 z_3.$$

This last term is the right side of (12).

Suppose that in the product zw, the first factor z is real. To be specific, let $z = a + 0i$ and $w = c + di$. Then

$$zw = (a+0i)(c+di) = (ac - 0d) + (ad + 0c)i$$
$$= ac + adi.$$

Now a is a common factor of both terms, so it seems reasonable to write

(14) $$a(c+di) = ac + adi.$$

In fact, we define the left side of (14) to be equal to the right side of (14). As a result we see that the product of the real number a and the complex number $c+di$ is the same whether we use the defining equation (14) or Definition 6 (equation (9)).

Theorems 1, 2, and 3 are rather routine and were to be expected. A more interesting and, perhaps, more surprising result is contained in

Theorem 4   If $z = a+bi \neq 0$, then z has a multiplicative inverse (reciprocal) denoted by z^{-1} or $1/z$ such that

(15) $$zz^{-1} = 1.$$

Proof: Since $z \neq 0$, it follows that either $a \neq 0$ or $b \neq 0$ and hence $a^2 + b^2 > 0$. Set

(16) $$z^{-1} = \frac{a}{a^2+b^2} + \frac{-b}{a^2+b^2} i \equiv \frac{1}{a^2+b^2}(a-bi).$$

Then, by Definition 6

$$zz^{-1} = (a+bi)\left(\frac{a}{a^2+b^2} + \frac{-b}{a^2+b^2}i\right)$$

$$= \left(\frac{a^2}{a^2+b^2} - \frac{-b^2}{a^2+b^2}\right) + \left(\frac{-ab}{a^2+b^2} + \frac{ba}{a^2+b^2}\right)i$$

$$= \frac{a^2+b^2}{a^2+b^2} + 0i = 1. \quad \blacksquare$$

EXAMPLE 1   Find the reciprocal of (a) $z_1 = 7-3i$, (b) $z_2 = -\dfrac{1}{2}+4i$.

Solution: By equation (16)

(a) $$z_1^{-1} = \frac{7}{7^2+3^2} + \frac{-(-3)}{7^2+3^2}i = \frac{7}{58} + \frac{3}{58}i = \frac{1}{58}(7+3i),$$

(b) $$z_2^{-1} = \frac{-1/2}{(1/2)^2+4^2} + \frac{-4}{(1/2)^2+4^2}i = \frac{-2}{65} - \frac{16}{65}i = \frac{-2}{65}(1+8i). \quad \blacktriangle$$

DEFINITION 7

Conjugate   *If $z = a + bi$ and $w = a - bi$, then w is called the conjugate of z.*

The conjugate of z is often indicated by placing a bar over z. Thus,

$$\bar{z} = \overline{a + bi} \equiv a - bi = w.$$

If w is the conjugate of z, then for the conjugate of w we have

$$\bar{w} = \overline{a - bi} = a - (-b)i = a + bi = z.$$

Hence z is the conjugate of w. It is worth noting that

$$z\bar{z} = (a + bi)(a - bi) = a^2 + b^2 + (-ab + ba)i$$

(17)
$$z\bar{z} = a^2 + b^2 + 0i = a^2 + b^2.$$

Consequently, $z\bar{z} \geqq 0$ and $z\bar{z} = 0$ if and only if $z = 0$.

This new symbolism allows us to condense the formula for the reciprocal of $z = a + bi$. Indeed, if $z \neq 0$, then

(18)
$$z^{-1} \equiv \frac{1}{z} = \frac{\bar{z}}{z\bar{z}} = \frac{a - bi}{a^2 + b^2} = \frac{a}{a^2 + b^2} - \frac{b}{a^2 + b^2}i.$$

DEFINITION 8

Division   *If $w \neq 0$, the quotient z/w is defined by*

(19)
$$\frac{z}{w} = z\left(\frac{1}{w}\right).$$

In other words, to divide z by w we merely multiply z by the reciprocal of w. In practice, one uses the conjugate of w. Suppose that, as usual, $z = a + bi$ and $w = c + di$. Then

$$\frac{z}{w} = z\left(\frac{1}{w}\right)\frac{\bar{w}}{\bar{w}} = (a + bi)\frac{c - di}{(c + di)(c - di)}$$

(20)
$$\frac{z}{w} = \frac{(a + bi)(c - di)}{c^2 + d^2} = \frac{ac + bd}{c^2 + d^2} + \frac{bc - ad}{c^2 + d^2}i.$$

EXAMPLE 2

Compute $(7 - 2i)/(3 + 5i)$.

Solution: Following the procedure indicated in equation (20) we have

$$\frac{7 - 2i}{3 + 5i} = \frac{7 - 2i}{3 + 5i} \cdot \frac{3 - 5i}{3 - 5i} = \frac{21 - 10 + (-35 - 6)i}{9 + 25}$$

$$= \frac{11 - 41i}{34} = \frac{11}{34} - \frac{41}{34}i. \quad \blacktriangle$$

Theorem 5   If $zw = 0$, then either $z = 0$ or $w = 0$.

Proof: If $z \neq 0$, then it has a reciprocal $1/z$. Hence

$$0 = \frac{1}{z} \cdot 0 = \frac{1}{z}(zw) = \left(\frac{1}{z} \cdot z\right)w = 1 \cdot w = w. \quad \blacksquare$$

Theorem 6   ***The Cancellation Law***   *If $wz_1 = wz_2$ and $w \neq 0$, then $z_1 = z_2$.*
PLE

Exercise 1

In problems 1 through 10 compute $z_1 + z_2$, $z_1 z_2$, and z_1/z_2.

1. $z_1 = 4 + 3i$, $z_2 = 5 + 12i$ 2. $z_1 = 4 + i$, $z_2 = 1 - 2i$

3. $z_1 = 5 - 12i$, $z_2 = 3 - 4i$ 4. $z_1 = -1 + i$, $z_2 = 1 - i$

5. $z_1 = -2 + 3i$, $z_2 = 2 + 3i$ 6. $z_1 = 5 + 5i$, $z_2 = 2 - 2i$

7. $z_1 = 4 + 3i$, $z_2 = 2i$ 8. $z_1 = -2 - 4i$, $z_2 = -4 + 2i$

9. $z_1 = -3 + 4i$, $z_2 = -5i$ 10. $z_1 = -7 + i$, $z_2 = -1 + 7i$

11. Compute the powers i^3, i^4, i^8, $(1+i)^4$, and $(1+i)^6$.

12. Given $z_1 = 1 + 2i$, $z_2 = 3 - i$, and $z_3 = 2 + 5i$, find (a) $(z_1 z_2)z_3$, (b) $z_1(z_2 z_3)$, and (c) $(z_3 z_1)z_2$.

13. Complete the proof of Theorem 3.

14. Prove that $1 \cdot z = z \cdot 1 = z$.

15. Theorem 4 shows that if $z \neq 0$, then z has a multiplicative inverse. Prove that this inverse is unique.

16. For the conjugate operation prove that

    (a) $\overline{z_1 + z_2} = \bar{z}_1 + \bar{z}_2$,   (b) $\overline{z_1 z_2} = \bar{z}_1 \bar{z}_2$,

    (c) $\overline{(z^n)} = (\bar{z})^n$,   (d) $\bar{\bar{z}} = z$,

    (e) $z = \bar{z}$ if and only if z is real,   (f) $z + \bar{z}$ is a real number,

    (g) $z - \bar{z}$ is an imaginary number.

17. If n is a positive integer, find a formula for (a) i^{4n}, (b) i^{4n+1}, (c) i^{4n+2}, (d) i^{4n+3}.

18. Use equation (20) to prove that if $w \neq 0$, then

$$w \cdot \left(\frac{z}{w}\right) = z.$$

19. Use the quadratic formula to find solutions of $x^2 - 4ix + 2 = 0$. Do the numbers obtained really satisfy this equation?

★20. Do problem 19 for $x^2 + (4+i)x + 2i = 0$.

21. Prove Theorem 6. *Hint:* Multiply both sides of the equation by w^{-1}.

★18.2   COMPLEX NUMBERS AS
ORDERED PAIRS

Now that we have examined the algebra of numbers of the form $a + bi$, we suppress the imaginary symbol i and use only pairs of real numbers (a, b) to form the complex number z.

DEFINITION 9
The system of complex numbers $\mathbf{C}^\star$ *is the set of all ordered pairs* (a, b) *where* a *and* b *are real numbers. We say that two complex numbers* (a, b) *and* (c, d) *are equal, and we write*

(21)
$$(a, b) = (c, d)$$

if and only if

(22)
$$a = c \quad \text{and} \quad b = d.$$

Further, addition and multiplication in $\mathbf{C}^\star$ *are defined by the two equations*

(23)
$$(a, b) + (c, d) \equiv (a+c, b+d)$$

and

(24)
$$(a, b)(c, d) \equiv (ac - bd, ad + bc).$$

The subset of all numbers in $\mathbf{C}^\star$ for which the second component b is zero forms a subset which can be identified with the real numbers. For example, in multiplication (the only difficult item)

$$(a, 0)(c, 0) = (ac - 0 \cdot 0, a0 + 0c) = (ac, 0),$$

so that it is parallel to the product of the real numbers $ac = ac$.

In the new system we have

(25)
$$(0, 1)(0, 1) = (0 \cdot 0 - 1 \cdot 1, 0 \cdot 1 + 1 \cdot 0)$$
$$= (-1, 0),$$

so that the number $(0, 1)$ behaves like $\sqrt{-1}$.

We have used the symbol $\mathbf{C}^\star$ for the set described in Definition 9 in order to distinguish it from the set $\mathbf{C}$ of numbers of the form $a + bi$. Indeed the two sets are different, but this difference is only superficial. Any theorem, definition, problem, or counterexample in either system can be transformed into an item with exactly the same status in the other system. One is tempted to say that $\mathbf{C}$ and $\mathbf{C}^\star$ are identical, but this is inaccurate. The correct phrase is "$\mathbf{C}$ and $\mathbf{C}^\star$ are *isomorphic* with respect to both addition and multiplication." Hence $\mathbf{C}$ and $\mathbf{C}^\star$ are isomorphic fields. To prove that $\mathbf{C}$ and $\mathbf{C}^\star$ are isomorphic we must show that there is a one-to-one correspondence (indicated by $\longleftrightarrow$) between the elements of $\mathbf{C}$ and $\mathbf{C}^\star$ and this correspondence must be preserved under addition and multiplication. We will avoid the details but merely note the important correspondences in the following table.

TABLE 1	$\mathbf{C}$	$\mathbf{C}^\star$		$\mathbf{C}$	$\mathbf{C}^\star$

$$a + bi \longleftrightarrow (a, b), \qquad\qquad (a + bi) + (c + di) \longleftrightarrow (a, b) + (c, d),$$
$$1 = 1 + 0i \longleftrightarrow (1, 0), \qquad\qquad (a + bi)(c + di) \longleftrightarrow (a, b)(c, d),$$
$$i = 0 + i \longleftrightarrow (0, 1). \qquad = ac - bd + (ad + bc)i \longleftrightarrow = (ac - bd,\ ad + bc),$$
$$i^2 = -1 \longleftrightarrow (0, 1)(0, 1) = (-1, 0).$$

18.3 GEOMETRIC REPRESENTATION OF COMPLEX NUMBERS

If we write the complex number z in the form $z = x + iy$, where x and y are real numbers, it is natural to represent the complex number z by the point with coordinates (x, y). Thus, in Figure 18.1 the points A, B, C, and D represent the

FIGURE 18.1

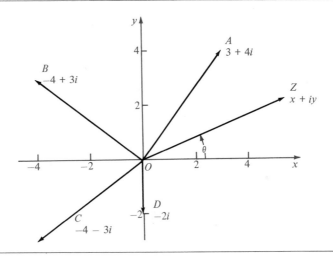

complex numbers $3+4i$, $-4+3i$, $-4-3i$, and $-2i$, respectively. When complex numbers are represented in this way by points on a plane with a rectangular coordinate system, the plane is called the *complex plane*. The horizontal or *x*-axis is called the *real axis*, because the real part of z is plotted along this axis. Similarly, the vertical or *y*-axis is called the *imaginary axis*. The complex plane is often called the *Argand plane* after the French mathematician Jean Argand (1768–1822).

SUMMARY

There is a one-to-one correspondence between the complex numbers **C** *and the points in the complex plane. Each number* $z = x+iy$ *corresponds to the unique point Z with coordinates* (x, y). *Conversely, each point* $Z(x, y)$ *corresponds to the unique complex number* $z = x+iy$.

We can also associate with each complex number a vector from the origin to the point that represents the complex number. For example, in Figure 18.1 the vector $\overrightarrow{OA}$ represents the complex number $3+4i$, the vector $\overrightarrow{OB}$ represents $-4+3i$, $\overrightarrow{OC}$ represents $-4-3i$, and $\overrightarrow{OD}$ represents $-2i$. In general the vector $\overrightarrow{OZ}$ from the origin to the point (x, y) represents the complex number $x+iy$. The length of this vector is called the *modulus* or the *absolute value* of the complex number $x+iy$, and is denoted by $|z|$ or $|x+iy|$. Obviously we have

(26) $$|z| = r = \sqrt{x^2 + y^2}.$$

If $z \neq 0$, then any angle θ from the positive *x*-axis to the vector $\overrightarrow{OZ}$ is called an *amplitude* of z or an *argument* of z and is denoted by amp z or arg z. It is not uniquely determined by the pair of numbers (x, y) but will be uniquely determined if we require that $0° \leq \theta < 360°$. In this case we can refer to θ as *the* amplitude or *the* argument rather than as an amplitude or an argument. If θ is an argument of $x+iy$, then from Figure 18.1 it is clear that

(27) $$\cos \theta = \frac{x}{r}, \quad \text{and} \quad \sin \theta = \frac{y}{r}.$$

From equation set (27) we have

(28) $$x = r\cos \theta, \quad \text{and} \quad y = r\sin \theta.$$

Hence,

(29) $$z = r\cos \theta + ir\sin \theta,$$

(30) $$z = r(\cos \theta + i\sin \theta).$$

Equation (30) is called the *polar form* of the complex number, in contrast to $z = x + iy$ which is the *rectangular form* of the complex number. In (30), r is the absolute value of z, and θ is any argument of z. If θ_0 denotes the argument of z ($0° \leq \theta_0 < 360°$), then there is some integer k such that $\theta = \theta_0 + 360°k$.

Finally, we note that the polar form of a complex number is used so frequently in certain branches of applied mathematics that still further abbreviation is useful, and one frequently finds r cis θ as a condensed version of $r(\cos\theta + i\sin\theta)$.

EXAMPLE 1   Find the modulus and an argument of $z = 1 + \sqrt{3}\,i$. Put z in polar form.

Solution: Clearly $r = \sqrt{1+3} = 2$. Since $\tan\theta = \sqrt{3}/1$ and the point Z is in Q.I, we have $\theta_0 = 60°$. Hence, $1 + \sqrt{3}\,i = 2(\cos 60° + i\sin 60°)$.   ▲

Exercise 2

1. For each of the complex numbers given in problems 1 through 5 of Exercise 1, find the corresponding point in the complex plane and draw the corresponding vector.

2. Find $|z_1|$ and $\arg z_1$ (to the nearest 10 minutes) for the complex number z_1 in problems 1 through 5 of Exercise 1.

3. Find $|z_1|$ and $\arg z_1$ (to the nearest 10 minutes) for the complex number z_1 in problems 6 through 10 of Exercise 1.

In problems 4 through 12 write the given complex number in polar form. In each case make a drawing showing the corresponding point in the complex plane.

4. $6\sqrt{3} - 6i$ 5. $3 - 3i$ 6. $1 + 2i$

7. -5 8. $-4\sqrt{3} + 4i$ 9. $2i$

10. $-3i$ 11. $5 + 12i$ 12. $-5 + 5i$

In problems 13 through 21 put z in rectangular form, given $|z|$ and $\arg z$.

13. $8, 90°$ 14. $15, 180°$ 15. $\sqrt{3}, 120°$

16. $5\sqrt{2}, -135°$ 17. $8, 510°$ 18. $1, 270°$

19. $0, 1000°$ 20. $100, 25°$ 21. $\sqrt{10}, -45°$

22. Prove that if two complex numbers have equal moduli and equal arguments, then they are equal.

23. Prove that if two complex numbers are equal, then they have equal moduli, and their arguments differ by an integral multiple of $360°$.

18.4 GEOMETRIC REPRESENTATION OF ADDITION

Let A, B, and C be the points in the complex plane that correspond to the complex numbers $z_1 = 6+i$, $z_2 = 3+4i$, and $z_1+z_2 = 9+5i$. If the vectors $\overrightarrow{OA}$, $\overrightarrow{OB}$, and $\overrightarrow{OC}$ are drawn (see Figure 18.2), then the figure seems to indicate

FIGURE 18.2

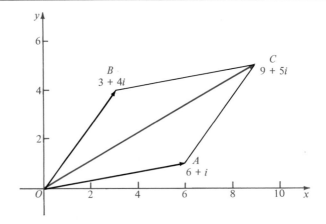

that $\overrightarrow{OC} = \overrightarrow{OA} + \overrightarrow{OB}$. In other words, we can add two complex numbers graphically by adding their vector representations. This relation between complex numbers and vectors under addition is stated accurately in

Theorem 7   Let $\overrightarrow{OA}$ be the position vector of z_1. Let $\overrightarrow{OB}$ be the position vector of z_2. Then the vector sum $\overrightarrow{OA} + \overrightarrow{OB}$ is the position vector of z_1+z_2.

Proof: As indicated in Figure 18.3 we let $\overrightarrow{OC}$ be the position vector of z_1+z_2. To prove that

(31) $$\overrightarrow{OC} = \overrightarrow{OA} + \overrightarrow{OB}$$

we must prove that no matter how the complex numbers are chosen, the quadrilateral $OACB$ is a parallelogram (which may collapse in special cases to a line segment). If $z_1 = x_1+iy_1$ and $z_2 = x_2+iy_2$, then by the distance formula Theorem 1 in Chapter 10

(32) $$|AC| = \sqrt{(x_1+x_2-x_1)^2 + (y_1+y_2-y_1)^2} = \sqrt{x_2^2 + y_2^2} = |OB|,$$

FIGURE 18.3

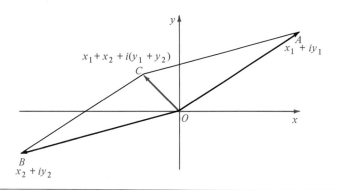

and

(33) $\quad |BC| = \sqrt{(x_1+x_2-x_2)^2 + (y_1+y_2-y_2)^2} = \sqrt{x_1^2 + y_1^2} = |OA|.$

Since the pairs of opposite sides of the quadrilateral are equal, the figure $OACB$ is a parallelogram. ∎

18.5 GEOMETRIC REPRESENTATION OF MULTIPLICATION

As in Section 18.4, we begin with an example. Again, let $z_1 = 6+i$, and let $z_2 = 3+4i$ be the complex numbers shown in Figure 18.2. In search of a geometrical relation among z_1, z_2 and the product $z_3 \equiv z_1 z_2$, we compute the modulus and argument of each of these numbers. We find that (to the nearest minute)

$$z_1 = 6 + i = \sqrt{37}(\cos 9°\,28' + i\sin 9°\,28') \equiv r_1(\cos\theta_1 + i\sin\theta_1)$$

$$z_2 = 3 + 4i = 5(\cos 53°\,8' + i\sin 53°\,8') \equiv r_2(\cos\theta_2 + i\sin\theta_2),$$

and

$$z_3 = (6+i)(3+4i) = (18-4) + i(24+3) = 14 + 27i$$
$$= \sqrt{925}(\cos 62°\,36' + i\sin 62°\,36') \equiv r_3(\cos\theta_3 + i\sin\theta_3).$$

We notice that

$$r_1 r_2 = 5\sqrt{37} = \sqrt{25 \times 37} = \sqrt{925} = r_3,$$

and

$$\theta_1 + \theta_2 = 9°\,28' + 53°\,8' = 62°\,36' = \theta_3.$$

It appears as though to multiply two complex numbers, we may multiply their moduli and add their arguments. This is the content of

Theorem 8   If $z_1 = r_1(\cos\theta_1 + i\sin\theta_1)$ and $z_2 = r_2(\cos\theta_2 + i\sin\theta_2)$, then

(34) $$z_1 z_2 = r_1 r_2[\cos(\theta_1 + \theta_2) + i\sin(\theta_1 + \theta_2)].$$

Proof: By direct multiplication we find that

$$\begin{aligned}
z_1 z_2 &= (r_1\cos\theta_1 + ir_1\sin\theta_1)(r_2\cos\theta_2 + ir_2\sin\theta_2)\\
&= (r_1 r_2\cos\theta_1\cos\theta_2 - r_1 r_2\sin\theta_1\sin\theta_2)\\
&\quad + i(r_1 r_2\sin\theta_1\cos\theta_2 + r_1 r_2\cos\theta_1\sin\theta_2)\\
&= r_1 r_2[(\cos\theta_1\cos\theta_2 - \sin\theta_1\sin\theta_2) + i(\sin\theta_1\cos\theta_2 + \cos\theta_1\sin\theta_2)]\\
&= r_1 r_2[\cos(\theta_1 + \theta_2) + i\sin(\theta_1 + \theta_2)],
\end{aligned}$$

by the addition formulas (13) and (19) of Chapter 9 (pages 203 and 207). ∎

Notice that $\theta_1 + \theta_2$ may be larger than 360°, so it may be necessary to subtract some multiple of 360° in order to find *the* argument of the product.

Since the quotient z_1/z_2 is a number z_3 such that $z_2 z_3 = z_1$, we have immediately

Theorem 9   If $z_1 = r_1(\cos\theta_1 + i\sin\theta_1)$ and $z_2 = r_2(\cos\theta_2 + i\sin\theta_2) \neq 0$, then

(35) $$\frac{z_1}{z_2} = \frac{r_1}{r_2}[\cos(\theta_1 - \theta_2) + i\sin(\theta_1 - \theta_2)].$$

Exercise 3

1. Prove that if $\overrightarrow{OA}$ represents z, then the vector that represents $-z$ has the same length as $\overrightarrow{OA}$ but the opposite direction.

2. Use the result of problem 1 to obtain a vector representation for $z_3 = z_1 - z_2$.

3. Prove that if z_1/z_2 is defined by equation (35), then the product of z_1/z_2 and z_2 is z_1.

In problems 4 through 13 express z_1 and z_2 in polar form, and compute $z_1 z_2$ and z_1/z_2. Compute the angle to the nearest 10 minutes. Wherever possible, without the use of tables, express the answer in rectangular form.

4. $z_1 = i$, $z_2 = 6\sqrt{3} + 6i$ 5. $z_1 = -4\sqrt{3} + 4i$, $z_2 = -i$

6. $z_1 = -\sqrt{2}+\sqrt{2}i, \; z_2 = 5+5i$ 7. $z_1 = -10+10i, \; z_2 = \sqrt{2}+\sqrt{2}i$

8. $z_1 = 1+2i, \qquad\qquad z_2 = 5i$ 9. $z_1 = -3i, \qquad\qquad z_2 = 4-2i$

10. $z_1 = 3+7i, \qquad\quad z_2 = 7+3i$ 11. $z_1 = 2+5i, \qquad\quad z_2 = 2-5i$

12. $z_1 = 2+3i, \qquad\quad z_2 = 4+6i$ 13. $z_1 = 3-2i, \qquad\quad z_2 = -4+6i$

In problems 14 through 17 perform the indicated operations and express the answer in rectangular form without using tables.

14. $2(\cos 18° + i\sin 18°) \cdot 3(\cos 27° + i\sin 27°)$

15. $6(\cos 42° + i\sin 42°) \cdot 3(\cos 78° + i\sin 78°)$

16. $5(\cos 1000° + i\sin 1000°) \div 4(\cos 790° + i\sin 790°)$

17. $12(\cos 2000° + i\sin 2000°) \div 3\sqrt{2}(\cos 1055° + i\sin 1055°)$

18.6 DE MOIVRE'S THEOREM

It is easy to extend Theorem 8 to the product of n complex numbers. If the n numbers are $z_1 = r_1(\cos\theta_1 + i\sin\theta_1)$, $z_2 = r_2(\cos\theta_2 + i\sin\theta_2)$, ..., and $z_n = r_n(\cos\theta_n + i\sin\theta_n)$, then clearly

$$(36) \quad z_1 z_2 \cdots z_n = r_1 r_2 \cdots r_n [\cos(\theta_1 + \theta_2 + \cdots + \theta_n)$$
$$+ i\sin(\theta_1 + \theta_2 + \cdots + \theta_n)].$$

In words, the modulus of the product of n complex numbers is the product of their moduli, and the argument of their product is the sum of the arguments of the numbers, or differs from it by some integer multiple of 360°.

If now we take the special case in which the n complex numbers are all equal, (36) gives a very simple formula for z^n.

Theorem 10   **De Moivre's Theorem** *If* $z = r(\cos\theta + i\sin\theta)$, *then for any positive integer* n

$$(37) \qquad\qquad z^n = r^n(\cos n\theta + i\sin n\theta).$$

The proof that we have just given is valid only for positive integer n. It is easy to see that (37) also holds when $n = 0$ or n is a negative integer as long as $z \neq 0$.

EXAMPLE 1   Compute $(2+2i)^{10}$.

Solution: Since $2+2i = 2\sqrt{2}(\cos 45° + i\sin 45°)$,

$$(2+2i)^{10} = (2\sqrt{2})^{10}(\cos 450° + i\sin 450°)$$
$$\doteq 2^{10}2^5(\cos 90° + i\sin 90°) = 2^{15}i = 32{,}768i. \quad \blacktriangle$$

De Moivre's Theorem can be used to compute roots as well as powers. Indeed, we can prove

Theorem 11   If $z = r(\cos\theta + i\sin\theta) \neq 0$, then there are precisely n, nth roots of z. These are given by the formula

$$(38) \quad z^{1/n} = r^{1/n}\left(\cos\frac{\theta + k360°}{n} + i\sin\frac{\theta + k360°}{n}\right), \quad k = 0, 1, 2, \cdots, n-1.$$

The corresponding points in the complex plane all lie on a circle of radius $r^{1/n}$ with center at the origin, and form the vertices of a regular polygon of n sides inscribed in this circle.

Proof: Let $w = \rho(\cos\alpha + i\sin\alpha)$ be an nth root of z, then by De Moivre's Theorem

$$w^n = \rho^n(\cos n\alpha + i\sin n\alpha) = z = r(\cos\theta + i\sin\theta).$$

Therefore $\rho^n = r$ or $\rho = r^{1/n}$. The angles $n\alpha$ and θ need not be equal, but they must differ at most by an integer multiple of 360°; that is

$$n\alpha = \theta + k360°.$$

But then

$$(39) \qquad\qquad \alpha = \frac{\theta + k360°}{n}.$$

If we use $\rho = r^{1/n}$ and $\alpha = (\theta + k360°)/n$ in $w = \rho(\cos\alpha + i\sin\alpha)$ with $k = 0, 1, 2, ..., n-1$, then we obtain the set **S** of n distinct complex numbers given by equation (38). Other integer values for k give complex numbers that are in **S**. We have proved that if w is an nth root of z, it is in the set **S**. Conversely, it is easy to see that if $w \in$ **S**, then $w^n = r(\cos\theta + i\sin\theta) = z$. ∎

EXAMPLE 2   Find the 4 fourth roots of $z = -16$.

Solution: Here $z = 16(\cos 180° + i\sin 180°)$. By Theorem 11 the roots are

$$w = 2\left(\cos\frac{180° + k360°}{4} + i\sin\frac{180° + k360°}{4}\right), \qquad k = 0, 1, 2, 3.$$

FIGURE 18.4

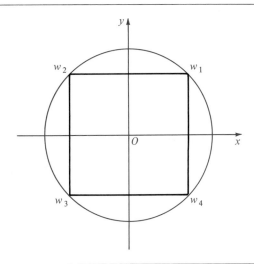

Using successively these values for k, we find

$$w_1 = 2(\cos 45° + i\sin 45°) = \sqrt{2} + i\sqrt{2}$$

$$w_2 = 2(\cos 135° + i\sin 135°) = -\sqrt{2} + i\sqrt{2}$$

$$w_3 = 2(\cos 225° + i\sin 225°) = -\sqrt{2} - i\sqrt{2}$$

$$w_4 = 2(\cos 315° + i\sin 315°) = \sqrt{2} - i\sqrt{2}. \quad \blacktriangle$$

The corresponding points are shown in Figure 18.4, together with the regular polygon of four sides, known as a square.

Exercise 4

1. Check the solution $w_2 = -\sqrt{2} + \sqrt{2}i$ in Example 2 by raising it to the fourth power.

In problems 2 through 11 use De Moivre's Theorem to compute the indicated power in rectangular form.

2. $(1 + \sqrt{3}i)^7$

3. $(1 - i)^{13}$

4. $(1 + i)^5$

5. $(\sqrt{3} - i)^6$

6. $(1 + 3i)^8$

7. $(-2 + i)^{10}$

8. $(\cos 27° + i\sin 27°)^{10}$

9. $(\cos 9° + i\sin 9°)^{20}$

10. $(\cos 4° + i\sin 4°)^8$

11. $(\cos 8° + i\sin 8°)^4$

In problems 12 through 23 find the indicated roots in rectangular form.

12. The fifth roots of 1

13. The sixth roots of -1

14. The square roots of $8i$

15. The fifth roots of $16\sqrt{2}-16\sqrt{2}i$

16. The cube roots of -1

17. The square roots of $-2+2\sqrt{3}i$

18. The cube roots of $-4\sqrt{3}+4i$

19. The sixth roots of -64

20. The eighth root of $256(\cos 48° + i\sin 48°)$ with the largest real part

21. The fourth root of $81(\cos 20° + i\sin 20°)$ with the largest imaginary part

22. The cube root of $27(\cos 30° + i\sin 30°)$ with the least imaginary part

23. The fourth root of $16(\cos 164° + i\sin 164°)$ with the least imaginary part

24. Solve the equation $x^4 + 1 = 0$.

25. Solve the equation $x^6 - 1 = 0$.

★ 26. If we use the Binomial Theorem to expand the right side of the equation $\cos n\theta + i\sin n\theta = (\cos\theta + i\sin\theta)^n$, we can then equate real and imaginary parts. This will give expressions for $\cos n\theta$ and $\sin n\theta$ in terms of $\cos\theta$ and $\sin\theta$. Use this method to prove that

$$(a)\ \cos 3\theta = \cos^3\theta - 3\cos\theta\sin^2\theta$$

and

$$(b)\ \sin 3\theta = 3\cos^2\theta\sin\theta - \sin^3\theta.$$

★ 27. Use the method of problem 26 to obtain $\cos 4\theta$ and $\sin 4\theta$ in terms of $\sin\theta$ and $\cos\theta$.

★ 28. Use the method of problem 26 for $\cos 5\theta$ and $\sin 5\theta$.

★ 29. Solve the equation $x^3 + x^2 + x + 1 = 0$. *Hint:* Multiply by $x-1$, solve the new equation, and reject the root $x = 1$.

Review Questions

Try to answer the following questions as accurately as possible before consulting the text.

1. What do we mean by the set **C** of all complex numbers?

2. In **C** what is the definition of
   (a) equality,
   (b) sum,
   (c) difference,
   (d) product,
   (e) reciprocal,
   (f) division.

3. Give the geometric interpretation of
   (a) addition,
   (b) subtraction,
   (c) multiplication,
   (d) division.

4. State De Moivre's Theorem.

5. Give the formula for the n, nth roots of a complex number $z \neq 0$.

Polynomials

We have already met polynomials in a casual way, and in Chapter 5 we studied quadratic polynomials thoroughly. In this chapter we prove a number of interesting theorems about polynomials of nth degree. To be precise we give

DEFINITION 1   ***Polynomial***   *An expression of the form*

(1) $$P(x) \equiv a_0 x^n + a_1 x^{n-1} + \cdots + a_{n-1} x + a_n, \qquad a_0 \neq 0,$$

is called a polynomial of nth degree. The equation $P(x) = 0$ is called a polynomial equation of nth degree.

If $n = 1$, the polynomial is linear; if $n = 2$, it is a quadratic polynomial; if $n = 3$, it is a cubic polynomial, etc. If all of the coefficients in (1) are zero, then the polynomial $P(x)$ is said to be identically zero and in this case we do *not* assign a degree to $P(x)$. In Theorem 3 in Chapter 5 we obtained a formula for finding the roots of an arbitrary quadratic equation. There is a similar formula for solving the general cubic equation called Cardan's formula although he stole it from Tartaglia (ca. 1499–1557). The formula for solving the general fourth degree equation (biquadratic) was discovered by Ferrari

469

(1522–1565) and is extremely complicated. We would be delighted to have a formula for solving the general fifth degree equation (quintic) using only a finite number of sums, products, and roots. However, Niels Abel (1802–1829) and Éveriste Galois (1811–1832) proved independently that no such formula exists.

We will not go into these matters here. Our purpose is to learn a scientific method of guessing at the roots of a polynomial equation, so that in an emergency we may be able to get by without the Tartaglia and Ferrari formulas. We will also learn a method for computing an approximate value of a root. This approach may seem unsatisfactory, but the problem of locating the roots of a polynomial equation is so complicated that even today there are still open questions in this branch of mathematics.

19.1 A COMPUTATION METHOD FOR POLYNOMIALS

If we wish to sketch the graph of a polynomial function

$$(2) \qquad\qquad y = P(x)$$

we must compute a few points and if the degree of $P(x)$ is high, this will place quite a strain upon our patience. To alleviate this strain we now give a method which will reduce the amount of computation. For simplicity, we give the method for a fourth degree polynomial, but it is applicable to any polynomial of nth degree.

Suppose that we wish to find the value of

$$(3) \qquad\qquad P(x) \equiv a_0 x^4 + a_1 x^3 + a_2 x^2 + a_3 x + a_4$$

when $x = r$. The computation of r^4 may be tedious, so we replace the direct assault by a gradual approach which consists of a sequence of multiplications and additions. We define b_0, b_1, b_2, b_3, and R by the equations

$$(4) \qquad\qquad b_0 = a_0,$$

$$(5) \qquad\qquad b_1 = b_0 r + a_1,$$

$$(6) \qquad\qquad b_2 = b_1 r + a_2,$$

$$(7) \qquad\qquad b_3 = b_2 r + a_3,$$

$$(8) \qquad\qquad R = b_3 r + a_4.$$

We claim that R is the value of $P(x)$ when $x = r$. To see this we use the expression for b_1 given by equation (5) in equation (6) and we find that

(9)
$$b_2 = (a_0 r + a_1)r + a_2$$
$$= a_0 r^2 + a_1 r + a_2.$$

We then use this expression for b_2 in equation (7) and find that

(10)
$$b_3 = (a_0 r^2 + a_1 r + a_2)r + a_3$$
$$= a_0 r^3 + a_1 r^2 + a_2 r + a_3.$$

Finally we use this expression for b_3 in equation (8) and find that

(11)
$$R = (a_0 r^3 + a_1 r^2 + a_2 r + a_3)r + a_4$$
$$= a_0 r^4 + a_1 r^3 + a_2 r^2 + a_3 r + a_4.$$

This is the value of $P(x)$ when $x = r$.

If we condense these steps into one expression, we obtain the identity

(12)
$$R = \{[(a_0 r + a_1)r + a_2]r + a_3\}r + a_4$$
$$= a_0 r^4 + a_1 r^3 + a_2 r^2 + a_3 r + a_4 = P(r).$$

Equation (12) may appear quite formidable, but the actual computation of R is very easy when we arrange the steps in the following systematic manner:

$$
\begin{array}{c|ccccc}
r & a_0 & a_1 & a_2 & a_3 & a_4 \\
 & & b_0 r & b_1 r & b_2 r & b_3 r \\
\hline
 & a_0 = b_0 & b_1 & b_2 & b_3 & R = P(r)
\end{array}
$$

In this scheme we place r in a small box to the left. The coefficients of $P(x)$ are placed in the first line. The arrows indicate that we multiply the element in the third row by r to obtain the next element in the second row. Finally, each b_k is obtained by adding the two elements in the column directly above b_k. Then $R = P(r)$ is the last such sum.

EXAMPLE 1   Sketch the graph of

(13)
$$y = P(x) \equiv x^3 - x^2 - 10x + 12$$

for $-4 \leq x \leq 4$.

Solution: We use the method just described to find y for various values of x. When $x = 4$ we have

$$
\begin{array}{r|rrrr}
4 & 1 & -1 & -10 & 12 \\
& & 4 & 12 & 8 \\
\hline
& 1 & 3 & 2 & 20 = R.
\end{array}
$$

Hence, if $x = 4$, then $y = 20$. When $x = 3$, we have

$$
\begin{array}{r|rrrr}
3 & 1 & -1 & -10 & 12 \\
& & 3 & 6 & -12 \\
\hline
& 1 & 2 & -4 & 0 = R.
\end{array}
$$

Hence, if $x = 3$, then $y = 0$. The method is the same when r is negative. For example, if $r = -3$ we have

$$
\begin{array}{r|rrrr}
-3 & 1 & -1 & -10 & 12 \\
& & -3 & 12 & -6 \\
\hline
& 1 & -4 & 2 & 6 = R.
\end{array}
$$

Consequently, the point $(-3, 6)$ is on the graph. Similar computations will give the points $(-4, -28)$, $(-2, 20)$, $(-1, 20)$, $(0, 12)$, $(1, 2)$, and $(2, -4)$ that are also on the graph. The graph of equation (13) is shown in Figure 19.1. We infer from this graph that the polynomial $P(x) = x^3 - x^2 - 10x + 12$ has a zero in each of the intervals $(-4, -3)$ and $(1, 2)$. It also has a zero at $x = 3$.

FIGURE 19.1

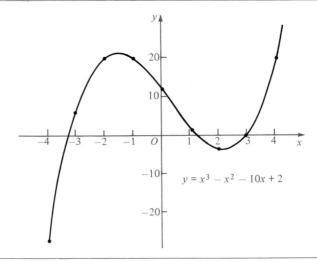

$$y = x^3 - x^2 - 10x + 2$$

If r is not an integer, we may expect that the computation of $P(r)$ will require more labor. For example, if $r = 1.5$, then we have

$$
\begin{array}{r|rrrr}
1.5 & 1 & -1 & -10 & 12 \\
   &   & 1.5 & 0.75 & 13.875 \\
\hline
   & 1 & 0.5 & -9.25 & -1.875 = R.
\end{array}
$$

Consequently, the point $(1.5, -1.875)$ is also on the graph of equation (13). ▲

This new method of computing $P(r)$ for a polynomial $P(x)$ is called *synthetic substitution*. We will see shortly that the arrangement of terms and the computation is exactly the same for synthetic division.

Of course, for some special values of x, direct substitution may be quicker than synthetic substitution. This is certainly true when $x = 0$ or $x = 1$. For example, if $x = 0$, equation (13) gives $y = 12$ immediately. When $x = 1$, equation (13) gives $y = 1 - 1 - 10 + 12 = 2$.

The curious student might try synthetic substitution with $r = 0$ and $r = 1$ to see if he obtains the same values for $P(0)$ and $P(1)$. He might also compute $P(4)$ by direct substitution in equation (13) and decide for himself if synthetic substitution really is quicker. The method of synthetic substitution is ideally suited for the evaluation of a polynomial by a computer.

Exercise 1

In problems 1 and 2 sketch the graph of the given function over the given interval. When computing coordinates use either direct substitution or synthetic substitution, but try to select the one that is quicker in each case.

1. $y = x^3 + 2x^2 - 3x - 1,$ $\quad -4 \leqq x \leqq 3$

2. $y = 2x^3 + 3x^2 - 16x - 9,$ $\quad -4 \leqq x \leqq 3$

In problems 3 through 6 use synthetic substitution to find the value of the given polynomial for the indicated values of x.

3. $P(x) = x^4 + 2x^3 - 3x^2 - 4x - 20,$ $\qquad x = 2, x = 3$

4. $P(x) = 3x^4 - 2x^3 + 4x^2 - 5x - 10,$ $\qquad x = 1, x = 2$

5. $P(x) = x^5 + 5x^4 + 3x^3 - 6x^2 - 9x + 11,$ $\quad x = -4, x = -5$

6. $P(x) = x^3 - (3 + a)x^2 + (3a + 5)x - 5a,$ $\quad x = a, x = 2$

7. If $P(x) = x^3 - bx^2 - 5x + 7b,$ find b so that $P(2) = 10$.

8. If $P(x) = cx^3 - 5x^2 + 2cx + 10$, find c so that $P(\sqrt{3}) = -20$.

9. For the polynomial given in problem 1 find suitable intervals in which the zeros of the polynomial must lie.

10. Do problem 9 for the polynomial given in problem 2.

11. Prove that the set of equations (4), (5), (6), (7), and (8) gives the first part of equation (12).

In problems 12 through 16 let $P(x) = a_0 x^n + a_1 x^{n-1} + \cdots + a_n$ be a polynomial of nth degree, and let $Q(x) = b_0 x^m + b_1 x^{m-1} + \cdots + b_m$ be a polynomial of mth degree.

12. Prove that $P(x)Q(x)$ is a polynomial of degree $m+n$.

13. If $m < n$, prove that $P(x) + Q(x)$ is a polynomial of degree n.

14. If $m = n$, prove that $P(x) + Q(x)$ may be of degree less than n or may even be identically zero.

★ 15. If $n \geq 1$, prove that $1/P(x)$ is not a polynomial.

★ 16. If $n > m$, prove that $Q(x)/P(x)$ is not a polynomial.

19.2 THE REMAINDER THEOREM

If we divide a polynomial $P(x)$ by a second polynomial $D(x)$, the standard procedure will yield two more polynomials $Q(x)$, the quotient, and $R(x)$, the remainder, such that

(14)
$$\frac{P(x)}{D(x)} = Q(x) + \frac{R(x)}{D(x)}.$$

This relation is true for all x for which the denominator $D(x)$ is not zero. We can remove this exceptional case if we multiply both sides of (14) by $D(x)$. This gives the identity

(15)
$$P(x) = D(x)Q(x) + R(x).$$

The division procedure can be continued until the degree of the remainder $R(x)$ is less than the degree of the divisor $D(x)$. This is illustrated in

EXAMPLE 1   Find $Q(x)$ and $R(x)$ when $P(x) \equiv x^4 - 8x^3 + 20x^2 - 8x + 1$ is divided by $D(x) \equiv x^2 - 2x + 5$.

Solution: The standard division method gives.

$$x^2 - 6x + 3 = Q(x). \quad \text{Quotient.}$$

Divisor. $D(x) = x^2 - 2x + 5 \,\overline{\smash{\big)}\,x^4 - 8x^3 + 20x^2 - 8x + 1} = P(x). \quad \text{Dividend.}$

$$x^2(x^2 - 2x + 5) \rightarrow x^4 - 2x^3 + 5x^2$$

(subtracting) $-6x^3 + 15x^2 - 8x$

$$-6x(x^2 - 2x + 5) \rightarrow -6x^3 + 12x^2 - 30x$$

(subtracting) $3x^2 + 22x + 1$

$$3(x^2 - 2x + 5) \rightarrow 3x^2 - 6x + 15$$

(subtracting) $28x - 14 = R(x). \quad \text{Remainder.}$

Hence, we have the identity (a particular case of equation (15))

$$x^4 - 8x^3 + 20x^2 - 8x + 1 = (x^2 - 2x + 5)(x^2 - 6x + 3) + 28x - 14,$$
$$P(x) \qquad = \qquad D(x) \qquad Q(x) \qquad + \quad R(x). \quad \blacktriangle$$

A careful study of the division process will yield a proof of

Theorem 1
PWO

Let $P(x)$ and $D(x)$ be polynomials where the degree of $D(x)$ is m. Then there are two polynomials $Q(x)$ and $R(x)$ such that:

(a) The equation

(15) $$P(x) = D(x)Q(x) + R(x)$$

is an identity in x.

(b) Either $R(x)$ is identically zero or the degree of $R(x)$ is less than m.
(c) $Q(x)$ and $R(x)$ are unique.

Our interest centers on the particular case in which the divisor is a linear polynomial, i.e., $D(x) = x - r$. In this case $R(x)$ must have degree less than the degree of $D(x)$. Hence, $R(x)$ is a constant (which may of course be zero). Then Theorem 1 gives

Theorem 2

The Remainder Theorem   *Let $P(x)$ be a polynomial, and let r be a number. Then there is a unique polynomial $Q(x)$ and a unique number R such that*

(16) $$P(x) = (x - r)Q(x) + R$$

is an identity in x.

A simple method for finding $Q(x)$ and R will be presented in the next section, and fortunately this method is identical with the method of synthetic substitution which we studied in Section 19.1. For the present we continue with the theory.

Theorem 3　　*The remainder R in equation (16) is the value of the polynomial at $x = r$.*

Proof: Let $x = r$ in equation (16). Then the first term on the right side is $(r-r)Q(r) = 0$. Hence, $P(r) = R$. ∎

Theorem 4　　*If r is a zero of the polynomial $P(x)$, then $x-r$ divides $P(x)$. Conversely, if $x-r$ divides $P(x)$, then r is a zero of $P(x)$.*

Proof: The phrase "$x-r$ divides $P(x)$" means that there is a polynomial $Q(x)$ such that

$$(17) \qquad P(x) = (x-r)Q(x).$$

Suppose now that r is a zero of $P(x)$. Then $P(r) = 0$, and by Theorem 3 $P(r) = R$. Hence, $R = 0$. Thus, equation (16) gives equation (17). Consequently, $(x-r)$ divides $P(x)$. Conversely, suppose that $x-r$ divides $P(x)$. Then equation (17) holds. Hence, $P(r) = (r-r)Q(r) = 0 \cdot Q(r) = 0$. Consequently, r is a zero of $P(x)$. ∎

Suppose now that we have located one zero of $P(x)$, and we have equation (17). Then any other zeros of $P(x)$ must be zeros of $Q(x)$. If s is a zero of $Q(x)$, then by Theorem 4 we can write $Q(x) = (x-s)Q_1(x)$. Using this in equation (17) we find

$$(18) \qquad P(x) = (x-r)(x-s)Q_1(x).$$

If we are able to continue this process, we will find n zeros which we denote by $r_1, r_2, \ldots, r_n$ (in place of $r, s, t, \ldots$). Then (18) will take the form

$$(19) \qquad P(x) = a_0(x-r_1)(x-r_2)(x-r_3)\cdots(x-r_n),$$

where n is the degree of $P(x)$, and a_0 is the leading coefficient (see equation (1)).

To convert these remarks into a theorem we must be certain that a polynomial *has* a zero. Although this may seem reasonably obvious, the proof is complicated. Further, if we restrict ourselves to only real numbers, the theorem is not even true. For example, the polynomial $P(x) \equiv x^2 + 4$ has no real zeros. It was Carl F. Gauss (1777–1855) who first gave a correct proof of

Theorem 5
PWO

The Fundamental Theorem of Algebra *If $P(x)$ is a polynomial of degree $n \geq 1$, then $P(x)$ has at least one zero in the field of complex numbers.*

Using this theorem, we can derive equation (19) where $r_1, r_2, ..., r_n$ are the zeros of $P(x)$. It may happen that some of the numbers $r_1, r_2, ..., r_n$ are the same. For example, we may have $r_1 = r_2 = r_3$, and if this is the case, we may combine the factors and write $(x - r_1)^3$ in equation (19) rather than $(x - r_1)(x - r_2)(x - r_3)$. If we make such combinations whenever possible, we arrive at

Theorem 6

Every polynomial $P(x)$ of the form (1) with $n \geq 1$ can be put in the form

(20) $$P(x) = a_0(x - r_1)^{e_1}(x - r_2)^{e_2} \cdots (x - r_k)^{e_k},$$

where $r_1, r_2, ..., r_k$ are the distinct zeros of $P(x)$ and each one of the exponents $e_1, e_2, ..., e_k$ is a positive integer.

DEFINITION 2

If $e_j = 1$ in equation (20), then the corresponding zero r_j is called a simple zero of $P(x)$ (or a simple root of the equation $P(x) = 0$). If $e_j > 1$, then r_j is called a multiple zero of $P(x)$ (or a multiple root of $P(x) = 0$). When $e_j > 1$, we call e_j the multiplicity of the zero (or root), and r_j is a zero (or root) of order e_j.

EXAMPLE 2

Discuss the nature of the zeros of the polynomial

(21) $$P(x) \equiv x^6 + 4x^5 - x^4 - 8x^3 + 7x^2 - 12x + 9.$$

Solution: We will study methods of locating zeros in the next three sections. For the present we merely note that if we know that r is a zero of $P(x)$, then $(x - r)$ is a factor of $P(x)$. After considerable labor (which we omit) we find that for this polynomial

(22) $$P(x) = (x - 1)^2 (x + 3)^2 (x + i)(x - i).$$

Hence, the zeros are 1, 1, -3, -3, i, and $-i$, where we list each zero in accordance with its multiplicity. Thus, 1 and -3 are each listed twice because each one is a zero of order 2. ▲

This example leads immediately to

Theorem 7
PLE

Each polynomial of degree $n \geq 1$ has precisely n zeros, when the zeros are counted in accordance with their multiplicities.

For example, the polynomial given by (21) is of sixth degree and has six zeros $(1, 1, -3, -3, i, -i)$, when the zeros are listed in accordance with their multiplicities.

Exercise 2

In problems 1 through 6 find the quotient, and the remainder when $P(x)$ is divided by $D(x)$, using the method illustrated in Example 1. A shorter method for the first 4 problems will be presented in the next section.

1. $P(x) = 2x^4 + 11x^3 - 31x^2 + 23x - 42$, $D(x) = 2x - 3$

2. $P(x) = 5x^5 - 11x^4 + 17x^3 - 3x^2 - 35x + 17$, $D(x) = 5x - 1$

3. $P(x) = 7x^5 - 5x^4 + 12x^3 - 17x^2 + 22x + 10$, $D(x) = 7x + 2$

4. $P(x) = 2x^5 - 3x^4 + 36x^2 - 29x + 15$, $D(x) = 2x + 5$

5. $P(x) = 2x^4 + x^3 - 16x^2 - 14x - 3$, $D(x) = 2x^2 - 5x - 3$

6. $P(x) = 2x^4 - x^3 - 19x^2 + 14x + 24$, $D(x) = x^2 + x - 6$

7. In problem 5, we have $D(x) = (2x+1)(x-3)$. Check your solution by dividing P first by $2x+1$ and then dividing the quotient by $x-3$.

8. Check your solution to problem 6 by dividing P first by $x+3$ and then by $x-2$.

19.3 SYNTHETIC DIVISION

We devote this section to a method of computing $Q(x)$ and R when $P(x)$ is divided by $x-r$. For simplicity we give the method when $P(x)$ is a fourth degree polynomial, but it is applicable to any polynomial of nth degree. Suppose that $Q(x) = b_0 x^3 + b_1 x^2 + b_2 x + b_3$, where we are to find the coefficients b_k in terms of r and the a_k. Since $Q(x)$ must satisfy equation (16), we have

$$(23) \quad a_0 x^4 + a_1 x^3 + a_2 x^2 + a_3 x + a_4$$
$$= (x-r)(b_0 x^3 + b_1 x^2 + b_2 x + b_3) + R$$
$$= b_0 x^4 + {}^{\bullet}b_1 x^3 + b_2 x^2 + b_3 x + R$$
$$\quad - rb_0 x^3 - rb_1 x^2 - rb_2 x - rb_3$$
$$= b_0 x^4 + (b_1 - rb_0)x^3 + (b_2 - rb_1)x^2 + (b_3 - rb_2)x + R - rb_3.$$

Since (23) is an identity in x, the corresponding coefficients in the first expression and the last expression must be equal. This leads to the set of equations:

$$a_0 = b_0 \qquad \text{(equating coefficients of } x^4\text{),}$$

$$a_1 = b_1 - rb_0 \qquad \text{(equating coefficients of } x^3\text{),}$$

$$a_2 = b_2 - rb_1 \qquad \text{(equating coefficients of } x^2\text{),}$$

$$a_3 = b_3 - rb_2 \qquad \text{(equating coefficients of } x\text{),}$$

$$a_4 = R - rb_3 \qquad \text{(equating the constant terms).}$$

If we solve this set of equations for the b_k we find

$$(24) \qquad\qquad b_0 = a_0,$$

$$(25) \qquad\qquad b_1 = rb_0 + a_1,$$

$$(26) \qquad\qquad b_2 = rb_1 + a_2,$$

$$(27) \qquad\qquad b_3 = rb_2 + a_3,$$

$$(28) \qquad\qquad R = rb_3 + a_4.$$

This set of equations is identical with the set (4), (5), (6), (7), and (8) in Section 19.1. Hence, the same scheme used in Section 19.1 will also give the coefficients of $Q(x)$ and R in the equation $P(x) = (x-r)Q(x) + R$. For convenience we reproduce the scheme used in Section 19.1.

$$
\begin{array}{c|ccccc}
r & a_0 & a_1 & a_2 & a_3 & a_4 \\
 & & b_0 r & b_1 r & b_2 r & b_3 r \\
\hline
a_0 = b_0 & b_1 & b_2 & b_3 & b_4 = R.
\end{array}
$$

EXAMPLE 1   Find the quotient and the remainder when

$$(21) \qquad P(x) \equiv x^6 + 4x^5 - x^4 - 8x^3 + 7x^2 - 12x + 9$$

is divided by $x - 1$.

Solution: For this sixth degree polynomial we have

$$
\begin{array}{c|ccccccc}
1 & 1 & 4 & -1 & -8 & 7 & -12 & 9 \\
 & & 1 & 5 & 4 & -4 & 3 & -9 \\
\hline
 & 1 & 5 & 4 & -4 & 3 & -9 & 0 = R.
\end{array}
$$

Hence, the quotient is $x^5 + 5x^4 + 4x^3 - 4x^2 + 3x - 9$ and $R = 0$. Therefore, $x = 1$ is a zero of this polynomial. ▲

EXAMPLE 2

Show that the polynomial given in Example 1 has the factorization indicated by equation (22).

Solution: We can use each successive quotient as the dividend in the next step. Starting with the quotient from Example 1 and dividing successively by $x-1$, $x+3$, and $x+3$, we have

$$
\begin{array}{r|rrrrrr}
\underline{1\rfloor} & 1 & 5 & 4 & -4 & 3 & -9 \\
& & 1 & 6 & 10 & 6 & 9 \\
\hline
\underline{-3\rfloor} & 1 & 6 & 10 & 6 & 9 & 0 = R, \\
& & -3 & -9 & -3 & -9 \\
\hline
\underline{-3\rfloor} & 1 & 3 & 1 & 3 & 0 = R, \\
& & -3 & 0 & -3 \\
\hline
& 1 & 0 & 1 & 0 = R.
\end{array}
$$

Hence, the quotient is x^2+1, and this is clearly the product $(x+i)(x-i)$. ▲

Now that we have a method for finding the quotient $Q(x)$, we can use this method to find bounds for the zeros of a polynomial. To be specific, suppose that we divide a fourth degree polynomial by $x-M$ (where $M>0$) and obtain

(29) $$P(x) = (x-M)(b_0 x^3 + b_1 x^2 + b_2 x + b_3) + R.$$

If all of the numbers b_0, b_1, b_2, b_3, and R are greater than or equal to zero, then $Q(x)>0$ whenever $x>0$. Consequently, if $x>M$ and $x>0$, then $P(x)>0$. Therefore, $P(x)$ has no zeros r such that $r>M$. Since the argument can be extended to a polynomial of arbitrarily high degree, we have

Theorem 8

Let M be a positive number such that when $P(x)$ is divided by $x-M$, the remainder and all of the coefficients in $Q(x)$ are nonnegative. Then $P(x)$ is not zero for any $x>M$.

Similar considerations lead to

Theorem 9
PLE

Let L be a negative number. Suppose that when $P(x)$ is divided by $x-L$, the sequence of coefficients and the remainder,

$$b_0, b_1, b_2, ..., b_{n-1}, R,$$

are alternately nonpositive and nonnegative. Then $P(x)$ is not zero for any $x < L$.

EXAMPLE 3   Find upper and lower bounds for the real zeros of

(30) $P(x) \equiv x^4 - 3x^3 - 10x^2 + x - 2.$

Solution: In searching for the proper divisor $x - M$ we try various values for M. When $M = 1, 2, 3,$ or 4, at least one b_k is negative. When $M = 5$, we have

$$
\begin{array}{r|rrrrr}
5 & 1 & -3 & -10 & 1 & -2 \\
  &   & 5 & 10 & 0 & 5 \\
\hline
  & 1 & 2 &   0 & 1 & 3.
\end{array}
$$

The last line has the property required in Theorem 8. Hence, the polynomial defined by equation (30) is never zero for $x \geq 5$.

To find a lower bound for the zeros of $P(x)$ we try $L = -1, -2,$ and -3. For $L = -3$ we have

$$
\begin{array}{r|rrrrr}
-3 & 1 & -3 & -10 &   1 & -2 \\
   &   & -3 & 18 & -24 & 69 \\
\hline
   & 1 & -6 &   8 & -23 & 67.
\end{array}
$$

The last line has the property required in Theorem 9. Hence, $P(x)$ is never zero for any $x \leq -3$.   ▲

EXAMPLE 4   Find all of the roots of the equation

$$P(x) \equiv x^3 - 9x^2 + 19x - 11 = 0.$$

Solution: If we can guess at a root of this equation, then by Theorem 4 we can write $P(x) = (x - r)Q(x)$, where $Q(x)$ is a quadratic polynomial. Then we can use the quadratic formula to find the other two roots. An examination of the coefficients gives $1 - 9 + 19 - 11 = 0$. Consequently, $x = 1$ is a root. Synthetic division gives

$$
\begin{array}{r|rrrr}
1 & 1 & -9 & 19 & -11 \\
  &   & 1 & -8 & 11 \\
\hline
  & 1 & -8 & 11 & 0 = R.
\end{array}
$$

Therefore, $Q(x) = x^2 - 8x + 11$. Applying the quadratic formula to this $Q(x)$ we find that $r = 4 + \sqrt{5}$ and $r = 4 - \sqrt{5}$ are also roots of the given equation. Thus, we have found three roots, and a cubic equation cannot have more than three roots.   ▲

Knowing the zeros of a polynomial, we can factor it (see Theorem 6). In this example we have

$$x^3 - 9x^2 + 19x - 11 = (x-1)(x-4-\sqrt{5})(x-4+\sqrt{5}).$$

Exercise 3

In problems 1 through 8 use synthetic division to find the quotient and the remainder when $P(x)$ is diveded by $x-r$.

1. $r = 2,$ $P(x) = x^4 - 3x^3 + 2x^2 + 4x + 5$
2. $r = 3,$ $P(x) = x^4 - 5x^3 - 3x^2 + 24x + 10$
3. $r = 2,$ $P(x) = x^6 - 5x^4 + 13x^2 - 25x$
4. $r = 4,$ $P(x) = 2x^6 - 35x^4 + 43x^2 + 21x - 11$
5. $r = -2,$ $P(x) = 3x^5 + 7x^4 - 3x^2 + 10x + 19$
6. $r = -3,$ $P(x) = 2x^5 - x^4 + 41x^2 - 7x + 7$
7. $r = 3/2,$ $P(x) = 2x^5 + 3x^4 - 17x^3 + 22x^2 - 17x + 3$
8. $r = -5/2,$ $P(x) = 2x^5 - x^4 - 15x^3 - 10x^2 - 9x + 40$

★ 9. Prove Theorem 9.

In problems 10 through 15 find suitable upper and lower bounds for the real zeros of the given polynomial.

10. $P(x) = x^3 + 2x^2 - 3x - 1$
11. $P(x) = 2x^3 + 3x^2 - 16x - 9$
12. $P(x) = x^4 + 2x^3 - 3x^2 - 4x - 20$
13. $P(x) = 3x^4 - 2x^3 + 4x^2 - 5x - 10$
14. $P(x) = x^5 + 5x^4 + 3x^3 - 6x^2 - 9x + 11$
15. $P(x) = x^6 - 7x^5 + 9x^4 - 57x^3 - 40x^2 - 13x - 5$

In problems 16 through 21 use the method of Example 4 to find all the roots of the given equation.

16. $x^3 - x^2 - 7x - 5 = 0$ 17. $2x^3 - 3x^2 - x + 2 = 0$
18. $x^3 + 4x^2 - 7x - 10 = 0$ 19. $x^3 - 2x^2 - 5x + 6 = 0$
20. $x^3 + x^2 - 13x + 3 = 0$ 21. $x^3 + 4x^2 - 2x - 15 = 0$

★22. If $A = 3$, $B = 2$, and $C = 4$, then the equation

$$A\cos^2 x + B\sin^2 x + C\cos 2x = 3\cos^2 x + 2\sin^2 x + 4\cos 2x$$

is an identity in x. Prove that the converse is not true by showing that if the equation is an identity in x, then A, B, and C may have other values; for example, $A = 4$, $B = 1$, and $C = 3$.

★23. Suppose that for the equation in problem 22 we select C to be any number. Prove that if we set $A = 7 - C$ and $B = C - 2$, then the equation is an identity in x.

19.4 POLYNOMIALS WITH REAL COEFFICIENTS

According to the fundamental theorem of algebra, every polynomial of degree $n \geq 1$ has at least one zero. This remarkable theorem includes the polynomials in which the coefficients are complex numbers.

When the coefficients are all real, the polynomial may have a complex zero $a + bi$ where $b \neq 0$. However, such zeros always occur in conjugate pairs, as stated in

Theorem 10   Let $a + bi$ be a zero of $P(x)$, where $b \neq 0$, and suppose that all of the coefficients of $P(x)$ are real. Then $a - bi$ is also a zero of $P(x)$.

Proof: We first recall some facts about complex numbers. The conjugate of $z = a + bi$ is $a - bi$ and is denoted by $\bar{z}$. The operation of conjugation has the following properties:

(31)
$$\overline{z_1 + z_2} = \bar{z}_1 + \bar{z}_2,$$

(32)
$$\overline{z_1 z_2} = \bar{z}_1 \bar{z}_2.$$

and if a_k is a real number, then

(33)
$$\overline{a_k} = a_k.$$

Further, repeated application of (32) will give

(34)
$$\overline{(z^k)} = (\bar{z})^k.$$

Now suppose that $P(a + bi) = 0$. In other words, suppose that

(35)
$$a_0(a + bi)^n + a_1(a + bi)^{n-1} + a_2(a + bi)^{n-2} + \cdots + a_n = 0.$$

We take the complex conjugate of both sides of (35) and use the relations (31) through (34). This gives

$$\overline{a_0(a+bi)^n + a_1(a+bi)^{n-1} + a_2(a+bi)^{n-2} + \cdots + a_n} = \overline{0},$$

(36) $$\overline{a_0}\overline{(a+bi)}^n + \overline{a_1}\overline{(a+bi)}^{n-1} + \overline{a_2}\overline{(a+bi)}^{n-2} + \cdots + \overline{a_n} = \overline{0}.$$

But $0, a_0, a_1, ..., a_n$ are all real numbers and $\overline{a+bi} = a-bi$. Hence, (36) yields

$$a_0(a-bi)^n + a_1(a-bi)^{n-1} + a_2(a-bi)^{n-2} + \cdots + a_n = 0.$$

Consequently, $P(a-bi) = 0$. Thus, $a-bi$ is a zero of $P(x)$. ∎

19.5 RATIONAL ROOTS OF POLYNOMIALS

We now give a scientific method of finding the rational zeros of a polynomial. The method is based on

Theorem 11   *Suppose that all of the coefficients of*

(1) $$P(x) = a_0 x^n + a_1 x^{n-1} + \cdots + a_{n-1} x + a_n, \qquad a_0 \neq 0$$

are integers and that p/q is a zero of $P(x)$. If p and q are integers with no common factor greater than 1, then p divides a_n and q divides a_0.

Proof:‡ If p/q is a zero of $P(x)$, we have

(37) $$a_0\left(\frac{p}{q}\right)^n + a_1\left(\frac{p}{q}\right)^{n-1} + \cdots + a_{n-1}\frac{p}{q} + a_n = 0.$$

We multiply both sides by q^n and transpose the last term to the right side. This gives

(38) $$a_0 p^n + a_1 p^{n-1} q + \cdots + a_{n-1} p q^{n-1} = -a_n q^n.$$

We can factor p from each term on the left side of (38). This gives

(39) $$p(a_0 p^{n-1} + a_1 p^{n-2} q + \cdots + a_{n-1} q^{n-1}) = -a_n q^n.$$

Now, $a_0, a_1, ..., a_{n-1}$, p, and q are integers. Hence, the left side of (39) is pC where C is an integer. Thus, p divides the left side of (39) and hence, p divides $-a_n q^n$. Since p and q have no common divisors other than 1, it follows that p divides a_n.

‡ This proof is probably the most complicated and difficult item in this text. The proof of Theorem 2 in Chapter 9 is a close rival.

To see that q divides a_0, we again multiply both sides of (37) by q^n, but this time we transpose the first term obtaining

(40) $$a_1 p^{n-1} q + \cdots + a_{n-1} pq^{n-1} + a_n q^n = -a_0 p^n,$$

(41) $$q(a_1 p^{n-1} + \cdots + a_{n-1} pq^{n-2} + a_n q^{n-1}) = -a_0 p^n.$$

Hence, $qD = -a_0 p^n$ where D is some suitable integer. Consequently q divides $-a_0 p^n$. Since p and q have no common divisors other than 1, it follows that q divides a_0. ∎

EXAMPLE 1

Find all the roots of

(42) $$P_1(x) \equiv 3x^3 + 17x^2 + 16x + 4 = 0.$$

Solution: We first try to locate the rational roots. If $r = p/q$ in lowest terms is a rational root, then (by Theorem 11) p must divide 4 and q must divide 3. Consequently, $p = \pm 1, \pm 2, \pm 4$ and $q = \pm 1, \pm 3$. We form all possible fractions with these values for p and q, and call the set of such fractions the test set

$$\mathbf{T} = \left\{ \pm 1, \pm 2, \pm 4, \pm \frac{1}{3}, \pm \frac{2}{3}, \pm \frac{4}{3} \right\}.$$

According to Theorem 11, if equation (42) has a rational root it must be in the set $\mathbf{T}$, but of course $\mathbf{T}$ may contain numbers that are not roots of (42). Now the set $\mathbf{T}$ has twelve elements and the methodical person will try all of these in equation (42). But if we notice that all of the coefficients in (42) are positive, then we see that $P_1(x) > 0$ whenever $x > 0$. Hence, any positive number is not a root, and we immediately eliminate from consideration all of the positive numbers from $\mathbf{T}$. Thus, we obtain a smaller test set

$$\mathbf{T}_1 = \left\{ -1, -2, -4, -\frac{1}{3}, -\frac{2}{3}, -\frac{4}{3} \right\}.$$

We test the numbers in $\mathbf{T}_1$ using synthetic division. The first four in the set $\mathbf{T}_1$ are not roots (we omit the computations), but the fifth one gives

$$
\begin{array}{r|rrrr}
-\frac{2}{3} & 3 & 17 & 16 & 4 \\
 & & -2 & -10 & -4 \\
\hline
 & 3 & 15 & 6 & 0 = R.
\end{array}
$$

The quotient is the quadratic $3x^2 + 15x + 6 = 3(x^2 + 5x + 2)$. By the quadratic formula the remaining roots of (42) are $(-5 \pm \sqrt{17})/2$. ▲

EXAMPLE 2   Find the rational zeros of

(43) $$P_2(x) \equiv x^5 + 3x^4 - 2x^3 - 5x^2 + x + 2.$$

Solution: By Theorem 11 if p/q is a zero of $P_2(x)$, then p divides 2 and q divides 1. In this case, the test set is $\mathbf{T} = \{2, 1, -1, -2\}$. Using synthetic division we test $r = 2$.

$2\rfloor$	1	3	-2	-5	1	2
		2	10	16	22	46
	1	5	8	11	23	$48 = R \neq 0$.

Clearly, $r = 2$ is not a zero of $P_2(x)$. We next test $r = 1$.

$1\rfloor$	1	3	-2	-5	1	2
		1	4	2	-3	-2
	1	4	2	-3	-2	$0 = R$.

Hence, $r = 1$ is a zero of $P_2(x)$. Further, the quotient $Q_1(x) \equiv P_2(x)/(x-1)$ is given by the last line, so we may use this in all further tests. Now $r = 1$ may be a repeated zero of $P_2(x)$, so we should test $Q_1(x)$ with $r = 1$. This gives

$1\rfloor$	1	4	2	-3	-2
		1	5	7	4
	1	5	7	4	$2 = R$.

Hence, 1 is a zero of $P_2(x)$, but it is not a repeated zero of $P_2(x)$. We next test $r = -1$.

$-1\rfloor$	1	4	2	-3	-2
		-1	-3	1	2
	1	3	-1	-2	$0 = R$.

Hence, $r = -1$ is a zero of $Q_1(x)$; therefore, it is a zero of $P_2(x)$. The scheme gives $Q_2(x) \equiv Q_1(x)/(x+1) = x^3 + 3x^2 - x - 2$. We test $Q_2(x)$ with $r = -1$ and $r = -2$,

$-1\rfloor$	1	3	-1	-2
		-1	-2	3
	1	2	-3	$1 = R$.

$-2\rfloor$	1	3	-1	-2
		-2	-2	6
	1	1	-3	$4 = R$.

We conclude that

$$P_2(x) = (x-1)(x+1)(x^3+3x^2-x-2) = (x-1)(x+1)Q_2(x)$$

where the last factor has no rational zeros. ▲

We will look at a method for approximating an irrational zero of $Q_2(x)$ in the next section.

EXAMPLE 3   Find the rational roots of the equation

(44) $x^5 + 3x^4 - 2x^3 - 5x^2 + x + 2 = 0.$

Solution: Since the polynomial in (44) is identical with the polynomial in Example 2, the roots of (44) are identical with the zeros of $P_2(x)$. Hence, the rational roots of equation (44) are $r_1 = 1$ and $r_2 = -1$. ▲

EXAMPLE 4   Find all the rational roots of

$$P_4(x) \equiv x^5 - 2x^4 - 3x^3 + 4x^2 - 2x + 1 = 0.$$

Solution: Here the test set is $\{1, -1\}$. Now $P_4(x) \neq 0$ for either of these values of x. Hence, the given equation has no rational roots. ▲

Exercise 4

In problems 1 through 10 find all of the rational roots of the given equation.

1. $x^3 - 8x^2 + 22x - 21 = 0$ 2. $x^3 + 4x^2 - 7x - 10 = 0$

3. $2x^3 - 7x^2 - 13x + 3 = 0$ 4. $x^3 - 4x^2 + 3x - 6 = 0$

5. $3x^4 - 14x^3 + 10x^2 + 11x + 2 = 0$ 6. $5x^4 + 19x^3 + 6x^2 - 7x + 1 = 0$

7. $x^4 - 5x^3 + 4x^2 + x - 15 = 0$ 8. $2x^4 + 11x^3 + 10x^2 - 21x - 18 = 0$

9. $x^5 - 5x^4 + 5x^3 + 5x^2 - 10x + 4 = 0$

10. $2x^5 - 7x^4 + 11x^3 - 8x^2 - 6x + 4 = 0$

In problems 11 and 12 give the test set **T** of possible rational roots in accordance with Theorem 11. Do not try to solve the equation.

11. $6x^4 - 16x^3 + 9x^2 + 11x - 35 = 0$

12. $10x^4 + 13x^3 - 39x^2 - 3x + 18 = 0$

In problems 13, 14, and 15 assume that $P(x)$ is a polynomial of nth degree for which all of the coefficients are real.

★13. Prove that if $x-(a+bi)$ divides $P(x)$, where a and b are real numbers and $b \neq 0$, then $x^2 - 2ax + a^2 + b^2$ divides $P(x)$.

★14. Prove that $P(x)$ can be factored into a product of linear and quadratic factors with real coefficients. *Hint:* You will need Theorem 5 and problem 13.

★15. Prove that if n is odd, then $P(x)$ has at least one real zero.

★16. Let r_1, r_2, and r_3 be distinct zeros of $a_0 x^3 + a_1 x^2 + a_2 x + a_3$. Prove that

$$a_3 = -a_0 r_1 r_2 r_3$$
$$a_2 = a_0 (r_1 r_2 + r_2 r_3 + r_3 r_1)$$
$$a_1 = -a_0 (r_1 + r_2 + r_3).$$

★17. Find formulas relating the zeros and coefficients of a fourth degree polynomial.

★**19.6 A METHOD FOR
APPROXIMATING AN
IRRATIONAL ROOT**

EXAMPLE 1

Find an approximate value for the largest irrational root of the equation

(45) $$P(x) \equiv x^5 + 3x^4 - 2x^3 - 5x^2 + x + 2 = 0.$$

Solution: This is the polynomial treated in Example 2 of the preceding section. There we found the factorization

(46) $$P(x) = (x-1)(x+1)Q(x),$$

where

(47) $$Q(x) \equiv x^3 + 3x^2 - x - 2,$$

and $Q(x)$ has no rational zeros. Thus, the zeros of $Q(x)$ are the irrational zeros of $P(x)$. Hence, we concentrate our attention on $Q(x)$.

A few simple computations will give the graph of $y = Q(x)$ shown in Figure 19.2. From this graph it is clear that $Q(x)$ has a zero in each of the intervals $(-4, -3)$, $(-1, 0)$, and $(0, 1)$. Since $Q(x)$ is a third degree polynomial, each of these zeros is a simple zero, and there are no others. Hence, the largest irrational root of $P(x) = 0$ is $r^\star$, the root of $Q(x) = 0$, which lies in $(0, 1)$.

FIGURE 19.2

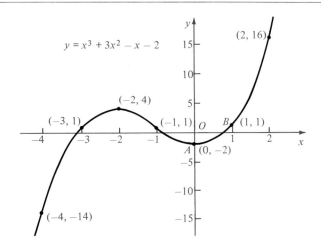

A first approximation to $r^\star$ can be obtained by using linear interpolation in the interval $(0, 1)$. To see the situation more clearly, we enlarge the graph and label the points of interest (see Figure 19.3). Then a first approximation to $r^\star$ is r_1, the x-coordinate of the point of intersection of the line AB with the x-axis. We compute r_1 using the same scheme for interpolation that we used for the trigonometric functions and for the logarithm function.

FIGURE 19.3

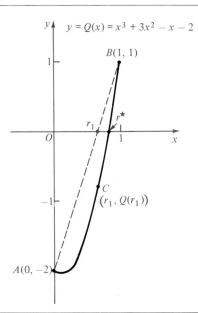

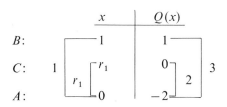

This gives $r_1/1 = 2/3$. Hence, $r_1 = 2/3 = 0.66\cdots$. Guided by the graph or rounding off to 1 significant figure, we set $r_1 = 0.7$ and compute $Q(0.7)$. If $Q(0.7) < 0$, then we have the situation shown in Figure 19.3, and $0.7 < r^\star < 1$. If $Q(0.7) > 0$, then the point $(r_1, Q(r_1))$ lies above the x-axis and in this case $0 < r^\star < 0.7$. To compute $Q(0.7)$, synthetic substitution gives

$$
\begin{array}{c|cccc}
0.7 & 1 & 3 & -1 & -2 \\
 & & 0.7 & 2.59 & 1.113 \\
\hline
 & 1 & 3.7 & 1.59 & -0.887 = Q(0.7)
\end{array}
$$

Thus, the coordinates of C are $(0.7, -0.887)$, and C lies below the x-axis as we suspected from Figure 19.2 or Figure 19.3.

The next step is to compute $Q(0.8)$, $Q(0.9)$, ... until we have a change in sign. The first step gives

$$
\begin{array}{c|cccc}
0.8 & 1 & 3 & -1 & -2 \\
 & & 0.8 & 3.04 & 1.632 \\
\hline
 & 1 & 3.8 & 2.04 & -0.368 = Q(0.8)
\end{array}
$$

Since $Q(0.8) < 0$, we must compute $Q(0.9)$. We find

$$
\begin{array}{c|cccc}
0.9 & 1 & 3 & -1 & -2 \\
 & & 0.9 & 3.51 & 2.259 \\
\hline
 & 1 & 3.9 & 2.51 & 0.259 = Q(0.9)
\end{array}
$$

Since Q changes sign as x increases from 0.8 to 0.9, we know that $0.8 < r^\star < 0.9$. We now use this interval as the basis for an interpolation to find an approximation of $r^\star$ to two decimal places.

$$
\frac{x}{0.1} = \frac{0.368}{0.627} = 0.586\cdots.
$$

Hence (to two decimal places) $r_2 = 0.8 + 0.06 = 0.86$. If we continue this process, we will find that $Q(0.860) \approx -0.0052$ and $Q(0.861) \approx 0.0012$. From this we infer that $r^\star = 0.861$ to three decimal places. ▲

This process may be continued by hand as long as our patience and energy hold up. If we have a computer available, this process can be programmed and the root $r^\star$ can be estimated (very quickly) to any degree of accuracy within the capacity of the computer.

The method used here may not be the quickest or the most convenient way to estimate $r^\star$. It does serve to introduce the reader to the techniques and problems of an important corner of applied mathematics. Of course, if the coefficients of the polynomial are complex numbers, or the root is a complex number (or both), then the problem and the solution are much more complicated.

Exercise 5

In problems 1 through 10 compute to two decimal places the root of the given equation that lies in the indicated interval.

1. $x^3 + 2x^2 - 3x - 1 = 0$, $\qquad r^\star \in (1, 2)$

2. $2x^3 + 3x^2 - 16x - 9 = 0$, $\qquad r^\star \in (2, 3)$

3. $x^3 + 2x^2 - 3x - 1 = 0$, $\qquad r^\star \in (-3, -2)$

4. $2x^3 + 3x^2 - 16x - 9 = 0$, $\qquad r^\star \in (-4, -3)$

5. $x^3 + 2x^2 - 3x - 1 = 0$, $\qquad r^\star \in (-1, 0)$

6. $2x^3 + 3x^2 - 16x - 9 = 0$, $\qquad r^\star \in (-1, 0)$

7. $x^3 + 10x - 80 = 0$, $\qquad r^\star \in (3, 4)$

8. $x^3 + x^2 + 2 = 0$, $\qquad r^\star \in (-2, -1)$

9. $x^4 + x^3 - 13x^2 - 14x + 15 = 0$, $\quad r^\star \in (3, 4)$

★10. $x^4 - x^3 - 5x + 3 = 0$, $\qquad r^\star \in (1, 2)$

In problems 11 through 14 compute to three decimal places the root of the given equation that lies in the indicated interval.

11. $x^3 - 2x^2 + 3x - 4 = 0$, $\quad r^\star \in (1, 2)$

12. $x^3 - 3x^2 + 5x - 7 = 0$, $\quad r^\star \in (2, 3)$

13. $x^4 - 7x^2 - 3x + 5 = 0$, $\quad r^\star \in (2, 3)$

14. $x^4 - 7x^2 - 3x + 5 = 0$, $\quad r^\star \in (0, 1)$

In problems 15 through 18 find the indicated root to two decimal places. Do not use logarithms or the binomial series.

15. $\sqrt{13}$ 16. $\sqrt[3]{7}$

17. $\sqrt[3]{11}$ 18. $\sqrt[4]{3}$

19. Using the three roots found in problems 1, 3, and 5, expand the product $(x-r_1)(x-r_2)(x-r_3)$, and round off the coefficients to two decimal places. Your answer should be reasonably close to $x^3 + 2x^2 - 3x - 1 = 0$.

20. Repeat problem 19 using the roots found in problems 2, 4, and 6.

Review Questions

Try to answer the following questions as accurately as possible before consulting the text.

1. If $P(x)$ and $D(x)$ are polynomials, what can we say about $Q(x)$ and $R(x)$ in the equation $P(x) = D(x)Q(x) + R(x)$? (See Theorem 1.)

2. State the Remainder Theorem.

3. Can a polynomial have complex numbers as coefficients?

4. Are Theorems 1, 2, 3, and 4 still true if the coefficients are complex numbers?

5. If all of the coefficients of $P(x)$ are real, what can be said about the complex zeros of $P(x)$?

6. If all of the coefficients of $P(x)$ are integers, what can be said about the rational zeros of $P(x)$?

7. State the fundamental theorem of algebra. What theorem follows from it? (See Theorems 5 and 6.)

Appendices

Appendix 1. Sets

The theory of sets forms one of the deepest and most interesting domains of modern mathematics. However, we are not ready to become involved in this fascinating study. Our purpose is more modest. We merely want to review the basic ideas and the notation, so that we can supply a precision to our ideas (theorems and definitions) that may otherwise be lacking.

A1.1 THE CONCEPT OF A SET

A set is any well-defined collection of objects. Each object in a set is called an *element* of the set. We will denote sets by bold letters such as $A, B, C, \ldots$. We write

$$x \in A, \qquad y \notin A$$

(read, "x is an element of A, y is not an element of A") to indicate that x is in the set A, and y is not.

The principal character of a set is that it is well-defined. Thus, given an element x, we must have some rule for determining whether or not x is in A. For example, the collection of all intelligent persons in the city of Tampa is

495

not a well-defined set because for some individuals their status as members or nonmembers may not be clear. If the collection is ill-defined (not well-defined), we merely reject it as a proper subject for mathematics‡.

If the set does not have too many elements, we can indicate the composition by enclosing the elements in braces. Thus, the notation

$$S = \{a, b, c, d, e\}$$

means that **S** is the set consisting of the five elements a, b, c, d, and e.

The set that has no elements is called the *empty set* or the void set and is denoted by the symbol $\varnothing$. A set may have infinitely many elements. The simplest example is

(1) $$N \equiv \{1, 2, 3, \ldots\},$$

the set of natural numbers (all the positive integers).

A set is often specified by setting conditions on the elements. For example, we may let **A** be the set of all integers greater than 11 and less than 19. Then

(2) $$A = \{12, 13, 14, 15, 16, 17, 18\}.$$

There is a convenient shorthand for writing a set that is formed by setting conditions on the elements. Thus, the example set just given could be written as

(3) $$A = \{x \mid 11 < x < 19,\ x \in N\}.$$

Here x is a letter which merely represents an element of the set. The vertical bar | can be translated as the phrase "such that," and after the bar we find the conditions on x that must be satisfied for x to be an element of the set **A**. Equation (3) would read "**A** is the set of all elements x such that x is greater than 11, x is less than 19, and x is a natural number." The reader may enjoy himself by thinking of a set **A** as a club or fraternity and the material after the vertical bar as the requirement on x for membership in the club.

Let $B = \{x \mid 1 < x < 25,\ x \in N\}$. Clearly, every element in **A** is also in **B**. We describe such an event by saying that **A** is contained in **B**, and we denote this by the symbol $A \subset B$.

DEFINITION 1

The set **A** *is said to be contained in* **B**, *and we write* $A \subset B$, *if every element of* **A** *is also an element of* **B**. *We also say that* **B** *contains* **A** *and write* $B \supset A$.

‡ We observe that much of the trouble with applying mathematics to the social sciences is that the sets involved are not well-defined. For example, the statements "Jones is a red-neck," "Smith is a liberal," involve the collections of red-necks and liberals, and these collections are certainly ill-defined.

If **A** is contained in **B**, we say that **A** is a *subset* of **B**, and **B** is a *superset* of **A**. The symbol **A** ⊂ **B** does not rule out the possibility that **A** and **B** are the same set. In fact, we have

DEFINITION 2   ***Equality of Sets*** *Two sets* **A** *and* **B** *are said to be equal if* **A** ⊂ **B** *and* **B** ⊂ **A**. *We symbolize this by writing* **A** = **B**.

This means that every element of **A** is an element of **B**, and every element of **B** is an element of **A**. Clearly, the equality of two sets means that they are identical. However, the two identical sets might arise from two different sources and at first glance appear to be quite different.

EXAMPLE 1   Let **A** = {0, 1, 2, 3}, and let **B** be the sets of roots of the equation $x^4 - 6x^3 + 11x^2 - 6x = 0$. Are these sets equal?

Solution: The reader can test that each element of **A** is also a root of the given equation (makes the left side equal to zero). In Chapter 19 we will prove that such an equation cannot have more than four roots. Hence, **A** = **B**. ▲

Theorem 1
PLE

For every set **A**, we have ∅ ⊂ **A**.

Theorem 2

If **A** ⊂ **B** and **B** ⊂ **C**, then **A** ⊂ **C**.

Proof: Let x be any element of **A**. Since **A** ⊂ **B**, we see that x is also an element of **B** (by the definition of **A** ⊂ **B**). Since **B** ⊂ **C** and x is an element of **B**, we see that x is also an element of **C**. Hence, $x \in$ **A** implies that $x \in$ **C**. It follows from the definition, that **A** is contained in **C**. ∎

A1.2 UNIONS AND INTERSECTIONS OF SETS

Given a number of sets, we may make larger sets by putting them together. When we put two sets **A** and **B** together, the resulting set is called the *union* of **A** and **B** and is denoted by the symbol **A** ∪ **B** (read "A union B"). Referring to Figure A.1, the elements of **A** and **B** are represented by the points inside the two circles, respectively. Then the set **A** ∪ **B** is represented by the points of the shaded region.

DEFINITION 3   ***Union*** *The set* **C** = **A** ∪ **B** *consists of all those elements that are either in* **A**, *or in* **B**, *or in both* **A** *and* **B**. *In symbols,* $x \in$ **C** *if either* $x \in$ **A**, *or* $x \in$ **B**, *or both. The set* **C** *is called the union of* **A** *and* **B**.

Notice that an element in both **A** and **B** is still counted just once in the set **C**.

FIGURE A.1

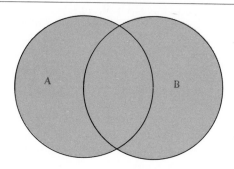

EXAMPLE 1   Find $A \cup B$ if

$$A = \{1, 2, 3, 4, 5, 6\} \qquad \text{and} \qquad B = \{4, 5, 6, 7, 8\}.$$

Solution: $A \cup B = \{1, 2, 3, 4, 5, 6, 7, 8\}$. Observe that the numbers 4, 5, and 6 occur in both sets, and yet when we form the union we list each of these numbers only once.   ▲

EXAMPLE 2   Let **D** be the set of all odd integers, and let **E** be the set of all even integers. Describe $D \cup E$.

Solution: $D \cup E$ is just the set of all integers. We denote this set by **Z**,

$$Z = D \cup E = \{0, \pm 1, \pm 2, \pm 3, \ldots\}. \quad ▲$$

Given two sets **A** and **B**, we may be interested in just those elements that belong to both of the sets. Such a set will, in general, be smaller than either one of the sets **A** and **B**. When we take the set of all elements that are common to the two sets **A** and **B**, the result is called the *intersection* of **A** and **B** and is denoted by the symbol $A \cap B$ (read, "A intersection B"). If we let the elements of **A** and **B** be represented by the points inside the two circles, as in Figure A.1, then the set $A \cap B$ is represented by the points of the shaded region in Figure A.2.

DEFINITION 4   ***Intersection***   *The set* $C = A \cap B$ *consists of all those elements that are in* **A** *and in* **B**. *In symbols,* $x \in C$ *if* $x \in A$ *and* $x \in B$. *The set* **C** *is called the intersection of* **A** *and* **B**. *If* $A \cap B$ *is empty, then the two sets are said to be disjoint.*

EXAMPLE 3   Find $A \cap B$, $D \cap E$, $A \cap D$, $A \cap E$, $B \cap D$, and $B \cap E$ for the sets of Examples 1 and 2.

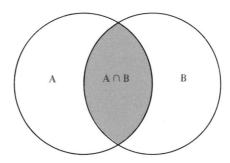

Solution: The elements common to **A** and **B** are 4, 5, and 6. Hence by the definition of intersection, **A** ∩ **B** = {4, 5, 6}. Since **D** consists of all of the odd integers and **E** consists of all the even integers, we see that **D** ∩ **E** has no elements, because there is no integer that is both odd and even (0 is even). Hence, **D** ∩ **E** = ∅, where we recall that ∅ is the symbol for the empty set.

It is easy to check each of the following

$$\textbf{A} \cap \textbf{D} = \{1, 3, 5\}, \qquad \textbf{A} \cap \textbf{E} = \{2, 4, 6\},$$

$$\textbf{B} \cap \textbf{D} = \{5, 7\}, \qquad \textbf{B} \cap \textbf{E} = \{4, 6, 8\}. \quad \blacktriangle$$

It is possible to unionize more than two sets or to take the intersection of more than two sets. For example, (**A** ∩ **B**) ∩ **D** means that one first forms a set by intersecting **A** and **B**, and then with this new set one takes the intersection with **D**. Using for **A**, **B**, and **D** the sets of Examples 1 and 2, we see that **A** ∩ **B** = {4, 5, 6}. Selecting from this set the odd integers we have

$$(\textbf{A} \cap \textbf{B}) \cap \textbf{D} = \{5\},$$

a set that consists of just one element. A set that has just one element is called a *singleton* set.

We have introduced an algebra of sets by defining two operations: ∪, union; and ∩, intersection. We can now state six simple theorems that are obviously true as soon as the meaning is clear.

Theorem 3
PLE

The union of any two sets contains either one of the sets. In symbols,

$$\textbf{A} \cup \textbf{B} \supset \textbf{A} \quad and \quad \textbf{A} \cup \textbf{B} \supset \textbf{B}.$$

Thus in general, the union of two sets is larger than either of its component parts. However, equality can occur. For example, if **Z** is the set of all integers and **D** is the set of all odd integers, then **Z** ∪ **D** = **Z**.

Theorem 4
PLE

The intersection of any two sets is contained in either one of the sets. In symbols,

$$A \cap B \subset A \quad \text{and} \quad A \cap B \subset B.$$

Equality can occur. For if Z and D denote the sets introduced above, then $Z \cap D = D$.

Theorem 5
PLE

Commutative Law for Intersections   *For any two sets A and B,*

$$A \cap B = B \cap A.$$

This means that in forming the intersection of two sets, the order is unimportant. The same is true when forming the union of two sets.

Theorem 6
PLE

Commutative Law For Unions   *For any two sets A and B,*

$$A \cup B = B \cup A.$$

If we have three sets, the symbol

$$(4) \qquad\qquad D = A \cap B \cap C$$

is at first glance confusing, because it is not clear which pair to consider first. That is, we might form $A \cap B$ and then intersect the resulting set with C. Calling the result D_1, this would be written

$$(5) \qquad\qquad D_1 = (A \cap B) \cap C.$$

If we first form $B \cap C$ and intersect the resulting set with A, we may get a set different from D_1; so to be safe we call it D_2. That is, we let

$$(6) \qquad\qquad D_2 = A \cap (B \cap C).$$

But the next theorem tells us that for set intersections such care is unnecessary, because $D_1 = D_2$.

Theorem 7
PLE

Associative Law For Intersections   *For any three sets A, B, and C,*

$$(7) \qquad\qquad (A \cap B) \cap C = A \cap (B \cap C).$$

Once this theorem is proved, the parentheses can be dropped and we can write the expression (4) to represent either (5) or (6) because, in fact, $D_1 = D_2$,

FIGURE A.3

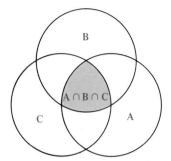

FIGURE A.4

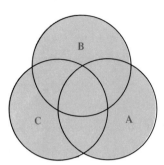

and we make this the definition of **D**. The perceptive reader will note at once that we could just define **D** in (4) to be the set of all elements that are in **A** and in **B**, and in **C**. The situation is illustrated in Figure A.3 where as usual circles represent the sets, and the intersection of the three sets is shaded.

A similar theorem holds for unions.

Theorem 8
PLE

Associative Law For Unions *For any three sets* **A**, **B**, *and* **C**,

(8) $$(A \cup B) \cup C = A \cup (B \cup C).$$

Once this theorem is proved, the parentheses can be dropped, and we can write $A \cup B \cup C$ for the set on either side of (8). This set is shaded in Figure A.4. Clearly this set consists of all elements that are in at least one of the three sets.

Diagrams such as those used in Figures A.1 through A.4 are known as *Venn diagrams*. As we have seen, they are very helpful in clarifying relations among sets.

Exercise 1

In problems 1 through 14 a collection is described by a statement. Which of these statements determine a set (the collection is well-defined) and which do not (the collection is ill-defined)?

1. The collection **N** of all positive integers

2. The collection **D** of all odd integers

3. The collection **E** of all even integers

4. The collection of all prime numbers

5. The collection of all good men living in the world

6. The collection of all integers that divide $(987, 654, 321)^{357} + 19$

7. The collection of all real numbers

8. The collection of all pairs (x, y), where x and y are real numbers

9. The collection of all pairs (x, y), where x is a straight line and y is a circle

10. The collection of all psychological concepts

11. The collection of all expert sociologists

12. The collection of all outstanding physicians

13. The collection of all graduates of a medical school approved by the AMA

14. The collection of all sets of real numbers

15. Prove that if $\mathbf{A} = \mathbf{B}$ and $\mathbf{B} = \mathbf{C}$, then $\mathbf{A} = \mathbf{C}$.

16. The negation of a symbol is denoted by putting a slanted line through the symbol. Thus, $x \notin \mathbf{A}$ means that x is not an element of $\mathbf{A}$. State clearly the meaning of $\mathbf{A} \not\subset \mathbf{B}$.

17. In each of the following list explicitly the elements in the given sets
    (a) $\{n \mid 3 \le n < 7, n \in \mathbf{N}\}$, (b) $\{n \mid -2 < n < 3, n \in \mathbf{N}\}$,
    (c) $\{n \mid -2 < n < 3, n \in \mathbf{Z}\}$, (d) $\{n \mid n^2 < 7, n \in \mathbf{Z}\}$,
    (e) $\{n \mid n \text{ divides } 1001, n \in \mathbf{N}\}$, (f) $\{n \mid n \text{ is a prime}, n \in \mathbf{N}, n < 17\}$.

18. Let $\mathbf{A} = \{1, 2, 3, 4\}$, $\mathbf{B} = \{1, 2, 3, 4, 5, 6\}$, $\mathbf{C} = \{2, 5, 7\}$, $\mathbf{D} = \{4, 5, 6, 7\}$, and $\mathbf{E} = \{1, 3, 4\}$. Find
    (a) $\mathbf{A} \cup \mathbf{B}$, (b) $\mathbf{A} \cap \mathbf{B}$, (c) $\mathbf{C} \cup \mathbf{D}$,
    (d) $\mathbf{C} \cap \mathbf{D}$, (e) $\mathbf{C} \cap \mathbf{D} \cap \mathbf{E}$, (f) $\mathbf{C} \cup \mathbf{D} \cup \mathbf{E}$,
    (g) $\mathbf{C} \cap (\mathbf{D} \cup \mathbf{E})$, (h) $\mathbf{A} \cap (\mathbf{C} \cup \mathbf{D})$, (i) $\mathbf{A} \cup \mathbf{B} \cup \mathbf{C} \cup \mathbf{D}$.

19. For the sets of problem 18 which of the following inclusion relations are true:
    (a) $\mathbf{A} \subset \mathbf{B}$, (b) $\mathbf{C} \subset \mathbf{B}$, (c) $\mathbf{C} \subset \mathbf{A} \cup \mathbf{D}$,
    (d) $\mathbf{A} \cap \mathbf{D} \subset \mathbf{E}$, (e) $\mathbf{C} \subset \mathbf{D} \cup \mathbf{E}$, (f) $\mathbf{D} \cap \mathbf{E} \subset \mathbf{B}$?

20. Prove that for any set $\mathbf{A}$
    (a) $\mathbf{A} \cup \mathbf{A} = \mathbf{A}$, (b) $\mathbf{A} \cup \varnothing = \mathbf{A}$,
    (c) $\mathbf{A} \cap \mathbf{A} = \mathbf{A}$, (d) $\mathbf{A} \cap \varnothing = \varnothing$.

21. (a) Prove that if $\mathbf{A} \subset \mathbf{C}$ and $\mathbf{B} \subset \mathbf{C}$, then $\mathbf{A} \cup \mathbf{B} \subset \mathbf{C}$.
    (b) Prove that if $\mathbf{A} \supset \mathbf{C}$ and $\mathbf{B} \supset \mathbf{C}$, then $\mathbf{A} \cap \mathbf{B} \supset \mathbf{C}$.

A1.3 THE COMPLEMENT OF A SET

Given a set **A**, we are often interested in the collection of elements that are not in **A**. This new set is called the *complement* of **A** and is denoted by **A'** (read, "**A** prime"). However, **A'** is not yet well-defined because we have not clearly stated which elements are eligible for membership in **A'**. For example, if $A = \{1, 2, 3, 4, 5, 6\}$ and we have in mind that **A** is a subset of the set of digits in the decimal system, then $A' = \{7, 8, 9, 0\}$. On the other hand, if **A** is regarded as a subset of **N**, then $A' = \{x \mid x \in N, \ x > 6\} = \{7, 8, 9, 10, 11, ...\}$. The solution to our difficulty is reasonably clear. In any sensible discussion there is usually some specified over-all set on which the discussion centers. This set is denoted by **U** and is called *the universal set for the discussion* or merely the *universal set*. It may change as the topic changes, but, in general, the composition of the universal set is obvious and, in fact, one often omits mentioning the set just because it *is* obvious. For example, if we are involved in a discussion of factoring numbers, then **U** would normally be the set of positive integers. On the other hand, if we are discussing the possibility of solving certain equations, then **U** would be the set of all real numbers or the set of all complex numbers.

It may seem that the changing nature of **U**, and, hence, **A'** might cause trouble. In actual practice it never does, because in any given problem the composition of **U** is either obvious, or is explicitly stated, and remains constant throughout the discussion of that problem.

DEFINITION 5

Complement *Let* **U** *be the universal set and let* **A** *be a subset of* **U**. *Then the complement of* **A** *is the set of all elements in* **U** *that are not in* **A**. *This set is symbolized by writing* **A'** *or* **U** − **A**.

EXAMPLE 1

Let **U** be the set of all positive integers, let **A** be the set of prime numbers, and let **B** be the set of perfect squares. Give the first 10 numbers in **A'** and **B'**.

Solution: $A = \{2, 3, 5, 7, 11, 13, 17, 19, 23, ...\}$, and $B = \{1, 4, 9, 16, ...\}$. Hence,

$$A' = \{1, 4, 6, 8, 9, 10, 12, 14, 15, 16, ...\}$$

and

$$B' = \{2, 3, 5, 6, 7, 8, 10, 11, 12, 13, ...\}. \quad \blacktriangle$$

In making a Venn diagram it is customary to let a box represent the universal set, with the usual circles representing certain subsets of the universal set. In the diagrams shown in Figures A.5 and A.6 the two circles representing the sets **A** and **B** divide the box into four regions. If x is any element in **U**, we can represent its member or nonmembership in the two sets **A** and **B** by a suitably selected region. For example, if x is not in **A** and not in **B**, then it is

FIGURE A.5

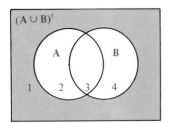

$(A \cup B)'$

FIGURE A.6

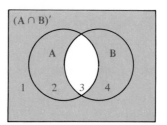

$(A \cap B)'$

represented by a point in the region labeled 1. Only three other cases are possible:

(II) $x \in A$, and $x \notin B$ is represented by the region labeled 2.
(III) $x \in A$, and $x \in B$ is represented by the region labeled 3.
(IV) $x \notin A$, and $x \in B$ is represented by the region labeled 4.

Since the Venn diagram in Figure A.5 represents all four possible cases, it can be used to prove certain theorems as we now show. The set $(A \cup B)'$ is shown shaded in Figure A.5. On the other hand, the set $A' \cap B'$ is represented by the same shaded region. Hence,

$$(9) \qquad\qquad (A \cup B)' = A' \cap B'.$$

Similarly, the shaded region in Figure A.6 represents the set $(A \cap B)'$ and it also represents the set $A' \cup B'$. Consequently,

$$(10) \qquad\qquad (A \cap B)' = A' \cup B'.$$

Theorem 9   ***De Morgan's Laws***   *For any two sets* **A** *and* **B** *contained in a universal set* **U**, *equations* (9) *and* (10) *hold.*

Some simpler relations are given in

Theorem 10   *If* **U** *is the universal set,* $\varnothing$ *is the empty set, and* **A** *is any set contained in* **U**, *then*
PLE

$$(11) \qquad\qquad U' = \varnothing, \qquad \varnothing' = U,$$

$$(12) \qquad\qquad A \cup A' = U, \qquad A \cap A' = \varnothing,$$

$$(13) \qquad\qquad (A')' = A.$$

A1.4 THE DISTRIBUTIVE LAWS
FOR SETS

All of the material in the first three sections is reasonably simple. The only difficulty might be the new symbols introduced. In this section we consider two laws that relate $\cup$ and $\cap$. Although relatively easy, these laws are slightly more complicated than those encountered up to now.

Theorem 11   ***The First Distributive Law for Sets***   *If* **A**, **B**, *and* **C** *are any three sets, then*

(14) $$A \cap (B \cup C) = (A \cap B) \cup (A \cap C).$$

Proof: The quickest way to see that Theorem 11 is correct is to use a Venn diagram. In Figure A.7 the three circles representing the three sets divide the box into 8 distinct regions which we have numbered $1, 2, \ldots, 8$. If x is any element in **U**, its relation as an element or nonelement of these three sets can be represented by a point in a suitably selected region. For example, if x is in **A**, **B**, and **C**, it is represented by a point in region 1. If x is in **B** and **C** but not in **A**, then it is represented by a point in region 3. If it is in **A** but not in **B** and not in **C**, then region 5 is the proper place for its representative point. If x is not in any of the three sets, then it is represented by a point in region 8. The remaining possibilities are left for the student to check. Since every possible case is represented by some region in Figure A.7, it is quite proper to use this figure as the basis for a valid proof of Theorem 11.

Now let **R** denote the set on the right side of equation (14), and **L** denote the set on the left side. The set **B** $\cup$ **C** is represented by the regions 1, 2, 3, 4, 6, and 7, and these are shown shaded with horizontal lines in Figure A.8. Intersecting this set with **A** gives the regions 1, 2, and 4. This is the set **L** and is shown as heavily shaded in Figure A.9.

For the set **R** we refer to Figure A.9. Here **A** $\cap$ **B** is shaded with horizontal lines, **A** $\cap$ **C** is shaded with vertical lines. Putting these regions together yields the region representing the union **R**. This is clearly the same region that we obtained for **L**, so **L** = **R**.   ∎

FIGURE A.7

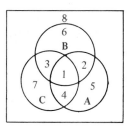

FIGURE A.8

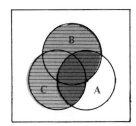

FIGURE A.9

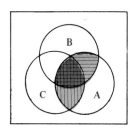

If we wish to avoid reference to a picture, the proof would go as follows: Let $x \in A \cap (B \cup C)$. Then x is in A, and it is either in B or in C. Hence, it is either in $A \cap B$ or it is in $A \cap C$. In either case, it is in the union of these two sets, so that $x \in (A \cap B) \cup (A \cap C)$. This proves that $L \subset R$. We leave it for the student to reverse the argument in order to show that $L \supset R$. It will then follow that $L = R$.

If we interchange the union and intersection signs in equation (14), we get a new equation which is also true.

Theorem 12   ***The Second Distributive Law for Sets***   *If* A, B, *and* C *are any three sets, then*

(15) $$A \cup (B \cap C) = (A \cup B) \cap (A \cup C).$$

We leave it for the student to prove this theorem by showing that both sides of (15) yield the shaded portion of Figure A.10.

FIGURE A.10

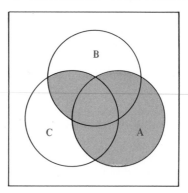

Exercise 2

1. Make a Venn diagram showing each of the following sets
   (a) $\{x \mid x \in A \text{ and } x \notin B\}$,
   (b) $\{x \mid x \in A, x \notin B, \text{ and } x \notin C\}$,
   (c) $\{x \mid x \in A, x \in B, \text{ and } x \notin C\}$,
   (d) $\{x \mid \text{either } x \in A \text{ or } x \in B, \text{ and } x \notin C\}$.

2. Complete the proofs of Theorems 10 and 12.

3. Subtraction of sets is defined as follows. If A and B are any two sets, then $A - B$ is the set of all x's that are in A, but not in B. In other words, subtracting B from A is effected by removing from A all of the elements that

are also in **B** (see problem 1a). Prove that $\mathbf{A} - \mathbf{B} = \mathbf{A}$ if and only if $\mathbf{A} \cap \mathbf{B} = \varnothing$.

4. Use a Venn diagram to prove that:

(a) $(\mathbf{A} - \mathbf{B}) - \mathbf{C} = \mathbf{A} - (\mathbf{B} \cup \mathbf{C})$,

(b) $\mathbf{A} \cap (\mathbf{B} - \mathbf{C}) = \mathbf{A} \cap \mathbf{B} - (\mathbf{A} \cap \mathbf{C})$,

(c) $(\mathbf{A} - \mathbf{B}) \cup (\mathbf{A} - \mathbf{C}) = \mathbf{A} - (\mathbf{B} \cap \mathbf{C})$,

(d) $(\mathbf{A} - \mathbf{B}) \cap (\mathbf{A} - \mathbf{C}) = \mathbf{A} - (\mathbf{B} \cup \mathbf{C})$.

5. Let $\mathbf{U} = \{0, 1, 2, 3, 4, 5, 6, 7, 8, 9\}$, $\mathbf{A} = \{0, 1, 2, 3\}$, $\mathbf{B} = \{0, 2, 4, 6, 8\}$, and $\mathbf{C} = \{1, 2, 3, 5, 6, 7, 9\}$. Find

(a) $(\mathbf{A} \cup \mathbf{B})'$,

(b) $\mathbf{A}' \cap \mathbf{B}'$,

(c) $(\mathbf{A} \cup \mathbf{C})'$,

(d) $\mathbf{A}' \cap \mathbf{C}'$,

(e) $(\mathbf{A} \cap \mathbf{C})'$,

(f) $\mathbf{A}' \cup \mathbf{C}'$.

6. Prove that if $\mathbf{A} \supset \mathbf{B}$, then $\mathbf{A}' \subset \mathbf{B}'$.

Appendix 2.   The Real Numbers

In this appendix we prove many of the theorems mentioned in Chapter 2, leaving a few of the easier ones for the student. However, proofs given in Chapter 2 will not be repeated here. To avoid unnecessary turning of pages, we will reproduce the eleven axioms for a field that were stated in Chapter 2. Because we go into much greater detail, the numbering of equations and theorems can not coincide with that used in Chapter 2.

We should mention that many of the eleven axioms given in Chapter 2 (and repeated here) are in fact theorems that can be derived from a smaller and simpler set of axioms, known as the Peano axioms. For the benefit of the curious reader, we have included a very brief account of the Peano axioms in the next appendix.

A2.1   THE AXIOMS FOR ADDITION

These are:

AXIOM 1

The Closure Property for Addition   *For every pair a, b in* **R**, *there is a uniquely determined element in* **R**, *called the sum of a and b and denoted by $a + b$.*

AXIOM 2   ***The Commutative Law for Addition***   *For every pair a, b in* **R**

(1) $$a + b = b + a.$$

AXIOM 3   ***The Associative Law for Addition***   *For every a, b, c in* **R**

(2) $$(a+b) + c = a + (b+c).$$

AXIOM 4   *There is a real number called zero (denoted by 0) in* **R** *such that for every a in* **R**

(3) $$a + 0 = a.$$

AXIOM 5   *For each a in* **R** *there is an element in* **R**, *denoted by $-a$ or by $(-a)$, such that*

(4) $$a + (-a) = 0.$$

Subtraction is defined in terms of addition. This is

DEFINITION 1   ***Subtraction***   *The difference $b-a$ is defined by*

(5) $$b - a = b + (-a).$$

We return to Axiom 4 and ask, "How many 'zeros' are there in **R**?" Of course Axiom 4 tells us that there is one such number denoted by 0, but it is possible that there are other "zeros" in **R**. By this we mean that there is some other number in **R** that has the same property that is expressed by equation (3). The word "unique" is used in mathematics to indicate that there is *exactly one* item of the type under consideration.

Theorem 1   *The identity element for addition is unique.*

Proof: Suppose that there are two identity elements. We denote these by 0 and $0^\star$. Then by the identity property (Axiom 4) for 0

(6) $$0^\star + 0 = 0^\star.$$

If $0^\star$ is also an identity for addition, then Axiom 4 applied to $0^\star$ gives

(7) $$0 + 0^\star = 0.$$

But by the Commutative Law (Axiom 2), the left sides of (6) and (7) are equal. Consequently, $0^\star = 0$. Hence, there is only one identity element. ∎

In a similar manner, we have

Theorem 2   For each a in **R**, the additive inverse is unique.

Proof: Suppose that $-a$ and b are two additive inverses for a. Then by the definition of an additive inverse, $a+b=0$. Adding $(-a)$ to both sides of $a+b=0$ we obtain

$$(-a)+(a+b) = (-a)+0 = (-a)$$
$$((-a)+a)+b = (-a) \qquad \text{(the Associative Law for Addition)}$$
$$0+b = (-a)$$
$$b = (-a). \quad \blacksquare$$

COROLLARY
PLE

If x,y are in **R** and $x+y=0$, then $x=-y$ and $y=-x$.

By Axiom 4, we know that $0+0=0$. If we use the corollary with $x=0$ and $y=0$, we obtain

Theorem 3   For the zero element in **R** we have

(8) $$-0 = 0.$$

Theorem 4   If $a+b=a+c$, then $b=c$.
PLE

Theorem 5   For every a,b in **R**
PLE
(9) $$-(-a) = a,$$

(10) $$-(a+b) = (-a)+(-b),$$

and

(11) $$-(a-b) = (-a)+b.$$

To illustrate the method, we prove that equation (10) is true. By the Associative and Commutative Laws for Addition

$$(a+b)+((-a)+(-b)) = a+(-a)+b+(-b) = 0+0 = 0.$$

Consequently, by the corollary to Theorem 2, we have

$$-(a+b) = (-a)+(-b).$$

A2.2 MULTIPLICATION AND THE DISTRIBUTIVE LAW

The five axioms of multiplication are similar to the ones for addition. Further, there is one axiom that relates addition and multiplication.

AXIOM 6   ***The Closure Property for Multiplication***   *For every pair a, b in* **R** *there is a uniquely determined element in* **R**, *called the product of a and b and denoted by ab (or by $a \cdot b$, or by $a \times b$).*

AXIOM 7   ***The Commutative Law for Multiplication***   *For every a, b in* **R**

(12) $$ab = ba.$$

AXIOM 8   ***The Associative Law for Multiplication***   *For every a, b, c in* **R**

(13) $$(ab)c = a(bc).$$

AXIOM 9   *There is a real number called one (denoted by 1) in* **R** *such that for every a in* **R**

(14) $$a \cdot 1 = a.$$

AXIOM 10   *For each a in* **R**, *such that $a \neq 0$, there is an element in* **R**, *denoted by $1/a$ $\left(\text{or by } \dfrac{1}{a} \right)$, such that*

(15) $$a\left(\frac{1}{a}\right) = 1.$$

Division is defined in terms of multiplication. This is

DEFINITION 2   ***Division***   *If $b \neq 0$, the quotient a/b is defined by*

(16) $$\frac{a}{b} = a\left(\frac{1}{b}\right).$$

In other words, to divide a by b, we multiply a by the reciprocal of b. If $b = 0$, then the quotient a/b is not defined. Following the method used in Section A2.1 we can prove

Theorem 6   *The identity element for multiplication is unique.*
PLE

Theorem 7   *For each a in* **R** *such that $a \neq 0$, the multiplicative inverse (reciprocal) is unique.*
PLE

COROLLARY
PLE

If x and y are in **R** *and xy = 1, then y = 1/x and x = 1/y.*

Addition and multiplication are related by

AXIOM 11

The Left Distributive Law   For every a, b, c *in* **R**,

(17)
$$a(b+c) = ab + ac.$$

Theorem 8

The Right Distributive Law   *For every* a, b, c *in* **R**

(18)
$$(a+b)c = ac + bc.$$

Proof: We use the Commutative Law (Axiom 7) and the Left Distributive Law (Axiom 11). These give

$$(a+b)c = c(a+b) = ca + cb = ac + bc. \quad \blacksquare$$

Theorem 9

For each a in **R**, *we have* $a \cdot 0 = 0 \cdot a = 0.$

Proof: Since $1+0 = 1$, the Left Distributive Law (Axiom 11) gives

$$a \cdot (1+0) = a \cdot 1$$
$$a \cdot 1 + a \cdot 0 = a$$
(19)
$$a + a \cdot 0 = a.$$

But $a+0 = a$, and the identity element for addition is unique. Hence, from (19) we have $a \cdot 0 = 0$. By the Commutative Law for Multiplication we also have $0 \cdot a = 0$. $\blacksquare$

If we use the definition of division and Axioms 7, 8, and 10, we can prove

Theorem 10
PLE

If a and b are in **R** *and* $b \neq 0$, *then*

(20)
$$\frac{b}{b} = 1, \qquad b\left(\frac{a}{b}\right) = a, \qquad \frac{a}{1} = a.$$

Note that in particular $1/1 = 1$. The second item in (20) is an important cancellation law that is extremely useful in algebraic manipulation. This same item also gives the

COROLLARY

If $a/b = 0$, then $a = 0$.

Theorem 11   *If a and b are in* **R** *and ab* $= 0$, *then either a* $= 0$ *or b* $= 0$.

Proof: If $a = 0$, there is nothing to prove. Assume that $a \neq 0$. Then it has a reciprocal $1/a$. If we multiply both sides of the equation $0 = ab$ by $1/a$ we have

$$(1/a) \cdot 0 = (1/a)(ab)$$
$$0 = ((1/a) \cdot a)b \qquad \text{(Associative Law for Multiplication)}$$
$$0 = 1 \cdot b = b. \quad \blacksquare$$

If we multiply both sides of $1 + 1 = 2$ by a and use the Distributive Law, we obtain

Theorem 12   *For any a in* **R**

(21) $$a + a = 2a.$$

Theorem 13   *If there are n terms in the sum* $a + a + \cdots + a$, *then* $a + a + \cdots + a = na$.
PLE

These last two theorems may seem trivial, but they are important. For example, we must have Theorem 12 in order to prove

Theorem 14   *If a and b are in* **R**, *then*

(22) $$(a+b)^2 = a^2 + 2ab + b^2.$$

Proof: Of course the notation a^2 means $a \cdot a$. Then

$$(a+b)^2 = (a+b)(a+b)$$
$$= (a+b)a + (a+b)b \qquad \text{(Left Distributive Law)}$$
$$= aa + ba + ab + bb \qquad \text{(Right Distributive Law)}$$
$$= a^2 + ab + ab + b^2 = a^2 + 2ab + b^2. \quad \blacksquare$$

Theorem 15   ***The Cancellation Law for Products***   *If* $a, b,$ *and* c *are in* **R**, $c \neq 0$, *and*

(23) $$ca = cb,$$

then (on cancellation)

(24) $$a = b.$$

Proof: We multiply both sides of equation (23) by $1/c$, and use the Associative Law for Multiplication. This gives

$$\frac{1}{c}(ca) = \frac{1}{c}(cb)$$

$$\left(\frac{1}{c} \cdot c\right)a = \left(\frac{1}{c} \cdot c\right)b.$$

Hence, $1 \cdot a = 1 \cdot b$ or $a = b$. ∎

Exercise 1

In problems 1 through 9 prove the given assertion using only the eleven axioms and any previously proved theorems. Naturally, all letters represent arbitrary real numbers.

1. $a(b+c+d) = ab + ac + ad$

2. $(a+b+c)d = ad + bd + cd$

3. If $d \neq 0$, then $\dfrac{a}{d} + \dfrac{b}{d} = \dfrac{a+b}{d}$ and $\dfrac{a}{d} + \dfrac{b}{d} + \dfrac{c}{d} = \dfrac{a+b+c}{d}$.

4. $(a+b)(c+d) = ac + ad + bc + bd$

5. $(a+b)^3 = a^3 + 3a^2b + 3ab^2 + b^3$

6. If $a \neq 0$ and $b \neq 0$, then $\dfrac{a}{b} \cdot \dfrac{b}{a} = 1$.

★7. If $a \neq 0$, then $1/(1/a) = a$. *Hint:* Use the corollary to Theorem 7.

★8. If $a \neq 0$ and $b \neq 0$, then $\dfrac{1}{a} \cdot \dfrac{1}{b} = \dfrac{1}{ab}$.

9. If $b \neq 0$, then $1/b \neq 0$.

**A2.3 MORE PROPERTIES OF
THE REAL NUMBERS**

We now derive the familiar rules for products and quotients that involve negative signs.

Theorem 16 For any a, b, c in **R**,

(25) $\qquad\qquad\qquad (-a)b = -(ab) = a(-b),$

(26) $$(-a)(-b) = ab,$$

(27) $$a(b-c) = ab - ac.$$

If $b \neq 0$, then

(28) $$\frac{-a}{b} = -\frac{a}{b}, \qquad \frac{a}{-b} = -\frac{a}{b}.$$

Proof: We multiply both sides of $a+(-a) = 0$ by b and use the Right Distributive Law (Theorem 8). This gives

(29) $$ab + (-a)b = 0 \cdot b = 0.$$

Thus, $(-a)b$ is the additive inverse of ab. In symbols, $(-a)b = -(ab)$. The second part of (25) now follows from the Commutative Law for Multiplication (Axiom 7), when applied to both sides of $(-b)a = -(ba)$.

We next apply (twice) the result just obtained to the product $(-a)(-b)$. This gives

(30) $$(-a)(-b) = -[a(-b)] = -[-(ab)].$$

By the first part of Theorem 5, the right side is ab.

To prove (27) we use the Left Distributive Law and the definition of subtraction. This yields

$$a(b-c) = a(b+(-c)) = ab + a(-c) = ab + [-(ac)] = ab - ac.$$

The definition of division and equation (25) give

(31) $$\frac{-a}{b} = (-a)\left(\frac{1}{b}\right) = -\left(a \cdot \frac{1}{b}\right) = -\frac{a}{b}.$$

If $b \neq 0$, equation (26) yields

(32) $$(-b)\left(-\frac{1}{b}\right) = b \cdot \frac{1}{b} = 1.$$

Consequently, by the corollary to Theorem 7 we see that $-(1/b)$ is the reciprocal of $-b$. In symbols,

(33) $$-\frac{1}{b} = \frac{1}{-b}.$$

Using (33) and the second part of (25) we obtain

$$\frac{a}{-b} = a\left(\frac{1}{-b}\right) = a\left(-\frac{1}{b}\right) = -\left(a\frac{1}{b}\right) = -\frac{a}{b}. \quad \blacksquare$$

The rules established in Theorem 16 are extremely useful in proving algebraic identities.

EXAMPLE 1   Prove that for all a, b in **R**

(34) $$a^2 - b^2 = (a+b)(a-b).$$

Solution: We start with the right side of (34). Then

$$
\begin{aligned}
(a+b)(a-b) &= (a+b)(a+(-b)) && \text{(Definition 1)}\\
&= (a+b)a + (a+b)(-b) && \text{(Left Distributive Law)}\\
&= aa + ba + a(-b) + b(-b) && \text{(Right Distributive Law)}\\
&= a^2 + ab + (-ab) + (-b^2) && \\
& && \text{(Commutative Law, Theorem 16)}\\
&= a^2 + 0 + (-b^2) = a^2 - b^2. \quad \blacktriangle
\end{aligned}
$$

Theorem 17   *If a, b, c, and d are in **R**, $b \neq 0$, and $d \neq 0$, then*

(35) $$\frac{a}{b} \cdot \frac{c}{d} = \frac{ac}{bd}.$$

Proof: We use the Associative and Commutative laws for multiplication together with problem 8 of Exercise 1. This gives

$$\frac{a}{b} \cdot \frac{c}{d} = \left(a \cdot \frac{1}{b}\right)\left(c \cdot \frac{1}{d}\right) = (ac)\left(\frac{1}{b} \cdot \frac{1}{d}\right) = (ac)\left(\frac{1}{bd}\right) = \frac{ac}{bd}. \quad \blacksquare$$

Since $d/d = 1$, equation (35) gives

Theorem 18   **The Cancellation Law for Fractions**   *If a, b, and d are in **R**, $b \neq 0$, and $d \neq 0$, then*

(36) $$\frac{ad}{bd} = \frac{a}{b}.$$

Although (36) is called a cancellation law, the equation can be read from right to left. Thus, equation (36) states that we can insert (instead of cancel) the common factor d on the right side to obtain the left side. This operation is used in the proof of the standard formula for adding two fractions.

Theorem 19   If a, b, c, and d are in $\mathbf{R}$, $b \neq 0$, and $d \neq 0$, then

(37)
$$\frac{a}{b} + \frac{c}{d} = \frac{ad + bc}{bd}.$$

Proof: Using (36) and the Left Distributive Law, we have

$$\frac{a}{b} + \frac{c}{d} = \frac{ad}{bd} + \frac{bc}{bd} = (ad)\frac{1}{bd} + (bc)\frac{1}{bd}$$

$$= (ad + bc)\frac{1}{bd} = \frac{ad + bc}{bd}. \quad \blacksquare$$

The Cancellation Law for Fractions is also used in the proof of

Theorem 20   Suppose that a, b, c, and d are in $\mathbf{R}$, $b \neq 0$, and $d \neq 0$. If
PLE

(38)
$$\frac{a}{b} = \frac{c}{d},$$

then

(39)
$$ad = bc.$$

Conversely, if (39) *is true, then* (38) *is also true.*

Theorem 21   If a, b, c, and d are in $\mathbf{R}$, $b \neq 0$, $c \neq 0$, and $d \neq 0$, then

(40)
$$\frac{\dfrac{a}{b}}{\dfrac{c}{d}} = \frac{ad}{bc}.$$

Proof: We use the criterion stated in Theorem 20. Indeed

(41)
$$\frac{a}{b} \cdot (bc) = \frac{c}{d} \cdot (ad)$$

since both sides give ac. $\blacksquare$

Exercise 2

In problems 1 through 5 prove the given assertion using only the eleven axioms and any previously proved theorems.

1. $(-1)a = a(-1) = -a$

2. If $b \neq 0$, then $\dfrac{-a}{-b} = \dfrac{a}{b}$.

3. $a^3 - b^3 = (a-b)(a^2 + ab + b^2)$

4. $a^3 + b^3 = (a+b)(a^2 - ab + b^2)$

5. $a^4 - b^4 = (a-b)(a+b)(a^2 + b^2)$

★A2.4 THE CONCEPT OF A FIELD

A *binary operation* defined on a set **S** is an operation that associates with every pair of elements of **S**, a uniquely determined third element.

Addition is an example of a binary operation defined on the set **R** of real numbers, because with every pair a, b in **R**, it associates the sum $a+b$. Multiplication is a second example of a binary operation, since it associates the product ab with every pair a, b in **R**.

We want to consider binary operations that resemble addition and multiplication but are not necessarily the same. For this purpose we introduce the two new symbols $\oplus$ and $\otimes$, which may be read, "abstract plus" and "abstract times," respectively. They may also be read, "circle plus" and "circle times." The "circles" are used to remind us that these operations are not necessarily the ordinary operations of addition and multiplication. If the set **S** is very large, then $a \oplus b$ and $a \otimes b$ are best given by a rule, but if **S** is a small set, then the abstract sum and the abstract product can be given by a table. For example, suppose that **S** is the set of digits $\{0, 1, 2, 3, 4\}$ and suppose further that $a \oplus b$ and $a \otimes b$ are defined by the entries in Tables 1 and 2, respectively.

TABLE 1 $a \oplus b$

a \ b	0	1	2	3	4
0	0	1	2	3	4
1	1	2	3	4	0
2	2	3	4	0	1
3	3	4	0	1	2
4	4	0	1	2	3

TABLE 2 $a \otimes b$

a \ b	0	1	2	3	4
0	0	0	0	0	0
1	0	1	2	3	4
2	0	2	4	1	3
3	0	3	1	4	2
4	0	4	3	2	1

The values for b and a are given in the top and left border, respectively. The result of the binary operation is given in the body of the table. As indicated by the colored bands, Table 1 gives $3 \oplus 4 = 2$, and Table 2 gives $2 \otimes 4 = 3$.

With considerable effort it can be proved that if $\mathbf{S} = \{0, 1, 2, 3, 4\}$ and $\oplus$ and $\otimes$ are defined by Tables 1 and 2, then all of the eleven axioms are satisfied. Consequently, the set of real numbers $\mathbf{R}$ is not the only set that satisfies the eleven axioms.

DEFINITION 3   ***Field***   *Let* $\mathbf{F}$ *be a set with at least two elements, and let* $\oplus$ *and* $\otimes$ *be two binary operations defined on* $\mathbf{F}$. *If* $\mathbf{F}$ *satisfies each of the Axioms 1 through 11 when* $+$ *is replaced by* $\oplus$ *and* $\times$ *is replaced by* $\otimes$, *then* $\mathbf{F}$ *is called a field with respect to these two operations.*

The real numbers form a field with respect to $+$ and $\times$. Rut $\mathbf{R}$ is not the only field. The set $\mathbf{F} = \{0, 1, 2, 3, 4\}$ is also a field with respect to the operations defined by Tables 1 and 2. This field has only a finite number of elements. The student who wishes to learn more about fields may consult any elementary book on abstract algebra.

Exercise 3

1. Explain why $\mathbf{Z}$ is not a field. Which of the eleven axioms does the set $\mathbf{Z}$ fail to satisfy?

2. Which of the eleven axioms does the set $\mathbf{N}$ fail to satisfy?

3. Is $\mathbf{Q}$ a field?

Appendix 3.   The Peano Axioms

The set of natural numbers $\mathbf{N}$ is just the set of counting numbers $\{1, 2, 3, \ldots\}$ and the discovery of this set predates the recorded history of man. The set was in common use for thousands of years before attempts were made to analyze the structure of $\mathbf{N}$ and to give a set of axioms for $\mathbf{N}$. The simplest set of axioms was the creation of Giuseppe Peano.‡

The fundamental character of the set $\mathbf{N}$ is that for each a in $\mathbf{N}$ there is a unique follower or successor, the number $a + 1$, that is also in $\mathbf{N}$. However, "addition" should be a consequence of the axioms, so that we wish to avoid the use of $+$ in the axioms. This is accomplished by introducing a neutral and inoffensive symbol, such as a' (read, "a prime"), for the follower of a. Further, the "first" element in $\mathbf{N}$ has a special position in the set, because it is not the follower of any other element in the set. As may be expected, we use the special symbol 1 for this particular element.

‡ Giuseppe Peano (1858–1932) was one of Italy's leading mathematicians. He also discovered a continuous curve that runs through every point in the interior of a square.

AXIOM 1   *The set* **N** *of natural numbers is not empty. It contains a particular element called one and denoted by* 1.

AXIOM 2   *For each a in* **N** *there is a unique element a′ in* **N**, *called the successor of a.*

AXIOM 3   *The number* 1 *is not the successor of any number in* **N**.

This means that if *a* is any element in **N**, then $a′ \neq 1$.

AXIOM 4   *Each element of* **N** *is the successor of at most one element in* **N**.

This means that if $a′ = b′$, then $a = b$.

AXIOM 5   *Let* **M** *be a set of natural numbers with the following properties:*

(A) 1 *is in* **M**.
(B) *If a is in* **M**, *then a′ is in* **M**.

Then **M** = **N**.

These are the five Peano axioms. Clearly, Axiom 5 is the *Principle of Mathematical Induction*. In Chapter 15 we give a detailed treatment of its many applications.

Of course, the notation for the elements of the set **N** is purely arbitrary, but on our planet the symbols

$$1′ = 2, \quad \text{the successor of 1 is 2,}$$

$$2′ = 3, \quad \text{the successor of 2 is 3,}$$

$$3′ = 4, \quad \text{the successor of 3 is 4, etc.}$$

have been adopted almost universally.

It is possible to define addition in a variety of ways, if we merely require that for each pair *a* and *b* in **N** there is associated an element of **N** called its *sum* and denoted by $a+b$. However, there is one and only one way of defining addition, if we require that for every pair *a* and *b* in **N**,

(1) $$a + 1 = a′$$

and

(2) $$a + b′ = (a+b)′.$$

Similar remarks apply to the product ab (also denoted by $a \cdot b$ and $a \times b$). It is completely arbitrary, but becomes uniquely determined if we require that for every pair a and b in $\mathbf{N}$,

(3)
$$a \cdot 1 = a$$

and

(4)
$$a \cdot b' = a \cdot b + a.$$

Naturally, in this text and in general, these uniquely determined definitions of addition and multiplication are the ones used. The nervous student may relax, because these are also just the definitions he has been practicing for most of his life.

Using the Peano axioms one can prove the following theorems.

Theorem 1   ***Commutative Law of Addition***   *If a and b are in $\mathbf{N}$, then*

(5)
$$a + b = b + a.$$

Theorem 2   ***Associative Law of Addition***   *If a, b, and c are in $\mathbf{N}$, then*

(6)
$$(a+b) + c = a + (b+c).$$

Theorem 3   ***Commutative Law of Multiplication***   *If a and b are in $\mathbf{N}$, then*

(7)
$$ab = ba.$$

Theorem 4   ***Associative Law of Multiplication***   *If a, b, and c are in $\mathbf{N}$, then*

(8)
$$(ab)c = a(bc).$$

Theorem 5   ***Distributive Law***   *If a, b, and c are in $\mathbf{N}$, then*

(9)
$$a(b+c) = ab + ac.$$

Starting with the natural numbers, one can enlarge the system to include (I) zero and the negative integers, (II) the rational numbers, and (III) the irrational numbers. The set of real numbers that one obtains is a complete ordered field. Each axiom stated in Chapters 2 and 3 is, in fact, a theorem that can be proved using only the Peano axioms. The reader may find this development of the real number system in *Foundations of Analysis*, by E. Landau, New York: Chelsea Publishing Co., 1951. However, the reader is advised to postpone this study. At this stage of his mathematical education, there are many other topics that are more important and more useful.

Exercise 1

1. Let $\mathbf{A} = \{1, 2, 3, 4, 5, 6\}$. Set $1' = 2$, $2' = 3$, $3' = 4$, $4' = 5$, $5' = 6$. This set, with the operation x', satisfies some of the Peano axioms, but not all. Which ones does the set $\mathbf{A}$ satisfy?

2. Suppose that in the set of problem 1 we put $6' = 1$, while keeping all of the other successors the same. Which of the Peano axioms are satisfied?

3. Suppose that in the set of problem 1 we put $6' = 2$, while keeping all of the other successors in problem 1 the same. Which of the Peano axioms are satisfied?

4. Suppose $\mathbf{N} = \{1, 2, 3, \ldots\}$, the set of natural numbers, but we define the successor relationship by $a' = a + 2$, for every a in $\mathbf{N}$. Which of the Peano axioms are satisfied?

5. The empty set satisfies some of the Peano axioms. Which ones?

6. Select natural numbers at random, and check that Theorems 1 through 5 are true for the particular numbers selected.

7. Suppose that we set $a + b = 5$ and $ab = 5$ for every a and b in $\mathbf{N}$. Naturally with these definitions, equations (1), (2), (3), and (4) are no longer true. What about Theorems 1, 2, 3, 4, and 5? With this new definition of sum and product, which of these theorems are still true?

8. Let $\mathbf{N}^\star = \mathbf{N} \cup \{a, b, c\}$. Define x' in the usual way for x in $\mathbf{N}$. For x in $\{a, b, c\}$ set $a' = b$, $b' = c$, and $c' = a$. Prove that with this definition of successor, the system $\mathbf{N}^\star$ satisfies the first four Peano axioms but not the fifth.

9. Using the results of problem 8, show that one cannot prove Axiom 5 (as a theorem) using only Axioms 1, 2, 3, and 4.

Appendix 4.   The Simplest Irrational Number

The simplest irrational number is $\sqrt{2}$. To prove that $\sqrt{2}$ is irrational we must prove

Theorem 1   *If x is a rational number, then $x^2 \neq 2$.*

Proof: Let $x = A/B$ where A and B are integers. If A and B have any common integer factors, we remove them by cancellation. Thus, if $A = ad$ and $B = bd$

where a, b, and d are integers, we have

(1)
$$x = \frac{A}{B} = \frac{ad}{bd} = \frac{a}{b}.$$

Consequently, we may assume that $x = a/b$, where a and b have no common divisor greater than one.

If $x^2 = 2$, then we have from (1)

$$\left(\frac{a}{b}\right)^2 = 2$$

or

(2)
$$a^2 = 2b^2.$$

Now the right side of equation (2) is an even integer and, hence, so is the left side. Thus, a^2 is divisible by 2. Consequently, a is also divisible by 2 (see problem 2 in the next exercise). Let $a = 2c$, where c is an integer. Using this expression for a in (2) we have

$$(2c)^2 = 2b^2$$

$$4c^2 = 2b^2$$

(3)
$$2c^2 = b^2.$$

Now the left side of (3) is an even integer and, hence, so is the right side. Thus, b^2 is divisible by 2. Consequently, b is also divisible by 2.

We have proved that if $x^2 = (a/b)^2 = 2$, where a and b have no common divisors greater than 1, then a and b do have the common divisor 2, which is greater than 1. This is a contradiction. Hence, $(a/b)^2$ cannot be 2. ∎

Exercise 1

1. Prove that if a is an odd integer, then a^2 is odd. *Hint:* For an appropriate n in **N**, we have $a = 2n + 1$. Hence, $a^2 = 2(2n^2 + 2n) + 1$ is also odd.

2. Prove that if a is an integer and a^2 is even, then a is an even integer. This was used in the proof of Theorem 1. *Hint:* Use problem 1.

3. Prove that if a is an integer and 3 divides a^2, then 3 divides a. *Hint:* If $a = 3k + 1$, then $a^2 = 9k^2 + 6k + 1 = 3(3k^2 + 2k) + 1$. If $a = 3k + 2$, then $a^2 = 9k^2 + 12k + 4 = 3(3k^2 + 4k + 1) + 1$. In each of these cases there is a remainder of one when a^2 is divided by 3.

4. Use the result of problem 3 to prove that the equation $x^2 = 3$ has no rational root. Hence, $\sqrt{3}$ is an irrational number.

5. Try to prove that $\sqrt{4}$ is an irrational number. Naturally you will fail. Why?

6. Suppose that n is an integer that is not the square of any other integer. What can you say about the nature of $\sqrt{n}$?

Appendix 5. The Order Axioms

Here we supply the proofs that were bypassed in Section 3.1. For convenience, we restate the three order axioms.

AXIOM 12

The Trichotomy Axiom *There is a set* **P** *called the set of positive numbers such that for each real number a exactly one of the following three assertions is true:*

(I) $a \in \mathbf{P}$,   (II) $a = 0$,   (III) $-a \in \mathbf{P}$.

The set of all real numbers a for which $-a \in \mathbf{P}$ is denoted by $\mathbf{P}^{(-)}$ and is called the set of *negative numbers*.

AXIOM 13

If a is in **P** *and b is in* **P**, *then* $a+b$ *is in* **P**.

AXIOM 14

If a is in **P** *and b is in* **P**, *then* ab *is in* **P**.

Theorem 1

If $a \neq 0$, *then* a^2 *is in* **P**.

Proof: We know that $(-a)(-a) = a \cdot a = a^2$. If $a \in \mathbf{P}$, then Axiom 14 implies that $a^2 \in \mathbf{P}$. If $a \notin \mathbf{P}$, then by Axiom 12 we have $-a \in \mathbf{P}$. In this case $(-a)(-a) = a^2$ is in **P**. ∎

COROLLARY

1 is in **P** *and* -1 *is in* $\mathbf{P}^{(-)}$.

Proof: Since $1^2 = 1$, we have $1 \in \mathbf{P}$. But if $1 \in \mathbf{P}$, then $-1 \notin \mathbf{P}$. Further, $-1 \neq 0$. Hence, $-1 \in \mathbf{P}^{(-)}$. ∎

Theorem 2

Every natural number is in **P**.

Proof: We know that $1 \in \mathbf{P}$. By adding 1 a sufficient number of times and using Axiom 13 each time, we see that each natural number is in **P**. ∎

Theorem 3   *If a is in $\mathbf{P}$ and b is in $\mathbf{P}^{(-)}$, then ab is in $\mathbf{P}^{(-)}$.*

Proof: If $b \in \mathbf{P}^{(-)}$, then $-b \in \mathbf{P}$. By Axiom 14 we have $a(-b) \in \mathbf{P}$. Consequently, $-(ab) \in \mathbf{P}$. Then by definition $ab \in \mathbf{P}^{(-)}$.   ∎

Theorem 4   *If a is in $\mathbf{P}^{(-)}$ and b is in $\mathbf{P}^{(-)}$, then ab is in $\mathbf{P}$.*

Proof: By hypothesis $-a \in \mathbf{P}$ and $-b \in \mathbf{P}$. Then by Axiom 14 we have $(-a)(-b) \in \mathbf{P}$. But $(-a)(-b) = ab$.   ∎

Theorem 5   *If a is in $\mathbf{P}$ and b is in $\mathbf{P}$, then a/b is in $\mathbf{P}$.*

Proof: Suppose to the contrary that $a/b \in \mathbf{P}^{(-)}$. Then $-(a/b) \in \mathbf{P}$. Hence, the product $b(-a/b) = -a$ is also in $\mathbf{P}$. But this is impossible (a contradiction), because we cannot have both a and $-a$ in $\mathbf{P}$ (see Axiom 12). Hence, $a/b \notin \mathbf{P}^{(-)}$. Since $a/b \neq 0$, we must have $a/b \in \mathbf{P}$.   ∎

Theorem 6   *If a is in $\mathbf{P}^{(-)}$ and b is in $\mathbf{P}^{(-)}$, then a/b is in $\mathbf{P}$.*

Proof: By hypothesis $-a \in \mathbf{P}$ and $-b \in \mathbf{P}$. By Theorem 5 we have $(-a)/(-b) \in \mathbf{P}$. But $(-a)/(-b) = a/b$.   ∎

Theorem 7   *If a is in $\mathbf{P}$ and b is in $\mathbf{P}^{(-)}$, then a/b is in $\mathbf{P}^{(-)}$ and b/a is in $\mathbf{P}^{(-)}$.*
PLE

Appendix 6.   Approximations and Scientific Notation

The number π is frequently given as 3.14, sometimes given as 3.1416, and occasionally as 3.14159265. Of course, not all of these values for π can be correct, and the truth is that none of them is. It can be proved that π is an irrational number and consequently, it can not be expressed in the decimal system with only a finite number of digits. However, we can use the symbol π to represent the ratio of the circumference of a circle to its diameter, and this symbol is used universally for this number.

It can be proved that

(1) $3.135 \leqq \pi \leqq 3.145,$

or better

(2) $$3.14155 \leqq \pi \leqq 3.14165,$$

or even better

(3) $$3.141592645 \leqq \pi \leqq 3.141592655.$$

When we write $\pi = 3.14$, what we really mean is that π satisfies the inequality (1). In this case we say that $\pi = 3.14$ to two decimal places. We also say $\pi = 3.14$ to three significant figures or with three-figure accuracy. Similarly,

Statement	Means	Degree of accuracy
$\pi = 3.1416$	Inequality (2)	5 significant figures
$\pi = 3.14159265$	Inequality (3)	9 significant figures

Approximate numbers also arise in making physical measurements. Thus, if we measure the thickness of a sheet of metal with a micrometer and record the thickness as 0.036 inch, we are *certainly not* claiming that this is the exact thickness. It is understood that the actual thcikness lies between 0.0355 inch and 0.0365 inch. Similarly, the claim that the average distance of the sun from the Earth is 92,900,000 miles is also to be taken as an approximation and that the correct value lies somewhere between 92,850,000 miles and 92,950,000 miles.

On the other hand, not all numbers are approximations. When we write that $x = 3$ is a solution of the equation $x^3 - 3x^2 + 2x - 6 = 0$, we do not mean that the solution is a number somewhere between 2.5 and 3.5. True enough 3 does lie in this interval, but 3 is the *exact* solution, and no amount of refinement in our computations will yield any closer estimation of the solution. Similarly, $\sqrt{16} = 4$ is an exact assertion, and not an approximate one.

Logically, we should have a special symbol to distinguish approximations from true equality. In this book we use $a \approx b$ to indicate that a is approximately equal to b. Thus, we write

$$\sqrt{3} \approx 1.732, \qquad \sqrt{5} \approx 2.236.$$

but we also write

$$\sqrt{4} = 2, \qquad \sqrt[3]{27} = 3.$$

However, we do not insist on using $\approx$ for every approximation. For example, we write $\sin 10° = 0.1736$, even though 0.1736 is really an approximation. The reason for this inconsistency is quite simple: We assume that the reader is sufficiently mature to know which assertions are approximations and which are exact. In this way, we can avoid any unnecessary distraction that may be caused by using $\approx$ too often. On the other hand, we will use the symbol $\approx$ whenever we feel that there is a good reason to emphasize that the assertion is an approximation.

To specify the number of significant figures in measurements such as 0.036 inch or 92,900,000 miles, we need to eliminate those zeros that are merely place holders. To this end we make

DEFINITION 1   **Scientific Notation**   *A number N is said to be written in scientific notation if it has the form*

(4) $$N = A \times 10^c,$$

where

(5) $$1 \leqq A < 10,$$

and c is an integer.

EXAMPLE 1   Put $x = 0.036$, $y = 92,900,000$, and $z = 0.0000123462$ in scientific notation.

Solution: We merely move the decimal point a suitable number of places and compensate by multiplying by an appropriate power of 10. Thus

$$x = 0.036 = 3.6 \times 10^{-2},$$
$$y = 92,900,000 = 9.29 \times 10^7,$$
$$z = 0.0000123462 = 1.23462 \times 10^{-5}. \quad \blacktriangle$$

It is clear from this example that the exponent c is determined by the

RULE   *Suppose that the decimal point is moved q places in converting the given number N to a number A such that $1 \leqq A < 10$. Then*

$c = q$ *if the decimal point is moved from right to left ($N \geqq 10$),*

$c = 0$ *if the decimal point is not moved,*

$c = -q$ *if the decimal point is moved from left to right ($N < 1$).*

The number of significant figures in N is indicated clearly by the following

AGREEMENT   *If $N = A \times 10^c$ is in scientific notation, then the number of significant figures in N is the number of digits in A.*

If A has k digits, then the true value of N satisfies the inequality

(6)
$$A - \frac{5}{10^k} \leqq \frac{N(\text{true})}{10^c} \leqq A + \frac{5}{10^k}.$$

Thus, the maximum error in the stated value of N is not more than $5 \times 10^c/10^k$.

EXAMPLE 2

Assuming that the numbers in Example 1 are measurements expressed properly in scientific notation, state the number of significant figures and the maximum error in each measurement.

Solution: Following the agreement we have:

x has 2 significant figures, max. error $= 5 \times 10^{-2}/10^2 = 0.0005$,

y has 3 significant figures, max. error $= 5 \times 10^7/10^3 = 50{,}000$,

z has 6 significant figures, max. error $= 5 \times 10^{-5}/10^6 = 0.00000000005$.

▲

It is not easy to determine the number of significant figures in a number that one obtains from a series of computations. We do not attempt to discuss this problem, but we give two examples to illustrate the difficulty.

EXAMPLE 3

Suppose that $x = 8.11$ and $y = 1.23$, and each number is a measurement that is correct to three significant figures. How many significant figures does the product xy have?

Solution: We compute the product and then "round off." We find that

$$(8.11)(1.23) = 9.9753$$

$$= 9.98 \text{ to three significant figures.}$$

But x could be 8.115, and y could be 1.235. Hence, the true value of the product xy could be as large as

$$(8.115)(1.235) = 10.022025$$

$$= 10.0 \text{ to three significant figures.}$$

In the other direction, we might have $x = 8.105$ and $y = 1.225$. Hence, the true value of the product xy could be as small as

$$(8.105)(1.225) = 9.928625$$

$$= 9.93 \text{ to three significant figures.}$$

In this particular example each of the factors is known to three significant figures, but the product is known to only two significant figures. Further, we note that although the difference

$$\text{maximum } xy - \text{minimum } xy = 0.0934,$$

we are not even certain about the first digit in the product. (It could be 1 or it could be 9.) ▲

Subtraction can lead to even "worse" results as indicated in

EXAMPLE 4   Suppose that $x = 54{,}326$ and $y = 54{,}321$, and each number is a measurement that is correct to five significant figures. How many significant figures does the difference $x - y$ have?

Solution: Computing with the numbers as given, we have

$$54{,}326 - 54{,}321 = 5.$$

The extreme possibilities are:

$$\text{largest } x - y = 54{,}326.5 - 54{,}320.5 = 6,$$
$$\text{smallest } x - y = 54{,}325.5 - 54{,}321.5 = 4.$$

Consequently, we have *no* significant figures in the difference. We can only claim that $4 \leq x - y \leq 6$, even though x and y were known to five significant figures. ▲

Although this example may seem artificial, there are many practical problems that take this form. For example, if we know the height (in inches) of a bridge over a roadway and the height (in inches) of a truck to three significant figures, the clearance between the two, as the truck passes under the bridge, may have no significant figures.

When we "round off" a number, such as the product $(8.11)(1.23) = 9.9753$ to three significant figures, we select that three digit number that is closest to 9.9753. In this case, it is obvious that we obtain 9.98 when we round off. The borderline case in which the number to be rounded off is "exactly in the middle," presents a problem.

EXAMPLE 5   Round off to two significant figures each of the following products: (a) $(3.8)(1.4)$, (b) $(4.2)(2.3)$, (c) $(2.3)(3.5)$, (d) $(3.5)(2.5)$.

Solution: Computation gives

(a) $(3.8)(1.4) = 5.32$ or 5.3 to two significant figures.

(b) $(4.2)(2.3) = 9.66$ or 9.7 to two significant figures.

(c) $(2.3)(3.5) = 8.05$.

(d) $(2.5)(3.5) = 8.75$.

Since 8.05 and 8.75 are "in the middle," there is some doubt as to whether we should increase or decrease the number when rounding off. There are a variety of rules for making a decision, but none of the rules have a solid theoretical foundation. For the sake of uniformity, in this text we adopt the convention that we round off so that the last digit is an even number. With this convention, 8.05 rounds off "downward" to 8.0, and 8.75 rounds off "upward" to 8.8. ▲

When computing the sides or angles of a triangle, we use the following table relating the accuracy in the computed or measured parts

Lengths of sides	Angles
Two significant figures	Nearest degree
Three significant figures	Nearest multiple of ten minutes
Four significant figures	Nearest minute
Five significant figures	Nearest tenth of a minute

For example, if the sides of a right triangle were given as $a = 4.00$ and $b = 7.50$ (three significant figures), we would compute the angles to the nearest multiple of ten minutes (unless we are given some different instructions).

Exercise 1

In problems 1 through 6 write the given number in scientific notation assuming that each number has three significant figures.

1. 34.5
2. 34,500
3. 0.0000345
4. 0.345
5. 0.0000000345
6. 345,000,000

In problems 7 through 9 put the given number in the usual form.

7. 6.42×10^{-3}
8. 6.42×10^9
9. 6.42×10^{-6}

In problems 10 through 18 round off the given number to three significant figures.

10. 5.678
11. 7.654
12. 1.11499
13. 1.115
14. 1.225001
15. 2.995
16. 2.985
17. 9.995
18. 9.990

Tables

TABLE A	N	N^2	$\sqrt{N}$	N	N^2	$\sqrt{N}$	N	N^2	$\sqrt{N}$	N	N^2	$\sqrt{N}$
SQUARES AND SQUARE ROOTS 1–200	1	1	1.0000	36	1,296	6.0000	71	5,041	8.4262	106	11,236	10.296
	2	4	1.4142	37	1,369	6.0828	72	5,184	8.4853	107	11,449	10.344
	3	9	1.7321	38	1,444	6.1644	73	5,329	8.5440	108	11,664	10.392
	4	16	2.0000	39	1,521	6.2450	74	5,476	8.6023	109	11,881	10.440
	5	25	2.2361	40	1,600	6.3246	75	5,625	8.6603	110	12,100	10.488
	6	36	2.4495	41	1,681	6.4031	76	5,776	8.7178	111	12,321	10.536
	7	49	2.6458	42	1,764	6.4807	77	5,929	8.7750	112	12,544	10.583
	8	64	2.8284	43	1,849	6.5574	78	6,084	8.8318	113	12,769	10.630
	9	81	3.0000	44	1,936	6.6333	79	6,241	8.8882	114	12,996	10.677
	10	100	3.1623	45	2,025	6.7082	80	6,400	8.9443	115	13,225	10.724
	11	121	3.3166	46	2,116	6.7823	81	6,561	9.0000	116	13,456	10.770
	12	144	3.4641	47	2,209	6.8557	82	6,724	9.0554	117	13,689	10.817
	13	169	3.6056	48	2,304	6.9282	83	6,889	9.1104	118	13,924	10.863
	14	196	3.7417	49	2,401	7.0000	84	7,056	9.1652	119	14,161	10.909
	15	225	3.8730	50	2,500	7.0711	85	7,225	9.2195	120	14,400	10.954
	16	256	4.0000	51	2,601	7.1414	86	7,396	9.2736	121	14,641	11.000
	17	289	4.1231	52	2,704	7.2111	87	7,569	9.3274	122	14,884	11.045
	18	324	4.2426	53	2,809	7.2801	88	7,744	9.3808	123	15,129	11.091
	19	361	4.3589	54	2,916	7.3485	89	7,921	9.4340	124	15,376	11.136
	20	400	4.4721	55	3,025	7.4162	90	8,100	9.4868	125	15,625	11.180
	21	441	4.5826	56	3,136	7.4833	91	8,281	9.5394	126	15,876	11.225
	22	484	4.6904	57	3,249	7.5498	92	8,464	9.5917	127	16,129	11.269
	23	529	4.7958	58	3,364	7.6158	93	8,649	9.6437	128	16,384	11.314
	24	576	4.8990	59	3,481	7.6811	94	8,836	9.6954	129	16,641	11.358
	25	625	5.0000	60	3,600	7.7460	95	9,025	9.7468	130	16,900	11.402
	26	676	5.0990	61	3,721	7.8103	96	9,216	9.7980	131	17,161	11.446
	27	729	5.1962	62	3,844	7.8740	97	9,409	9.8489	132	17,424	11.489
	28	784	5.2915	63	3,969	7.9373	98	9,604	9.8995	133	17,689	11.533
	29	841	5.3852	64	4,096	8.0000	99	9,801	9.9499	134	17,956	11.576
	30	900	5.4772	65	4,225	8.0623	100	10,000	10.0000	135	18,225	11.619
	31	961	5.5678	66	4,356	8.1240	101	10,201	10.050	136	18,496	11.662
	32	1,024	5.6568	67	4,489	8.1854	102	10,404	10.100	137	18,769	11.705
	33	1,089	5.7446	68	4,624	8.2462	103	10,609	10.149	138	19,044	11.747
	34	1,156	5.8310	69	4,761	8.3066	104	10,816	10.198	139	19,321	11.790
	35	1,225	5.9161	70	4,900	8.3656	105	11,025	10.247	140	19,600	11.832
	N	N^2	$\sqrt{N}$	N	N^2	$\sqrt{N}$	N	N^2	$\sqrt{N}$	N	N^2	$\sqrt{N}$

TABLE A
SQUARES AND
SQUARE ROOTS
(CONTINUED)

N	N^2	$\sqrt{N}$	N	N^2	$\sqrt{N}$	N	N^2	$\sqrt{N}$	N	N^2	$\sqrt{N}$
141	19,881	11.874	156	24,336	12.490	171	29,241	13.077	186	34,596	13.638
142	20,164	11.916	157	24,649	12.530	172	29,584	13.115	187	34,969	13.675
143	20,449	11.958	158	24,964	12.570	173	29,929	13.153	188	35,344	13.711
144	20,736	12.000	159	25,281	12.610	174	30,276	13.191	189	35,721	13.748
145	21,025	12.042	160	25,600	12.649	175	30,625	13.229	190	36,100	13.784
146	21,316	12.083	161	25,921	12.689	176	30,976	13.266	191	36,481	13.820
147	21,609	12.124	162	26,244	12.728	177	31,329	13.304	192	36,864	13.856
148	21,904	12.166	163	26,569	12.767	178	31,684	13.342	193	37,249	13.892
149	22,201	12.207	164	26,896	12.806	179	32,041	13.379	194	37,636	13.928
150	22,500	12.247	165	27,225	12.845	180	32,400	13.416	195	38,025	13.964
151	22,801	12.288	166	27,556	12.884	181	32,761	13.454	196	38,416	14.000
152	23,104	12.329	167	27,889	12.923	182	33,124	13.491	197	38,809	14.036
153	23,409	12.369	168	28,224	12.961	183	33,489	13.528	198	39,204	14.071
154	23,716	12.410	169	28,561	13.000	184	33,856	13.565	199	39,601	14.107
155	24,025	12.450	170	28,900	13.038	185	34,225	13.601	200	40,000	14.142
N	N^2	$\sqrt{N}$	N	N^2	$\sqrt{N}$	N	N^2	$\sqrt{N}$	N	N^2	$\sqrt{N}$

TABLE B

VALUES OF THE
TRIGONOMETRIC
FUNCTIONS TO
FOUR PLACES

Radians	Degrees	Sin	Tan	Cot	Cos		
.0000	0° 00′	.0000	.0000	——	1.0000	90° 00′	1.5708
.0029	10′	.0029	.0029	343.8	1.0000	89° 50′	1.5679
.0058	20′	.0058	.0058	171.9	1.0000	40′	1.5650
.0087	30′	.0087	.0087	114.6	1.0000	30′	1.5621
.0116	40′	.0116	.0116	85.94	.9999	20′	1.5592
.0145	50′	.0145	.0145	68.75	.9999	10′	1.5563
.0175	1° 00′	.0175	.0175	57.29	.9998	89° 00′	1.5533
.0204	10′	.0204	.0204	49.10	.9998	88° 50′	1.5504
.0233	20′	.0233	.0233	42.96	.9997	40′	1.5475
.0262	30′	.0262	.0262	38.19	.9997	30′	1.5446
.0291	40′	.0291	.0291	34.37	.9996	20′	1.5417
.0320	50′	.0320	.0320	31.24	.9995	10′	1.5388
.0349	2° 00′	.0349	.0349	28.64	.9994	88° 00′	1.5359
.0378	10′	.0378	.0378	26.43	.9993	87° 50′	1.5330
.0407	20′	.0407	.0407	24.54	.9992	40′	1.5301
.0436	30′	.0436	.0437	22.90	.9990	30′	1.5272
.0465	40′	.0465	.0466	21.47	.9989	20′	1.5243
.0495	50′	.0494	.0495	20.21	.9988	10′	1.5213
.0524	3° 00′	.0523	.0524	19.08	.9986	87° 00′	1.5184
.0553	10′	.0552	.0553	18.07	.9985	86° 50′	1.5155
.0582	20′	.0581	.0582	17.17	.9983	40′	1.5126
.0611	30′	.0610	.0612	16.35	.9981	30′	1.5097
.0640	40′	.0640	.0641	15.60	.9980	20′	1.5068
.0669	50′	.0669	.0670	14.92	.9978	10′	1.5039
.0698	4° 00′	.0698	.0699	14.30	.9976	86° 00′	1.5010
.0727	10′	.0727	.0729	13.73	.9974	85° 50′	1.4981
.0756	20′	.0756	.0758	13.20	.9971	40′	1.4952
.0785	30′	.0785	.0787	12.71	.9969	30′	1.4923
.0814	40′	.0814	.0816	12.25	.9967	20′	1.4893
.0844	50′	.0843	.0846	11.83	.9964	10′	1.4864
.0873	5° 00′	.0872	.0875	11.43	.9962	85° 00′	1.4835
.0902	10′	.0901	.0904	11.06	.9959	84° 50′	1.4806
.0931	20′	.0929	.0934	10.71	.9957	40′	1.4777
.0960	30′	.0958	.0963	10.39	.9954	30′	1.4748
.0989	40′	.0987	.0992	10.08	.9951	20′	1.4719
.1018	50′	.1016	.1022	9.788	.9948	10′	1.4690
.1047	6° 00′	.1045	.1051	9.514	.9945	84° 00′	1.4661
.1076	10′	.1074	.1080	9.255	.9942	83° 50′	1.4632
.1105	20′	.1103	.1110	9.010	.9939	40′	1.4603
.1134	30′	.1132	.1139	8.777	.9936	30′	1.4573
.1164	40′	.1161	.1169	8.556	.9932	20′	1.4544
.1193	50′	.1190	.1198	8.345	.9929	10′	1.4515
.1222	7° 00′	.1219	.1228	8.144	.9925	83° 00′	1.4486
		Cos	Cot	Tan	Sin	Degrees	Radians

TABLE B

TRIGONOMETRIC
FUNCTIONS
(CONTINUED)

Radians	Degrees	Sin	Tan	Cot	Cos		
.1222	**7° 00′**	.1219	.1228	8.144	.9925	**83° 00′**	1.4486
.1251	10′	.1248	.1257	7.953	.9922	82° 50′	1.4457
.1280	20′	.1276	.1287	7.770	.9918	40′	1.4428
.1309	30′	.1305	.1317	7.596	.9914	30′	1.4399
.1338	40′	.1334	.1346	7.429	.9911	20′	1.4370
.1367	50′	.1363	.1376	7.269	.9907	10′	1.4341
.1396	**8° 00′**	.1392	.1405	7.115	.9903	**82° 00′**	1.4312
.1425	10′	.1421	.1435	6.968	.9899	81° 50′	1.4283
.1454	20′	.1449	.1465	6.827	.9894	40′	1.4254
.1484	30′	.1478	.1495	6.691	.9890	30′	1.4224
.1513	40′	.1507	.1524	6.561	.9886	20′	1.4195
.1542	50′	.1536	.1554	6.435	.9881	10′	1.4166
.1571	**9° 00′**	.1564	.1584	6.314	.9877	**81° 00′**	1.4137
.1600	10′	.1593	.1614	6.197	.9872	80° 50′	1.4108
.1629	20′	.1622	.1644	6.084	.9868	40′	1.4079
.1658	30′	.1650	.1673	5.976	.9863	30′	1.4050
.1687	40′	.1679	.1703	5.871	.9858	20′	1.4021
.1716	50′	.1708	.1733	5.769	.9853	10′	1.3992
.1745	**10° 00′**	.1736	.1763	5.671	.9848	**80° 00′**	1.3963
.1774	10′	.1765	.1793	5.576	.9843	79° 50′	1.3934
.1804	20′	.1794	.1823	5.485	.9838	40′	1.3904
.1833	30′	.1822	.1853	5.396	.9833	30′	1.3875
.1862	40′	.1851	.1883	5.309	.9827	20′	1.3846
.1891	50′	.1880	.1914	5.226	.9822	10′	1.3817
.1920	**11° 00′**	.1908	.1944	5.145	.9816	**79° 00′**	1.3788
.1949	10′	.1937	.1974	5.066	.9811	78° 50′	1.3759
.1978	20′	.1965	.2004	4.989	.9805	40′	1.3730
.2007	30′	.1994	.2035	4.915	.9799	30′	1.3701
.2036	40′	.2022	.2065	4.843	.9793	20′	1.3672
.2065	50′	.2051	.2095	4.773	.9787	10′	1.3643
.2094	**12° 00′**	.2079	.2126	4.705	.9781	**78° 00′**	1.3614
.2123	10′	.2108	.2156	4.638	.9775	77° 50′	1.3584
.2153	20′	.2136	.2186	4.574	.9769	40′	1.3555
.2182	30′	.2164	.2217	4.511	.9763	30′	1.3526
.2211	40′	.2193	.2247	4.449	.9757	20′	1.3497
.2240	50′	.2221	.2278	4.390	.9750	10′	1.3468
.2269	**13° 00′**	.2250	.2309	4.331	.9744	**77° 00′**	1.3439
.2298	10′	.2278	.2339	4.275	.9737	76° 50′	1.3410
.2327	20′	.2306	.2370	4.219	.9730	40′	1.3381
.2356	30′	.2334	.2401	4.165	.9724	30′	1.3352
.2385	40′	.2363	.2432	4.113	.9717	20′	1.3323
.2414	50′	.2391	.2462	4.061	.9710	10′	1.3294
.2443	**14° 00′**	.2419	.2493	4.011	.9703	**76° 00′**	1.3265
		Cos	Cot	Tan	Sin	Degrees	Radians

TABLE B

TRIGONOETMRIC
FUNCTIONS
(CONTINUED)

Radians	Degrees	Sin	Tan	Cot	Cos		
.2443	**14° 00′**	.2419	.2493	4.011	.9703	**76° 00′**	1.3265
.2473	10′	.2447	.2524	3.962	.9696	75° 50′	1.3235
.2502	20′	.2476	.2555	3.914	.9689	40′	1.3206
.2531	30′	.2504	.2586	3.867	.9681	30′	1.3177
.2560	40′	.2532	.2617	3.821	.9674	20′	1.3148
.2589	50′	.2560	.2648	3.776	.9667	10′	1.3119
.2618	**15° 00′**	.2588	.2679	3.732	.9659	**75° 00′**	1.3090
.2647	10′	.2616	.2711	3.689	.9652	74° 50′	1.3061
.2676	20′	.2644	.2742	3.647	.9644	40′	1.3032
.2705	30′	.2672	.2773	3.606	.9636	30′	1.3003
.2734	40′	.2700	.2805	3.566	.9628	20′	1.2974
.2763	50′	.2728	.2836	3.526	.9621	10′	1.2945
.2793	**16° 00′**	.2756	.2867	3.487	.9613	**74° 00′**	1.2915
.2822	10′	.2784	.2899	3.450	.9605	73° 50′	1.2886
.2851	20′	.2812	.2931	3.412	.9596	40′	1.2857
.2880	30′	.2840	.2962	3.376	.9588	30′	1.2828
.2909	40′	.2868	.2994	3.340	.9580	20′	1.2799
.2938	50′	.2896	.3026	3.305	.9572	10′	1.2770
.2967	**17° 00′**	.2924	.3057	3.271	.9563	**73° 00′**	1.2741
.2996	10′	.2952	.3089	3.237	.9555	72° 50′	1.2712
.3025	20′	.2979	.3121	3.204	.9546	40′	1.2683
.3054	30′	.3007	.3153	3.172	.9537	30′	1.2654
.3083	40′	.3035	.3185	3.140	.9528	20′	1.2625
.3113	50′	.3062	.3217	3.108	.9520	10′	1.2595
.3142	**18° 00′**	.3090	.3249	3.078	.9511	**72° 00′**	1.2566
.3171	10′	.3118	.3281	3.047	.9502	71° 50′	1.2537
.3200	20′	.3145	.3314	3.018	.9492	40′	1.2508
.3229	30′	.3173	.3346	2.989	.9483	30′	1.2479
.3258	40′	.3201	.3378	2.960	.9474	20′	1.2450
.3287	50′	.3228	.3411	2.932	.9465	10′	1.2421
.3316	**19° 00′**	.3256	.3443	2.904	.9455	**71° 00′**	1.2392
.3345	10′	.3283	.3476	2.877	.9446	70° 50′	1.2363
.3374	20′	.3311	.3508	2.850	.9436	40′	1.2334
.3403	30′	.3338	.3541	2.824	.9426	30′	1.2305
.3432	40′	.3365	.3574	2.798	.9417	20′	1.2275
.3462	50′	.3393	.3607	2.773	.9407	10′	1.2246
.3491	**20° 00′**	.3420	.3640	2.747	.9397	**70° 00′**	1.2217
.3520	10′	.3448	.3673	2.723	.9387	69° 50′	1.2188
.3549	20′	.3475	.3706	2.699	.9377	40′	1.2159
.3578	30′	.3502	.3739	2.675	.9367	30′	1.2130
.3607	40′	.3529	.3772	2.651	.9356	20′	1.2101
.3636	50′	.3557	.3805	2.628	.9346	10′	1.2072
.3665	**21° 00′**	.3584	.3839	2.605	.9336	**69° 00′**	1.2043
		Cos	Cot	Tan	Sin	Degrees	Radians

TABLE B	Radians	Degrees	Sin	Tan	Cot	Cos		
TRIGONOMETRIC FUNCTIONS (CONTINUED)	.3665	**21° 00′**	.3584	.3839	2.605	.9336	**69° 00′**	1.2043
	.3694	10′	.3611	.3872	2.583	.9325	68° 50′	1.2014
	.3723	20′	.3638	.3906	2.560	.9315	40′	1.1985
	.3752	30′	.3665	.3939	2.539	.9304	30′	1.1956
	.3782	40′	.3692	.3973	2.517	.9293	20′	1.1926
	.3811	50′	.3719	.4006	2.496	.9283	10′	1.1897
	.3840	**22° 00′**	.3746	.4040	2.475	.9272	**68° 00′**	1.1868
	.3869	10′	.3773	.4074	2.455	.9261	67° 50′	1.1839
	.3898	20′	.3800	.4108	2.434	.9250	40′	1.1810
	.3927	30′	.3827	.4142	2.414	.9239	30′	1.1781
	.3956	40′	.3854	.4176	2.394	.9228	20′	1.1752
	.3985	50′	.3881	.4210	2.375	.9216	10′	1.1723
	.4014	**23° 00′**	.3907	.4245	2.356	.9205	**67° 00′**	1.1694
	.4043	10′	.3934	.4279	2.337	.9194	66° 50′	1.1665
	.4072	20′	.3961	.4314	2.318	.9182	40′	1.1636
	.4102	30′	.3987	.4348	2.300	.9171	30′	1.1606
	.4131	40′	.4014	.4383	3.282	.9159	20′	1.1577
	.4160	50′	.4041	.4417	2.264	.9147	10′	1.1548
	.4189	**24° 00′**	.4067	.4452	2.246	.9135	**66° 00′**	1.1519
	.4218	10′	.4094	.4487	2.229	.9124	65° 50′	1.1490
	.4247	20′	.4120	.4522	2.211	.9112	40′	1.1461
	.4276	30′	.4147	.4557	2.194	.9100	30′	1.1432
	.4305	40′	.4173	.4592	2.177	.9088	20′	1.1403
	.4334	50′	.4200	.4628	2.161	.9075	10′	1.1374
	.4363	**25° 00′**	.4226	.4663	2.145	.9063	**65° 00′**	1.1345
	.4392	10′	.4253	.4699	2.128	.9051	64° 50′	1.1316
	.4422	20′	.4279	.4734	2.112	.9038	40′	1.1286
	.4451	30′	.4305	.4770	2.097	.9026	30′	1.1257
	.4480	40′	.4331	.4806	2.081	.9013	20′	1.1228
	.4509	50′	.4358	.4841	2.066	.9001	10′	1.1199
	.4538	**26° 00′**	.4384	.4877	2.050	.8988	**64° 00′**	1.1170
	.4567	10′	.4410	.4913	2.035	.8975	63° 50′	1.1141
	.4596	20′	.4436	.4950	2.020	.8962	40′	1.1112
	.4625	30′	.4462	.4986	2.006	.8949	30′	1.1083
	.4654	40′	.4488	.5022	1.991	.8936	20′	1.1054
	.4683	50′	.4514	.5059	1.977	.8923	10′	1.1025
	.4712	**27° 00′**	.4540	.5095	1.963	.8910	**63° 00′**	1.0996
	.4741	10′	.4566	.5132	1.949	.8897	62° 50′	1.0966
	.4771	20′	.4592	.5169	1.935	.8884	40′	1.0937
	.4800	30′	.4617	.5206	1.921	.8870	30′	1.0908
	.4829	40′	.4643	.5243	1.907	.8857	20′	1.0879
	.4858	50′	.4669	.5280	1.894	.8843	10′	1.0850
	.4887	**28° 00′**	.4695	.5317	1.881	.8829	**62° 00′**	1.0821
			Cos	Cot	Tan	Sin	Degrees	Radians

	Radians	Degrees	Sin	Tan	Cot	Cos		
TABLE B	.4887	**28° 00′**	.4695	.5317	1.881	.8829	**62° 00′**	1.0821
TRIGONOMETRIC	.4916	10′	.4720	.5354	1.868	.8816	61° 50′	1.0792
FUNCTIONS	.4945	20′	.4746	.5392	1.855	.8802	40′	1.0763
(CONTINUED)	.4974	30′	.4772	.5430	1.842	.8788	30′	1.0734
	.5003	40′	.4797	.5467	1.829	.8774	20′	1.0705
	.5032	50′	.4823	.5505	1.816	.8760	10′	1.0676
	.5061	**29° 00′**	.4848	.5543	1.804	.8746	**61° 00′**	1.0647
	.5091	10′	.4874	.5581	1.792	.8732	60° 50′	1.0617
	.5120	20′	.4899	.5619	1.780	.8718	40′	1.0588
	.5149	30′	.4924	.5658	1.767	.8704	30′	1.0559
	.5178	40′	.4950	.5696	1.756	.8689	20′	1.0530
	.5207	50′	.4975	.5735	1.744	.8675	10′	1.0501
	.5236	**30° 00′**	.5000	.5774	1.732	.8660	**60° 00′**	1.0472
	.5265	10′	.5025	.5812	1.720	.8646	59° 50′	1.0443
	.5294	20′	.5050	.5851	1.709	.8631	40′	1.0414
	.5323	30′	.5075	.5890	1.698	.8616	30′	1.0385
	.5352	40′	.5100	.5930	1.686	.8601	20′	1.0356
	.5381	50′	.5125	.5969	1.675	.8587	10′	1.0327
	.5411	**31° 00′**	.5150	.6009	1.664	.8572	**59° 00′**	1.0297
	.5440	10′	.5175	.6048	1.653	.8557	58° 50′	1.0268
	.5469	20′	.5200	.6088	1.643	.8542	40′	1.0239
	.5498	30′	.5225	.6128	1.632	.8526	30′	1.0210
	.5527	40′	.5250	.6168	1.621	.8511	20′	1.0181
	.5556	50′	.5275	.6208	1.611	.8496	10′	1.0152
	.5585	**32° 00′**	.5299	.6249	1.600	.8480	**58° 00′**	1.0123
	.5614	10′	.5324	.6289	1.590	.8465	57° 50′	1.0094
	.5643	20′	.5348	.6330	1.580	.8450	40′	1.0065
	.5672	30′	.5373	.6371	1.570	.8434	30′	1.0036
	.5701	40′	.5398	.6412	1.560	.8418	20′	1.0007
	.5730	50′	.5422	.6453	1.550	.8403	10′	.9977
	.5760	**33° 00′**	.5446	.6494	1.540	.8387	**57° 00′**	.9948
	.5789	10′	.5471	.6536	1.530	.8371	56° 50′	.9919
	.5818	20′	.5495	.6577	1.520	.8355	40′	.9890
	.5847	30′	.5519	.6619	1.511	.8339	30′	.9861
	.5876	40′	.5544	.6661	1.501	.8323	20′	.9832
	.5905	50′	.5568	.6703	1.492	.8307	10′	.9803
	.5934	**34° 00′**	.5592	.6745	1.483	.8290	**56° 00′**	.9774
	.5963	10′	.5616	.6787	1.473	.8274	55° 50′	.9745
	.5992	20′	.5640	.6830	1.464	.8258	40′	.9716
	.6021	30′	.5664	.6873	1.455	.8241	30′	.9687
	.6050	40′	.5688	.6916	1.446	.8225	20′	.9657
	.6080	50′	.5712	.6959	1.437	.8208	10′	.9628
	.6109	**35° 00′**	.5736	.7002	1.428	.8192	**55° 00′**	.9599
			Cos	Cot	Tan	Sin	Degrees	Radians

Radians	Degrees	Sin	Tan	Cot	Cos		
TABLE B							
.6109	**35° 00′**	.5736	.7002	1.428	.8192	**55° 00′**	.9599
.6138	10′	.5760	.7046	1.419	.8175	54° 50′	.9570
.6167	20′	.5783	.7089	1.411	.8158	40′	.9541
.6196	30′	.5807	.7133	1.402	.8141	30′	.9512
.6225	40′	.5831	.7177	1.393	.8124	20′	.9483
.6254	50′	.5854	.7221	1.385	.8107	10′	.9454
.6283	**36° 00′**	.5878	.7265	1.376	.8090	**54° 00′**	.9425
.6312	10′	.5901	.7310	1.368	.8073	53° 50′	.9396
.6341	20′	.5925	.7355	1.360	.8056	40′	.9367
.6370	30′	.5948	.7400	1.351	.8039	30′	.9338
.6400	40′	.5972	.7445	1.343	.8021	20′	.9308
.6429	50′	.5995	.7490	1.335	.8004	10′	.9279
.6458	**37° 00′**	.6018	.7536	1.327	.7986	**53° 00′**	.9250
.6487	10′	.6041	.7581	1.319	.7969	52° 50′	.9221
.6516	20′	.6065	.7627	1.311	.7951	40′	.9192
.6545	30′	.6088	.7673	1.303	.7934	30′	.9163
.6574	40′	.6111	.7720	1.295	.7916	20′	.9134
.6603	50′	.6134	.7766	1.288	.7898	10′	.9105
.6632	**38° 00′**	.6157	.7813	1.280	.7880	**52° 00′**	.9076
.6661	10′	.6180	.7860	1.272	.7862	51° 50′	.9047
.6690	20′	.6202	.7907	1.265	.7844	40′	.9018
.6720	30′	.6225	.7954	1.257	.7826	30′	.8988
.6749	40′	.6248	.8002	1.250	.7808	20′	.8959
.6778	50′	.6271	.8050	1.242	.7790	10′	.8930
.6807	**39° 00′**	.6293	.8098	1.235	.7771	**51° 00′**	.8901
.6836	10′	.6316	.8146	1.228	.7753	50° 50′	.8872
.6865	20′	.6338	.8195	1.220	.7735	40′	.8843
.6894	30′	.6361	.8243	1.213	.7716	30′	.8814
.6923	40′	.6383	.8292	1.206	.7698	20′	.8785
.6952	50′	.6406	.8342	1.199	.7679	10′	.8756
.6981	**40° 00′**	.6428	.8391	1.192	.7660	**50° 00′**	.8727
.7010	10′	.6450	.8441	1.185	.7642	49° 50′	.8698
.7039	20′	.6472	.8491	1.178	.7623	40′	.8668
.7069	30′	.6494	.8541	1.171	.7604	30′	.8639
.7098	40′	.6517	.8591	1.164	.7585	20′	.8610
.7127	50′	.6539	.8642	1.157	.7566	10′	.8581
.7156	**41° 00′**	.6561	.8693	1.150	.7547	**49° 00′**	.8552
.7185	10′	.6583	.8744	1.144	.7528	48° 50′	.8523
.7214	20′	.6604	.8796	1.137	.7509	40′	.8494
.7243	30′	.6626	.8847	1.130	.7490	30′	.8465
.7272	40′	.6648	.8899	1.124	.7470	20′	.8436
.7301	50′	.6670	.8952	1.117	.7451	10′	.8407
.7330	**42° 00′**	.6691	.9004	1.111	.7431	**48° 00′**	.8378
		Cos	Cot	Tan	Sin	Degrees	Radians

TABLE B
TRIGONOMETRIC
FUNCTIONS
(CONTINUED)

	Radians	Degrees	Sin	Tan	Cot	Cos		
TABLE B	.7330	**42° 00′**	.6691	.9004	1.111	.7431	**48° 00′**	.8378
TRIGONOMETRIC	.7359	10′	.6713	.9057	1.104	.7412	47° 50′	.8348
FUNCTIONS	.7389	20′	.6734	.9110	1.098	.7392	40′	.8319
(CONTINUED)	.7418	30′	.6756	.9163	1.091	.7373	30′	.8290
	.7447	40′	.6777	.9217	1.085	.7353	20′	.8261
	.7476	50′	.6799	.9271	1.079	.7333	10′	.8232
	.7505	**43° 00′**	.6820	.9325	1.072	.7314	**47° 00′**	.8203
	.7534	10′	.6841	.9380	1.066	.7294	46° 50′	.8174
	.7563	20′	.6862	.9435	1.060	.7274	40′	.8145
	.7592	30′	.6884	.9490	1.054	.7254	30′	.8116
	.7621	40′	.6905	.9545	1.048	.7234	20′	.8087
	.7650	50′	.6926	.9601	1.042	.7214	10′	.8058
	.7679	**44° 00′**	.6947	.9657	1.036	.7193	**46° 00′**	.8029
	.7709	10′	.6967	.9713	1.030	.7173	45° 50′	.7999
	.7738	20′	.6988	.9770	1.024	.7153	40′	.7970
	.7767	30′	.7009	.9827	1.018	.7133	30′	.7941
	.7796	40′	.7030	.9884	1.012	.7112	20′	.7912
	.7825	50′	.7050	.9942	1.006	.7092	10′	.7883
	.7854	**45° 00′**	.7071	1.0000	1.000	.7071	**45° 00′**	.7854
			Cos	Cot	Tan	Sin	Degrees	Radians

TABLE C
COMMON LOGARITHMS TO FOUR PLACES

N	0	1	2	3	4	5	6	7	8	9
1.0	.0000	.0043	.0086	.0128	.0170	.0212	.0253	.0294	.0334	.0374
1.1	.0414	.0453	.0492	.0531	.0569	.0607	.0645	.0682	.0719	.0755
1.2	.0792	.0828	.0864	.0899	.0934	.0969	.1004	.1038	.1072	.1106
1.3	.1139	.1173	.1206	.1239	.1271	.1303	.1335	.1367	.1399	.1430
1.4	.1461	.1492	.1523	.1553	.1584	.1614	.1644	.1673	.1703	.1732
1.5	.1761	.1790	.1818	.1847	.1875	.1903	.1931	.1959	.1987	.2014
1.6	.2041	.2068	.2095	.2122	.2148	.2175	.2201	.2227	.2253	.2279
1.7	.2304	.2330	.2355	.2380	.2405	.2430	.2455	.2480	.2504	.2529
1.8	.2553	.2577	.2601	.2625	.2648	.2672	.2695	.2718	.2742	.2765
1.9	.2788	.2810	.2833	.2856	.2878	.2900	.2923	.2945	.2967	.2989
2.0	.3010	.3032	.3054	.3075	.3096	.3118	.3139	.3160	.3181	.3201
2.1	.3222	.3243	.3263	.3284	.3304	.3324	.3345	.3365	.3385	.3404
2.2	.3424	.3444	.3464	.3483	.3502	.3522	.3541	.3560	.3579	.3598
2.3	.3617	.3636	.3655	.3674	.3692	.3711	.3729	.3747	.3766	.3784
2.4	.3802	.3820	.3838	.3856	.3874	.3892	.3909	.3927	.3945	.3962
2.5	.3979	.3997	.4014	.4031	.4048	.4065	.4082	.4099	.4116	.4133
2.6	.4150	.4166	.4183	.4200	.4216	.4232	.4249	.4265	.4281	.4298
2.7	.4314	.4330	.4346	.4362	.4378	.4393	.4409	.4425	.4440	.4456
2.8	.4472	.4487	.4502	.4518	.4533	.4548	.4564	.4579	.4594	.4609
2.9	.4624	.4639	.4654	.4669	.4683	.4698	.4713	.4728	.4742	.4757
3.0	.4771	.4786	.4800	.4814	.4829	.4843	.4857	.4871	.4886	.4900
3.1	.4914	.4928	.4942	.4955	.4969	.4983	.4997	.5011	.5024	.5038
3.2	.5051	.5065	.5079	.5092	.5105	.5119	.5132	.5145	.5159	.5172
3.3	.5185	.5198	.5211	.5224	.5237	.5250	.5263	.5276	.5289	.5302
3.4	.5315	.5328	.5340	.5353	.5366	.5378	.5391	.5403	.5416	.5428
3.5	.5441	.5453	.5465	.5478	.5490	.5502	.5514	.5527	.5539	.5551
3.6	.5563	.5575	.5587	.5599	.5611	.5623	.5635	.5647	.5658	.5670
3.7	.5682	.5694	.5705	.5717	.5729	.5740	.5752	.5763	.5775	.5786
3.8	.5798	.5809	.5821	.5832	.5843	.5855	.5866	.5877	.5888	.5899
3.9	.5911	.5922	.5933	.5944	.5955	.5966	.5977	.5988	.5999	.6010
N	0	1	2	3	4	5	6	7	8	9

TABLE C

COMMON LOGARITHMS TO FOUR PLACES (CONTINUED)

N	0	1	2	3	4	5	6	7	8	9
4.0	.6021	.6031	.6042	.6053	.6064	.6075	.6085	.6096	.6107	.6117
4.1	.6128	.6138	.6149	.6160	.6170	.6180	.6191	.6201	.6212	.6222
4.2	.6232	.6243	.6253	.6263	.6274	.6284	.6294	.6304	.6314	.6325
4.3	.6335	.6345	.6355	.6365	.6375	.6385	.6395	.6405	.6415	.6425
4.4	.6435	.6444	.6454	.6464	.6474	.6484	.6493	.6503	.6513	.6522
4.5	.6532	.6542	.6551	.6561	.6571	.6580	.6590	.6599	.6609	.6618
4.6	.6628	.6637	.6646	.6656	.6665	.6675	.6684	.6693	.6702	.6712
4.7	.6721	.6730	.6739	.6749	.6758	.6767	.6776	.6785	.6794	.6803
4.8	.6812	.6821	.6830	.6839	.6848	.6857	.6866	.6875	.6884	.6893
4.9	.6902	.6911	.6920	.6928	.6937	.6946	.6955	.6964	.6972	.6981
5.0	.6990	.6998	.7007	.7016	.7024	.7033	.7042	.7050	.7059	.7067
5.1	.7076	.7084	.7093	.7101	.7110	.7118	.7126	.7135	.7143	.7152
5.2	.7160	.7168	.7177	.7185	.7193	.7202	.7210	.7218	.7226	.7235
5.3	.7243	.7251	.7259	.7267	.7275	.7284	.7292	.7300	.7308	.7316
5.4	.7324	.7332	.7340	.7348	.7356	.7364	.7372	.7380	.7388	.7396
5.5	.7404	.7412	.7419	.7427	.7435	.7443	.7451	.7459	.7466	.7474
5.6	.7482	.7490	.7497	.7505	.7513	.7520	.7528	.7536	.7543	.7551
5.7	.7559	.7566	.7574	.7582	.7589	.7597	.7604	.7612	.7619	.7627
5.8	.7634	.7642	.7649	.7657	.7664	.7672	.7679	.7686	.7694	.7701
5.9	.7709	.7716	.7723	.7731	.7738	.7745	.7752	.7760	.7767	.7774
6.0	.7782	.7789	.7796	.7803	.7810	.7818	.7825	.7832	.7839	.7846
6.1	.7853	.7860	.7868	.7875	.7882	.7889	.7896	.7903	.7910	.7917
6.2	.7924	.7931	.7938	.7945	.7952	.7959	.7966	.7973	.7980	.7987
6.3	.7993	.8000	.8007	.8014	.8021	.8028	.8035	.8041	.8048	.8055
6.4	.8062	.8069	.8075	.8082	.8089	.8096	.8102	.8109	.8116	.8122
6.5	.8129	.8136	.8142	.8149	.8156	.8162	.8169	.8176	.8182	.8189
6.6	.8195	.8202	.8209	.8215	.8222	.8228	.8235	.8241	.8248	.8254
6.7	.8261	.8267	.8274	.8280	.8287	.8293	.8299	.8306	.8312	.8319
6.8	.8325	.8331	.8338	.8344	.8351	.8357	.8363	.8370	.8376	.8382
6.9	.8388	.8395	.8401	.8407	.8414	.8420	.8426	.8432	.8439	.8445
N	0	1	2	3	4	5	6	7	8	9

TABLE C
COMMON LOGARITHMS TO FOUR PLACES (CONTINUED)

N	0	1	2	3	4	5	6	7	8	9
7.0	.8451	.8457	.8463	.8470	.8476	.8482	.8488	.8494	.8500	.8506
7.1	.8513	.8519	.8525	.8531	.8537	.8543	.8549	.8555	.8561	.8567
7.2	.8573	.8579	.8585	.8591	.8597	.8603	.8609	.8615	.8621	.8627
7.3	.8633	.8639	.8645	.8651	.8657	.8663	.8669	.8675	.8681	.8686
7.4	.8692	.8698	.8704	.8710	.8716	.8722	.8727	.8733	.8739	.8745
7.5	.8751	.8756	.8762	.8768	.8774	.8779	.8785	.8791	.8797	.8802
7.6	.8808	.8814	.8820	.8825	.8831	.8837	.8842	.8848	.8854	.8859
7.7	.8865	.8871	.8876	.8882	.8887	.8893	.8899	.8904	.8910	.8915
7.8	.8921	.8927	.8932	.8938	.8943	.8949	.8954	.8960	.8965	.8971
7.9	.8976	.8982	.8987	.8993	.8998	.9004	.9009	.9015	.9020	.9025
8.0	.9031	.9036	.9042	.9047	.9053	.9058	.9063	.9069	.9074	.9079
8.1	.9085	.9090	.9096	.9101	.9106	.9112	.9117	.9122	.9128	.9133
8.2	.9138	.9143	.9149	.9154	.9159	.9165	.9170	.9175	.9180	.9186
8.3	.9191	.9196	.9201	.9206	.9212	.9217	.9222	.9227	.9232	.9238
8.4	.9243	.9248	.9253	.9258	.9263	.9269	.9274	.9279	.9284	.9289
8.5	.9294	.9299	.9304	.9309	.9315	.9320	.9325	.9330	.9335	.9340
8.6	.9345	.9350	.9355	.9360	.9365	.9370	.9375	.9380	.9385	.9390
8.7	.9395	.9400	.9405	.9410	.9415	.9420	.9425	.9430	.9435	.9440
8.8	.9445	.9450	.9455	.9460	.9465	.9469	.9474	.9479	.9484	.9489
8.9	.9494	.9499	.9504	.9509	.9513	.9518	.9523	.9528	.9533	.9538
9.0	.9542	.9547	.9552	.9557	.9562	.9566	.9571	.9576	.9581	.9586
9.1	.9590	.9595	.9600	.9605	.9609	.9614	.9619	.9624	.9628	.9633
9.2	.9638	.9643	.9647	.9652	.9657	.9661	.9666	.9671	.9675	.9680
9.3	.9685	.9689	.9694	.9699	.9703	.9708	.9713	.9717	.9722	.9727
9.4	.9731	.9736	.9741	.9745	.9750	.9754	.9759	.9763	.9768	.9773
9.5	.9777	.9782	.9786	.9791	.9795	.9800	.9805	.9809	.9814	.9818
9.6	.9823	.9827	.9832	.9836	.9841	.9845	.9850	.9854	.9859	.9863
9.7	.9868	.9872	.9877	.9881	.9886	.9890	.9894	.9899	.9903	.9908
9.8	.9912	.9917	.9921	.9926	.9930	.9934	.9939	.9943	.9948	.9952
9.9	.9956	.9961	.9965	.9969	.9974	.9978	.9983	.9987	.9991	.9996
N	**0**	**1**	**2**	**3**	**4**	**5**	**6**	**7**	**8**	**9**

TABLE D

LOGARITHMS
OF THE
TRIGONOMETRIC
FUNCTIONS TO
FOUR PLACES

Angle θ	L. sin θ	L. tan θ	L. cot θ	L. cos θ	
0° 00′	No value	No value	No value	10.0000	**90° 00′**
10′	7.4637	7.4637	12.5363	10.0000	89° 50′
20′	7.7648	7.7648	12.2352	10.0000	40′
30′	7.9408	7.9409	12.0591	10.0000	30′
40′	8.0658	8.0658	11.9342	10.0000	20′
50′	8.1627	8.1627	11.8373	10.0000	10′
1° 00′	8.2419	8.2419	11.7581	9.9999	**89° 00′**
10′	8.3088	8.3089	11.6911	9.9999	88° 50′
20′	8.3668	8.3669	11.6331	9.9999	40′
30′	8.4179	8.4181	11.5819	9.9999	30′
40′	8.4637	8.4638	11.5362	9.9998	20′
50′	8.5050	8.5053	11.4947	9.9998	10′
2° 00′	8.5428	8.5431	11.4569	9.9997	**88° 00′**
10′	8.5776	8.5779	11.4221	9.9997	87° 50′
20′	8.6097	8.6101	11.3899	9.9996	40′
30′	8.6397	8.6401	11.3599	9.9996	30′
40′	8.6677	8.6682	11.3318	9.9995	20′
50′	8.6940	8.6945	11.3055	9.9995	10′
3° 00′	8.7188	8.7194	11.2806	9.9994	**87° 00′**
10′	8.7423	8.7429	11.2571	9.9993	86° 50′
20′	8.7645	8.7652	11.2348	9.9993	40′
30′	8.7857	8.7865	11.2135	9.9992	30′
40′	8.8059	8.8067	11.1933	9.9991	20′
50′	8.8251	8.8261	11.1739	9.9990	10′
4° 00′	8.8436	8.8446	11.1554	9.9989	**86° 00′**
10′	8.8613	8.8624	11.1376	9.9989	85° 50′
20′	8.8783	8.8795	11.1205	9.9988	40′
30′	8.8946	8.8960	11.1040	9.9987	30′
40′	8.9104	8.9118	11.0882	9.9986	20′
50′	8.9256	8.9272	11.0728	9.9985	10′
5° 00′	8.9403	8.9420	11.0580	9.9983	**85° 00′**
10′	8.9545	8.9563	11.0437	9.9982	84° 50′
20′	8.9682	8.9701	11.0299	9.9981	40′
30′	8.9816	8.9836	11.0164	9.9980	30′
40′	8.9945	8.9966	11.0034	9.9979	20′
50′	9.0070	9.0093	10.9907	9.9977	10′
6° 00′	9.0192	9.0216	10.9784	9.9976	**84° 00′**
10′	9.0311	9.0336	10.9664	9.9975	83° 50′
20′	9.0426	9.0453	10.9547	9.9973	40′
30′	9.0539	9.0567	10.9433	9.9972	30′
40′	9.0648	9.0678	10.9322	9.9971	20′
50′	9.0755	9.0786	10.9214	9.9969	10′
7° 00′	9.0859	9.0891	10.9109	9.9968	**83° 00′**
	L. cos θ	L. cot θ	L. tan θ	L. sin θ	Angle θ

Subtract 10 from all logarithms.

TABLE D

LOGARITHMS
OF THE
TRIGONOMETRIC
FUNCTIONS TO
FOUR PLACES
(CONTINUED)

Angle θ	L. sin θ	L. tan θ	L. cot θ	L. cos θ	
7° 00′	9.0859	9.0891	10.9109	9.9968	**83° 00′**
10′	9.0961	9.0995	10.9005	9.9966	82° 50′
20′	9.1060	9.1096	10.8904	9.9964	40′
30′	9.1157	9.1194	10.8806	9.9963	30′
40′	9.1252	9.1291	10.8709	9.9961	20′
50′	9.1345	9.1385	10.8615	9.9959	10′
8° 00′	9.1436	9.1478	10.8522	9.9958	**82° 00′**
10′	9.1525	9.1569	10.8431	9.9956	81′ 50°
20′	9.1612	9.1658	10.8342	9.9954	40′
30′	9.1697	9.1745	10.8255	9.9952	30′
40′	9.1781	9.1831	10.8169	9.9950	20′
50′	9.1863	9.1915	10.8085	9.9948	10′
9° 00′	9.1943	9.1997	10.8003	9.9946	**81° 00′**
10′	9.2022	9.2078	10.7922	9.9944	80° 50′
20′	9.2100	9.2158	10.7842	9.9942	40′
30′	9.2176	9.2236	10.7764	9.9940	30′
40′	9.2251	9.2313	10.7687	9.9938	20′
50′	9.2324	9.2389	10.7611	9.9936	10′
10° 00′	9.2397	9.2463	10.7537	9.9934	**80° 00′**
10′	9.2468	9.2536	10.7464	9.9931	79° 50′
20′	9.2538	9.2609	10.7391	9.9929	40′
30′	9.2606	9.2680	10.7320	9.9927	30′
40′	9.2674	9.2750	10.7250	9.9924	20′
50′	9.2740	9.2819	10.7181	9.9922	10′
11° 00′	9.2806	9.2887	10.7113	9.9919	**79° 00′**
10′	9.2870	9.2953	10.7047	9.9917	78° 50′
20′	9.2934	9.3020	10.6980	9.9914	40′
30′	9.2997	9.3085	10.6915	9.9912	30′
40′	9.3058	9.3149	10.6851	9.9909	20′
50′	9.3119	9.3212	10.6788	9.9907	10′
12° 00′	9.3179	9.3275	10.6725	9.9904	**78° 00′**
10′	9.3238	9.3336	10.6664	9.9901	77° 50′
20′	9.3296	9.3397	10.6603	9.9899	40′
30′	9.3353	9.3458	10.6542	9.9896	30′
40′	9.3410	9.3517	10.6483	9.9893	20′
50′	9.3466	9.3576	10.6424	9.9890	10′
13° 00′	9.3521	9.3634	10.6366	9.9887	**77° 00′**
10′	9.3575	9.3691	10.6309	9.9884	76° 50′
20′	9.3629	9.3748	10.6252	9.9881	40′
30′	9.3682	9.3804	10.6196	9.9878	30′
40′	9.3734	9.3859	10.6141	9.9875	20′
50′	9.3786	9.3914	10.6086	9.9872	10′
14° 00′	9.3837	9.3968	10.6032	9.9869	**76° 00′**
	L. cos θ	L. cot θ	L. tan θ	L.sin θ	Angle θ

Subtract 10 from all logarithms.

TABLE D	Angle θ	L. sin θ	L. tan θ	L. cot θ	L. cos θ	
LOGARITHMS	**14° 00′**	9.3837	9.3968	10.6032	9.9869	**76° 00′**
OF THE	10′	9.3887	9.4021	10.5979	9.9866	75° 50′
TRIGONOMETRIC	20′	9.3937	9.4074	10.5926	9.9863	40′
FUNCTIONS TO	30′	9.3986	9.4127	10.5873	9.9859	30′
FOUR PLACES	40′	9.4035	9.4178	10.5822	9.9856	20′
(CONTINUED)	50′	9.4083	9.4230	10.5770	9.9853	10′
	15° 00′	9.4130	9.4281	10.5719	9.9849	**75° 00′**
	10′	9.4177	9.4331	10.5669	9.9846	74° 50′
	20′	9.4223	9.4381	10.5619	9.9843	40′
	30′	9.4269	9.4430	10.5570	9.9839	30′
	40′	9.4314	9.4479	10.5521	9.9836	20′
	50′	9.4359	9.4527	10.5473	9.9832	10′
	16° 00′	9.4403	9.4575	10.5425	9.9828	**74° 00′**
	10′	9.4447	9.4622	10.5378	9.9825	73° 50′
	20′	9.4491	9.4669	10.5331	9.9821	40′
	30′	9.4533	9.4716	10.5284	9.9817	30′
	40′	9.4576	9.4762	10.5238	9.9814	20′
	50′	9.4618	9.4808	10.5192	9.9810	10′
	17° 00′	9.4659	9.4853	10.5147	9.9806	**73° 00′**
	10′	9.4700	9.4898	10.5102	9.9802	72° 50′
	20′	9.4741	9.4943	10.5057	9.9798	40′
	30′	9.4781	9.4987	10.5013	9.9794	30′
	40′	9.4821	9.5031	10.4969	9.9790	20′
	50′	9.4861	9.5075	10.4925	9.9786	10′
	18° 00′	9.4900	9.5118	10.4882	9.9782	**72° 00′**
	10′	9.4939	9.5161	10.4839	9.9778	71′ 50′
	20′	4.4977	9.5203	10.4797	9.9774	40′
	30′	9.5015	9.5245	10.4755	9.9770	30′
	40′	9.5052	9.5287	10.4713	9.9765	20′
	50′	9.5090	9.5329	10.4671	9.9761	10′
	19° 00′	9.5126	9.5370	10.4630	9.9757	**71° 00′**
	10′	9.5163	9.5411	10.4589	9.9752	70° 50′
	20′	9.5199	9.5451	10.4549	9.9748	40′
	30′	9.5235	9.5491	10.4509	9.9743	30′
	40′	9.5270	9.5531	10.4469	9.9739	20′
	50′	9.5306	9.5571	10.4429	9.9734	10′
	20° 00′	9.5341	9.5611	10.4389	9.9730	**70° 00′**
	10′	9.5375	9.5650	10.4350	9.9725	69° 50′
	20′	9.5409	9.5689	10.4311	9.9721	40′
	30′	9.5443	9.5727	10.4273	9.9716	30′
	40′	9.5477	9.5766	10.4234	9.9711	20′
	50′	9.5510	9.5804	10.4196	9.9706	10′
	21° 00′	9.5543	9.5842	10.4158	9.9702	**69° 00′**
		L. cos θ	L. cot θ	L. tan θ	L. sin θ	Angle θ

Subtract 10 from all logarithms.

TABLE D	Angle θ	L. sin θ	L. tan θ	L. cot θ	L. cos θ	
LOGARITHMS	**21° 00′**	9.5543	9.5842	10.4158	9.9702	**69° 00′**
OF THE	10′	9.5576	9.5879	10.4121	9.9697	68° 50′
TRIGONOMETRIC	20′	9.5609	9.5917	10.4083	9.9692	40′
FUNCTIONS TO	30′	9.5641	9.5954	10.4046	9.9687	30′
FOUR PLACES	40′	9.5673	9.5991	10.4009	9.9682	20′
(CONTINUED)	50′	9.5704	9.6028	10.3972	9.9677	10′
	22° 00′	9.5736	9.6064	10.3936	9.9672	**68° 00′**
	10′	9.5767	9.6100	10.3900	9.9667	67° 50′
	20′	9.5798	9.6136	10.3864	9.9661	40′
	30′	9.5828	9.6172	10.3828	9.9656	30′
	40′	9 5859	9.6208	10.3792	9.9651	20′
	50′	9.5889	9.6243	10.3757	9.9646	10′
	23° 00′	9.5919	9.6279	10.3721	9.9640	**67° 00′**
	10′	9.5948	9.6314	10.3686	9.9635	66° 50′
	20′	9.5978	9.6348	10.3652	9.9629	40′
	30′	9.6007	9.6383	10.3617	9.9624	30′
	40′	9.6036	9.6417	10.3583	9.9618	20′
	50′	9.6065	9.6452	10.3548	9.9613	10′
	24° 00′	9.6093	9.6486	10.3514	9.9607	**66° 00′**
	10′	9.6121	9.6520	10.3480	9.9602	65° 50′
	20′	9.6149	9.6553	10.3447	9.9596	40′
	30′	9.6177	9.6587	10.3413	9.9590	30′
	40′	9.6205	9.6620	10.3380	9.9584	20′
	50′	9.6232	9.6654	10.3346	9.9579	10′
	25° 00′	9.6259	9.6687	10.3313	9.9573	**65° 00′**
	10′	9.6286	9.6720	10.3280	9.9567	64° 50′
	20′	9.6313	9.6752	10.3248	9.9561	40′
	30′	9.6340	9.6785	10.3215	9.9555	30′
	40′	9.6366	9.6817	10.3183	9.9549	20′
	50′	9.6392	9.6850	10.3150	9.9543	10′
	26° 00′	9.6418	9.6882	10.3118	9.9537	**64° 00′**
	10′	9.6444	9.6914	10.3086	9.9530	63° 50′
	20′	9.6470	9.6946	10.3054	9.9524	40′
	30′	9.6495	9.6977	10.3023	9.9518	30′
	40′	9.6521	9.7009	10.2991	9.9512	20′
	50′	9.6546	9.7040	10.2960	9.9505	10′
	27° 00′	9.6570	9.7072	10.2928	9.9499	**63° 00′**
	10′	9.6595	9.7103	10.2897	9.9492	62° 50′
	20′	9.6620	9.7134	10.2866	9.9486	40′
	30′	9.6644	9.7165	10.2835	9.9479	30′
	40′	9.6668	9.7196	10.2804	9.9473	20′
	50′	9.6692	9.7226	10.2774	9.9466	10′
	28° 00′	9.6716	9.7257	10.2743	9.9459	**62° 00′**
		L. cos θ	L. cot θ	L. tan θ	L. sin θ	Angle θ

Subtract 10 from all logarithms.

TABLE D	Angle θ	L. sin θ	L. tan θ	L. cot θ	L. cos θ	
LOGARITHMS	**28° 00′**	9.6716	9.7257	10.2743	9.9459	**62° 00′**
OF THE	10′	9.6740	9.7287	10.2713	9.9453	61° 50′
TRIGONOMETRIC	20′	9.6763	9.7317	10.2683	9.9446	40′
FUNCTIONS TO	30′	9.6787	9.7348	10.2652	9.9439	30′
FOUR PLACES	40′	9.6810	9.7378	10.2622	9.9432	20′
(CONTINUED)	50′	9.6833	9.7408	10.2592	9.9425	10′
	29° 00′	9.6856	9.7438	10.2562	9.9418	**61° 00′**
	10′	9.6878	9.7467	10.2533	9.9411	60° 50′
	20′	9.6901	9.7497	10.2503	9.9404	40′
	30′	9.6923	9.7526	10.2474	9.9397	30′
	40′	9.6946	9.7556	10.2444	9.9390	20′
	50′	9.6968	9.7585	10.2415	9.9383	10′
	30° 00′	9.6990	9.7614	10.2386	9.9375	**60° 00′**
	10′	9.7012	9.7644	10.2356	9.9368	59° 50′
	20′	9.7033	9.7673	10.2327	9.9361	40′
	30′	9.7055	9.7701	10.2299	9.9353	30′
	40′	9.7076	9.7730	10.2270	9.9346	20′
	50′	9.7097	9.7759	10.2241	9.9338	10′
	31° 00′	9.7118	9.7788	10.2212	9.9331	**59° 00′**
	10′	9.7139	9.7816	10.2184	9.9323	58° 50′
	20′	9.7160	9.7845	10.2155	9.9315	40′
	30′	9.7181	9.7873	10.2127	9.9308	30′
	40′	9.7201	9.7902	10.2098	9.9300	20′
	50′	9.7222	9.7930	10.2070	9.9292	10′
	32° 00′	9.7242	9.7958	10.2042	9.9284	**58° 00′**
	10′	9.7262	9.7986	10.2014	9.9276	57° 50′
	20′	9.7282	9.8014	10.1986	9.9268	40′
	30′	9.7302	9.8042	10.1958	9.9260	30′
	40′	9.7322	9.8070	10.1930	9.9252	20′
	50′	9.7342	9.8097	10.1903	9.9244	10′
	33° 00′	9.7361	9.8125	10.1875	9.9236	**57° 00′**
	10′	9.7380	9.8153	10.1847	9.9228	56° 50′
	20′	9.7400	9.8180	10.1820	9.9219	40′
	30′	9.7419	9.8208	10.1792	9.9211	30′
	40′	9.7438	9.8235	10.1765	9.9203	20′
	50′	9.7457	9.8263	10.1737	9.9194	10′
	34° 00′	9.7476	9.8290	10.1710	9.9186	**56° 00′**
	10′	9.7494	9.8317	10.1683	9.9177	55° 50′
	20′	9.7513	9.8344	10.1656	9.9169	40′
	30′	9.7531	9.8371	10.1629	9.9160	30′
	40′	9.7550	9.8398	10.1602	9.9151	20′
	50′	9.7568	9.8425	10.1575	9.9142	10′
	35° 00′	9.7586	9.8452	10.1548	9.9134	**55° 00′**
		L. cos θ	L. cot θ	L. tan θ	L. sin θ	Angle θ

Subtract 10 from all logarithms.

	Angle θ	L. sin θ	L. tan θ	L. cot θ	L. cos θ	
TABLE D	**35° 00′**	9.7586	9.8452	10.1548	9.9134	**55° 00′**
LOGARITHMS	10′	9.7604	9.8479	10.1521	9.9125	54° 50′
OF THE	20′	9.7622	9.8506	10.1494	9.9116	40′
TRIGONOMTERIC	30′	9.7640	9.8533	10.1467	9.9107	30′
FUNCTIONS TO	40′	9.7657	9.8559	10.1441	9.9098	20′
FOUR PLACES	50′	9.7675	9.8586	10.1414	9.9089	10′
(CONTINUED)	**36° 00′**	9.7692	9.8613	10.1387	9.9080	**54° 00′**
	10′	9.7710	9.8639	10.1361	9.9070	53° 50′
	20′	9.7727	9.8666	10.1334	9.9061	40′
	30′	9.7744	9.8692	10.1308	9.9052	30′
	40′	9.7761	9.8718	10.1282	9.9042	20′
	50′	9.7778	9.8745	10.1255	9.9033	10′
	37° 00′	9.7795	9.8771	10.1229	9.9023	**53° 00′**
	10′	9.7811	9.8797	10.1203	9.9014	52° 50′
	20′	9.7828	9.8824	10.1176	9.9004	40′
	30′	9.7844	9.8850	10.1150	9.8995	30′
	40′	9.7861	9.8876	10.1124	9.8985	20′
	50′	9.7877	9.8902	10.1098	9.8975	10′
	38° 00′	9.7893	9.8928	10.1072	9.8965	**52° 00′**
	10′	9.7910	9.8954	10.1046	9.8955	51° 50′
	20′	9.7926	9.8980	10.1020	9.8945	40′
	30′	9.7941	9.9006	10.0994	9.8935	30′
	40′	9.7957	9.9032	10.0968	9.8925	20′
	50′	9.7973	9.9058	10.0942	9.8915	10′
	39° 00′	9.7989	9.9084	10.0916	9.8905	**51° 00′**
	10′	9.8004	9.9110	10.0890	9.8895	50° 50′
	20′	9.8020	9.9135	10.0865	9.8884	40′
	30′	9.8035	9.9161	10.0839	9.8874	30′
	40′	9.8050	9.9187	10.0813	9.8864	20′
	50′	9.8066	9.9212	10.0788	9.8853	10′
	40° 00′	9.8081	9.9238	10.0762	9.8843	**50° 00′**
	10′	9.8096	9.9264	10.0736	9.8832	49° 50′
	20′	9.8111	9.9289	10.0711	9.8821	40′
	30′	9.8125	9.9315	10.0685	9.8810	30′
	40′	9.8140	9.9341	10.0659	9.8800	20′
	50′	9.8155	9.9366	10.0634	9.8789	10′
	41° 00′	9.8169	9.9392	10.0608	9.8778	**49° 00′**
	10′	9.8184	9.9417	10.0583	9.8767	48° 50′
	20′	9.8198	9.9443	10.0557	9.8756	40′
	30′	9.8213	9.9468	10.0532	9.8745	30′
	40′	9.8227	9.9494	10.0506	9.8733	20′
	50′	9.8241	9.9519	10.0481	9.8722	10′
	42° 00′	9.8255	9.9544	10.0456	9.8711	**48° 00′**
		L. cos θ	L. cot θ	L. tan θ	L. sin θ	Angle θ

Subtract 10 from all logarithms.

TABLE D	Angle θ	L. sin θ	L. tan θ	L. cot θ	L. cos θ	
LOGARITHMS	**42° 00'**	9.8255	9.9544	10.0456	9.8711	**48° 00'**
OF THE	10'	9.8269	9.9570	10.0430	9.8699	47° 50'
TRIGONOMETRIC	20'	9.8283	9.9595	10.0405	9.8688	40'
FUNCTIONS TO	30'	9.8297	9.9621	10.0379	9.8676	30'
FOUR PLACES	40'	9.8311	9.9646	10.0354	9.8665	30'
(CONTINUED)	50'	9.8324	9.9671	10.0329	9.8653	10'
	43° 00'	9.8338	9.9697	10.0303	9.8641	**47° 00'**
	10'	9.8351	9.9722	10.0278	9.8629	46° 50'
	20'	9.8365	9.9747	10.0253	9.8618	40'
	30'	9.8378	9.9772	10.0228	9.8606	30'
	40'	9.8391	9.9798	10.0202	9.8594	20'
	50'	9.8405	9.9823	10.0177	9.8582	10'
	44° 00'	9.8418	9.9848	10.0152	9.8569	**46° 00'**
	10'	9.8431	9.9874	10.0126	9.8557	45° 50'
	20'	9.8444	9.9899	10.0101	9.8545	40'
	30'	9.8457	9.9924	10.0076	9.8532	30'
	40'	9.8469	9.9949	10.0051	9.8520	20'
	50'	9.8482	9.9975	10.0025	9.8507	10'
	45° 00'	9.8495	10.0000	10.0000	9.8495	**45° 00'**
		L. cos θ	L. cot θ	L. tan θ	L. sin θ	Angle θ

Subtract 10 from all logarithms.

Answers

Chapter 2, Exercise 1, page 23

29. 4	**30.** $-(7/17)$	**31.** -99
32. 3	**33.** 4/5	**34.** 20
35. 9/25	**36.** 441/25	**37.** 317/140
38. 93/140	**39.** $-107/140$	**40.** 841
41. 5,832	**42.** 5/6	**43.** 4/3
44. 6/17	**45.** 5/4	**46.** 7/10
47. -1	**48.** 12	**49.** 12/55

Chapter 3, Exercise 1, page 35

1. (a) -8,   (c) $\sqrt[3]{9}\sqrt[3]{15} = 3\sqrt[3]{5}$,   (e) 5/18,   (g) $9/7 - 2/3 = 13/21$

3. $\{x \mid x \leqq 4\}$ **5.** $\{x \mid -3 \leqq x \leqq 1\}$

7. $\{x \mid -5(\sqrt{2}+1) \leqq x \leqq 5(\sqrt{2}-1)\}$ **9.** $\{x \mid -3 \leqq x \leqq 7/3\}$

19. If $a > 0$ and $b < 0$, then $ab < 0$, $a/b < 0$ and $b/a < 0$.

20. If $a < 0$ and $b < 0$, then $ab > 0$ and $a/b > 0$.

21. Let $a = -2$, $b = 1$, $c = -1$, $d = 1$. Then $a < b$ and $c < d$.

But, $ac = 2 > 1 = bd$.

Chapter 3, Exercise 2, page 42

1. $1/6$

3. $\sqrt{15} - 2\sqrt{3}$

5. $3\sqrt{10} - 5\sqrt{3}$

9. $[-3, 13]$

11. $(-8, -2)$

13. $x = 71$ (one point)

15. $(-\sqrt{7}, \sqrt{7})$

17. $[-4, -1) \cup (1, 4]$

19. $(-5, \infty)$

21. $(2.9, 3.1)$

23. $(-\infty, -103) \cup (97, \infty)$

27. If $b < a$, then $[a, b]$ is the empty set.

Chapter 3, Exercise 3, page 45

5. (a) 5 (c) -6 (e) $15/2$

6. (a) $17, 20$ (c) $11/10, 9/5$

7. $23/6, 7, 61/6, 40/3$

Chapter 4, Exercise 1, page 54

1. $-243/32$

3. 2

5. $1/2^3 = 1/8$

7. $-ab^5$

9. c/d^2

11. $-a^5 b^9/2$

13. y^{23}/z^{22}

15. $2z^5/x^4 y$

17. $1/c^2 d^6 e^{10}$

19. $a(a+1)$

21. $1/(x+y)$

23. $(r^2 + s^2)/rs(r+s)$

25. $1/x$

27. $a^4 b^8$

29. 1

Chapter 4, Exercise 2, page 63

1. $(x+1)(x+2)$

3. $(y-3)(y-5)$

5. Irreducible

7. $a(2a+1)(a-5)$

9. $(3x-4)(5x+6)$

11. $(x^2+4)(x+1)$

13. $x^4(x-2y)(x^2+2xy+4y^2)$

15. $(3a+4)(2x+y)$

17. $(4x-3y)(3x+10y)$

19. $(2x+2y+z)(x+y+3z)$

21. $-(3x^k-1)(x^k-1)$

23. Irreducible

25. $4(4x+1)(x-1)$

27. $9(a+1)(a^2+3a+3)$

29. $bc(5a^2-b)(a+2c)$

31. $(3u-v+2w)^2$

33. $(x+y)(3x-y)(5x^2-2xy+y^2)$

35. $(a+b+2c)(x+3z)$

37. $(x+1)^2(x^2+2x-1)$

39. $(a+3b-c)(2a-b+5c)$

41. $4(2u-v)^2$

43. $(x+3y+7)(x+3y-2)$

45. $(u+v-w)(2u-3v+5w)$

47. $16, \quad (x+4)^2$

49. $25/2, \quad 2(x-5/2)^2$

51. $1/3, \quad 3(y+1/3)^2$

53. $9/20, \quad 5(x^2-3/10)^2$

55. $(x+3)^2 + 10$

57. $2(y+3/4)^2 + 31/8$

59. $5(x+3/2)^2 + 7/4$

Chapter 4, Exercise 3, page 70

1. x/y

3. $(a+b)^2/(a-b)$

5. $(x-3)/(2x+3)$

7. $(a+b)/2(2a+3b)$

9. $-10/(x+y)$

11. $(2x+1)(x+4)/(x+2)(x+3)$

13. $3/x$

15. $4a(a+3b)$

17. $z/7(2z-1)$

19. $(x+3)/(x+2)$

21. $15/(x-1)(x+3)$

23. $(x^2+y^2)/xy = x/y + y/x$

25. $(x-3)/(7x-1)$

27. $(x+2y)(x+3y)/(4x^2+y^2)$

29. $(1+x+x^2+x^3+2x^4+x^5+x^6)/(1+x^2+x^3+x^4+x^6)$

Chapter 4, Exercise 4, page 74

1. $2\sqrt{2}$

3. $25\sqrt{5}$

5. $-3\sqrt[3]{12}/4$

7. $6\sqrt[5]{54}$

9. $-10\sqrt[3]{10}$

11. $2\sqrt[3]{3}$

13. $xy^3\sqrt{xy}$

15. $2a^2\sqrt{ab}/b^4$

17. $a^4b^4\sqrt[3]{ab}$

19. $a(1+a)$

21. $b\sqrt[3]{21ab}/7a$

23. $ab^2\sqrt[3]{4a}$

25. $3\sqrt{3}+2\sqrt{5}$

27. $\sqrt{x}-\sqrt{2x}$

29. $9\sqrt{3}$

31. $s^3t^2(3+2st)\sqrt[3]{s^2t}$

Chapter 4, Exercise 5, page 79

1. -243

3. $1/81$

5. 0.04

7. $27/8$

9. 4

11. $1/y^{3/20}$

13. $b^{7/4}$

15. $y^{7/12}/x^{1/15}$

17. $a^{1/15}b^{2/15}c^{11/30}$

19. $9a^{7/2}b/7$

21. $x^{1/2}+2x^{1/4}y^{1/2}+y$

23. $x^{8/5}-y^3$

25. a

27. $a^2/(a^2-x^2)^{1/2}$

29. $a^3/x^{1/2}(a^3-x^3)^{2/3}$

31. $4/\sqrt{y}$

33. $(x-1)/(4x-3)^{2/3}$

35. $R-\{7, 13\}$

37. $[-4, \infty)$

39. R

41. $\left[-\sqrt{7}, \sqrt{7}\right]$

43. $(-\infty, -3] \cup [5, \infty)$

45. $[-7, 5]$

47. $\varnothing$

49. $[-1, \infty)$

Chapter 5, Exercise 1, page 88

1. $\{-4, 7\}$ **3.** $\{-9, -3, 1\}$ **5.** $\{-5, 6\}$

7. $\{-3/2, 4\}$ **9.** $\{-9, 9\}$ **11.** $\{-8/15\}$

13. $\{0\}$ **15.** $\{-2\}$ **17.** $\varnothing$

19. $\{-3, -2\}$ **21.** $\{-9/4, 1\}$ **23.** $\{b+2a\}$

25. $b = -56$ **27.** $b = 2$ **29.** $b = c$

Chapter 5, Exercise 2, page 93

1. 165 miles/hour, 205 miles/hour   **3. (a)** 13 gallons, **(b)** 52 gallons, no

5. 1 hour 30 minutes   **7.** 82

9. 271   **11.** 23, 27

13. 73, 438   **15.** 75 miles

17. 71 miles/hour   **19.** $7,000

21. 70 feet × 140 feet   **23.** Arvin, $120

25. 4 hours   **27.** 8 years, 19 years

29. $9\dfrac{3}{13}$ days   **31.** 16 minutes

Chapter 5, Exercise 3, page 100

1. $3 \pm \sqrt{6}$ **3.** $(-7 \pm \sqrt{29})/2$

5. 1, 4   **7.** No real roots

9. $(-7 \pm \sqrt{17})/8$ **11.** 1/2, 11/2

13. $(-3 \pm \sqrt{33})/12$ **15.** 1/3, 5/2

17. $-3, -2, 2, 3$ **19.** $\pm \sqrt{5 \pm \sqrt{20}}$

21. Two real roots, $\pm \sqrt{-3 + 3\sqrt{2}}$

23. No real roots   **25.** Two real roots, $-2 \pm \sqrt{6}$

27. $-2, 1, 4, 7$ **29.** $-9, -3, -1, 3$

31. By Theorem 4, $ax^2 + bx + c = a(x - r_1)(x - r_2) = ax^2 - a(r_1 + r_2)x + ar_1 r_2$. Hence, $b = -a(r_1 + r_2)$ and $c = ar_1 r_2$.

33. $x^2 - 7x + 4 = 0$ **35.** $12x^2 - 21x + 4 = 0$

37. $x^2 + 5x - 1 = 0$ **39.** 25

41. ± 30 **43.** 9/8

45. 11, 13 or $-13, -11$

47. $(\sqrt{33} - 5)/2, (\sqrt{33} + 5)/2$ or $(-\sqrt{33} - 5)/2, (-\sqrt{33} + 5)/2$

49. $(9 - \sqrt{17})/2$ inches   **51.** $6 + 5\sqrt{2}$ inches by $6 + 5\sqrt{2}$ inches

53. $-2, \pm\sqrt{3}/2$ **55.** $-7, -2, 2$

Chapter 5, Exercise 4, page 105

1. -6 3. 5   5. $-5/3$

7. 4   9. No solution   11. -7

13. 5   15. 1, 4/5   17. No solution

19. (a) $2 \pm \sqrt{3}$,   (b) $5 \pm 2\sqrt{6}$ 21. $-4 \pm 2\sqrt{2}$

23. -12

Chapter 5, Exercise 5, page 108

1. 3   3. 5, -2 5. -7

7. 8, 4   9. 17/4   11. -1

13. 2, 3/2   15. 16/9   17. 1/5, 4/5

19. 0   21. 1/12   23. 1, 3

Chapter 5, Exercise 6, page 110

1. $\sqrt{19} + \sqrt{21}$ 3. $5\sqrt{7}$ 5. $2\sqrt{3} + 3\sqrt{2}$

9. $a = 1$ 11. $c = d$ 13. $a = b$

15. $x = 2y$ 17. $c = d$ 19. $c = d$

21. Problems 12, 14, 15, 16, 17

Chapter 6, Exercise 1, page 117

1. $\mathbf{D = R, G = R}$ 3. $\mathbf{D = R - \{-5\}, G = R - \{0\}}$

5. $\mathbf{D} = (-\infty, -3] \cup [3, \infty), \mathbf{G} = [0, \infty)$

7. $\mathbf{D} = (-2, 2), \mathbf{G} = [3, \infty)$

9. $\mathbf{D} = (-\infty, -2] \cup [-1, 1] \cup [2, \infty), \mathbf{G} = [0, \infty)$

13. 0, 0, 6, 24, -6, -24 15. -2, $s^2 - 3$, $t^2 - 3$, $y^2 - 3$

17. $x^2 + x + 3$, $4x^2 - 2x + 3$, $x^2 + x + 3$, $x^2 + 2xy + y^2 - x - y + 3$,

     $x^2 - 2xy + y^2 - x + y + 3$

19. $V = x^3$ 21. $A = \sqrt{3} x^2/4$

23. 3, 4, 9, 11, 4, 4

Chapter 6, Exercise 2, page 121

3. $f(g(x)) = abx + ab - 1$ 9. $2x^2 - 2x - 1$

11. $2x^2 - 2x + 1$ 13. None

15. $(4x^2 - 12x + 11)^2$, $(2x^2 + 8x + 5)^2$ 17. $2x^2 + 2x + 5$

Chapter 6, Exercise 3, page 124

1. $5, 5, \sqrt{13}, 3\sqrt{3}$ **3.** $y = -4$

5. $y = 2$ **7.** $x = -3$

9. $x = \sqrt{3}$ **11.** II or III

13. I or IV   **15.** II or IV

17. I or II   **19.** III or IV

21. Horizontal line 4 units above x-axis

23. Vertical line 5 units to left of the y-axis

25. Two horizontal lines each 2 units from x-axis

27. The empty set (or figure)

29. A circle, center at origin, radius 6

31. The origin

33. Two vertical lines each 3 units from the y-axis

35. No

36. $\sqrt{(x_2 - x_1)^2 + (y_2 - y_1)^2}$, see Theorem 1, Chapter 9, page 199.

Chapter 6, Exercise 4, page 130

7. Minimum $f = -4$ at $x = 0$ **9.** Maximum $f = 14$ at $x = 3$

11. Minimum $f = -7$ at $x = -1$ **13.** Maximum $f = 21/4$ at $x = -3/2$

33. (a) R,   **(c)** $\emptyset$,   **(e)** $[-1, \infty)$,   **(g)** $(-\infty, -3/2,]$ **(i)** R

35. $f(3+d) = f(3-d) = 14 - d^2$ **37.** 10, 10

Chapter 6, Exercise 5, page 136

1. P and Q are symmetric with respect to the y-axis. P and R are symmetric with respect to the x-axis. P and S are symmetric with respect to the origin. P and T are symmetric with respect to the line $y = x$.

3. Symmetric with respect to the x-axis

5. Circle, all symmetries

7. No symmetries

9. Symmetric with respect to the x-axis, y-axis, and origin

11. Symmetric with respect to the x-axis

13. Symmetric with respect to the x-axis

15. Symmetric with respect to the x-axis

17. Symmetric with respect to the origin

19. All symmetries

21. Symmetric with respect to the x-axis, y-axis, and origin

23. At origin

Chapter 6, Exercise 6, page 139

1. 10

3. 2/5

5. 4/3

7. 64/243

9. 42/55

11. (a) 144 feet, **(b)** $s = 16t^2$ feet

13. 90 cubic inches

15. 300 pounds

17. 40 ohms

19. 184.32 tons

21. $y = Ct$

23. $C = (ak^3 + 1)/(ak^3 - b)$

Chapter 6, Exercise 7, page 144

1. $S., T.$, no

3. $N.$, yes

5. $S.$, no

7. $T.$, no

9. $T.$, no

11. $R., S., T.$, no

13. $R., S., T.$, no

15. $R., T.$, no

17. mn

Chapter 7, Exercise 1, page 152

1. 1, 2, 3, 4, 5, 6, 8, 9, 10, 12, 15, 18, 20, 24, 30, 36, 40, 45, 60, 72, 90, 120, 180, 360

3. (a) $\pi/3,$ **(b)** $4\pi/3,$ **(c)** $4\pi,$
(d) $-3\pi/4,$ **(e)** $\pi/15,$ **(f)** $11\pi/15,$
(g) $\pi/5,$ **(h)** $-5\pi/6$

5. (a) $45°,$ **(b)** $-405°,$ **(c)** $450°,$
(d) $16° 30'$ **(e)** $-40°,$ **(f)** $95°,$
(g) $1260°,$ **(h)** $-1440°$

9. (a) $(2n+1)\pi, n \in \mathbf{Z};$ **(b)** $(2n+1/2)\pi, n \in \mathbf{Z};$ **(c)** $(2n+3/2)\pi, n \in \mathbf{Z}$

11. Given a curve C, select a finite number of points on C and join successive pairs of these points with chords to form an inscribed polygonal path. (See the figure in problem 12.) Compute the length of this polygon. The length of C is the limit of the lengths of these polygons as the number of points of division grows indefinitely, and the length of each chord tends to zero.

12.

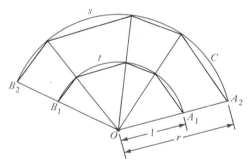

From similar triangles in the two approximating polygonal paths we infer that $s/r = t/1$ or $s = rt$.

13. 6.91 inches

15. (**a**) 11 radians, (**b**) 630°

17. (**a**) 69.1 miles (**b**) 1.15 miles

19. 2,920 miles

21. (**a**) 3.57 miles, (**b**) 1.68 feet

Chapter 7, Exercise 2, page 157

3.

θ	sin	cos	tan	cot	sec	csc
$\dfrac{2\pi}{3} = 120°$	$\dfrac{\sqrt{3}}{2}$	$-\dfrac{1}{2}$	$-\sqrt{3}$	$-\dfrac{\sqrt{3}}{3}$	-2	$\dfrac{2\sqrt{3}}{3}$
$-\dfrac{\pi}{4} = -45°$	$-\dfrac{\sqrt{2}}{2}$	$\dfrac{\sqrt{2}}{2}$	-1	-1	$\sqrt{2}$	$-\sqrt{2}$
$\dfrac{7\pi}{6} = 210°$	$-\dfrac{1}{2}$	$-\dfrac{\sqrt{3}}{2}$	$\dfrac{\sqrt{3}}{3}$	$\sqrt{3}$	$-\dfrac{2\sqrt{3}}{3}$	-2
$4\pi = 720°$	0	1	0	Und.	1	Und.
$-\dfrac{3\pi}{2} = -270°$	1	0	Und.	0	Und.	1

5. See Section 7.4, Table 2, page 160.

7. (**a**) 5/13, 12/13; (**b**) 3/5, −4/5;

(**c**) $-5\sqrt{41}/41, -4\sqrt{41}/41$; (**d**) $-2\sqrt{5}/5, \sqrt{5}/5$

Chapter 7, Exercise 3, page 159

1. For sin and cos, $\mathbf{D} = \mathbf{R}$, $\mathbf{G} = [-1, 1]$.
For tan, $\mathbf{D} = \mathbf{R} - \{(n+1/2)\pi \,|\, n \in \mathbf{Z}\}$, $\mathbf{G} = \mathbf{R}$.
For cot, $\mathbf{D} = \mathbf{R} - \{n\pi \,|\, n \in \mathbf{Z}\}$, $\mathbf{G} = \mathbf{R}$.
For sec, $\mathbf{D}$ is the same as for tan, $\mathbf{G} = (-\infty, -1] \cup [1, \infty)$.
For csc, $\mathbf{D}$ is the same as for cot, $\mathbf{G}$ is the same as for sec.

3. Tangent, secant, and cosecant are increasing.

Chapter 7, Exercise 4, page 163

1. Q. I, $\rho = \pi/7$ **3.** Q. II, $\rho = 2\pi/5$
5. Q. IV, $\rho = \pi/8$ **7.** Q. I, $\rho = 80°$
9. Q. II, $\rho = 19°$ **11.** Q. II, $\rho = 26°$

Problem	13	15	17	19	21	23
θ	$\dfrac{25\pi}{6}$	$-\dfrac{20\pi}{3}$	$\dfrac{9\pi}{4}$	$540°$	$-240°$	$-1110°$
ρ	$\dfrac{\pi}{6}$	$\dfrac{\pi}{3}$	$\dfrac{\pi}{4}$	0	$60°$	$30°$
sin	$\dfrac{1}{2}$	$-\dfrac{\sqrt{3}}{2}$	$\dfrac{\sqrt{2}}{2}$	0	$\dfrac{\sqrt{3}}{2}$	$-\dfrac{1}{2}$
cos	$\dfrac{\sqrt{3}}{2}$	$-\dfrac{1}{2}$	$\dfrac{\sqrt{2}}{2}$	-1	$-\dfrac{1}{2}$	$\dfrac{\sqrt{3}}{2}$
tan	$\dfrac{\sqrt{3}}{3}$	$\sqrt{3}$	1	0	$-\sqrt{3}$	$-\dfrac{\sqrt{3}}{3}$

Chapter 7, Exercise 5, page 165

1. 0.0523 **3.** 0.3249 **5.** 0.7412
7. 1.991 **9.** 11.83 **11.** 0.9293
13. 0.7050 **15.** 0.6009 **17.** $A = 0° \, 10'$
19. $C = 8° \, 0'$ **21.** $E = 13° \, 30'$ **23.** $G = 23° \, 30'$
25. $\beta = 0.4102$ **27.** $\theta = 0.6370$ **29.** $\alpha = 0.0902$
31. $\gamma = 0.3025$

Chapter 7, Exercise 6, page 168

1. Yes	**3.** 1.303	**5.** 0.6472
7. 4.989	**9.** 0.4067	**11.** 5.769
13. 0.2278	**15.** $B = 79°\,30'$	**17.** $D = 72°\,0'$
19. $F = 56°\,10'$	**21.** $\alpha = 1.3701$	**23.** $\gamma = 0.8901$
25. $\psi = 1.2392$		

Chapter 7, Exercise 7, page 173

5. $[0, \pi]$ **7.** $n\pi,\ n \in \mathbf{Z}$

9.

Interval	$\sin\theta$	$\cos\theta$	$\tan\theta$	$\cot\theta$	$\sec\theta$	$\csc\theta$
$0 < \theta < \dfrac{\pi}{2}$, Q. I	Inc.	Dec.	Inc.	Dec.	Inc.	Dec.
$\dfrac{\pi}{2} < \theta < \pi$, Q. II	Dec.	Dec.	Inc.	Dec.	Inc.	Inc.
$\pi < \theta < \dfrac{3\pi}{2}$, Q. III	Dec.	Inc.	Inc.	Dec.	Dec.	Inc.
$\dfrac{3\pi}{2} < \theta < 2\pi$, Q. IV	Inc.	Inc.	Inc.	Dec.	Dec.	Dec.

17. The sum is neither odd nor even unless one of the functions is the zero function (zero for all values of x in **D**).

Chapter 8, Exercise 1, page 180

11. $n^2 + 6n + 5 = (n+5)(n+1)$

13. If a person has fun proving trigonometric identities, then he has a normal, healthy, inquisitive mind.

Chapter 8, Exercise 4, page 194

1. 50°, 110°	**3.** 25°, 205°
5. 30°, 150°	**7.** 0°
9. 35° 50′, 144° 10′	

11. 27° 20′, 62° 40′, 117° 20′, 152° 40′, 207° 20′, 242° 40′, 297° 20′, 332° 40′

13. 107° 10′, 142° 10′, 287° 10′, 322° 10′

15. 270°, 306° 50' **17.** 22° 40', 143° 10'
19. 0°, 30°, 90°, 150°, 180°, 210°, 270°, 330°
21. 63° 30', 108° 30', 243° 30', 288° 30'
23. 30°, 45°, 150°, 225° **25.** 0°, 30°, 150°, 210°, 330°

Chapter 9, Exercise 1, page 199

1. 13 **3.** 5 **5.** 10
7. 6 **9.** $\sqrt{74}$ **11.** $\sqrt{146}$
13. $\sqrt{2}$ **15.** Yes **17.** $d = \sqrt{x_2^2 + y_2^2} = r$

Chapter 9, Exercise 2, page 206

1. $\cos 10°$ **3.** $1/2$ **5.** $-\sqrt{3}/2$ **7.** $\cos 5A$
9. $-(\sqrt{6}+\sqrt{2})/4$ **11.** $-(\sqrt{6}-\sqrt{2})/4$
29. $(\sqrt{35}-6)/12$, Q. II **31.** $\sqrt{13}/65$, Q. I
35. No, Theorem 5 of Chapter 8 was proved only for $0 \leqq \alpha \leqq \pi/2$ (or $0° \leqq A \leqq 90°$).

Chapter 9, Exercise 3, page 209

1. $1/2$ **3.** 0 **5.** 1 **7.** $\tan 2C$
9. (a) $(\sqrt{6}+\sqrt{2})/4$, (b) $-(2+\sqrt{3})$ **11.** (a) $(\sqrt{2}+\sqrt{6})/4$, (b) $2+\sqrt{3}$
13. (a) $49/37$, (b) $-7/61$ **15.** (a) $49/\sqrt{3770}$, (b) $-7/\sqrt{3770}$
17. (a) 0.034899, (b) 0.999391

Chapter 9, Exercise 4, page 214

1. $\tan 40°$ **3.** $\sin 28°$ **5.** $\cos 80°$ **7.** $\frac{1}{2}\tan 2A$
9. (a) $(\sqrt{6}-\sqrt{2})/4$, (b) $(\sqrt{6}+\sqrt{2})/4$
11. (a) $\sqrt{10}/10$, (b) $3\sqrt{10}/10$ **13.** (a) $24/25$, (b) $7/25$
15. (a) $3\sqrt{10}/10$, (b) $-\sqrt{10}/10$
17. (a) $(n-1)/\sqrt{2(n^2+1)}$, (b) $(n+1)/\sqrt{2(n^2+1)}$
21. $\cos 3C = 4\cos^3 C - 3\cos C$

Chapter 9, Exercise 5, page 218

1. $(\sin 20° + \sin 10°)/2$ **3.** $(\cos 4A + \cos 2A)/2$
5. $(\cos C + \cos 3C + \cos 5C + \cos 9C)/4$
7. $2\sin 2\theta \cos \theta$ **9.** $2\cos 4\theta \cos 2\theta$

23. $\cos A = (3\sqrt{2}+\sqrt{3})/6$, $\cos B = (3\sqrt{2}-\sqrt{3})/6$

25. $\cos A = \left(\sqrt{(1+x)(1+y)} + \sqrt{(1-x)(1-y)}\right)/2$,

$\cos B = \left(\sqrt{(1+x)(1+y)} - \sqrt{(1-x)(1-y)}\right)/2$,

$\sin A = \left(\sqrt{(1-x)(1+y)} - \sqrt{(1+x)(1-y)}\right)/2$,

$\sin B = \left(\sqrt{(1-x)(1+y)} + \sqrt{(1+x)(1-y)}\right)/2$

Chapter 9, Exercise 6, page 222

1. $0°$, $180°$

3. $30°$, $150°$, $270°$

5. $30°$, $60°$, $210°$, $240°$

7. $22°\,30'$, $112°\,30'$, $202°\,30'$, $292°\,30'$

9. $17°+n36°$, $n = 0, 1, 2, 3, 4, 5, 6, 7, 8, 9$

11. $90°$, $270°$; $22°\,30'+n45°$, $n = 0, 1, 2, 3, 4, 5, 6, 7$

13. $45°$, $135°$, $225°$, $315°$

15. $30°+n60°$, $n = 0, 1, 2, 3, 4, 5$

17. $0°$, $180°$; $10°+n20°$, $n = 0, 1, 2, ..., 17$

19. $36°\,50'$, $53°\,10'$

21. $82°\,50'$

23. $36°\,50'$

25. $19°\,20'$

Chapter 10, Exercise 1, page 230

1. $u = (v+5)/3$, $\mathbf{D} = \mathbf{R}$, $\mathbf{G} = \mathbf{R}$

3. $u = 5-2v$, $\mathbf{D} = \mathbf{R}$, $\mathbf{G} = \mathbf{R}$

5. $u = (v+3)/(v-2)$, $\mathbf{D} = \mathbf{R}-\{2\}$, $\mathbf{G} = \mathbf{R}-\{1\}$

7. $u = -1+\sqrt{v+16}$, $\mathbf{D} = [-16, \infty)$, $\mathbf{G} = [-1, \infty)$

9. $x = 2+\sqrt{9-y}$, $\mathbf{D} = (-\infty, 9]$, $\mathbf{G} = [2, \infty)$

11. $t = \sqrt[3]{z}$, $\mathbf{D} = \mathbf{R}$, $\mathbf{G} = \mathbf{R}$

15. $x = (y+\sqrt{y^2-4})/2$, $\mathbf{D} = [2, \infty)$, $\mathbf{G} = [1, \infty)$

17. $x = (-dy+b)/(cy-a)$

Chapter 10, Exercise 3, page 239

1. $90°$

3. $-30°$

5. $135°$

7. $13°\,30'$

9. $139°\,50'$

11. $-\pi/2$

13. $\pi/3$

15. $\pi/4$

17. $2\pi/3$

19. $4/5$

21. $12/13$

23. -0.9827

25. a

27. c

29. $e/\sqrt{1+e^2}$

31. $1/\sqrt{1-g^2}$

33. $2u\sqrt{1-u^2}$

35. $(1-u^2)/(1+u^2)$

37. $\pm\sqrt{(1-\sqrt{1-u^2})/2}$

39. $\sqrt{1-u^2}/(1+u)$

41. False

43. False

45. True

47. True

49. True

Chapter 10, Exercise 4, page 242

1. 63/65, 16/65, Q. I

3. $(12\sqrt{5}+\sqrt{10})/30$, $(4\sqrt{5}-3\sqrt{10})/30$, Q. II

5. $x(1-x^2+2\sqrt{1-x^2})/(1+x^2)$

9.

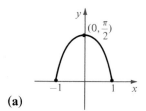

(a)

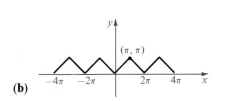

(b)

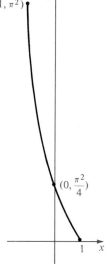

(c)

11. (a) Odd, **(b)** neither, **(c)** even, **(d)** neither, **(e)** odd, **(f)** even

Chapter 10, Exercise 5, page 248

1. Each of the numbers is a period of $f(x)$

3. Not periodic   **5.** Periodic, $p = 1$

7. Periodic, $p = 2\pi$ **9.** Periodic, $p = 2\pi/5$

11. $p = \pi/2$, amplitude $= 3$ **13.** $p = 3\pi$

15. $p = 2\pi$, amplitude $= \sqrt{5}$ **17.** $p = \pi$, amplitude $= \sqrt{10}$

19. $p = \pi/3$, amplitude $= 1/2$

21. (a) $-\operatorname{Arc\,sin}(2/\sqrt{5})$,   **(c)** $\dfrac{1}{2}\operatorname{Arc\,sin}(3/\sqrt{10})$

Chapter 10, Exercise 6, page 252

1.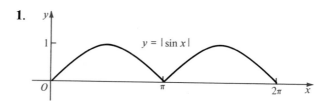

$y = |\sin x|$

3.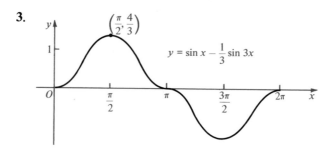

$\left(\dfrac{\pi}{2}, \dfrac{4}{3}\right)$

$y = \sin x - \dfrac{1}{3}\sin 3x$

5.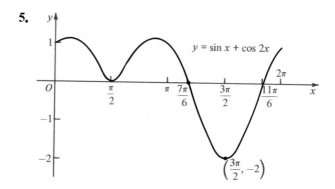

$y = \sin x + \cos 2x$

$\left(\dfrac{3\pi}{2}, -2\right)$

7.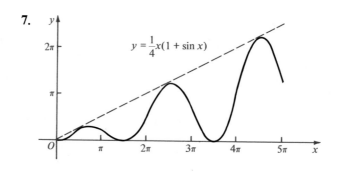

$y = \dfrac{1}{4}x(1 + \sin x)$

9.

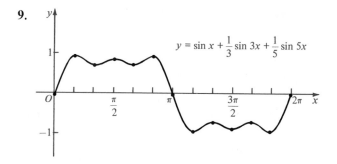

$$y = \sin x + \frac{1}{3} \sin 3x + \frac{1}{5} \sin 5x$$

Chapter 11, Exercise 2, page 259

1. 1	**3.** 4	**5.** -2
7. 5	**9.** 4	**11.** 1/3
13. 5/3	**15.** -4	**17.** $-1/2$
19. -2	**21.** 2	**23.** 3
25. 5/2	**27.** 6	**29.** 81
31. 81	**33.** 1/64	**35.** 8
37. 1	**39.** 256	

41 (*a*) or (*b*), **D** $= (0, \infty)$, **G** $=$ **R**

Chapter 11, Exercise 3, page 261

1. 0.602	**3.** 0.778	**5.** 0.954
7. 1.908	**9.** -0.301	**11.** 0.211
13. 1.845	**15.** 0.669	**17.** -0.669
19. 1.738	**21.** False	**23.** False
25. True	**27.** False	

Chapter 11, Exercise 4, page 266

1. 0.8837	**3.** 0.9542	**5.** 0.1303
7. 0.8075	**9.** 2.3617	**11.** 5.2810
13. $0.3617 - 2$	**15.** $0.9243 - 5$	**17.** 1.43
19. 2.06	**21.** 3.09	**23.** 7.60
25. 34,000,000	**27.** 378,000	**29.** 0.000735
31. 0.00000055		

Chapter 11, Exercise 5, page 271

1. $8.7348 - 10$	**3.** $4.8971 - 10$	**5.** 0.836
7. 0.00000175	**9.** 1.7350	**11.** $9.9428 - 10$
13. 0.3702	**15.** $6.4782 - 10$	**17.** 2.5228
19. 3.9541	**21.** $270,400$	**23.** 17.16
25. 0.0003678	**27.** 83.02	**29.** 0.01367

Chapter 11, Exercise 6, page 274

1. 3.51	**3.** $151,000$	**5.** 0.0000972
7. 0.289	**9.** $8,730$	**11.** 0.0000288
13. $37,000$	**15.** 2.13	**17.** 0.572
19. 0.609	**21.** 0.02054	**23.** 0.0001919
25. 17.63	**27.** $62,600,000$	**29.** 0.5120
31. 142.9	**33.** $32,660,000$	**35.** 11.97
37. $286,900$	**39.** 0.7552	**41.** 0.7552
43. 0.07422	**45.** 0.8382	**47.** $635,000,000,000$
49. 690 days		

Chapter 11, Exercise 7, page 277

1. $7/3$ **3.** $1/5$

5. $(3 \pm \sqrt{5})/2$ **7.** 3.32

9. -1.01 **11.** -1.13

13. **(a)** $23.5 \approx 24$ years,   **(b)** $11.9 \approx 12$ years,   **(c)** $8.05 \approx 9$ years

15. -0.000189 **17.** 29.8 years

19. 6.7%

Chapter 12, Exercise 1, page 284

1. $\sin Q = q/r,$ $\cos Q = p/r,$ $\tan Q = q/p,$ $\cot Q = p/q,$ $\sec Q = r/p,$ $\csc Q = r/q$

3. $\sin W = w/v,$ $\cos W = u/v,$ $\tan W = w/u,$ $\cot W = u/w,$ $\sec W = v/u,$ $\csc W = v/w$

Chapter 12, Exercise 2, page 288

1. $a = 905.8,$ $b = 1,528,$ $B = 59° 20'$

3. $a = 1,934,$ $b = 440.6,$ $A = 77° 10'$

5. $A = 41°50'$, $B = 48°10'$, $b = 2,235$

7. $b = 2,309$, $c = 3,000$, $B = 50°20'$

9. $A = 43°30'$, $B = 46°30'$, $c = 2,757$

Chapter 12, Exercise 3, page 292

1. 0.3368	**3.** 0.9442	**5.** 0.4752
7. 0.0995	**9.** 0.7296	**11.** 4.309
13. 9°19'	**15.** 61°7'	**17.** 79°4'
19. 79°4'	**21.** 63°37'	**23.** 81°28'
25. 9.6350 − 10	**27.** 9.1789 − 10	**29.** 10.3666 − 10
31. 11.0494 − 10	**33.** 9.9598 − 10	**35.** 9.9616 − 10
37. 29°1'	**39.** 8°59'	**41.** 57°56'
43. 9°44'	**45.** 21°4'	**47.** 55°29'

Chapter 12, Exercise 4, page 294

1. $A = 39°14'$, $B = 50°46'$, $c = 2.562$

3. $a = 6.780$, $b = 7.183$, $B = 46°39'$

5. $a = 0.02355$, $c = 0.04941$, $B = 61°32'$

7. $A = 33°23'$, $b = 20.60$, $B = 56°37'$

Chapter 12, Exercise 5, page 296

1. 273 feet

3. (a) 17.63, (b) 12.14

5. 31°

7. 6.7 feet, no

9. (a) 270 miles/hour, (b) 15 miles **11.** S 53° W

13. 13.85 feet, 19.60 feet

15. 4,980 feet

Chapter 12, Exercise 6, page 302

1. $c = a + b$

3. 13

5. 13

7. 0.013

9. $A = 19°40'$, $B = 28°10'$, $C = 132°10'$

11. (16, 14, 6), (16, 14, 10), (16, 16, 16), (16, 19, 21), (16, 26, 30), (16, 49, 55)

13. 3,800 feet

15. 27 miles

Chapter 12, Exercise 7, page 305

1. $b = 90$, $c = 73$ **3.** $a = 140$, $c = 170$
5. $a = 0.02401$, $c = 0.03184$ **7.** $a = 17.50$, $b = 20.76$
9. 192 feet   **11.** 1,680 feet
13. 220 feet

Chapter 12, Exercise 8, page 310

1. No triangles   **3.** $B = 14°\,29'$
5. $B = 61°\,3'$, $B' = 118°\,57'$ **7.** No triangles
9. $B = 59°\,43'$
11. $\triangle ABC$: $a = 8,015$, $C = 74°\,0'$, $A = 72°\,27'$;
        $\triangle ABC'$: $a = 5,454$, $C' = 106°\,0'$, $A = 40°\,27'$
13. No triangles   **15.** $12°\,10'$
17. 3.2×10^7 miles, 1.3×10^8 miles   **19.** 24.1 inches

Chapter 13, Exercise 1, page 318

1. $(3, -2)$ **3.** $(-3, -8)$
5. $(10, -3)$ **7.** $A = 0$, $B = 5$
9. No solution   **11.** $F = 35$, $S = 5$
13. $W = 8$, $B = 7$

Chapter 13, Exercise 2, page 321

1. $(1, 2)$, $(5, 2)$ **3.** $(1, 10)$, $(4, 1)$
5. $(0, 1)$, $(-2, 5)$ **7.** $(3, 2)$
9. $(-2, 4)$, $(1, 1)$, $(2, 0)$ **11.** $(-1, 1)$, $(3, 3)$
13. $(\pm 3, \pm 2)$ **15.** $(1, 0)$, $(0, 1)$

Chapter 13, Exercise 3, page 329

1. $(1, 2, -3)$ **3.** $(3, 3, 4)$
5. $(2, 5, -3)$ **7.** $(1, -1, 1)$
9. $x = 1$, $y = 2$, $z = 3$, $u = -3$
11. $V = 10$, $E = 15$, $F = 7$ **19.** $a = 1$, $b = -2$, $c = 3$
21. $x + 2y - 3z = 4$

Chapter 13, Exercise 4, page 336

3. $(-1, 1, -1)$ **5.** $(3, 0, -5)$

7. No solution   **9.** $(1/2, 1/2, 1/2)$

11. $(-2, 0, 2, 1)$ **13.** $(1, 2, 3, 4)$

15. Eight solutions $(\pm 2, \pm 3, \pm 1)$ **17.** $(-1, -11/3, 3)$

Chapter 13, Exercise 5, page 347

1. 207   **3.** -6 **5.** -8

7. 133   **9.** 1

Chapter 13, Exercise 6, page 352

1. $(2, -1, 0)$ **3.** $(1/2, 2, -3/2)$ **5.** $(2, -1, 1)$

Chapter 13, Exercise 7, page 354

1. 80   **3.** 0   **5.** 16

7. 1   **9.** $(1, 0, 0, 0)$ **11.** $(-3/2, 3, 4, 1/2)$

Chapter 13, Exercise 8, page 361

1. $\begin{bmatrix} 6 & -1 \\ 2 & 6 \end{bmatrix}$ **3.** $\begin{bmatrix} 4 & 3 \\ 3 & 1 \end{bmatrix}$

5. $\begin{bmatrix} -23 & 19 \\ 11 & -2 \end{bmatrix}$ **7.** $\begin{bmatrix} 3 & 1 \\ 11 & -1 \end{bmatrix}$

9. $\begin{bmatrix} -4 & -2 \\ 5 & 6 \end{bmatrix}$ **11.** $\begin{bmatrix} -5 & 7 \\ -23 & 49 \end{bmatrix}$

13. $\begin{bmatrix} 3 & -5 \\ 9 & -9 \end{bmatrix}$ **15.** $\begin{bmatrix} 0 & 0 \\ 0 & 0 \end{bmatrix}$

17. $\begin{bmatrix} 4 & -5 \\ -4 & -2 \end{bmatrix}$

21. See problems 7 and 9 (or 8 and 10).

Chapter 14, Exercise 1, page 371

29. $n = 1, \quad n \geqq 10$

Chapter 14, Exercise 2, page 376

7. 98	**9.** 330	**11.** -24
13. True	**15.** True	**17.** True
19. False	**21.** True	**23.** False
25. True	**27.** True	

Chapter 14, Exercise 3, page 380

1. 7, 9, 11, 13, 15, $S_5 = 55$ **3.** 6, 4, 2, 0, -2, $S_5 = 10$

5. $-4, -2, -1, -1/2, -1/4, S_5 = -31/4$

7. 2, 3, 5, 8, 12 **9.** 1, 0, 4, -5, 11

11. $d = \dfrac{2}{n(n-1)} [S_n - na_1]$ **13.** na_1

15. (a) 8, 11, 14, 17; (b) 12, 5, -2, -9

17. 459 **19.** 60,000

21. 5 **25.** 75

27. 4, 9, 14 **29.** 3, 6, 12

Chapter 15, Exercise 1, page 386

1. (a) 720, (b) 5,040, (c) 40,320

5. $n = 1$ **7.** $(m, 1)$, $(1, n)$, $(0, 0)$

Chapter 15, Exercise 2, page 389

1. (a) 1, 7, 21, 35, 35, 21, 7, 1;

 (b) 1, 8, 28, 56, 70, 56, 28, 8, 1;

 (c) 1, 9, 36, 84, 126, 126, 84, 36, 9, 1

3. (a) 210, (b) 165, (c) 210, (d) 165

9. (a) 20, (b) 190, (c) 7, (d) 595

Chapter 15, Exercise 3, page 393

1. (a) 165, (b) 330, (c) 0

3. (a) $-1,080$, (b) 720

5. Set $a = 1$, $b = 1$ in equation (6).

7. Sum is 3^n, set $a = 1$, $b = 2$ in equation (6).

9. (a) 135, (b) 0, (c) 540

11. (a) -8, **(b)** 24, **(c)** -48, **(d)** -4

13. (a) $x^{27} - 9x^{26}y + 39x^{25}y^2 - 325x^{24}y^3/3$,

$\quad$ **(b)** $1 - 100x + 4{,}750x^2 - 142{,}500x^3$

15. (a) 96,059,601, **(b)** 1.771561

17. $\sqrt{x}(x^5 + 5x^4 + 10x^3 + 10x^2 + 5x + 1)/x^3$

19. 194

Chapter 15, Exercise 4, page 397

1. $1 - \dfrac{1}{2}x - \dfrac{1}{8}x^2 - \dfrac{1}{16}x^3 - \dfrac{5}{128}x^4$

3. $1 + \dfrac{1}{5}x - \dfrac{2}{25}x^2 + \dfrac{6}{125}x^3 - \dfrac{21}{625}x^4$

5. $1 - x + x^2 - x^3 + x^4 - \cdots$ $\qquad$ **7.** $(-1)^n$

9. It is legitimate. $\qquad$ **11.** $1 - x + 0x^2 + 0x^3 + 0x^4$

13. 1.016 $\qquad\qquad$ **15.** 1.066

17. 3.873 $\qquad\qquad$ **19.** 3.915

21. 2.091

Chapter 16, Exercise 1, page 406

1. 210 $\qquad\qquad\qquad$ **3.** 504

5. 6,720 $\qquad\qquad\qquad$ **9.** 12

11. $5^5 = 3{,}125$ $\qquad\qquad$ **13.** 36

15. 200 $\qquad\qquad\qquad$ **17.** 625,000

19. $6! = 720$ $\qquad\qquad$ **21.** $P(21, 3) = 7{,}980$

23. $(5 - 1)! = 24$

25. (a) $(4!)(5!) = 2{,}880$, **(b)** The same

27. (a) 24, **(b)** 60, **(c)** 504

29. (a) $7!/(2!\,2!)$, **(b)** $8!/3!$, **(c)** $6!/(2!\,3!)$, **(d)** $11!/(4!\,4!\,2!)$

31. (a) 2, **(b)** 36, **(c)** 10,800, **(d)** 90,720

33. (a) 6, **(b)** 18, **(c)** 24, **(d)** 10

Chapter 16, Exercise 2, page 410

1. 56 $\qquad\qquad\qquad$ **3.** 56

5. 286 $\qquad\qquad\qquad$ **7.** 286

9. $n(n-1)(n-2)(n-3)/24$

11. (a) $C(9,3) = 84$, (b) $n(n-1)(n-2)/6$

13. $C(9,3) = 84$, $2^9/2 = 256$ **15.** $C(52,3) = 22{,}100$

17. $4C(13,3) = 1{,}144$ **19.** $C(24,3) = 2{,}024$

21. $2^7 = 128$ **23.** $13C(4,3)\cdot 48\cdot 44/2 = 54{,}912$

25. $4C(13,5) - 40 = 5{,}108$

27. $44C(13,2)\,C(4,2)\,C(4,2) = 123{,}552$

29. $C(12,2)\,C(10,2) = 2{,}970$

31. $C(n,4) = n(n-1)(n-2)(n-3)/24$

33. (a) 10, (b) 16

Chapter 16, Exercise 3, page 415

1. $n(\mathbf{U}) = 4$; (a) 1/4, (b) 1/2, (c) 3/4, (d) 1/4

3. $n(\mathbf{U}) = P(52,2) = 2{,}652$, if the order is regarded as important. If not, $n(\mathbf{U}) = C(52,2) = 1{,}326$.

   (a) $C(16,2)/C(52,2) = 20/221$, (b) $C(32,2)/C(52,2) = 248/663$,

   (c) $C(4,2)/C(52,2) = 1/221$, (d) $13C(4,2)/C(52,2) = 1/17$

5. $n(\mathbf{U}) = 36$; (a) 1/6, (b) 1/6, (c) 1/6, (d) 1/3, (e) 7/36

7. $n(\mathbf{U}) = 6^3 = 216$; (a) 1/72, (b) 1/36, (c) 5/108, (d) 1/36

9. $n(\mathbf{U}) = 72$; (a) 1/12, (b) 1/12, (c) 1/2, (d) 13/72

11. $n(\mathbf{U}) = 15\cdot 14 = 210$; (a) 1/35, (b) 34/105, (c) 3/7, (d) 1/5

13. $n(\mathbf{U}) = 100$; (a) 1/2, (b) 1/5, (c) 1/10, (d) 3/5

15. $n(\mathbf{U}) = C(52,4) = 270{,}725$; (a) $C(13,4)/C(52,4) = 11/4{,}165$,

   (b) $C(13,2)\,C(13,2)/C(52,4) = \dfrac{468}{20{,}825}$,

   (c) $13/C(52,4) = 2/20{,}825$,

   (d) $C(4,2)\,C(4,2)\,C(13,2)/C(52,4) = 216/20{,}825$

17. $n(\mathbf{U}) = 132$; (a) 1/11, (b) 13/22, (c) 1/6, (d) 47/66

Chapter 16, Exercise 4, page 418

1. 1/3

3. (a) 1/4, (b) 3/4, (c) 7/24, (d) 11/12

5. $d = 1/30$; (a) 3/20, (b) 11/20, (c) 4/15

7. 1/11, 6/11, 2/11, 2/11, $P[B \text{ or } C] = 8/11$

15. $A \cap B \neq \varnothing$

Chapter 16, Exercise 5, page 423

1. (a) 1/16, (b) 1/4, (c) 5/16, (d) 3/8
3. (a) 1/81, (b) 32/81, (c) 48/81, (d) 24/81
5. (a) 25/72, (b) 91/216, (c) 125/1,728, (d) 1,603/1,728
7. (a) 3/5, (b) 2/5, (c) 1/5, (d) 4/5
9. 1/8   **11.** 5/16
13. (a) 0.0837, (b) $13/4^4 \approx 0.0508$ **15.** $n = 8$

Chapter 16, Exercise 6, page 428

1. 5 to 8   **3.** $3.50
5. $-\$3/13 \approx -23$ cents   **7.** 3
9. 19.5 cents, no   **11.** 7/2
13. 7 to 5   **15.** 8 to 5
19. 7/12

Chapter 17, Exercise 1, page 440

1. $(10, 10)$ **3.** $(-5, 9)$
5. $(9, -5)$ **7.** $(-8, 5)$
9. $r = 7\sqrt{2}, \theta = 45°$ **11.** $r = 2\sqrt{2}, \theta = 120°$
13. $r = 130, \theta = 292° 40'$ **15.** $[-5\sqrt{3}, 5]$
17. $[-25\sqrt{2}, -25\sqrt{2}]$ **19.** $[37.59, -13.68]$
21. $[-\sqrt{11}, 0]$

Chapter 17, Exercise 2, page 445

1. $|R| = 41$ pounds, S 52° W   **3.** -60 pounds, 27 pounds
5. $|R| = 219$ pounds, S 38° 0' W   **7.** $|R| = 493$ pounds, $\theta = 27°$
9. 97 pounds, 541 pounds   **11.** 200 pounds
13. 200 pounds
15. 186 miles per hour, 188 miles per hour
17. 38 miles per hour, 471 miles per hour
19. 86 miles per hour, N 57° W
21. $T_1 = 815$ pounds, $T_2 = 922$ pounds
23. $T_1 = 174$ pounds, $T_2 = 985$ pounds

Chapter 18, Exercise 1, page 455

1. (a) $9+15i$, (b) $-16+63i$, (c) $(56-33i)/169$

3. (a) $8-16i$, (b) $-33-56i$, (c) $(63-16i)/25$

5. (a) $6i$, (b) -13, (c) $(5+12i)/13$

7. (a) $4+5i$, (b) $-6+8i$, (c) $(3-4i)/2$

9. (a) $-3-i$, (b) $5(4+3i)$, (c) $-(4+3i)/5$

11. (a) $-i$, (b) 1, (c) 1, (d) -4, (e) $-8i$

17. (a) 1, (b) i, (c) -1, (d) $-i$

19. $(2\pm\sqrt{6})\,i$

Chapter 18, Exercise 2, page 459

3. (a) $5\sqrt{2}$, $45°$, (b) 5, $36°\,50'$, (c) $2\sqrt{5}$, $243°\,30'$, (d) 5, $126°\,50'$
  (e) $5\sqrt{2}$, $171°\,50'$

5. $3\sqrt{2}(\cos 315° + i\sin 315°)$
7. $5(\cos 180° + i\sin 180°)$

9. $2(\cos 90° + i\sin 90°)$
11. $13(\cos 67°\,20' + i\sin 67°\,20')$

13. $8i$
15. $(-\sqrt{3}+3i)/2$

17. $-4\sqrt{3}+4i$
19. 0

21. $\sqrt{5}-\sqrt{5}i$

Chapter 18, Exercise 3, page 462

5. (a) $4(1+\sqrt{3}i)$, (b) $-4(1+\sqrt{3}i)$

7. (a) $-20\sqrt{2}$, (b) $5\sqrt{2}i$

9. (a) $6\sqrt{5}(\cos 243°\,30' + i\sin 243°\,30')$,
  (b) $3\sqrt{5}(\cos 296°\,30' + i\sin 296°\,30')/10$

11. (a) 29, (b) $\cos 136°\,20' + i\sin 136°\,20'$

13. (a) $26i$, (b) $(\cos 202°\,40' + i\sin 202°\,40')/2$

15. $9(-1+\sqrt{3}i)$
17. $-2(1+i)$

Chapter 18, Exercise 4, page 465

3. $64(-1+i)$
5. -64

7. $-236.2+3{,}117i$
9. -1

11. $0.8480+0.5299i$
13. $\pm i$, $(\pm\sqrt{3}\pm i)/2$

15. $0.9080+1.782i$, $\sqrt{2}(-1+i)$, $-1.782-0.9080i$, $0.3128-1.975i$,
  $1.975-0.3128i$

17. $\pm(1+\sqrt{3}i)$ **19.** $\pm 2i,\ \pm\sqrt{3}\pm i$

21. $-0.2616+2.989i$ **23.** $1.312-1.509i$

25. $(\pm 1\pm\sqrt{3}i)/2,\ \pm 1$

27. (a) $\cos 4\theta = \cos^4\theta - 6\cos^2\theta\sin^2\theta + \sin^4\theta,$

   (b) $\sin 4\theta = 4(\cos^3\theta\sin\theta - \cos\theta\sin^3\theta)$

29. $-1,\ \pm i$

Chapter 19, Exercise 1, page 473

3. $P(2)=-8,\ P(3)=76$ **5.** $P(-4)=15,\ P(-5)=-469$

7. $b=4$

9. One root lies in each of the intervals $(-3,-2),\ (-1,0),$ and $(1,2).$

Chapter 19, Exercise 2, page 478

1. $Q(x)=x^3+7x^2-5x+4,\ R=-30$

3. $Q(x)=x^4-x^3+2x^2-3x+4,\ R=2$

5. $Q(x)=x^2+3x+1,\ R=0$

Chapter 19, Exercise 3, page 482

1. $Q(x)=x^3-x^2+4,\ R=13$

3. $Q(x)=x^5+2x^4-x^3-2x^2+9x-7,\ R=-14$

5. $Q(x)=3x^4+x^3-2x^2+x+8,\ R=3$

7. $Q(x)=2x^4+6x^3-8x^2+10x-2,\ R=0$

11. $L=-4,\ M=3$ **13.** $L=-1,\ M=2$

15. $L=-1,\ M=7$ **17.** $1,\ (1\pm\sqrt{17})/4$

19. $-2,1,3$ **21.** $-3,\ (-1\pm\sqrt{21})/2$

Chapter 19, Exercise 4, page 487

1. 3 **3.** $-3/2$

5. $2,\ -1/3$ **7.** No rational roots

9. $1,1,1$ (a root of 3rd order)

11. $T=\{\pm 1,\ \pm 5,\ \pm 7,\ \pm 35,\ \pm 1/2,\ \pm 5/2,\ \pm 7/2,\ \pm 35/2,\ \pm 1/3,\ \pm 5/3,\ \pm 7/3,$
$\pm 35/3,\ \pm 1/6,\ \pm 5/6,\ \pm 7/6,\ \pm 35/6\}.$

The equation has no rational roots.

17. $a_4 = a_0 r_1 r_2 r_3 r_4,\quad a_3 = -a_0(r_1 r_2 r_3 + r_1 r_2 r_4 + r_1 r_3 r_4 + r_2 r_3 r_4),$
$a_2 = a_0(r_1 r_2 + r_1 r_3 + r_1 r_4 + r_2 r_3 + r_2 r_4 + r_3 r_4),\quad a_1 = -a_0(r_1+r_2+r_3+r_4)$

Chapter 19, Exercise 5, page 491

1. 1.20   **3.** -2.91 **5.** -0.29

7. 3.54   **9.** 3.50   **11.** 1.651

13. 2.725   **15.** 3.61   **17.** 2.22

19. $x^3 + 2.00x^2 - 2.9961x - 1.01268$

Appendix 1, Exercise 1, page 501

1. W **2.** W **3.** W

4. W **5.** I **6.** W

7. W **8.** W **9.** W

10. I **11.** I **12.** I

13. W **14.** W

16. There is at least one x in **A** that is not in **B**.

17. (a) $\{3, 4, 5, 6\}$,   (b) $\{1, 2\}$,   (c) $\{-1, 0, 1, 2\}$,   (d) $\{-2, -1, 0, 1, 2\}$,
   (e) $\{1, 7, 11, 13, 77, 91, 143, 1{,}001\}$,   (f) $\{2, 3, 5, 7, 11, 13\}$

18. (a) $\{1, 2, 3, 4, 5, 6\}$,   (b) $\{1, 2, 3, 4\}$,   (c) $\{2, 4, 5, 6, 7\}$,   (d) $\{5, 7\}$,
   (e) $\varnothing$,   (f) $\{1, 2, 3, 4, 5, 6, 7\}$,   (g) $\{5, 7\}$,   (h) $\{2, 4\}$,
   (i) $\{1, 2, 3, 4, 5, 6, 7\}$

19. (a), (c), (d), and (f) are true.

Appendix 1, Exercise 2, page 506

1.

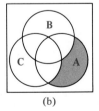

(a)   (b)

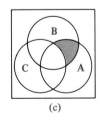 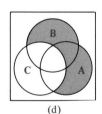

(c)   (d)

5. (a) and (b) $\{5, 7, 9\}$, (c) and (d) $\{4, 8\}$, (e) and (f) $\{0, 4, 5, 6, 7, 8, 9\}$

Appendix 2, Exercise 3, page 518

1. Axiom 10   **2.** Axioms 4, 5, and 10   **3.** Yes

Appendix 3, Exercise 1, page 521

1. Axioms 1, 3, 4, 5 (replace **N** by **A**)

2. Axioms 1, 2, 4, 5 (replace **N** by **A**)

3. Axioms 1, 2, 3, 5 (replace **N** by **A**)

4. Axioms 1, 2, 3, 4

5. Axioms 2, 3, 4, 5 (replace **N** by the empty set)

7. All of these theorems are still true

Appendix 4, Exercise 1, page 522

6. It is an irrational number.

Appendix 6, Exercise 1, page 529

1. 3.45×10	**2**. 3.45×10^4	**3**. 3.45×10^{-5}
4. 3.45×10^{-1}	**5**. 3.45×10^{-8}	**6**. 3.45×10^8
7. 0.00642	**8**. 6,420,000,000	**9**. 0.00000642
10. 5.68	**11**. 7.65	**12**. 1.11
13. 1.12	**14**. 1.23	**15**. 3.00
16. 2.98	**17**. 10.0	**18**. 9.99

Credits

Index of Special Symbols

SIMPLE SYMBOLS	*Symbol*	*Meaning or Name*	*Page*
	$\equiv$	Identically equal	5, 179
	$\equiv$	Is defined to be	5
	$<$	Less than	31
	$>$	Greater than	31
	$\approx$	Approximately equal	164, 525
	$\in$	Element of the set	495
	$\varnothing$	The empty set	496
	$\cup$	Union	497
	$\cap$	Intersection	498
	$\subset$	Is contained in	496
	$\supset$	Contains	496
	$\ldots$	Et cetera	10, 366
	$\blacksquare$	End of proof	9
	$\blacktriangle$	End of solution of an example	9
	∞	Infinity	12
	$n!$	Factorial n	385, 386
	$\vec{0}$	Zero vector	437
	0	Zero matrix	332, 357
	$\oplus$	Abstract sum	517
	$\otimes$	Abstract product	517
	$\Rightarrow$	Implies	8
	ρ	Reference angle (Greek "rho")	160
	$\sim$	Is equivalent to	335

COMPOUND SYMBOLS	*Symbol*	*Meaning or Name*	*Page*
	PLE	Proof is left as an exercise	9
	PWO	Proof will be omitted	9
	$|P_1 P_2|$	Length of the line segment $P_1 P_2$	39, 198
	$\overline{P_1 P_2}$	Directed distance from P_1 to P_2	43
	$\overrightarrow{AB}$	Vector from A to B	436
	$|AB|$	Length of the vector $\overrightarrow{AB}$	437
	$|a|$	Absolute value of a	36
	$|z|$	Absolute value of the complex number z	458
	$|OP|$	Distance from the origin to P	123, 458
	$[a, b]$	Closed interval	41
	(a, b)	Open interval	41

Index